Advanced Lectures in Mathematics
Volume II

Asymptotic Theory in Probability and Statistics with Applications

Editors: Tze-Leung Lai, Lianfen Qian, Qi-Man Shao

International Press
www.intlpress.com

高等教育出版社
HIGHER EDUCATION PRESS

Tze-Leung Lai
Department of Statistics
Stanford University

Qi-Man Shao
Department of Mathematics
University of Oregon

Lianfen Qian
Department of Mathematical Sciences
Florida Atlantic University

2000 Mathematics Subject Classification. 46N30.

ISBN 978-1-57146-169-8

Typeset using the LaTeX system.
Printed in the USA on acid-free paper.

ADVANCED LECTURES IN MATHEMATICS

Professor Zhengyan Lin

June 19-21, 2006, Center of Mathematical Sciences, Zhejiang university

Preface

To celebrate the 65th birthday of Professor Zhengyan Lin, an International Conference on Asymptotic Theory in Probability and Statistics was held at the Center of Mathematical Sciences and Department of Mathematics of Zhejiang University, Hangzhou, China, in the summer of 2006. One of the aims of the conference was to provide a platform for the exchange of new ideas and recent developments in asymptotic theory and applications. Many speakers of the conference were invited to contribute to this volume, which consists of expository papers based on their invited talks or related research areas. All papers were carefully peer refereed. We would like to dedicate this book to Professor Zhengyan Lin and wish him continuing success in many years to come!

Professor Lin is a leading probabilist in China. He has made significant contributions to the development of asymptotic theory, especially, limit theorems for mixing dependent random variables and self-normalized sums, and sample path properties of Gaussian processes. Professor Lin has published over 140 papers and 7 books. He and Professor Chuanrong Lu have supervised over 160 graduate students at Hangzhou University (now merged with Zhejiang University) since 1982.

An objective of the present volume of 18 papers by the invited speakers and contributors to the conference is to introduce graduate students to some active research areas in probability and statistics. Most papers are survey papers so that the present volume can provide readers with a valuable resource in probability, statistics and their applications. Obviously, we cannot cover all of the important topics of current research.

The volume consists of three parts: (I) Limit Theorems, (II) Statistics and Applications, and (III) Mathematical Finance and Insurance.

Part I has 8 papers, focusing on limit theory through various angles. It starts with the probability theory of self-normalization (Lai and Shao), followed by random partitions (Su), adaptive designs (Zhang), Gaussian processes (Wang), Gaussian random fields (Xiao), large deviations theory for two-parameter Gaussian processes (Chen and Csörgő), intersection local times (Chen), and ends with limit theorems for U-statistics (JING) to serve as a link between probability and statistics.

Part II contains 7 papers, exploring various research areas in statistics and its applications. It starts with the classical inverse problem for the t-Statistics (Yang, Fang and Kotz), followed by statistical issues in rounded data (Bai, Zheng, Zhang and Hu). Then it introduces a variety of useful statistical models such as piecewise regression models (Qian), partially linear models (Wang, Liang and Jin) and nonlinear time series models (H. Yu). Finally it reviews recent work on classification and illustrates a probabilistic classifier with environmental and remote sensing applications (J. Yu and Ranneby), mixed linear model approaches for complex trait analysis (Yang and Zhu).

Part III consists of 3 papers. It begins with introducing inference and computation for stochastic volatility models related to option pricing (Ji), and then gives an overview of Choquet integrals and their applications to risk theory (Wang and Yan). The volume ends with a paper by Yang on actuarial science and its recent developments.

We thank all authors for their superb contributions and the referees for their thorough and timely work. Thanks also go to the Center of Mathematical Sciences at Zhejiang University for providing conference facilities and financial support, to the executive editors, S. T. Yau, K. F. Liu and L. Z. Ji, for their interests in publishing the volume for the Advanced Lecture in Mathematics, and to the Chinese National Science Foundation for financial support. Last but not the least, we thank the local organizers for their tremendous contributions to the conference.

T. L. Lai, L. F. Qian and Q. M. Shao
March 2007

List of contributors:

Zhidong Bai	Northeast Normal University, China
Ranneby Bo	Swedish University of Agricultural Sciences, Sweden
Bin Chen	Carleton University, Canada
Maklós Csörgő	Carleton University, Canada
Xia Chen	The University of Tennessee, USA
Kai-Tai Fang	Zhuhai Campus of Beijing Normal University, China
Guorong Hu	Northeast Normal University, China
Chuanshu Ji	University of North Carolina, USA
Lei Jin	Texas A&M University, USA
Bing-Yi JING	Hong Kong University of Science and Technology, Hong Kong, China
Samuel Kotz	George Washington University, USA
Tze Leung Lai	Stanford University, USA
Hua Liang	University of Rochester Medical Center, USA
Lianfen Qian	Florida Atlantic University, USA
Qi-Man Shao	Hong Kong University of Science and Technology, Hong Kong, China
Zhonggen Su	Zhejiang University, China
Suojin Wang	Texas A&M University, USA
Wensheng Wang	East China Normal University and Hangzhou Normal University, China
Zengwu Wang	Academy of Mathematics and Systems Science, China
Baoxue Zhang	Northeast Normal University, China
Li-xin Zhang	Zhejiang University, China
Shurong Zheng	Northeast Normal University, China
Yimin Xiao	Michigan State University, USA
Jia-An Yan	Academy of Mathematics and Systems Science, China
Hailiang Yang	The University of Hong Kong, Hong Kong, China
Jian Yang	Zhejiang University, China
Zhenhai Yang	Beijing University of Technology, China
Hao Yu	University of Western Ontario, Canada
Jun Yu	Swedish University of Agricultural Sciences, Sweden
Jun Zhu	Zhejiang University, China

Contents

Part I: Limit Theorems

Part II: Statistics and Applications

Part III: Mathematical Finance and Insurance

PartI: Limit Theorems

Self-normalized Limit Theorems in Probability and Statistics

Tze Leung Lai
Department of Statistics, Stanford University
Stanford, CA 94305-4065, USA
E-mail: lait@stanford.edu

Qi-Man Shao
Department of Mathematics
Hong Kong University of Science and Technology
Clear Water Bay, Kowloon, Hong Kong, China
and
Department of Mathematics, University of Oregon
Eugene, OR 97403, USA
and
Department of Mathematics, Zhejiang University
Hangzhou 310027, China
E-mail: maqmshao@ust.hk

Abstract

The normalizing constants in classical limit theorems are usually sequences of real numbers. Moment conditions or other related assumptions are necessary and sufficient for many classical limit theorems. However, the situation becomes very different when the normalizing constants are sequences of random variables. A self-normalized large deviation shows that no any moment condition is needed for such large deviation type results. A self-normalized law of the iterated logarithm remains valid for all distributions in the domain of attraction of a normal or stable law. This reveals that self-normalization preserves essential properties much better than deterministic normalization does. In this chapter we review some important developments on self-normalized limit theorems in the last decade, especially on self-normalized large deviations, self-normalized saddle point approximations, self-normalized moderate deviations, self-normalized Cramér type large deviations for independent random variables, self-normalized law of the iterated logarithm and increments, self-normalized large and moderate deviations in $\mathbb{R}^d$, large and moderate deviations of Hotelling's T^2 statistic, large and moderate deviations for self-normalized empirical processes; limiting distributions of self-normalized sums, weak invariance principle for self-normalized partial sum processes, Darling-Erdős type theorem, asymptotic distributions of non-central self-normalized sums, the pseudo-maximization approach for self-normalized stochastic processes.

1 Introduction

Let X, $X_1, X_2, \cdots$ be independent and identically distributed random variables. Put

$$S_n = \sum_{i=1}^{n} X_i, \quad V_n^2 = \sum_{i=1}^{n} X_i^2. \tag{1.1}$$

The standardized sum usually means $b_n(S_n - a_n)$, where a_n and b_n are non-random sequences, while self-normalized sum refers to S_n/V_n. It is well-known that moment conditions or other related conditions are necessary and sufficient for many classical limit theorems. For example, the strong law of large numbers holds if and only if the mean of X is finite; the central limit theorem holds if and only if $EX^2I(|X| \leqslant x)$ is slowing varying as $x \to \infty$; and a necessary and sufficient condition for the large deviation is that the moment generating function of X is finite in a neighborhood of zero. On the other hand, limit theorems for self-normalized sums S_n/V_n put a totally new countenance upon the classical limit theorems. In contrast to the well-known Hartman-Wintner law of the iterated logarithm and its converse by Strassen (1966), Griffin and Kuelbs (1989) obtained a self-normalized law of the iterated logarithm for all distributions in the domain of attraction of a normal or stable law. Shao (1997) showed that no moment conditions are needed for a self-normalized large deviation result $P(S_n/V_n \geqslant x\sqrt{n})$ and that the tail probability of S_n/V_n is Gaussian like when X_1 is in the domain of attraction of the normal law and sub-Gaussian like when X is in the domain of attraction of a stable law, while Giné, Götze and Mason (1997) proved that the tails of S_n/V_n are uniformly sub-Gaussian when the sequence is stochastically bounded. Shao (1999) established a Cramér type result for self-normalized sums only under a finite third moment condition. These results strongly show that self-normalized partial sums preserve desirable properties much better than non-randomized partial sums. Self-normalization is also commonly used in statistics. Many statistical inferences require the use of classical limit theorems. However, these classical results often involve some unknown parameters, one needs to first estimate the unknown parameters and then substitute the estimators into the classical limit theorems. This commonly used practice is exactly the self-normalization. A typical case is the Student t-statistic.

The close relationship between the Student t-statistic T_n and the self-normalized sum S_n/V_n can be seen below:

$$T_n := \frac{\bar{X}}{s/\sqrt{n}} = \frac{S_n}{V_n}\Big(\frac{n-1}{n-(S_n/V_n)^2}\Big)^{1/2} \tag{1.2}$$

and

$$\{T_n \geqslant t\} = \Big\{\frac{S_n}{V_n} \geqslant t\Big(\frac{n}{n+t^2-1}\Big)^{1/2}\Big\}, \tag{1.3}$$

where $\bar{X}$ is the sample mean and s is the sample standard deviation.

In this chapter we review some important developments in the self-normalized limit theorems in the last 20 years. We will focus on the following topics:

1. Self-normalized large deviations;
2. Self-normalized saddlepoint approximations;
3. Self-normalized moderate deviations;
4. Cramér type large deviations for independent random variables;
5. Self-normalized law of the iterated logarithm and increments;
6. Self-normalized large and moderate deviations in $\mathbb{R}^d$;
7. Large and moderate deviations of Hotelling's T^2 statistic;
8. Large and moderate deviations for self-normalized empirical processes;
9. Limiting distributions of self-normalized sums;
10. Weak invariance principle for self-normalized partial sum processes;
11. Darling-Erdős type theorem;
12. Asymptotic distributions of non-central self-normalized sums;
13. Self-normalized processes: a pseudo-maximization approach;
14. Applications to statistics.

The review is based on three survey papers by Shao (1998), Shao (2004) and de la Pena, Klass and Lai (2007).

2 Self-normalized large deviations

Let $X, X_1, X_2, \cdots$ be a sequence of independent and identically distributed (i.i.d.) random variables. The classical Cramér-Chernoff large deviation [8] states that if

A) $Ee^{t_0 X} < \infty$ for some $t_0 > 0$,

then for every $x > EX$,

$$\lim_{n\to\infty} P\left(\frac{S_n}{n} \geqslant x\right)^{1/n} = \rho(x), \tag{2.1}$$

where $\rho(x) = \inf_{t\geqslant 0} e^{-tx} Ee^{tX}$.

Roughly speaking, this type of large deviation shows that the convergence rate in the law of large numbers is exponential if the moment generating function is finite in a right neighbourhood of zero. The latter is also necessary for an exponential scale (see, e.g. citePS73). Essentially built on condition **A**), the theory of large deviations in finite dimensional spaces, as well as in abstract spaces, has been well developed, and various applications in statistics (c.f., e.g., Bahadur (1971)), engineering, statistical mechanics and applied probability have been found in recent years. We refer to de Acosta (1988), Stroock (1984), Donsker and Varadhan (1987) and the book by Dembo and Zeitouni (1992), and references therein, for more details.

On the other hand, the following result shows that a self-normalized large deviation remains valid for arbitrary random variables without any moment conditions.

Theorem 2.1. [Shao (1997)] *Assume that either $E(X) \geqslant 0$ or $E(X^2) = \infty$. Then*

$$\lim_{n\to\infty} P\left(S_n \geqslant x\, n^{1/2} V_n\right)^{1/n} = \sup_{b\geqslant 0}\inf_{t\geqslant 0} Ee^{t(bX - x(X^2+b^2)/2)} \tag{2.2}$$

for $x > E(X)/(E(X^2))^{1/2}$, where $E(X)/(E(X^2))^{1/2} = 0$ if $E(X^2) = \infty$.

Note that for any random variable X, $E(X^2)$ is either finite or infinite. When $E(X^2)$ is finite, of course, $E(|X|)$ is finite. It is reasonable to assume $E(X) \geqslant 0$. In fact, when $E(X) \leqslant 0$, (2.2) remains true for $x > 0$.

More generally, using $V_{n,p} := \left(\sum_{i=1}^{n} |X_i|^p\right)^{1/p} n^{1-1/p}$, $p > 1$ as normalizing constants, we have

Theorem 2.2. [Shao (1997)] *Let $p > 1$ and assume that either $EX \geqslant 0$ or $E|X|^p = \infty$. Then*

$$\lim_{n\to\infty} P\left(S_n \geqslant x\, n^{1-1/p} V_{n,p}\right)^{1/n} = \sup_{b\geqslant 0}\inf_{t\geqslant 0} Ee^{t(bX - x(\frac{1}{p}|X|^p + \frac{p-1}{p} b^{p/(p-1)}))} \tag{2.3}$$

for $x > EX/(E|X|^p)^{1/p}$, where $EX/(E|X|^p)^{1/p} = 0$ if $E|X|^p = \infty$.

A key observation of the proof of Theorem 2.1 is the following identity

$$\forall \ x \geqslant 0, y \geqslant 0, \quad x^{1/2} y^{1/2} = (1/2) \inf_{b>0} (x/b + yb). \tag{2.4}$$

Thus, one can write

$$\sqrt{n} V_n = (1/2) \inf_{c \geqslant 0} (bn + V_n^2/b)$$

and

$$\begin{aligned} P(S_n/V_n \geqslant x\sqrt{n}) &= P\Big(S_n \geqslant (x/2) \inf_{b \geqslant 0} (b\, n + V_n^2 b)\Big) \\ &= P\Big(\sup_{b \geqslant 0} \sum_{i=1}^{n} (bX_i - x(X_i^2 + b^2)/2) \geqslant 0\Big). \quad (2.5) \end{aligned}$$

Now for each fixed $b \geqslant 0$, $\{bX_i - x(X_i^2 + b^2)/2, i \geqslant 1\}$ is a sequence of i.i.d. random variables with finite moment generating function in a right neighborhood of zero. Applying the Cramér-Chernoff large deviation (2.1) yields

$$\liminf_{n \to \infty} P(S_n/V_n \geqslant x\sqrt{n})^{1/n} \geqslant \sup_{b \geqslant 0} \inf_{t \geqslant 0} E\left(e^{t(bX - x(X^2 + b^2)/2)}\right).$$

As to the upper bound, intuitively, the order of the probability of the union of events should be close to the maximum of the probability of the individual event. The original proof is quite complicated, Shao (2006) gives a much simpler proof. For completeness, we restate the main steps of the proof. Let $A \geqslant 1$ and $m = [2A/x] + 1$, where $[y]$ denotes the integer part of y. Observe that

$$P\left(S_n \geqslant x\, n^{1/2} V_n\right) = I_1 + I_2, \tag{2.6}$$

where

$$I_1 = P\left(S_n \geqslant x\, n^{1/2} V_n, V_n > m\, n^{1/2}\right),$$
$$I_2 = P\left(S_n \geqslant x\, n^{1/2} V_n, V_n \leqslant m\, n^{1/2}\right).$$

Noting that $nA \leqslant xn^{1/2} V_n/2$ for $V_n > m\, n^{1/2}$, We have

$$\begin{aligned} I_1 &\leqslant P\Big(\sum_{i=1}^{n} X_i I(|X_i| > A) + nA \geqslant xn^{1/2} V_n, V_n > m\, n^{1/2}\Big) \\ &\leqslant P\Big(\sum_{i=1}^{n} X_i I(|X_i| > A) \geqslant (1/2) x n^{1/2} V_n\Big) \\ &\leqslant P\Big(\sum_{i=1}^{n} I(|X_i| > A) \geqslant x^2 n/4\Big) \\ &\leqslant \left(12 x^{-2} P(|X| > A)\right)^{x^2 n/4}, \end{aligned}$$

As to I_2, note that

$$n^{1/2}V_n = \inf_{0\leqslant b\leqslant V_n/n^{1/2}} (bn + V_n^2/b)/2, \tag{2.7}$$

we have

$$\begin{aligned}
I_2 &\leqslant P\left(\sup_{0\leqslant b\leqslant m} \{b\,S_n - x(V_n^2 + nb^2)/2\} \geqslant 0\right) \\
&\leqslant \sum_{i=1}^{nm} P\left(\sup_{(i-1)/n\leqslant b\leqslant i/n} \{b\,S_n - x(V_n^2 + nb^2)/2\} \geqslant 0\right) \\
&\leqslant \sum_{i=1}^{mn} P\left((i/n)S_n - x(V_n^2 + n((i-1)/n)^2)/2 \geqslant 0\right) \\
&\leqslant \sum_{i=1}^{mn} P\left((i/n)S_n - x(V_n^2 + n(i/n)^2)/2 \geqslant -xm\right) \\
&\leqslant \sum_{i=1}^{mn} \inf_{0\leqslant t\leqslant k} e^{xmt}\Big\{E\exp\Big(t((i/n)X - (x/2)(X^2 + (i/n)^2)\Big)\Big\}^n \\
&\leqslant mn\,e^{xmk} \sup_{0\leqslant b\leqslant m}\ \inf_{0\leqslant t\leqslant k} \Big\{E\exp\Big(t(bX - (x/2)(X^2 + b^2))\Big)\Big\}^n
\end{aligned}$$

for any $k \geqslant 1$. Therefore

$$\limsup_{n\to\infty} I_2^{1/n} \leqslant \sup_{0\leqslant b\leqslant m}\ \inf_{0\leqslant t\leqslant k} E\exp\Big(t(bX - (x/2)(X^2 + b^2))\Big). \tag{2.8}$$

It is easy to see that $\inf_{0\leqslant t\leqslant k} E\exp\Big(t(bX - (x/2)(X^2 + b^2))\Big)$ is a non-increasing sequence of k and a continuous function of b for each k. Thus, by Lemma 4 of Chernoff (1952)

$$\begin{aligned}
&\lim_{k\to\infty} \sup_{0\leqslant b\leqslant m}\ \inf_{0\leqslant t\leqslant k} E\exp\Big(t(bX - (x/2)(X^2 + b^2))\Big) \\
&= \sup_{0\leqslant b\leqslant m+1}\ \inf_{0\leqslant t<\infty} E\exp\Big(t(bX - (x/2)(X^2 + b^2))\Big)
\end{aligned}$$

and hence

$$\begin{aligned}
\limsup_{n\to\infty} I_2^{1/n} &\leqslant \sup_{0\leqslant b\leqslant m}\ \inf_{t\geqslant 0} E\exp\Big(t(bX - (x/2)(X^2 + b^2))\Big) \qquad (2.9)\\
&\leqslant \sup_{0\leqslant b<\infty}\ \inf_{t\geqslant 0} E\exp\Big(t(bX - (x/2)(X^2 + b^2))\Big).
\end{aligned}$$

Letting $A \to \infty$ together with (2.6) and (2.9) completes the proof of the upper bound.

Theorem 2.1 has been extended to high dimension by Dembo and Shao (1998a) and to self-normalized empirical processes by Bercu, Gassiat and Rio (2002).

3 Self-normalized saddlepoint approximations

Let $X, X_1, X_2, \cdots$ be i.i.d. random variables and denote $\bar{X}$ the sample mean of $\{X_i, 1 \leqslant i \leqslant n\}$. Large deviation result provides an exponential rate of convergence for tail probability. However, a more fine-tuned approximation can be offered by saddlepoint approximations. Daniels (1954) showed that the density function of $\bar{X}$ satisfies

$$f_{\bar{X}}(x) = e^{-n(\tau x - \kappa(\tau))}\Big(\frac{n}{2\pi\kappa''(\tau)}\Big)^{1/2}(1 + O(n^{-1})),$$

where τ is the saddlepoint satisfying $\kappa'(\tau) = x$. Lugannani and Rice (1980) obtained the tail probability of $\bar{X}$:

$$P(\bar{X} \geqslant x) = 1 - \Phi(\sqrt{n}\hat{w}) + \frac{\phi(\sqrt{n}\hat{w})}{\sqrt{n}}\Big(\frac{1}{\hat{u}} - \frac{1}{\hat{w}} + O(n^{-1})\Big),$$

where $\kappa'(\tau) = x$, $\hat{w} = \{2[\tau\kappa'(\tau) - \kappa(\tau)]\}^{1/2}\mathrm{sign}\{\tau\}$, $\hat{u} = \tau[\kappa''(\tau)]^{1/2}$, Φ and ϕ denote the standard normal distribution function and density function, respectively. So, the error incurred by the saddlepoint approximation is $O(n^{-1})$ as against the more usual $O(n^{-1/2})$ associated with the normal approximation. Another desirable feature of saddlepoint approximation is that the approximation is quite satisfactory even when the sample size n is small. The book by Jensen (1995) gives a detailed account of saddlepoint approximations and related techniques. However, a finite moment generating function is an essential requirement for saddlepoint expansions. Daniels and Young (1991) derived saddlepoint approximations for the tail probability of the Student t-statistic under the assumption that the moment generating function of X^2 exists. Note that Theorem 2.1 holds without any moment assumption. It is natural to ask whether the saddle point approximation is still valid without any moment condition for the t statistic or equivalently, for the self-normalized sum S_n/V_n. Jing, Shao and Zhou (2004) recently give an affirmative answer to this question. Let $K(s,t) = \ln Ee^{sX+tX^2}$,

$$K_{11}(s,t) = \frac{\partial^2 K(s,t)}{\partial s^2}, K_{12}(s,t) = \frac{\partial^2 K(s,t)}{\partial s \partial t}, K_{22}(s,t) = \frac{\partial^2 K(s,t)}{\partial t^2}.$$

For $0 < x < 1$, let $\hat{t}_0$ and a_0 be solutions t and a to the equations

$$\frac{EXe^{t(-2aX/x^2+X^2)}}{Ee^{t(-2aX/x^2+X^2)}} = a, \quad \frac{EX^2e^{t(-2aX/x^2+X^2)}}{Ee^{t(-2aX/x^2+X^2)}} = \frac{a^2}{x^2}.$$

It is proved in [43] that $\hat{t}_0 < 0$. Put $\hat{s}_0 = -2a_0\hat{t}_0/x^2$ and define

$$\Lambda_0(x) = \hat{s}_0 a_0 + \hat{t}_0 a_0^2/x^2 - K(\hat{s}_0, \hat{t}_0),$$
$$\Lambda_1(x) = 2\hat{t}_0/x^2 + (1, 2a_0/x^2)\Delta^{-1}(1, 2a_0/x^2)',$$

where

$$\Delta = \begin{pmatrix} K_{11}(\hat{s}_0, \hat{t}_0) & K_{12}(\hat{s}_0, \hat{t}_0) \\ K_{12}(\hat{s}_0, \hat{t}_0) & K_{22}(\hat{s}_0, \hat{t}_0) \end{pmatrix}.$$

Theorem 3.1. [Jing, Shao and Zhou (2004)]. *Assume* $EX = 0$ *or* $EX^2 = \infty$ *and that*

$$\int_{-\infty}^{\infty} \int_{-\infty}^{\infty} |Ee^{isX+itX^2}|^r dsdt < \infty \ for \ some r > 1.$$

Then for $0 < x < 1$

$$P(S_n/V_n \geqslant x\sqrt{n}) = 1 - \Phi(\sqrt{n}w) - \frac{\phi(\sqrt{n}w)}{\sqrt{n}} \Big(\frac{1}{w} - \frac{1}{v} + O(n^{-1}) \Big),$$

where $w = \sqrt{2\Lambda_0(x)}$, *and* $v = (-\hat{t}_0/2)^{1/2} x^{3/2} a_0^{-1} (\det \Delta)^{1/2} \Lambda_1(x)^{1/2}$.

4 Self-normalized moderate deviations

Let $X, X_1, X_2, \cdots$ be i.i.d. random variables and let $\{x_n, n \geqslant 1\}$ be a sequence of positive numbers with $x_n \to \infty$ as $n \to \infty$. It is known that

$$\lim_{n\to\infty} x_n^{-2} \ln P \left(\frac{|S_n|}{\sqrt{n}} \geqslant x_n \right) = -\frac{1}{2}$$

holds for any sequence $\{x_n\}$ with $x_n \to \infty$ and $x_n = o(\sqrt{n})$ if and only if $EX = 0$, $EX^2 = 1$ and $Ee^{t_0|X|} < \infty$ for some $t_0 > 0$. The sufficient part follows from the Cramér large deviation (c.f. Petrov (1975)). As to the necessary part, assume that for any $x_n \to \infty$ with $x_n = o(\sqrt{n})$, there exist $c_0 > 0$ and $c_1 \geqslant 2$ such that

$$P(|S_n| \geqslant x_n\sqrt{n}) \leqslant c_1 \exp(-c_0\, x_n^2) \text{ for } n \geqslant 1. \tag{4.1}$$

We prove that there is $t_0 > 0$ such that $Ee^{t_0|X|} < \infty$. If this is not true, that is, $Ee^{t|X|} = \infty$ for every $t > 0$, by the fact that

$$\sum_{k=1}^{\infty} e^{2t2^k} P(|X| \geqslant 2^k) \geqslant Ee^{t|X|},$$

there exists $t_k \to 0$ such that $t_k 2^k \geqslant 2^{k/2}$ and

$$\sum_{k=1}^{\infty} e^{t_k 2^k} P(|X| \geqslant 2^k) = \infty. \tag{4.2}$$

Let $n = [2^k/t_k^{1/2}]$ and $x_n = 2^k$. Then by (4.1)

$$\begin{aligned} P(|X| \geqslant 2^k) &\leqslant P(|S_n| \geqslant 2^{k-1}) + P(|S_{n-1}| \geqslant 2^{k-1}) \\ &\leqslant 2c_1 \exp(-c_0 2^{2(k-1)}/n) \\ &\leqslant 2c_1 \exp(-c_0 2^k t_k^{1/2}/4) \\ &= 2c_1 \exp(-c_0 2^k t_k/(4t_k^{1/2})) \\ &\leqslant 2c_1 \exp(-2^{k+1} t_k) \end{aligned}$$

for sufficient large k. Hence

$$\sum_{k=1}^{\infty} e^{t_k 2^k} P(|X| \geqslant 2^k) < \infty,$$

which contradicts (4.2). This proves that there must exist a $t_0 > 0$ such that $Ee^{t_0|X|} < \infty$.

The next result shows again that the situation is quite different in the case of self-normalized limit theorems. It tells us that the main term of the asymptotic probability of $P(S_n \geqslant x_n V_{n,2})$ is distribution free as long as X is in the domain of attraction of a normal law and $x_n = o(\sqrt{n})$.

Theorem 4.1. [Shao (1997)] *Let $\{x_n, n \geqslant 1\}$ be a sequence of positive numbers with $x_n \to \infty$ and $x_n = o(\sqrt{n})$ as $n \to \infty$. If $EX = 0$ and $EX^2 I\{|X| \leqslant x\}$ is slowly varying as $x \to \infty$, then*

$$\lim_{n\to\infty} x_n^{-2} \ln P\left(\frac{S_n}{V_{n,2}} \geqslant x_n\right) = -\frac{1}{2}.$$

Similar results to that of Theorem 4.1 remain valid when X is in the domain of attraction of a stable law.

Theorem 4.2. [Shao (1997)] *Assume that there exist $0 < \alpha < 2$, $c_1 \geqslant 0$, $c_2 \geqslant 0$, $c_1 + c_2 > 0$ and a slowly varying function $h(x)$ such that*

$$P(X \geqslant x) = \frac{c_1 + o(1)}{x^\alpha} h(x), \quad P(X \leqslant -x) = \frac{c_2 + o(1)}{x^\alpha} h(x) \quad \textit{as } x \to \infty.$$

Moreover, assume that $EX = 0$ if $1 < \alpha < 2$, X is symmetric if $\alpha = 1$ and that $c_1 > 0$ if $0 < \alpha < 1$. Let $p > \max(1, \alpha)$, and let $\{x_n, n \geqslant 1\}$ be a sequence of positive numbers with $x_n \to \infty$ and $x_n = o(n^{(p-1)/p})$ as $n \to \infty$. Then, we have

$$\lim_{n\to\infty} x_n^{-p/(p-1)} \ln P\left(\frac{S_n}{V_{n,p}} \geqslant x_n\right) = -(p-1)\beta_p(\alpha, c_1, c_2),$$

where $V_{n,p} = (\sum_{i=1}^n |X_i|^p)^{1/p}$ *and* $\beta_p(\alpha, c_1, c_2)$ *is the solution of* $\Gamma_p(\beta, \alpha, c_1, c_2) = 0$ *and* $\Gamma_p(\beta, \alpha, c_1, c_2) =$

$$\begin{cases} c_1 \int_0^\infty \frac{1 + px - e^{px - x^p/\beta^{p-1}}}{x^{\alpha+1}} dx \\ \qquad + c_2 \int_0^\infty \frac{1 - px - e^{-px - x^p/\beta^{p-1}}}{x^{\alpha+1}} dx, & \text{if } 1 < \alpha < 2, \\ c_1 \int_0^\infty \frac{2 - e^{px - x^p/\beta^{p-1}} - e^{-px - x^p/\beta^{p-1}}}{x^2} dx, & \text{if } \alpha = 1, \\ c_1 \int_0^\infty \frac{1 - e^{px - x^p/\beta^{p-1}}}{x^{\alpha+1}} dx + c_2 \int_0^\infty \frac{1 - e^{-px - x^p/\beta^{p-1}}}{x^{\alpha+1}} dx, & \text{if } 0 < \alpha < 1. \end{cases}$$

In particular, if X *is symmetric, then*

$$\lim_{n\to\infty} x_n^{-p/(p-1)} \ln P\left(\frac{S_n}{V_{n,p}} \geqslant x_n\right) = -(p-1)\beta_p(\alpha),$$

where $\beta_p(\alpha)$ *is the solution of* $\int_0^\infty \frac{2 - e^{px - x^p/\beta^{p-1}} - e^{-px - x^p/\beta^{p-1}}}{x^{\alpha+1}} dx = 0.$

5 Cramér type large deviations for independent random variables

Let $X_1, X_2, \cdots$ be a sequence of independent random variables with $EX_i = 0$ and $0 < EX_i^2 < \infty$ for $i \geqslant 1$. Set

$$S_n = \sum_{i=1}^n X_i, \qquad B_n^2 = \sum_{i=1}^n EX_i^2, \qquad V_n^2 = \sum_{i=1}^n X_i^2.$$

It is well-known that the central limit theorem holds if the Lindeberg condition is satisfied. There are mainly two approaches for estimating the error of the normal approximation. One approach is to study the absolute error in the central limit theorem via Berry-Esseen bounds or Edgeworth expansions. Another approach is to estimate the relative error of $P(S_n \geqslant xB_n)$ to $1 - \Phi(x)$. One of the typical results in this direction is the so-called Cramér large deviation. If $X_1, X_2, \cdots$ are a sequence of i.i.d. random variables with zero means and a finite moment-generating function $Ee^{tX_1} < \infty$ for t in a neighborhood of zero, then for $x \geqslant 0$ and $x = o(n^{1/2})$

$$\frac{P(S_n \geqslant xB_n)}{1 - \Phi(x)} = \exp\left\{x^2 \lambda(\frac{x}{\sqrt{n}})\right\}\left[1 + O(\frac{1+x}{\sqrt{n}})\right],$$

where $\lambda(t)$ is the so-called Cramér's series (see [Petrov (1975), Chapter VIII] for details). In particular, if $Ee^{t|X_1|^{1/2}} < \infty$ for some $t > 0$, then

$$\frac{P(S_n \geqslant xB_n)}{1 - \Phi(x)} \to 1 \tag{5.1}$$

holds uniformly for $x \in (0, o(n^{1/6}))$. Note that the moment condition $Ee^{t|X_1|^{1/2}} < \infty$ is necessary. Similar results are also available for independent but not necessarily identically distributed random variables under a finite moment-generating function condition.

Shao (1999) established a (5.1) type result for self-normalized sums only under a finite third-moment condition. More precisely, he showed that if $E|X_1|^{2+\delta} < \infty$ for $0 < \delta \leqslant 1$, then

$$\frac{P(S_n \geqslant xV_n)}{1 - \Phi(x)} \to 1 \tag{5.2}$$

holds uniformly for $x \in (0, o(n^{\delta/(2(2+\delta))}))$. Recently, several papers have focused on the self-normalized limit theorems for independent but not necessarily identically distributed random variables. Bentkus, Bloznelis and Götze (1996) obtained the following Berry-Esseen bound:

$$\begin{aligned}&|P(S_n/V_n \geqslant x) - (1 - \Phi(x))| \\ &\leqslant A\Big(B_n^{-2}\sum_{i=1}^{n} EX_i^2 I\{|X_i| > B_n\} + B_n^{-3}\sum_{i=1}^{n} E|X_i|^3 I\{|X_i| \leqslant B_n\}\Big),\end{aligned}$$

where A is an absolute constant. Assuming only finite third moments, Wang and Jing (1999) derived exponential non-uniform Berry-Esseen bounds. Chistyakov and Götze (2003) refined Wang and Jing's results and obtained the following result among others: If $X_1, X_2, \cdots$ are *symmetric* independent random variables with finite third moments, then

$$P(S_n/V_n \geqslant x) = (1 - \Phi(x))\Big(1 + O(1)(1 + x)^3 B_n^{-3}\sum_{i=1}^{n} E|X_i|^3\Big) \tag{5.3}$$

for $0 \leqslant x \leqslant B_n/(\sum_{i=1}^{n} E|X_i|^3)^{1/3}$, where $O(1)$ is bounded by an absolute constant.

Result (5.3) is useful because it provides not only the relative error but also a Berry-Esseen rate of convergence. It should be noted that if X_i is symmetric, then

$$S_n/V_n \text{ and } \Big(\sum_{i=1}^{n} \varepsilon_i X_i\Big)/V_n$$

have the same distribution, where $\{\varepsilon_i, i \geqslant 1\}$ is a Rademacher sequence that is independent of $\{X_i, i \geqslant 1\}$. Hence given $\{X_i, 1 \leqslant i \leqslant n\}$, the problem reduces to estimate the tail probability of the partial sum of independent random variables $\varepsilon_i X_i / V_n, 1 \leqslant i \leqslant n$. So, the assumption of symmetry not only takes away the main difficulty in proving a self-normalized limit theorem but also limits its potential applications. Jing, Shao and Wang (2003) recently obtained a Cramér-type large deviation for general independent random variables. In particular, they show that (5.3) remains valid for non-symmetric independent random variables. Let

$$\Delta_{n,x} = \frac{(1+x)^2}{B_n^2} \sum_{i=1}^n EX_i^2 I\{|X_i| > B_n/(1+x)\} + \frac{(1+x)^3}{B_n^3} \sum_{i=1}^n E|X_i|^3 I\{|X_i| \leqslant B_n/(1+x)\}$$

for $x \geqslant 0$.

Theorem 5.1. [Jing, Shao and Wang (2003)]. *There is an absolute constant A (>1) such that*

$$\frac{P(S_n \geqslant xV_n)}{1 - \Phi(x)} = e^{O(1)\Delta_{n,x}}$$

for all $x \geqslant 0$ satisfying

$$x^2 \max_{1 \leqslant i \leqslant n} EX_i^2 \leqslant B_n^2$$

and

$$\Delta_{n,x} \leqslant (1+x)^2 / A,$$

where $|O(1)| \leqslant A$.

Theorem 5.1 provides a very general framework. The following result is a direct consequence of the above general theorem.

Theorem 5.2. [Jing, Shao and Wang (2003)]. *Let $\{a_n, n \geqslant 1\}$ be a sequence of positive numbers. Assume that*

$$a_n^2 \leqslant B_n^2 / \max_{1 \leqslant i \leqslant n} EX_i^2 \tag{5.4}$$

and

$$\forall\, \varepsilon > 0, \quad B_n^{-2} \sum_{i=1}^n EX_i^2 I\{|X_i| > \varepsilon B_n/(1+a_n)\} \to 0, \quad \text{as } n \to \infty. \tag{5.5}$$

Then

$$\frac{\ln P(S_n/V_n \geqslant x)}{\ln(1-\Phi(x))} \to 1 \tag{5.6}$$

holds uniformly for $x \in (0, a_n)$.

When the X_i's have a finite $(2+\delta)$th moment for $0 < \delta \leqslant 1$, we obtain (5.3) without assuming any symmetric condition.

Theorem 5.3. [Jing, Shao and Wang (2003)]. *Let* $0 < \delta \leqslant 1$ *and set*

$$L_{n,\delta} = \sum_{i=1}^{n} E|X_i|^{2+\delta}, \ d_{n,\delta} = B_n/L_{n,\delta}^{1/(2+\delta)}.$$

Then,

$$\frac{P(S_n/V_n \geqslant x)}{1-\Phi(x)} = 1 + O(1)\Big(\frac{1+x}{d_{n,\delta}}\Big)^{2+\delta}, \tag{5.7}$$

for $0 \leqslant x \leqslant d_{n,\delta}$, *where* $O(1)$ *is bounded by an absolute constant. In particular, if* $d_{n,\delta} \to \infty$ *as* $n \to \infty$, *we have*

$$\frac{P(S_n \geqslant xV_n)}{1-\Phi(x)} \to 1 \tag{5.8}$$

uniformly in $0 \leqslant x \leqslant o(d_{n,\delta})$.

By the fact that $1-\Phi(x) \leqslant 2e^{-x^2/2}/(1+x)$ for $x \geqslant 0$, it follows from (5.7) that the following exponential non-uniform Berry-Esseen bound

$$|P(S_n/V_n \geqslant x) - (1-\Phi(x))| \leqslant A(1+x)^{1+\delta}e^{-x^2/2}/d_{n,\delta}^{2+\delta} \tag{5.9}$$

holds for $0 \leqslant x \leqslant d_{n,\delta}$.

Theorem 5.1 has been successfully applied to study the studentized bootstrap and the self-normalized law of the iterated logarithm. The proof of Theorem 5.1 is quite complicated. Shao (2004) refined Theorem 5.2 which only requires condition (5.5).

Theorem 5.4. [Shao (2004)]. *Let* x_n *be a sequence of real numbers such that* $x_n \to \infty$ *and* $x_n = o(B_n)$. *Assume*

$$\forall\, \varepsilon > 0, \ B_n^{-2}\sum_{i=1}^{n} EX_i^2 I\{|X_i| > \varepsilon B_n/x_n\} \to 0. \tag{5.10}$$

Then we have

$$\ln P(S_n/V_n \geqslant x_n) \sim -x_n^2/2. \tag{5.11}$$

6 Self-normalized laws of the iterated logarithm and increments

The classical law of the iterated logarithm holds if and only if the second moment condition of X is finite, while the general path sample property of increments of the partial sum depends also on the moment condition. In particular,

$$\limsup_{n\to\infty} \max_{0\leqslant k\leqslant n} \max_{1\leqslant i\leqslant a_n} \frac{|S_{k+i}-S_k|}{\Big(2\, a_n(\log(n/a_n)+\log\log n)\Big)^{1/2}} = 1 \quad \text{a.s.}$$

holds for any non-decreasing sequence of integers $\{a_n,\ n\geqslant 1\}$ satisfying

$$\text{(i) } 1\leqslant a_n\leqslant n;\ \text{(ii) } a_{2n}=O(a_n);\ \text{(iii) } a_n/\log n\to\infty$$

if and only if $E(X)=0,\ EX^2=1$ and $Ee^{t_0|X|}<\infty$ for some $t_0>0$ (see, for example, Csörgő and Révész (1981)). In contrast to the well-known Hartman-Wintner law of the iterated logarithm and its converse by Strassen (1966), Griffin and Kuelbs (1989) established a self-normalized law of the iterated logarithm for all distributions in the domain of attraction of a normal or stable law:

Theorem 6.1. [Griffin and Kuelbs (1989)].

(a) *If $EX=0$ and $EX^2 I\{|X|\leqslant x\}$ is slowly varying as $x\to\infty$, then*

$$\limsup_{n\to\infty} \frac{S_n}{V_{n,2}(2\log\log n)^{1/2}} = 1 \quad \text{a.s..}$$

(b) *Under the conditions of Theorem 4.2, there is a positive constant C such that*

$$\limsup_{n\to\infty} \frac{S_n}{V_{n,2}(2\log\log n)^{1/2}} = C \quad \text{a.s..}$$

Csörgő, Lin and Shao (1994) obtained an analogue for the increments.

Theorem 6.2. [Csörgő, Lin and Shao (1994)]. *If $EX=0$ and $EX^2I\{|X| \leqslant x\}$ is slowly varying as $x\to\infty$, then*

$$\limsup_{n\to\infty} \max_{0\leqslant k\leqslant n} \max_{1\leqslant i\leqslant a_n} \frac{|S_{k+i}-S_k|}{(\sum_{k<j\leqslant k+a_n} X_j^2)^{1/2}\Big(2(\log(n/a_n)+\log\log n)\Big)^{1/2}}$$
$$= 1 \quad \text{a.s.}$$

for any non-decreasing sequence of integers $\{a_n,\ n\geqslant 1\}$ satisfying (i), (ii) *and* (iii).

By applying the self-normalized moderate deviation Theorem 4.2 and the subsequence method, the precise constant C in Griffin and Kuelbs's LIL can be determined. Furthermore, the result can be extended to a more general case.

Theorem 6.3. [Shao (1997)]. *Under the conditions of Theorem 4.2, we have*

$$\limsup_{n\to\infty} \frac{S_n}{(\log\log n)^{(p-1)/p} V_{n,p}} = ((p-1)\beta_p(\alpha, c_1, c_2))^{(1-p)/p} \quad \text{a.s..}$$

Similar to Theorem 6.2, we have

Theorem 6.4. [Shao (1998)]. *Let* $\{a_n,\ n \geqslant 1\}$ *be a non-decreasing sequence of integers satisfying* (i), (ii) *and* (iii). *Under the conditions of Theorem 4.2, we have*

$$\limsup_{n\to\infty} \max_{0\leqslant k\leqslant n} \max_{1\leqslant i\leqslant a_n} \frac{S_{k+i} - S_k}{(\sum_{k<j\leqslant k+a_n} |X_j|^p)^{1/p}\Big(\log(n/a_n) + \log\log n\Big)^{1-1/p}}$$
$$= ((p-1)\beta_p(\alpha, c_1, c_2))^{(1-p)/p} \quad \text{a.s.,}$$

for $p > \max(1, \alpha)$, *and*

$$\limsup_{n\to\infty} \max_{0\leqslant k\leqslant n} \max_{1\leqslant i\leqslant a_n} \frac{S_{k+i} - S_k}{\max_{k<j\leqslant k+a_n} |X_j|\Big(\log(n/a_n) + \log\log n\Big)}$$
$$= \frac{1}{\tau(\alpha, c_1, c_2)} \quad \text{a.s..}$$

7 Self-normalized moderate and large deviations in $\mathbb{R}^d$

It is natural to ask if the results stated in previous sections can be generalized to high dimensional space. Dembo and Shao (1998a, b) initiated the research in this direction.

Let $\mathcal{Z}$ be a topological space equipped with a σ-field $\mathcal{F}$, $\mathcal{K} = \{K : K \in \mathcal{F} \text{ is pre-compact}\}$, $\mathcal{S}$ a subset of $\mathcal{F}$, $I : \mathcal{Z} \to [0, \infty]$ a lower semi-continuous function, and $a_n \to \infty$. We say that the $\mathcal{Z}$-valued random variables $\boldsymbol{Z}_n$ satisfy a Partial Large Deviation Principle (PLDP) of speed a_n and rate function $I(\boldsymbol{z})$ with respect to $\mathcal{S}$ if the lower bound

$$-\inf_{\boldsymbol{z}\in A^o} I(\boldsymbol{z}) \leqslant \liminf_{n\to\infty} a_n^{-1} \log P(\boldsymbol{Z}_n \in A)\,,$$

holds for every $A \in \mathcal{F}$ and the upper bound

$$\limsup_{n\to\infty} a_n^{-1} \log P(\boldsymbol{Z}_n \in A) \leqslant - \inf_{\boldsymbol{z}\in\bar{A}} I(\boldsymbol{z}) ,$$

holds for every $A \in \mathcal{S}$.

Let $\boldsymbol{X}$, $\boldsymbol{X}_1, \boldsymbol{X}_2, \cdots$ be i.i.d. $\mathbb{R}^d$-valued random variables and let

$$\widetilde{SS}_n = n^{-1}\sum_{i=1}^{n} \boldsymbol{X}_i, \quad \widetilde{V}_{n,p} = n^{-1}\sum_{i=1}^{n} \|\boldsymbol{X}_i\|^p, \quad p > 1.$$

Define

$$\Lambda^*(\boldsymbol{s}, v) = \sup_{\boldsymbol{\lambda}\in\mathbb{R}^d,\theta\in\mathbb{R}} \left\{ \langle\boldsymbol{\lambda}, \boldsymbol{s}\rangle + \theta v - \log E(\exp(\langle\boldsymbol{\lambda}, \boldsymbol{X}\rangle + \theta||\boldsymbol{X}||^p)) \right\}.$$

Theorem 7.1. [Dembo and Shao (1998a, b)].

(a) $(\widetilde{SS}_n, \widetilde{V}_{n,p})$ *satisfies a Partial Large Deviation Principle (PLDP) of speed* $a_n = n$ *and rate function* $\Lambda^*(\boldsymbol{s}, v)$ *with respect to the collection of sets* $A \subset \mathbb{R}^{d+1}$ *such that*

$$\liminf_{v\to\infty,\ (\boldsymbol{s},v)\in A} \{||\boldsymbol{s}||^p)/v\} > 0,$$

(b) *Assume that* $E\boldsymbol{X} = 0$, $h(s) := E\|\boldsymbol{X}\|^2 I\{\|\boldsymbol{X}\| \leqslant s\}$ *is slowly varying and*

$$\sum = \lim_{s\to\infty} E\boldsymbol{X}\boldsymbol{X}' I\{\|\boldsymbol{X}\| \leqslant s\}/h(s)$$

exists. Let $a_n \to \infty$, $a_n = o(n)$ *and* $z_n \to \infty$ *be any sequence such that* $a_n = (1 + o(1))nz_n^{-2}h(z_n)$. *Let*

$$\boldsymbol{T}_n = a_n^{-1} z_n^{-1} \sum_{i=1}^{n} \boldsymbol{X}_i,$$

$$U_{n,p} = a_n^{-1} z_n^{-p} \sum_{i=1}^{n} \|\boldsymbol{X}_i\|^p.$$

Set $J_2(\boldsymbol{x}, 1) = \sup_{\boldsymbol{\lambda}}\{\langle\boldsymbol{\lambda}, \boldsymbol{x}\rangle - \langle\boldsymbol{\lambda}, \Sigma\boldsymbol{\lambda}\rangle/2\}$ *and* $J_2(\boldsymbol{x}, y) = \infty$ *for* $y \neq 1$. *Then,* $(\boldsymbol{T}_n, U_{n,2})$ *satisfy a PLDP of speed* $a_n = o(n)$ *and rate function* $J_2(\boldsymbol{x}, y)$ *with respect to the sets* $A \subset \mathbb{R}^{d+1}$ *such that*

$$(\boldsymbol{x}, y) \in A, y > 1 \implies (\boldsymbol{x}, 1) \in A \ ; \quad \liminf_{y\to\infty,\ (\boldsymbol{x},y)\in A} \{\|\boldsymbol{x}\|^2/y\} > 0.$$

(c) *Assume that the following conditions are satisfied for some* $0 < \alpha < 2$:

(i) $EX = 0$ *for* $1 < \alpha < 2$ *and* $\boldsymbol{X} \overset{d.}{=} -\boldsymbol{X}$ *for* $\alpha = 1$,

(ii) $P(\|\boldsymbol{X}\| > x) = l(x)x^{-\alpha}$,

(iii) $\boldsymbol{X}/\|\boldsymbol{X}\|$ *given condition* $\|\boldsymbol{X}\| > x$ *converges in distribution to a random variable* $\boldsymbol{Y}$ *on the unit sphere of* $\mathbb{R}^d$ *as* $x \to \infty$,

(iv) $\inf_{\boldsymbol{\lambda} \neq 0} P(\langle \boldsymbol{\lambda}, \boldsymbol{Y} \rangle > 0) > 0$.

Let $J_\alpha(\boldsymbol{x}, y) = \sup_{\boldsymbol{\lambda},\theta}\{\langle \boldsymbol{\lambda}, \boldsymbol{x}\rangle + \theta y - L_\alpha(\boldsymbol{\lambda}, \theta)\}$, *where*

$$L_\alpha(\boldsymbol{\lambda}, \theta) = \alpha \int_0^\infty \frac{E(\exp(y\langle \boldsymbol{\lambda}, \boldsymbol{Y}\rangle + \theta y^p)) - y\langle \boldsymbol{\lambda}, E\boldsymbol{Y}\rangle I\{\alpha > 1\} - 1}{y^{\alpha+1}} dy,$$

for $\theta < 0$, *while* $L_\alpha(\mathbf{0}, 0) = 0$ *and* $L_\alpha(\boldsymbol{\lambda}, \theta) = \infty$ *for* $\theta > 0$ *or* $\theta = 0$, $\boldsymbol{\lambda} \neq \mathbf{0}$. *Let* $p > \max(\alpha, 1)$, $z_n \to \infty$ *such that* $a_n = (1 + o(1))nz_n^{-\alpha} l(z_n)$. *Then, the random vectors* $(\boldsymbol{T}_n, U_{n,p})$ *satisfy the PLDP of speed* a_n *and rate function* $J_\alpha(\boldsymbol{x}, y)$ *with respect to sets* $A \subset \mathbb{R}^{d+1}$ *such that*

$$\liminf_{y \to \infty,\ (\boldsymbol{x}, y) \in A} \{\|\boldsymbol{x}\|^p / y\} > 0.$$

Theorem 7.1 has many interesting consequences, among which are the self-normalized laws of the iterated type results.

Theorem 7.2. [Demdo and Shao (1998a, b)].

(I) *Under the condition of* [b] *in Theorem 7.1,*

$$\lim_{n\to\infty} a_n^{-1} \log P\left(\frac{\|\sum_{i=1}^n \boldsymbol{X}_i\|_2}{(\sum_{i=1}^n \|\boldsymbol{X}_i\|_2^2)^{1/2}} \geqslant \sqrt{a_n}\right) = -\frac{1}{2\,\sigma^2}$$

and

$$\limsup_{n\to\infty} \frac{\|\sum_{i=1}^n \boldsymbol{X}_i\|_2}{(2\,\sigma^2\,(\log\log n)\sum_{i=1}^n \|\boldsymbol{X}_i\|_2^2)^{1/2}} = 1 \quad \text{a.s..}$$

where $\sigma^2 = \sup_{\|\boldsymbol{x}\|_2 = 1}\langle \boldsymbol{x}, \Sigma \boldsymbol{x}\rangle$.

(II) *Let* $\boldsymbol{X} = (X_1, ..., X_d)$, *where* $X_1, ..., X_d$ *are i.i.d. real valued random variables such that* $EX_1 = 0$ *and* $h_1(s) := EX_1^2 I\{|X_1| \leqslant s\}$ *is slowly varying at* $s = \infty$. *For* $q \geqslant 1$, *define* $\sigma_q^2 = d^{-1+(2-q)/qI\{1\leqslant q\leqslant 2\}}$. *Then*

$$\lim_{n\to\infty} a_n^{-1} \log P\left(\frac{\|\sum_{i=1}^n \boldsymbol{X}_i\|_q}{(\sum_{i=1}^n \|\boldsymbol{X}_i\|_2^2)^{1/2}} \geqslant \sqrt{a_n}\right) = -\frac{1}{2\sigma_q^2}$$

and

$$\limsup_{n\to\infty} \frac{\|\sum_{i=1}^n \boldsymbol{X}_i\|_q}{(2\sigma_q^2(\log\log n)\sum_{i=1}^n \|\boldsymbol{X}_i\|_2^2)^{1/2}} = 1 \quad \text{a.s..}$$

(III) *Under the assumptions of* (c) *in Theorem 7.1,*

$$\lim_{n\to\infty} a_n^{-1} \log P\left(\frac{\|\sum_{i=1}^n \boldsymbol{X}_i\|}{(\sum_{i=1}^n \|\boldsymbol{X}_i\|^p)^{1/p}} \geqslant a_n^{1-1/p} \right) = -\inf_{\boldsymbol{x}} J_\alpha(\boldsymbol{x}, \|\boldsymbol{x}\|^p)$$

and

$$\limsup_{n\to\infty} \frac{\|\sum_{i=1}^n \boldsymbol{X}_i\|}{(\log\log n)^{1-1/p}(\sum_{i=1}^n \|\boldsymbol{X}_i\|^p)^{1/p}} = \left(\inf_{\boldsymbol{x}} J_\alpha(\boldsymbol{x}, \|\boldsymbol{x}\|^p)\right)^{(1-p)/p} \quad \text{a.s..}$$

We note that $(\sum_{i=1}^n \|\boldsymbol{X}_i\|^p)^{1/p}$ may not be a useful self-normalizing sequence as a lot of commonly used statistics do not have this form. In next section, we study the large and moderate deviation of the Hotelling T^2 statistic.

8 Large and moderate deviations of Hotelling's T^2 statistic

Let $\boldsymbol{X}, \boldsymbol{X}_1, \boldsymbol{X}_2, \ldots$ be a sequence of independent and identically distributed (i.i.d.) nondegenerate $\mathbb{R}^d$-valued random vectors with mean $\boldsymbol{\mu}$, where $d \geqslant 1$. Let

$$SS_n = \sum_{i=1}^n \boldsymbol{X}_i, \ \boldsymbol{V}_n = \sum_{i=1}^n (\boldsymbol{X}_i - SS_n/n)(\boldsymbol{X}_i - SS_n/n)'.$$

Define Hotelling's T^2 statistic by

$$T_n^2 = (SS_n - n\boldsymbol{\mu})' \boldsymbol{V}_n^{-1} (SS_n - n\boldsymbol{\mu}). \tag{8.1}$$

Here when $\boldsymbol{V}_n$ is not full rank, i.e., $\boldsymbol{V}_n$ is degenerate, $\boldsymbol{x}'\boldsymbol{V}_n^{-1}\boldsymbol{x}$ is defined as

$$\boldsymbol{x}'\boldsymbol{V}_n^{-1}\boldsymbol{x} = \sup_{||\boldsymbol{\theta}||=1, \boldsymbol{\theta}'\boldsymbol{x}>0} \frac{(\boldsymbol{\theta}'\boldsymbol{x})^2}{\boldsymbol{\theta}'\boldsymbol{V}_n\boldsymbol{\theta}}$$

and $0/0$ is interpreted as ∞.

The T^2-statistic is used for testing hypotheses about the mean $\boldsymbol{\mu}$ and for obtaining confidence regions for the unknown $\boldsymbol{\mu}$. When $\boldsymbol{X}$ has a normal distribution $N(\boldsymbol{\mu}, \Sigma)$, it is known that $(n-d)T_n^2/(dn)$ is distributed as an F-distribution with d and $n-d$ degrees of freedom (see, e.g., Anderson (1984)). When the distribution of $\boldsymbol{X}$ is not normal, it was proved by Sepanski (1994) that the limiting distribution of T_n^2 as $n \to \infty$ is a χ^2-distribution with d degrees of freedom. Parallel to Theorems 2.1 and 4.1, Dembo and Shao (2006) proved large and moderate deviations for the T^2-statistic.

Theorem 8.1. [Dembo and Shao (2006)].

(i) *Assume that* $\boldsymbol{\mu} = 0$. *For* $\alpha \in (0,1)$, *let*

$$K(\alpha) = \sup_{b \geqslant 0} \sup_{||\boldsymbol{\theta}||=1} \inf_{t \geqslant 0} E \exp\Big(t\big(b\boldsymbol{\theta}'\boldsymbol{X} - \alpha((\boldsymbol{\theta}'\boldsymbol{X})^2 + b^2)/2\big)\Big).$$

Then, for all $x > 0$,

$$\lim_{n\to\infty} P\Big(T_n^2 \geqslant x\,n\Big)^{1/n} = K(\sqrt{x/(1+x)}).$$

(ii) *Let* $\{x_n,\ n \geqslant 1\}$ *be a sequence of positive numbers with* $x_n \to \infty$ *and* $x_n = o(n)$ *as* $n \to \infty$. *Assume that* $h(x) := E||\boldsymbol{X}||^2 I\{||\boldsymbol{X}|| \leqslant x\}$ *is slowly varying and*

$$\liminf_{x\to\infty} \inf_{\boldsymbol{\theta}\in\mathbb{R}^d, ||\boldsymbol{\theta}||=1} E(\boldsymbol{\theta}'\boldsymbol{X})^2 I\{||\boldsymbol{X}|| \leqslant x\}/h(x) > 0\,. \tag{8.2}$$

If $\boldsymbol{\mu} = 0$, *then*

$$\lim_{n\to\infty} x_n^{-1} \ln P\Big(T_n^2 \geqslant x_n\Big) = -\frac{1}{2}.$$

(iii) *Assume that* $h(x) := E||\boldsymbol{X}||^2 I\{||\boldsymbol{X}|| \leqslant x\}$ *is slowly varying and* (8.2) *is satisfied. If* $\boldsymbol{\mu} = 0$, *then*

$$\limsup_{n\to\infty} \frac{T_n^2}{2\log\log n} = 1 \quad \text{a.s.}.$$

9 Large and moderate deviations for self-normalized empirical processes

Let $\{X, X_i, i \geqslant 1\}$ be a sequence of i.i.d. random variables with values in some measurable space $(\boldsymbol{X}, \mathbf{C})$. Let $\mathbf{F}$ be a permissible class of real measurable functions on $(\boldsymbol{X}, \mathbf{C})$. Consider all functions f in $\mathbf{F}$ satisfy

$$Ef(X) = 0 \ \text{ and } \ Ef^2(X) = 1,$$

and define

$$W_n(f) = \frac{P_n(f)}{\sqrt{P_n(f^2)}},$$

where P_n denotes the empirical measure $P_n = \frac{1}{n}\sum_{k=1}^n \delta_{X_k}$. For each fixed function f in $\mathbf{F}$, $W_n(f)$ is a self-normalized sum. Bercu, Gassiat and Rio (2002) recently studied large and moderate deviations for the self-normalized empirical processes $(W_n(f))_{f\in\mathbf{F}}$ as well as various exponential bounds for the deviation of $W_n = \sup_{f\in\mathbf{F}} W_n(f)$.

Theorem 9.1. [Bercu, Gassiat and Rio (2002)]. *Assume that* $\mathbf{F}$ *is a countable P-Dansker class with finite bracketing numbers in* $L^2(P)$ *satisfying concordance of signs. Then, for any sequence* (x_n) *such that* $x_n \to \infty$, $x_n = o(\sqrt{n})$

$$\lim_{n\to\infty} \frac{1}{x_n^2} \ln P(W_n \geqslant \frac{x_n}{\sqrt{n}}) = -\frac{1}{2}.$$

They also proved a large deviation principle for W_n. Consider $\mathbf{F} = \{f_\gamma : \gamma \in [m, 0^-] \cap [0^+, M]\}$ where $m < 0$, $M > 0$,

$$f_\gamma(x) = \begin{cases} \exp(\gamma t(x) - l(\gamma)) - 1, & \text{if } \gamma \neq 0, \\ t(x), & \text{if } \gamma = 0^+, \\ -t(x), & \text{if } \gamma = 0^-. \end{cases}$$

The function t is continuous real measurable with moment generating function $l(\gamma) = \ln E\exp(\gamma t(X))$. Assume that $[2m, 2M]$ is included in the domain of l so that the functions of $\mathbf{F}$ are square integrable. Define the rate function I by

$$I = \begin{cases} \inf_{f\in\mathbf{F}} I_f(x), & \text{if } x \geqslant 0, \\ \inf_{f\in\mathbf{F}} I_{-f}(x), & \text{otherwise.} \end{cases}$$

where $I_f(x) = -\ln\sup_{a\geqslant 0}\inf_{t\geqslant 0} E\exp(t(af(X) - x(f^2(X) + a^2)/2))$.

Theorem 9.2. [Bercu, Gassiat and Rio (2002)]. *Assume that* $t(X)$ *has a continuous distribution function and* $Et(X) = 0$. *Then,* (W_n) *satisfies a large deviation principle with continuous rate function* I. *In particular, for any* $x > 0$

$$\lim_{n\to\infty} \frac{1}{n} \ln P(W_n \geqslant x) = -I(x).$$

The restrictions on $\mathbf{F}$ in the above theorems seem too strong. It would be interesting to obtain large and moderate deviations for general self-normalized empirical processes.

10 Limiting distributions of self-normalized sums

Let $X, X_1, X_2, \cdots$ be i.i.d. random variables. It is well-known that $S_n/b_n - a_n$ has a non-degenerate limiting distribution for some suitably chosen real constants a_n and $b_n > 0$ if and only if X is in the domain of attraction of a stable law with index α $(0 < \alpha \leqslant 2)$. When $\alpha = 2$, it is equivalent to $EX^2 I(|X| \leqslant x)$ is slowly varying as $x \to \infty$, while for $0 < \alpha < 2$, it is equivalent to

$$P(X \geqslant x) = \frac{(c_1 + o(1))h(x)}{x^\alpha}, \quad P(X \leqslant -x) = \frac{(c_2 + o(1))h(x)}{x^\alpha}$$

as $x \to \infty$, where $c_1 \geqslant 0, c_2 \geqslant 0, c_1 + c_2 > 0$ and $h(x)$ is slowly varying at ∞. It is also known that the normalizing constants a_n and b_n are complicatedly determined by the slowly varying function l. On the other hand, self-normalization shows again its neatness. Logan, Mallows, Rice and Shepp (LMRS)(1973) proved that all limit laws of $S_n/V_{n,2}$ for X in the domain of attraction of a stable law with index $\alpha \in (0,\ 2)$ have a sub-Gaussian tail which depends in a complicated way on the parameter α, and posed a conjecture which was solved in [56].

Theorem 10.1. [Logan, Mallows, Rice and Shepp (1973) and Shao (1997)].

Under the conditions of Theorem 4.2, the limiting density function $p(x)$ of $S_n/V_{n,2}$ exists and satisfies as $x \to \infty$

$$p(x) \sim \frac{1}{\alpha}\left(\frac{2}{\pi}\right)^{1/2} \sqrt{2\beta(\alpha, c_1, c_2)}e^{-x^2\beta(\alpha,c_1,c_2)}.$$

Logan, Mallows, Rice and Shepp (1973) also stated that S_n/V_n is asymptotically normal if [and perhaps only if] X is in the domain of attraction of the normal law and that it seems worthy of conjecture that the only possible nontrivial limiting distributions of S_n/V_n are those obtained when X follows a stable law. The first conjecture of LMRS was proved by Giné, Götze and Mason (1997) and the second conjecture was recently confirmed by Chistyakov and Götze (2004).

Theorem 10.2. [Gine, Götze and Mason (1997)].

$$S_n/V_{n,2} \xrightarrow{d.} N(0,1)$$

if and only if $EX = 0$ and $EX^2 I\{|X| \leqslant x\}$ is slowly varying.

Theorem 10.3. [Chistyakov and Götze (2004)]. *S_n/V_n converges weakly to a random variable Z such that $P(|Z| = 1) < 1$ if and only if*

(i) *X is in the domain of attraction of a stable law with index $\alpha \in (0, 2]$;*

(ii) *$EX = 0$ if $1 < \alpha \leqslant 2$;*

(iii) *if* $\alpha = 1$, *then* X *is in the domain of attraction of the Cauchy law and Feller's condition holds, that is,* $\lim_{n\to\infty} nE\sin(X/a_n)$ *exists and is finite, where* $a_n = \inf\{x > 0 : nx^{-2}(1 + EX^2 I\{|X| \leqslant x\}) \leqslant 1\}$.

Chistyakov and Götze (2004) also proved that S_n/V_n converges weakly to Z with $P(|Z| = 1)$ if and only if $P(|X| > x)$ is a slowly varying function at $+\infty$.

For the asymptotic distribution of self-normalized censored sums and trimmed sums, we refer to Hahn, Kuelbs and Weiner (1990) and Griffin and Pruitt (1991).

11 Weak invariance principle for self-normalized partial sum processes

Let $X, X_1, X_2, \cdots$ be i.i.d. random variables. As a generalization of the central limit theorem, the classical weak invariance principle states that on an appropriate probability space,

$$\sup_{0\leqslant t\leqslant 1} |\frac{1}{\sqrt{n}\sigma} S_{[nt]} - \frac{1}{\sqrt{n}} W(nt)| = o(1) \text{ in probability}$$

if and only if $EX = 0$ and $\mathrm{Var}(X) = \sigma^2$, where $\{W(t), t \geqslant 0\}$ is a standard Wiener process. The weak invariance principle is a stronger version of Donsker's classical functional central limit theorem. In view of the self-normalized central limit theorem, Csörgő, Szyszkowicz and Wang (2003) proved a self-normalized version of the weak invariance principle.

Theorem 11.1. [Csörgő, Szyszkowicz and Wang (2003)]. *As* $n \to \infty$, *the following statements are equivalent:*

(a) $EX = 0$ *and* X *is in the domain of attraction of the normal law;*

(b) $S_{[nt]}/V_n \to W(t)$ *weakly on* $(D[0,1], \rho)$, *where* ρ *is the sup-norm metric for functions in* $D[0,1]$, *and* $\{W(t), 0 \leqslant t \leqslant 1\}$ *is a standard Wiener process;*

(c) *On an appropriate probability space, we can construct a standard Wiener process* $\{W(t), t \geqslant 0\}$ *such that*

$$\sup_{0\leqslant t\leqslant 1} |S_{[nt]}/V_n - W(nt)/\sqrt{n}| = o(1) \quad \textit{in probability.}$$

The following are immediate corollaries of the above weak invariance principle. If $EX = 0$ and X is in the domain of attraction of the normal law, then

$$P(\max_{1\leqslant i\leqslant n} S_i/V_n \leqslant x) \to P(\sup_{0\leqslant t\leqslant 1} W(t) \leqslant x),$$

$$P(\max_{1\leqslant i\leqslant n} |S_i|/V_n \leqslant x) \to P(\sup_{0\leqslant t\leqslant 1} |W(t)| \leqslant x),$$

$$P(n^{-1}\sum_{i=1}^n S_i^2/V_n^2 \leqslant x) \to P\Big(\int_0^1 W^2(t)dt \leqslant x\Big),$$

$$P(n^{-1}\sum_{i=1}^n |S_i|/V_n \leqslant x) \to P\Big(\int_0^1 |W(t)|dt \leqslant x\Big).$$

When X is in the domain of attraction of a stable law, Hu and Shao (2005) obtained a similar invariance principle.

Let N_1, N_2 be two independent Poisson random measures with intensity L on σ-finite measure space $(\mathbb{R}^2, \mathcal{B}^2, L)$, where $\mathcal{B}^2$ is the Borel σ-algebra of $\mathbb{R}^2$ and L is the Lebesgue measure on $\mathbb{R}^2$. For $i = 1, 2$ put

$$N_i(t, y) := N_i([0, t] \times [0, y]) \text{ for } t \geqslant 0,\ y \geqslant 0$$

and define the following processes:

$$\Delta_{\alpha,i}(t) = \begin{cases} \alpha^{-1}\int_0^\infty N_i(t,y)y^{-1-1/\alpha}dy, & \text{if } 0 < \alpha < 1, \\ \int_e^\infty (N_i(t,y) - ty)y^{-2}dy + \int_0^e N_i(t,y)y^{-2}dy, & \text{if } \alpha = 1, \\ \alpha^{-1}\int_0^\infty (N_i(t,y) - ty)y^{-1-1/\alpha}dy, & \text{if } 1 < \alpha \leqslant 2. \end{cases}$$

Theorem 11.2. [Hu and Shao (2005)]. *Assume that there exists $\alpha \in (0,2)$, $c_1 \geqslant 0, c_2 \geqslant 0, c_1 + c_2 > 0$ and a slowly varying function $h(x)$ such that*

$$P(X \geqslant x) = \frac{c_1 + o(1)}{x^\alpha} h(x) \quad \text{and } P(X \leqslant -x) = \frac{c_2 + o(1)}{x^\alpha} h(x)$$

as $x \to \infty$. Moreover, assume that $EX = 0$ if $1 < \alpha < 2$ and X is symmetric if $\alpha = 1$. Write

$$\Delta_\alpha(t) = -\omega_1\Delta_{\alpha,1}(t) + \omega_2\Delta_{\alpha,2}(t),$$
$$\widetilde{\Delta}_\alpha(t) = \omega_1^2\Delta_{\alpha/2,1}(t) + \omega_2^2\Delta_{\alpha/2,2}(t), \quad 0 \leqslant t \leqslant 1,$$

where $w_1 = \left(\frac{c_2}{c_1+c_2}\right)^{\frac{1}{\alpha}}$, $w_2 = \left(\frac{c_1}{c_1+c_2}\right)^{\frac{1}{\alpha}}$, then in $D[0,1]$,

$$S_{[nt]}/V_n \Rightarrow \Delta_\alpha(t)/\sqrt{\widetilde{\Delta}_\alpha(1)}.$$

Theorem 11.2 can be applied to obtain the limiting distribution of a unit-root test statistic when the data is in the domain of attraction of a stable law.

12 Asymptotic distributions of non-central self-normalized sums

Let $X, X_1, X_2, \ldots$ be a sequence of i.i.d. random variables with $\sigma^2 = \mathrm{Var}(X) < \infty$. Define $\mathbb{T}_n = \sqrt{n}\bar{X}/S$, where $\bar{X}$ and S^2 are the sample mean and the sample variance, respectively. We refer to $\mathbb{T}_n$ as the central or non-central (Student's) t-statistic, depending on $E(X) = 0$ or $E(X) \neq 0$, respectively. The non-central t-statistic arises naturally in the calculation of powers for t-tests. The central t-statistic have been well studied while there is a very limited literature on the non-central t-statistic. Bentkus, Jing, Shao and Zhou (2006) studied the limiting behavior of the non-central t-statistic, which turns out to be much more complicated. For instance, it is well known that, under finite second moment conditions, the limiting distributions for the central t-statistic are normal while those for the non-central t-statistic can be non-normal, and critically depend on whether or not $E(X^4) = \infty$.

Assume that $E|X| < \infty$ and $\mu \neq 0$. Define $\nu = \mu/\sigma$ and let Z_τ follow a stable law with an index $\tau \in (0, 2]$ whose characteristic function is given by

$$E\exp(itZ_\tau) = \begin{cases} \exp\{-t^2/2\}, & \text{if } \tau = 2, \\ \exp\{-|t|^\tau(1 - i(\mathrm{sign} t)\tan(\pi\tau/2))\}, & \text{if } \tau \in (1,2)\cup(0,1), \\ \exp\{-|t|(1 + 2(i/\pi)(\mathrm{sign} t)\ln|t|)\}, & \text{if } \tau = 1. \end{cases}$$

When $\tau = 2$, write $Z = Z_2$, which is a standard normal r.v. Write $X \in DA(\tau)$ if X belongs to the domain of attraction of the stable law of Z_τ. Denote $Y \in \mathcal{B}(p, 0, 1)$ $(0 < p < 1)$ if Y is a standardized Bernoulli random variable and write $X \in \mathcal{B}(p, \mu, \sigma^2)$ if $X = \sigma Y + \mu$, where $Y \in \mathcal{B}(p, 0, 1)$. It is easy to see that $Y \in \mathcal{B}(p, 0, 1)$ if and only if

$$P\left\{Y = -\frac{p}{\sqrt{pq}}\right\} = q, \quad P\left\{Y = \frac{q}{\sqrt{pq}}\right\} = p,$$

where $q = 1 - p$.

Next result gives the limiting distribution when $E(X^4) < \infty$.

Theorem 12.1. [Bentkus, Jing, Shao and Zhou (2006)]. *Assume that* $E(X^4) < \infty$.

(i) *For* $X \sim \mathcal{B}(p, \mu, \sigma^2)$ *such that* $\mu/\sigma = 2\sqrt{pq}/(q - p)$ *and* $p \neq 1/2$, *we have*

$$a_n(\mathbb{T}_n - \nu_0\sqrt{n-1}) \xrightarrow{\mathcal{D}} N^2(0, 1),$$

where $\nu_0 = 2\sqrt{pq}/(q - p)$ *and* $a_n = 2\nu_0\sqrt{n}/(1 + \nu_0^2)$.

(ii) *For any r.v. X other than the one given in* (i), *we have*

$$\sigma_0^{-1}(\mathbb{T}_n - \nu\sqrt{n}) \xrightarrow{\mathcal{D}} N(0,1),$$

where $\sigma_0^2 = 1 - \nu\alpha_3 + \nu^2(\alpha_4 - 1)/4$, *and* $\alpha_k = E(X-\mu)^k/\sigma^k$, $k = 3, 4$.

The limiting distribution is quite different when $E(X^4) = \infty$. Standardize $\{X, X_k, k \geqslant 1\}$ by $Y = (X-\mu)/\sigma$ and $Y_k = (X_k - \mu)/\sigma$ for $k \geqslant 1$. Assume that $X^2 \in DA(\tau)$ with $\tau \in [1,2]$. If $\tau \in (1,2]$, then $Y^2 \in DA(\tau)$ and

$$c_n n^{-1/\tau}(Y_1^2 + \cdots + Y_n^2 - n) \xrightarrow{\mathcal{D}} Z_\tau, \quad \text{as } n \to \infty, \tag{12.1}$$

where c_n is a slowly varying sequence (see Feller, 1971 or Ibragimov and Linnik, 1971). In the case $\tau = 1$, we have

$$c_n n^{-1}(Y_1^2 + \cdots + Y_n^2) - d_n \xrightarrow{\mathcal{D}} Z_1, \quad \text{as } n \to \infty, \tag{12.2}$$

with slowly varying $c_n > 0$. The centering constants d_n satisfy $d_n \to \infty$.

Theorem 12.2. *Assume that $\mu \neq 0$, $E(X^2) < \infty$, and $E(X^4) = \infty$. Further assume that $X^2 \in DA(\tau)$ for $1 \leqslant \tau \leqslant 2$ and that $(d_n - c_n)^2 = o(c_n)$ when $\tau = 1$. Define*

$$a_n = 2c_n n^{1/2-1/\tau}/\nu, \; b_n = \begin{cases} \nu\sqrt{n}, & \text{if } 1 < \tau \leqslant 2, \\ \nu\sqrt{n}(3/2 - d_n/(2c_n)), & \text{if } \tau = 1. \end{cases} \tag{12.3}$$

(i) *If $\tau = 2$, then*

$$a_n(\mathbb{T}_n - b_n) \xrightarrow{\mathcal{D}} Z;$$

(ii) *If $\tau \in [1,2)$, then*

$$a_n(\mathbb{T}_n - b_n) \xrightarrow{\mathcal{D}} -Z_\tau.$$

Bentkus, Jing, Shao and Zhou (2006) also studied the limiting distributions of the non-central t-statistic when $\mu = \mu_n$.

13 Self-normalized processes: a pseudo-maximization approach

Consider a pair of random variables A, B, with $B > 0$ such that

$$E\exp\{\lambda A - \lambda^2 B^2/2\} \leqslant 1, \tag{13.1}$$

for all $-\infty < \lambda < \infty$. This is the "canonical assumption" in the pseudo-maximization approach of de la Pena, Klass and Lai (2000, 2004) to self-normalized processes. They note that if the integrand $\exp\{\lambda A - \lambda^2 B^2/2\}$ in (13.1) can be maximized over λ inside the expectation (as can be done if A/B^2 is non-random), taking the value $\lambda = A/B^2$ one would get $E\exp\{\frac{A^2}{2B^2}\} \leqslant 1$. This in turn would give the optimal Chebyshev-type bound $P(\frac{A}{B} > x) \leqslant \exp\{\frac{-x^2}{2}\}$. Since A/B^2 can not (in general) be taken to be non-random, we need to find an alternative method for dealing with this maximization. One approach for attaining a similar effect involves integrating over a probability measure F, and using Fubini's theorem to interchange the order of integration with respect to P and F. To be effective for all possible pairs (A, B), the F chosen would need to be as uniform as possible so as to "pick" the maximum value of $T(\lambda) = \exp\{\lambda A - \lambda^2 B^2/2\}$ regardless of where it might occur. Thereby some mass is certain to be assigned to and near the random value $\lambda = A/B^2$ which maximizes $\lambda A - \lambda^2 B^2/2$. Since all uniform measures are multiples of Lebesgue measure (which is infinite), we construct a finite measure (or a sequence of finite measures) which tapers off to zero as slowly as we can to manage, at $\lambda = \infty$. This approach will be used in what follows to provide exponential and moment inequalities for A/B, $A/\sqrt{B^2 + (EB)^2}$, $A/\{B\sqrt{\log\log(B \vee e^2)}\}$ as well as LIL's. We begin with the second case where the proof is more transparent. The approach, pioneered by Robbins and Siegmund (1970) and commonly known as the method of mixtures, was used by de la Peña, Klass and Lai (2004) to prove the following.

Theorem 13.1. *Let A, B with $B > 0$ be random variables satisfying the canonical assumption* (13.1) *for all $\lambda \in \mathbb{R}$. Then*

$$P\Big(\frac{|A|}{\sqrt{B^2 + (EB)^2}} > x\Big) \leqslant \sqrt{2}\exp(-x^2/4) \tag{13.2}$$

for all $x > 0$.

Proof. Multiplying both sides of (13.1) by $(2\pi)^{-1/2} y \exp(-\lambda^2 y^2/2)$ (with $y > 0$) and integrating over λ, we obtain by using Fubini's theorem that

$$1 \geqslant \int_{-\infty}^{\infty} E\frac{y}{\sqrt{2\pi}} \exp\Big\{\lambda A - \frac{\lambda^2}{2} B^2\Big\} \exp\Big\{-\frac{\lambda^2 y^2}{2}\Big\} d\lambda \tag{13.3}$$

$$= E\Big[\frac{y}{\sqrt{B^2 + y^2}} \exp\Big\{\frac{A^2}{2(B^2 + y^2)}\Big\} \times \int_{-\infty}^{\infty} \frac{\sqrt{B^2 + y^2}}{\sqrt{2\pi}}$$

$$\exp\Big\{-\frac{B^2+y^2}{2}\Big(\lambda^2-2\frac{A}{B^2+y^2}\lambda+\frac{A^2}{(B^2+y^2)^2}\Big)\Big\}d\lambda\Big]$$

$$=E\Big[\frac{y}{\sqrt{B^2+y^2}}\exp\Big\{\frac{A^2}{2(B^2+y^2)}\Big\}\Big].$$

By the Cauchy-Schwarz inequality and (13.3),

$$E\exp\Big\{\frac{A^2}{4(B^2+y^2)}\Big\} \tag{13.4}$$

$$\leqslant\Big\{\Big(E\frac{y\exp\{\frac{A^2}{2(B^2+y^2)}\}}{B^2+y^2}\Big)\Big(E\sqrt{\frac{B^2+y^2}{y^2}}\Big)\Big\}^{1/2}$$

$$\leqslant\Big(E\sqrt{\frac{B^2}{y^2}+1}\Big)^{1/2}.$$

Since $E\sqrt{\frac{B^2}{y^2}+1}\leqslant E(\frac{B}{y}+1)$, the special case $y=EB$ in (13.4) gives

$$E\exp(A^2/[4(B^2+(EB)^2])\leqslant\sqrt{2}.$$

Then, using Markov's inequality and this we get

$$P\{\frac{|A|}{\sqrt{B^2+(EB)^2}}\geqslant x\}P\{\frac{A^2}{4(B^2+(EB)^2)}\geqslant\frac{x^2}{4}\}\leqslant\sqrt{2}\ \exp(-x^2/4).$$

In what follows we discuss the analysis of other boundary crossing probabilities by using the method of mixtures, first introduced in Robbins and Siegmund (1970), under the following variant of the canonical assumption:

$$\Big\{\exp(\lambda A_t-\lambda^2B_t^2/2),t\geqslant 0\Big\}\text{ is a supermartingale with mean }\leqslant 1 \tag{13.5}$$

for $0<\lambda\leqslant\lambda_0$. We begin by introducing the so-called "Robbins-Siegmund" (R-S) boundaries; see Lai (1976). Let F be a finite positive measure on $(0,\lambda_0)$ and assume that $F(0,\lambda_0)>0$. Let $\Psi(u,v^2)=\int\exp(\lambda u-\lambda^2v^2/2)dF(\lambda)$. Given $c>0$ and $v^2>0$, the equation

$$\Psi(u,v^2)=c,$$

has a unique solution

$$u=\beta_F(v^2,c).$$

Moreover, $\beta_F(v^2,c)$ is a concave function of v^2 and

$$\lim_{v^2\to\infty}\frac{\beta_F(v^2,c)}{v^2}=b/2,$$

where

$$b = \sup\{y > 0 : \int_0^y F(d\lambda) = 0\},$$

with sup over the empty set equal to zero. As shown by de la Pena, Klass and Lai (2004), the R-S boundaries $\beta_F(v^2, c)$ can be used for analyzing the boundary crossing probability

$$P(A_t \geqslant g(B_t) \text{ for some } t \geqslant 0)$$

when $g(B_t) = \beta_F(B_t^2, c)$ for some F and $c > 0$. This probability equals

$$\begin{aligned} &P(A_t \geqslant \beta_F(B_t^2, c) \text{ for some } t \geqslant 0) \\ &= P(\Psi(A_t, B_t^2) \geqslant c \text{ for some } t \geqslant 0) \leqslant F(0, \lambda_0)/c, \end{aligned} \tag{13.6}$$

using Doob's inequality on the supermartingale $\Psi(A_t, B_t)$, $t \geqslant 0$. In particular, let $\delta > 0$ and

$$dF(\lambda) = \frac{1}{\lambda(\log(1/\lambda))(\log\log(1/\lambda)^{1+\delta}} d\lambda$$

for $0 < \lambda < e^{-e}$. Let $\log_2(x) = \log\log(x)$ and $\log_3(x) = \log\log_2(x)$. As shown in Example 4 of Robbins and Siegmund (1970),

$$\beta_F(v^2, c) = \sqrt{2v^2[\log_2(v^2) + (3/2 + \delta)\log_3(v^2) + \log(c/2\sqrt{\pi}) + o(1)]},$$

as $v^2 \to \infty$. With this choice of F, the probability in (13.6) is bounded by $F(0, e^{-e})/c$ for all $c > 0$. Given $\epsilon > 0$, take δ small enough and and $c = c(\delta)$ large enough so that $F_\delta(0, e^{-e})/c(\delta) < \epsilon$. Since ϵ can be arbitrarily small and for fixed $c = c(\delta)$, $\beta_F(v^2, c) \sim \sqrt{2v^2 \log\log(v)}$ as $v^2 \to \infty$,

$$\limsup_{t\to\infty} \frac{A_t}{\sqrt{2B_t \log\log(B_t)}} \leqslant 1 \text{ on } \{\lim_{t\to\infty} B_t = \infty\}. \tag{13.7}$$

13.1 LILs for self-normalized martingales

By making use of the following lemma, de la Pena, Klass and Lai (2004) proved a universal upper LIL for self-normalized and suitably centered sums of random variables, under no assumptions on their joint distributions, providing a far-reaching generalization of (13.7) that requires assumption (13.5).

Lemma 13.1. *Let $0 < \gamma < 1$ and define $a_\gamma = -\{\gamma + \log(1-\gamma)\}/\gamma^2$. Let Y_n be random variables adapted to the filtration $\{\mathcal{F}_n\}$. Let $0 \leqslant \gamma_n < 1$ and $0 < \lambda_n \leqslant 1/a_{\gamma_n}$ be $\mathcal{F}_{n-1}$-measurable random variables. Let $\mu_n = E(Y_n \mathbf{1}_{\{-\gamma_n \leqslant Y_n < \lambda_n\}} | \mathcal{F}_{n-1})$. Then $\{\exp[\Sigma_{i=1}^n (Y_i - \mu_i - \lambda_i^{-1} Y_i^2)]$, $\mathcal{F}_n$, $n \geqslant 1\}$ is a supermartingale whose expectation is $\leqslant 1$.*

Theorem 13.2. *Let X_n be random variables adapted to a filtration $\{\mathcal{F}_n\}$. Let $S_n = X_1 + \cdots + X_n$ and $V_n^2 = X_1^2 + \cdots + X_n^2$. Then, given any $\lambda > 0$, there exist positive constants c_λ and b_λ such that $\lim_{\lambda \to 0} b_\lambda = \sqrt{2}$ and*

$$\limsup \frac{S_n - \sum_{i=1}^n \mu_i(-\lambda v_n, c_\lambda v_n)\}}{V_n(\log\log V_n)^{1/2}} \leqslant b_\lambda \text{ a.s. on } \{\lim V_n = \infty\}, \tag{13.8}$$

where $v_n = V_n(\log\log V_n)^{-1/2}$ and $\mu_i(c,d) = E(X_i I\{c \leqslant X < d\}|\mathcal{F}_{i-1})$ for $c < d$.

The constants c_λ and b_λ in Theorem 13.2 can be determined as follows. For $\lambda > 0$, let $h(\lambda)$ be the positive solution of $h - \log(1+h) = \lambda^2$. Let $b_\lambda = h(\lambda)/\lambda$, $\gamma = h(\lambda)/\{1 + h(\lambda)\}$ and $c_\lambda = \lambda/(\gamma a_\gamma)$, where a_γ is given in Lemma 13.1. Then $\lim_{\lambda\to 0} b_\lambda = \sqrt{2}$. Let $e_k = \exp\{k/\log k\}$. The basic idea underlying the proof of Theorem 13.2 pertains to upper-bounding the probability of an event of the form $E_k = \{t_{k-1} \leqslant \tau_k < t_k\}$, in which t_j and τ_j are stopping times defined by

$$\begin{aligned} t_j &= \inf\{n : V_n \geqslant e_j\}, \\ \tau_j &= \inf\{n \geqslant t_j : S_n - \sum_{i=1}^n \mu_i(-\lambda v_n, c_\lambda v_n) \geqslant (1+3\epsilon) b_\lambda V_n (\log\log V_n)^{1/2}\}, \end{aligned} \tag{13.9}$$

letting $\inf \emptyset = \infty$. Note that for $i < n$, the centering constants $\mu_i(-\lambda v_n, c_\lambda v_n)$ involve v_n that is not determined until time n, so the centered sums that result do not form a martingale. Sandwiching τ_k between t_{k-1} and t_k enables us to replace both the random exceedance and truncation levels in (13.9) by constants. Then the event E_k can be re-expressed in terms of two simultaneous inequalities, one involving centered sums and the other involving a sum of squares. These inequalities are combined with Lemma 13.1 to conclude that $P(E_k)$ can be upper-bounded by the probability of an event involving the supremum of a nonnegative supermartingale with mean $\leqslant 1$, to which Doob's maximal inequality can be applied; see de la Pena, Klass and Lai (2004, pp.1924-1925) for details.

Although Theorem 13.2 gives an upper LIL for any adapted sequence of random variables X_n, the upper bound in (13.8) may not be attained. Example 6.4 of de la Pena, Klass and Lai (2004) suggests that one way to sharpen the bound is to center X_n at its conditional median before applying Theorem 13.2 to $\widetilde{X}_n = X_n - \text{med}(X_n|\mathcal{F}_{n-1})$. On the other hand, if X_n is a martingale difference sequence such that $|X_n| \leqslant m_n$ a.s. for some $\mathcal{F}_{n-1}$-measurable m_n with $m_n = o(v_n)$ and $v_n \to \infty$ a.s., then

Theorem 6.1 of de la Pena, Klass and Lai (2004) shows that (13.8) is sharp in the sense that

$$\limsup \frac{S_n}{V_n(\log\log V_n)^{1/2}} = \sqrt{2} \text{ a.s..} \tag{13.10}$$

Stout (1970, 1973) proved an analogous LIL in which V_n in (13.10) is replaced by the sum $s_n^2 = \Sigma_{i=1}^n E(X_i^2|\mathcal{F}_{i-1})$ of conditional variances, under $m_n = o(s_n(\log\log s_n)^{-1/2})$ and $s_n \to \infty$ a.s..

13.2 Moment inequalities for self-normalized processes

Caballero, Fernandez and Nualart (1998) provide moment inequalities for a continuous martingale divided by its quadratic variation, and use these results to show that if $\{M_t, t \geqslant 0\}$ is a continuous martingale null at zero, then for every $1 \leqslant p < q$, there exists a universal constant $C = C(p,q)$ such that

$$\left\|\frac{M_t}{\langle M_t\rangle}\right\|_p \leqslant C \left\|\frac{1}{\langle M_t\rangle^{\frac{1}{2}}}\right\|_q. \tag{13.11}$$

Related work in Revuz and Yor (1999 page 167) for continuous local martingales yields for all $p > q > 0$ the existence of a constant C_{pq} such that

$$E\frac{(\sup_{s<\infty}|M_s|)^p}{\langle M_\infty\rangle^{q/2}} \leqslant C_{pq}E(\sup_{s<\infty}|M_s|)^{p-q}. \tag{13.12}$$

The inequality (13.11) is used by Caballero, Fernandez and Nualart (1998) is used by them to establish the continuity (through Kolmogorov's criteria) and uniqueness of the solutions of a nonlinear stochastic partial differential equation with additive white noise. The basic idea behind their application is that the problems at hand translate into (negative) moment bounds on the quadratic variation of a martingale. A natural question that arises in connection with (13.11) is what about the case when the normalization is done by $\sqrt{\langle M\rangle_t}$. The following result from de la Peña, Klass and Lai (2000, 2004) provides information in this direction.

Theorem 13.3. *Let $A, B > 0$ be two random variables. Assume $E\exp\{\lambda A - \lambda^2B^2/2\} \leqslant 1$ for all $\lambda > 0$. Let $\log^+(x) = 0 \vee \log x$. Then for all $p > 0$,*

$$E\left|\frac{A^+}{B}\right|^p \leqslant c_{1,p} + c_{2,p}E(\log^+\log(B \vee \frac{1}{B}))^{p/2}, \tag{13.13}$$

$$E\Big(\frac{A^+}{B\sqrt{1\vee\log^+\log(B\vee\frac{1}{B})}}\Big)^p\leqslant C_p, \tag{13.14}$$

for some universal constants $0<c_{1,p},c_{2,p},C_p<\infty$.

The following example shows the optimality properties of Theorem 13.3. Let $\{Y_i\}$, $i=1,...$ be a sequence of i.i.d. Bernoulli random variables $Y_i=1$ with probability $1/2$ and $Y_i=-1$ with equal probability. Let

$$T=\inf\{n\geqslant e^e:\sum_{i=1}^n Y_i\geqslant\sqrt{2n\log\log n}\},$$

with $T=\infty$ if no such n exists. From the delicate LIL by Erdős (1942), it follows that $P(T<\infty)=1$. Let $d_{n,j}=Y_j1(T\geqslant j)$ if $1\leqslant j\leqslant n$ and $d_{n,j}=0$ if $j>n$. Then almost surely,

$$\frac{d_{n,1}+...+d_{n,n}}{\sqrt{d_{n,1}^2+...+d_{n,n}^2}}\frac{\sum_{j=1}^{\min\{T,n\}}Y_j}{\sqrt{\min\{T,n\}}}\to\sqrt{2\log\log T},$$

as $n\to\infty$. Therefore, the second term in (13.13) can not be removed.

The proof of Theorem 13.3 is based on the following result. Let $L:(0,\infty)\to(0,\infty)$ be a nondecreasing function such that

$$L(cy)\leqslant 3cL(y)\text{ for all }c\geqslant 1\text{ and }y>0, \tag{13.15}$$

$$L(y^2)\leqslant 3L(y)\text{ for all }y\geqslant 1, \tag{13.16}$$

$$\int_1^\infty\frac{dx}{xL(x)}=\frac{1}{2}. \tag{13.17}$$

Then

$$f(\lambda)=\frac{1}{\lambda L(\max(\lambda,1/\lambda))},\ \lambda>0,$$

is a density. An example satisfying (13.15) — (13.17) is

$$L(x)=2(\log xe^e)(\log\log xe^e)^2I(x\geqslant 1).$$

Let $g(x)=\frac{\exp\{x^2/2\}}{x}I(x\geqslant 1)$. Then

$$E\left[\frac{g(\frac{A}{B})}{L(\frac{A}{B})\vee L(B\vee\frac{1}{B})}\right]\leqslant\frac{3}{\int_0^1 e^{-x^2/2}dx}. \tag{13.18}$$

13.3 An expectation LIL for self-normalized processes

We next study the case of self-normalized inequalities when there is no possibility of explosion at the origin by shifting the denominator away from zero. An important result in this direction comes from Graversen and Peskir (2000).

Theorem 13.4. *Let $\{M_t, \mathcal{F}_t,\ t \geqslant 0\}$ be a continuous local martingale with quadratic variation process $\langle M\rangle_t$, $t \geqslant 0$. Let*

$$h(x) = \sqrt{\log(1 + \log(1 + x)}.$$

Then there exist universal constants D_1, $D_2 > 0$ such that

$$D_1 E[h(\langle M\rangle_\tau)] \leqslant E\left[\frac{\sup_{0\leqslant t\leqslant \tau} |M_s|}{\sqrt{1 + \langle M\rangle_t}}\right] \leqslant D_2 E[h(\langle M\rangle_\tau)],$$

for all stopping times τ of M.

The proof of this result was obtained by making use of Lenglart's inequality, the optional sampling theorem and Itô's formula. Shortly after this result appeared, de la Peña, Klass and Lai (2000) introduced the non-shifted moment bounds presented in last section. In particular, in de la Peña, Klass and Lai (2000), it is shown how to extend the above inequality to the case of higher moments and to more general of increasing functions including exponentials. Subsequently, de la Peña, Klass and Lai (2004) proved part (i) of the following theorem for more general processes than continuous local martingales. Part (ii) of the theorem can be proved by arguments similar to those of Theorem 2 of de la Peña, Klass and Lai (2000).

Theorem 13.5. *Let $T = \{0, 1, 2, \cdots\}$ or $T = [0, \infty)$, $1 < q \leqslant 2$, and $\Phi_q(\theta) = \theta^q/q$. Let A_t and $B_t > 0$ be stochastic processes satisfying*

$$\{\exp(\lambda A_t - \Phi_q(\lambda B_t)), t \in T\} \text{ is a supermartingale with mean } \leqslant 1 \tag{13.19}$$

for $0 < \lambda < \lambda_0$ and such that $A_0 = 0$ and B_t is nondecreasing in $t > 0$. In the case $T = [0, \infty)$, assume furthermore that A_t and B_t are right-continuous. Let $L : [1, \infty) \to (0, \infty)$ be a nondecreasing function satisfying (13.15) — (13.17). Let $\eta > 0, \lambda_0\eta > \delta > 0$, and $h : [0, \infty) \to [0, \infty)$ be a nondecreasing function such that $h(x) \leqslant e^{\delta x}$ for all large x.

(i) *There exists a constant κ depending only on $\lambda_0, \eta, \delta, q, h$ and L such that*

$$Eh\Big(\sup_{t\geqslant 0}\Big\{A_t(B_t\vee\eta)^{-1}[1\vee\log^+ L(B_t\vee\eta)]^{-(q-1)/q}\Big\}\Big) \leqslant \kappa. \tag{13.20}$$

(ii) *There exists* $\widetilde{\kappa}$ *such that for any stopping time* τ,

$$Eh\Big(\sup_{0\leqslant t\leqslant\tau}|(B_t\vee\eta)^{-1}A_t|\Big)\leqslant\widetilde{\kappa}+Eh(|\widetilde{\kappa}\vee\log^+L(B_t\vee\eta)|^{(q-1)/q}). \tag{13.21}$$

13.4 Lemmas on the canonical assumptions

The following lemmas provide a wide range of random variables and stochastic processes that satisfy the canonical assumption (13.1) or (13.5) or (13.19); see de la Pena, Klass and Lai (2004).

Lemma 13.2. *Let* W_t *be a standard Brownian motion. Assume that* T *is a stopping time such that* $T<\infty$ *a.s.. Then*

$$E\exp\{\lambda W_T-\lambda^2T/2\}\leqslant 1,$$

for all $\lambda\in\mathbb{R}$.

Lemma 13.3. *Let* M_t *be a continuous, square-integrable martingale, with* $M_0=0$. *Then*

$$E\exp\{\lambda M_t-\lambda^2\langle M\rangle_t/2\}\leqslant 1,$$

for all $t\geqslant 0$. *Moreover,* $\exp\{\lambda M_t-\lambda^2\langle M\rangle_t/2\}$ *is a supermartingale for all* $\lambda\in\mathbb{R}$. *If* M_t *is only assumed to be a continuous local martingale, then the inequality is also valid by application of Fatou's lemma.*

Lemma 13.4. *Let* $\{M_t:t\geqslant 0\}$ *be a locally square-integrable martingale, with* $M_0=0$. *Let* $\{V_t\}$ *be an increasing process, which is adapted, purely discontinuous and locally integrable; let* $V^{(p)}$ *be its dual predictable projection. Set* $X_t=M_t+V_t$, $C_t=\sum_{s\leqslant t}((\Delta X_s)^+)^2$, $D_t=\{\sum_{s\leqslant t}((\Delta X_s)^-)^2\}_t^{(p)}$, $H_t=\langle M\rangle_t^c+C_t+D_t$. *Then* $\exp\{X_t-V_t^{(p)}-\frac{1}{2}H_t\}$ *is a supermartingale and hence*

$$E\exp\{\lambda(X_t-V_t^{(p)})-\lambda^2H_t/2\}\leqslant 1 \text{ for all } \lambda\in\mathbb{R}.$$

Lemma 13.5. *Let* $\{d_i\}$ *be a sequence of variables adapted to an increasing sequence of* σ*-fields* $\{\mathcal{F}_i\}$. *Assume that the* d_i*'s are conditionally symmetric (that is* $\mathcal{L}(d_i|\mathcal{F}_{i-1})=\mathcal{L}(-d_i|\mathcal{F}_{i-1})$). *Then* $\exp\{\lambda\Sigma_{i=1}^nd_i-\lambda^2\Sigma_{i=1}^nd_i^2/2\}$, $n\geqslant 1$, *is a supermartingale with mean* $\leqslant 1$, *for all* $\lambda\in\mathbb{R}$.

Lemma 13.6. *Let $\{d_n\}$ be a sequence of random variables adapted to an increasing sequence of σ-fields $\{\mathcal{F}_n\}$ such that $E(d_n|\mathcal{F}_{n-1}) \leqslant 0$ and $d_n \leqslant M$ a.s. for all n and some nonrandom positive constant M. Let $0 < \lambda_0 \leqslant M^{-1}$, $A_n = \sum_{i=1}^n d_i$, $B_n^2 = (1 + \frac{1}{2}\lambda_0 M)\sum_{i=1}^n E(d_i^2|\mathcal{F}_{i-1})$, $A_0 = B_0 = 0$. Then $\{\exp(\lambda A_n - \frac{1}{2}\lambda^2 B_n^2),\ \mathcal{F}_n,\ n \geqslant 0\}$ is a supermartingale for every $0 \leqslant \lambda \leqslant \lambda_0$.*

Lemma 13.7. *Let $\{d_n\}$ be a sequence of random variables adapted to an increasing sequence of σ-fields $\{\mathcal{F}_n\}$ such that $E(d_n|\mathcal{F}_{n-1}) = 0$ and $\sigma_n^2 = E(d_n^2|\mathcal{F}_{n-1}) < \infty$. Assume that there exists a positive constant M such that $E(|d_n|^k|\mathcal{F}_{n-1}) \leqslant (k!/2)\sigma_n^2 M^{k-2}$ a.s. or $P(|d_n| \leqslant M|\mathcal{F}_{n-1}) = 1$ a.s. for all $n \geqslant 1$, $k > 2$. Let $A_n = \sum_{i=1}^n d_i$, $V_n^2 = \sum_{i=1}^n E(d_i^2|\mathcal{F}_{i-1})$, $A_0 = V_0 = 0$. Then $\{\exp(\lambda A_n - \frac{1}{2(1-M\lambda)}\lambda^2 V_n^2),\ \mathcal{F}_n,\ n \geqslant 0\}$ is a supermartingale for every $0 \leqslant \lambda \leqslant 1/M$.*

Lemma 13.8. *Let $\{d_n\}$ be a sequence of random variables adapted to an increasing sequence of σ-fields $\{\mathcal{F}_n\}$ such that $E(d_n|\mathcal{F}_{n-1}) \leqslant 0$ and $d_n \geqslant -M$ a.s. for all n and some non-random positive constant M. Let $A_n = \sum_{i=1}^n d_i$, $B_n^r r c_{\gamma,r} \sum_{i=1}^n |d_i|^r$, $A_0 = B_0 = 0$. Then $\{\exp(\lambda A_n - (\lambda B_n)^r/r), \mathcal{F}_n,\ n \geqslant 0\}$ is a supermartingale for every $0 \leqslant \lambda \leqslant \gamma M^{-1}$, where $c_{\gamma,r} = -\{\gamma + \log(1-\gamma)\}/\gamma^r$.*

14 Applications to statistics

Self-normalization is ubiquitous in statistical applications. Self-normalized statistics have been used, out of necessity, for quite some time by now. The celebrated "Student t-statistic" t_n, a classical object familiar to virtually everyone who does data analysis, is closely related to $S_n/V_{n,2}$ via the following identities

$$t_n = \frac{S_n}{V_{n,2}} \left(\frac{n-1}{n - (S_n/V_{n,2})^2} \right)^{1/2}$$

and

$$\{t_n \geqslant t\} = \left\{ \frac{S_n}{V_{n,2}} \geqslant t \left(\frac{n}{n + t^2 - 1} \right)^{1/2} \right\}.$$

Thus, appropriate properties of self-normalized partial sums can be easily transformed to the t-statistic. For example, under the conditions of Theorem 4.1,

$$\lim_{n\to\infty} x_n^{-2} \ln P(t_n \geqslant x_n) = -\frac{1}{2}$$

for any $x_n \to \infty$ with $x_n = o(\sqrt{n})$.

The above result enables one to evaluate the exact Bahadur slope of all score tests for any univariate model. He and Shao (1996) derived the exact Bahadur slopes of studentized score tests and showed that under mild conditions the likelihood score is locally optimal in Bahadur efficiency. The self-normalized technique was successfully applied in Chen and Shao (1997) to study the performance of Monte Carlo methods for estimating ratios of normalizing constants.

Cao (2007) recently studied the moderate deviations for two-sample t-statistics. Let $X_1, \cdots, X_{n_1}$ be a random sample from a population with mean μ_1 and variance σ_1^2, and $Y_1, \cdots, Y_{n_2}$ be a random sample from another population with mean μ_2 and variance σ_2^2. Define the two sample t-statistic

$$T = \frac{\bar{X} - \bar{Y} - (\mu_1 - \mu_2)}{\sqrt{s_1^2/n_1 + s_2^2/n_2}},$$

where

$$s_1^2 = \frac{1}{n_1 - 1} \sum_{i=1}^{n_1} (X_i - \bar{X})^2, s_2^2 = \frac{1}{n_2 - 1} \sum_{i=1}^{n_2} (Y_i - \bar{Y})^2.$$

Theorem 14.1. [Cao (2007)]. *Let $n = n_1 + n_2$. Assume that there are $0 < c_1 \leqslant c_2 < \infty$ such that $c_1 \leqslant n_1/n_2 \leqslant c_2$. Then for any $x := x(n_1, n_2)$ satisfying $x \to \infty$, $x = o(n^{1/2})$*

$$\ln P(T \geqslant x) \sim -x^2/2$$

as $n_1, n_2 \to \infty$. If in addition, $E|X_1|^3 < \infty$ and $E|Y_1|^3 < \infty$, then

$$\frac{P(T \geqslant x)}{1 - \Phi(x)} = 1 + O(1)(1+x)^3 n^{-1/2} d^3$$

for $0 \leqslant x \leqslant n^{1/6}/d$, where $d^3 = (E|X_1 - \mu_1|^3 + E|Y_1 - \mu_2|^3)/(\sigma_1^2 + \sigma_2^2)^{3/2}$ and $O(1)$ is a finite constant depending only on c_1 and c_2. In particular,

$$\frac{P(T \geqslant x)}{1 - \Phi(x)} \to 1$$

uniformly in $x \in (0, o(n^{1/6}))$.

The t-statistic $\sqrt{n}(\bar{X}_n - \mu)/s_n$ is a special case of more general *Studentized statistics* $(\widehat{\theta}_n - \theta)/\widehat{\text{se}}_n$ that are of fundamental importance in statistical inference on an unknown parameter θ of an underlying distribution from which the sample observations $X_1, \cdots, X_n$ are drawn. In nonparametric inference, θ is a functional $g(F)$ of the underlying distribution function F and $\widehat{\theta}_n$ is usually chosen to be $g(\widehat{F}_n)$, where $\widehat{F}_n$ is the empirical distribution. The standard deviation of $\widehat{\theta}_n$ is often called its *standard error*, which is typically unknown, and $\widehat{\mathrm{se}}_n$ denotes a consistent estimate of the standard error of $\widehat{\theta}_n$. For the t-statistic, μ is the mean of F and $\bar{X}_n$ is the mean of $\widehat{F}_n$. Since $\mathrm{Var}(\bar{X}_n) = \mathrm{Var}(X_1)/n$, we estimate the standard error of $\bar{X}_n$ by $s_n/\sqrt{n}$, where s_n^2 is the sample variance. An important property of a Studentized statistic is that it is an *approximate pivot*, which means that its distribution is approximately the same for all θ; see Efron and Tibshirani (1993, Section 12.5) who make use of this pivotal property to construct bootstrap-t confidence intervals and tests. For parametric problems, θ is usually a multidimensional vector and $\widehat{\theta}_n$ is an asymptotically normal estimate (e.g., by maximum likelihood). Moreover, the asymptotic covariance matrix $\Sigma_n(\theta)$ of $\widehat{\theta}_n$ depends on the unknown parameter θ, so $\Sigma_n^{-1/2}(\widehat{\theta}_n)(\widehat{\theta}_n - \theta)$ is the self-normalized (Studentized) statistic that can be used an approximate pivot for tests and confidence regions.

The theoretical basis for the approximate pivotal property of Studentized statistics lies in the limiting standard normal distribution, or in some other limiting distribution that does not involve θ (or F in the nonparametric case). To derive the asymptotic normality of $\widehat{\theta}_n$, one often uses a martingale M_n associated with the data and approximates $\Sigma_n^{-1/2}(\widehat{\theta}_n)(\widehat{\theta}_n - \theta)$ by $\langle M\rangle_n^{-1/2} M_n$. For example, in the asymptotic theory of the maximum likelihood estimator $\widehat{\theta}_n, \Sigma_n(\theta)$ is the inverse of the observed Fisher information matrix $I_n(\theta)$, and the asymptotic normality of $\widehat{\theta}_n$ follows by using Taylor's theorem to derive

$$-I_n(\theta)(\widehat{\theta}_n - \theta) \doteq \sum_{i=1}^{n} \nabla \log f_\theta(X_i|X_1, \cdots, X_{i-1}), \tag{14.1}$$

The right-hand side of (14.1) is a martingale whose predictable variation is $-I_n(\theta)$. Therefore the Studentized statistic associated with the maximum likelihood estimate can be approximated by a self-normalized martingale.

The pseudo-maximization technique is closely related to likelihood and Bayesian inference. Let $X_1, \cdots, X_n$ be observations from a distribution with joint density function $f_\theta(x_1, \cdots, x_n)$. Likelihood inference is based on the likelihood function $L_n(\theta) = f_\theta(X_1, \cdots, X_n)$, whose maximization leads to the maximum likelihood estimator $\widehat{\theta}_n$. Bayesian inference is based on the posterior distribution of θ, which is the conditional distribution of θ given $X_1, \cdots, X_n$ when θ is assumed to have a prior distribution with density function π. Under squared error loss, the Bayes estimator $\widetilde{\theta}_n$ is the mean of the posterior distribution whose density function is proportional to $L_n(\theta)\pi(\theta)$, i.e.,

$$\widetilde{\theta}_n = \int \theta\pi(\theta)L_n(\theta)d\theta \Big/ \int \pi(\theta)L_n(\theta)d\theta. \tag{14.2}$$

Recall that $L_n(\theta)$ is maximized at the maximum likelihood estimator $\widehat{\theta}_n$. Applying Laplace's asymptotic formula to the integrals in (14.2) shows that $\widetilde{\theta}_n$ is asymptotically equivalent to $\widehat{\theta}_n$, so integrating θ over the posterior distribution in (14.2) amounts to pseudo-maximization.

Let θ_0 denote the true parameter value. A fundamental quantity in likelihood inference is the likelihood ratio martingale

$$\frac{L_n(\theta)}{L_n(\theta_0)} = e^{\ell_n(\theta)}, \text{ where } \ell_n(\theta) = \sum_{i=1}^n \log \frac{f_\theta(X_i|X_1, \cdots, X_{i-1})}{f_{\theta_0}(X_i|X_1, \cdots, X_{i-1})}. \tag{14.3}$$

Note that $\nabla\ell_n(\theta_0)$ is also a martingale; it is the martingale in the right-hand side of (14.1). Clearly $\int e^{\ell_n(\theta)}\pi(\theta)d\theta$ is also a martingale for any probability density function π of θ. Lai (2004) shows how the pseudo-maximization approach can be applied to $e^{\ell_n(\theta)}$ to derive boundary crossing probabilities for the generalized likelihood ratio statistics $\ell_n(\widehat{\theta}_n)$ that lead to efficient procedures in sequential analysis.

References

[1] A. de Acosta, Large deviations for vector-valued functionals of a Markov chain: lower bounds. *Ann. Probab.* **16** (1988), 925–960.

[2] R. R. Bahadur, Some limit theorems in statistics. Regional Conference Series in Appl. Math. No. 4, *SIAM*, Philadelphia, 1971.

[3] V. Bentkus and F. Götze, The Berry-Esséen bound for Student's statistic. *Ann. Probab.* **24** (1996), 491–503.

[4] B. Bercu, E. Gassiat and E. Rio, Concentration inequalities, large and moderate deviations for self-normalized empirical processes. *Ann. Probab.* **30** (2002), 1576–1604.

[5] M. E. Caballero, B. Fernández and D. Nualart, Estimation of densities and applications. *J. Theoret. Probab.* **11** (1998), 831–851.

[6] H. Y. Cao, Moderate deviations for two sample t-statistics. *ESAIM-P&S* (2007) (to appear).

[7] M. H. Chen and Q. M. Shao, On Monte Carlo methods for estimating ratios of normalizing constants. *Ann. Stat.* **25** (1997), 1563–1594.

[8] H. Chernoff, A measure of asymptotic efficiency for tests of a hypothesis based on the sum of observations. *Ann. Math. Stat.* **23** (1952), 493–507.

[9] G. P. Chistyakov and F. Götze, Moderate deviations for self-normalized sums. *Theory Probab. Appl.* **47** (2003), 415–428.

[10] G. P. Chistyakov and F. Götze, Limit distributions of Studentized means. *Ann. Probab.* **32** (2004), 28–77.

[11] H. Cramér, Sur un nouveaux théoréme limite de la théorie des probabilités. *Actualités Sci. Indust.* **736** (1938), 5–23, Hermann, Paris.

[12] S. Csörgő, E. Haeusler and D. M. Mason, The asymptotic distribution of trimmed sums. *Ann. Probab.* **16** (1988), 672–699.

[13] S. Csörgő, E. Haeusler and D. E. Mason, The quantile-transform approach to the asymptotic distribution of modulus trimmed sums. Sums, Trimmed Sums and Extremes (Hahn, M. G., Mason, D.M., Weiner, D.C., eds) 337–354, *Progr. Probab.* **23** (1991), Birkhäuser, Boston.

[14] M. Csörgő and L. Horváth, Asymptotic representations of self-normalized sums. *Probab. Math. Stat.* **9** (1988), 15–24.

[15] M. Csörgő and L. Horváth, *Weighted Approximations in Probability and Statistics.* Wiley & Sons, New York, 1993.

[16] M. Csörgő, B. Szyszkowicz and Q. Wang, Donsker's theorem for self-normalized partial sums processes. *Ann. Probab.* **31** (2003a), 1228–1240.

[17] M. Csörgő, B. Szyszkowicz and Q. Wang, Darling-Erdős theorems for self-normalized sums. *Ann. Probab.* **31** (2003b), 676–692.

[18] H. E. Daniels, Saddlepoint approximations in statistics. *Ann. Math. Stat.* **25**, 631–650 (1954).

[19] H. E. Daniels and G. A. Young, Saddlepoint approximation for the studentized mean, with an application to the bootstrap. *Biometrika* **78**, 169–179 (1991).

[20] D. A. Darling, The influence of the maximum term in the addition of independent random variables. *Trans. Am. Math. Soc.* **73** (1952), 95–107.

[21] V. H. de la Peña, M. J. Klass and T. L. Lai, Moment bounds for self-normalized martingales. In *High Dimensional Probability II* (E. Giné, D. M. Mason and J. A. Wellner, eds.) (2000), 1–11. Birkhauser, Boston.

[22] V. de la Pena, M. Klass and T. L. Lai, Self-normalized processes: exponential inequalities, moment bounds and iterated logarithm laws. *Ann. Probab.* **32** (2004), 1902–1933.

[23] V. H. de la Peña, M. J. Klass and T. L. Lai. Pseudo-maximization and self-normalized processes. *Probability Surveys* (2007) (to appear)

[24] A. Dembo and Q. M. Shao, Self-normalized large deviations in vector spaces. In: *Progress in Probability* (Eberlein, Hahn, Talagrand, eds) Vol. **43** (1998a), 27–32

[25] A. Dembo and Q. M. Shao, Self-normalized moderate deviations and LILs *Stochastic Process Appl.* **75** (1998b), 51–65.

[26] A. Dembo and Q. M. Shao. Large and moderate deviations for Hotelling's t^2-statistic. *Elect. Comm. Probab.* **11** (2006), 149–159.

[27] A. Dembo and O. Zeitouni, *Large Deviations Techniques and Applications.* Jones and Bartlett, Boston, 1992.

[28] M. D. Donsker and S. R. S Varadhan, Large deviations for noninteracting particle systems. *J. Stat. Physics.* **46** (1987), 1195–1232.

[29] B. Efron and R. J. Tibshirani, *An Introduction to the Bootstrap.* Chapman & Hall, New York, 1993.

[30] P. Erdős. On the law of the iterated logarithm. *Ann. Math.* **43**(1942), 419–436.

[31] E. Gine, F. Götze and D. M. Mason, When is the Student t-statistic asymptotically standard normal. *Ann. Probab.* **25** (1997), 1514–1531.

[32] S. E. Graversen and G. Peskir, Maximal inequalities for the ornstein-Uhlenbeck process. *Proc. Am. Math. Soc.* **128** (10) (2000), 3035–3041.

[33] P. S. Griffin and J. Kuelbs, Self-normalized laws of the iterated logarithm. *Ann. Probab.* **17** (1989), 1571–1601.

[34] P. S. Griffin and D.M. Mason, On the asymptotic normality of self-normalized sums. *Math. Proc. camb. Phil. Soc.* **109** (1991), 597–610.

[35] P. S. Griffin and W. E. Pruitt, Asymptotic normality and subsequential limits of trimmed sums. *Ann. Probab.* **17** (1989), 1186–1219.

[36] P. S. Griffin and W. E. Pruitt, Weak convergence of trimmed sums. Sums, Trimmed Sums and Extremes (Hahn, M. G., Mason, D.M., Weiner, D.C., eds) 55–80, *Progr. Probab.* **23** (1991), Birkhäuser, Boston.

[37] M. G. Hahn, J. Kuelbs and D. C. Weiner, The asymptotic joint distribution of self-normalized censored sums and sums of squares. *Ann. Probab.* **18** (1990), 1284–1341.

[38] M. G. Hahn, J. Kuelbs and D. C. Weiner, Asymptotic behavior of partial sums: a more robust approach via trimming and self-normalization. Sums, Trimmed Sums and Extremes (Hahn, M. G., Mason, D.M., Weiner, D.C., eds) 1–54, *Progr. Probab.* **23** (1991), Birkhäuser, Boston.

[39] X. He and Q. M. Shao, Bahadur efficiency and robustness of studentized score tests. *Ann. Inst. Stat. Math.* **48** (1996), 295–314.

[40] Z. S. Hu and Q. M. Shao (2005), Self-normalized weak invariance principle in the domain of a stable law (in preparation)

[41] J. L. Jensen, *Saddlepoint Approximations.* Oxford University Press, New York, 1995.

[42] B. Y. Jing, Q. M. Shao and Q. Y. Wang, Self-normalized Cramér type large deviations for independent random variables. *Ann. Probab.* (with B. Y. Jing and Q.Y. Wang) **31** (2003), 2167–2215.

[43] B. Y. Jing, Q. M. Shao and W. Zhou, Saddlepoint approximation for Student's t-statistic with no moment conditions. *Ann. Stat.* **32** (2004), 2679–2711.

[44] T. L. Lai. Likelihood ratio identities and their applications to sequential analysis (with Discussion). *Sequential Anal.* **23** (2004), 467–556.

[45] R. LePage, M. Woodroofe and J. Zinn, Convergence to a stable distribution via order statistics. *Ann. Probab.* **9** (1981), 624–632.

[46] Yu. V. Linnik, Limit theorems for sums of independent random variables, taking account of large deviations. *Theory Probab. Appl.* **7** (1962), 121–134.

[47] B. F. Logan, C. L. Mallows, S. O. Rice and L. A. Shepp, Limit distributions of self-normalized sums. *Ann. Probab.* **1** (1973), 788–809.

[48] R. Lugannani and S. Rice, Saddlepoint approximations for the distribution of the sum of independent random variables. *Adv. Appl. Probab.* **12**, 475–490 (1980).

[49] R. Maller, Asymptotic normality of trimmed means in higher dimensions. *Ann. Probab.* **16** (1988), 1608–1622.

[50] R. Maller, A review of some asymptotic properties of trimmed sums of multivariate data. Sums, Trimmed Sums and Extremes (Hahn, M. G., Mason, D.M., Weiner, D.C., eds) 179–214, *Progr. Probab.* **23** (1991), Birkhäuser, Boston.

[51] V. V. Petrov, *Sums of Independent Random Variables.* Springer-Verlag, New York, 1975.

[52] D. Revuz and M. Yor. *Continuous Martingales and Brownian Motion*, 3rd edition. Springer, New York, 1999.

[53] H. Robbins and D. Siegmund. Boundary crossing probabilities for the Wiener process and sample sums. *Ann. Math. Stat.* **41** (1970), 1410–1429.

[54] Q. M. Shao, On a problem of Csörgő and Révész. *Ann. Probab.* **17** (1989), 809–812.

[55] Q. M. Shao, Cramér-type large deviation for Student's t statistic. *J. Theorect. Probab.* **12**, 387–398 (1999).

[56] Q. M. Shao, Self-normalized large deviations. *Ann. Probab.* **25** (1997), 285–328.

[57] W. F. Stout, A martingale analogue of Kolmogorov's law of the iterated logarithm. *Z. Wahrsch. verw. Geb.* **15** (1970), 279–290.

[58] W. F. Stout. Maximal inequalities and the law of the iterated logarithm. *Ann. Probab.* **1** (1973), 322–328.

[59] V. Strassen, A converse to the law of the iterated logarithm. Z. Wahrsch. *verw. Gebiete.* **4** (1966), 265–268.

[60] D. W. Stroock, *An Introduction to the Theory of Large Deviations.* Springer-Verlag, Berlin, 1984.

[61] Q. Wang and B. Y. Jing, An exponential non-uniform Berry-Esseen bound for self-normalized sums. *Ann. Probab.* **27**, 2068–2088 (1999).

Asymptotic Analysis of Random Partitions*

Zhonggen Su
Department of Mathematics, Zhejiang University
Hangzhou 310027, China
E-mail: suzhonggen@zju.edu.cn

Abstract

In this paper, we aim to review some works on the asymptotic behaviors of random partitions. The focus is upon two basic probability models— uniform partitions and Plancherel partitions. One of fundamental results is the existence of limit shapes, which corresponds to the classic law of large numbers. The fluctuations around the limit shape are also considered; the well-known Gumbel distribution, Tracy-Widom distribution and normal distribution will be used to describe the asymptotic fluctuations at the edge and in the bulk of the spectrum of random partitions. The technique used in asymptotic analysis is the so-called poissonization and depoissonization method, a peculiar conditioning argument. We are content with describing some basic facts and remarkable results, no complete proof is given.

1 Introduction

The theory of partitions is one of the very few branches of mathematics that can be appreciated by anyone who is endowed with little more than a lively interest in the subject. Its applications are found wherever discrete objects are to be counted or classified, whether in the molecular and the atomic studies of matter, in the theory of numbers, or in combinatorial problems from all sources.

The theory of partitions has an interesting history. Certain special problems in partitions certainly date back to the Middle Ages; however, the first discovery of any depth was made in the eighteenth century when Euler proved many beautiful and significant partition theorems. Euler indeed laid the foundations of the theory of partitions. Many of the other great mathematician—Cayley, Gauss, Hardy, Jacobi, Lagrange, Legendre, Littlewood, Rademacher, Ramanujan, Schur, and Sylverster-

*The author is partly supported by NSF of China (Grant No. 10371109, No.10671176) and the Royal Society K. C. Wong Education Foundation International Incoming Fellowship (2006).

have contributed to the development of the theory. Andrews [2] is a first thorough survey of this field, which contains many wonderful results and informative historic notes.

Random partitions occur in mathematics and physics in a wide variety of contexts. A partition can record a state of some random growth process. More often it happens that a certain quantity of interest is expressed, explicitly or implicitly, as a sum over partitions. A simple but fruitful perspective is to treat sums over partitions probabilistically, that is, treat them as expectations of some functions of a random partition. See Okounkov [31], Vershik [46] for more.

In the past decades there have been an intensive activity around the theory of distribution and applications of random partitions. The wealth of applications and connections to many areas of different mathematics is such that it is utterly impossible to paint a whole picture in one short paper. Instead, I will give a few illustrative examples, selected according to my own limited expertise and taste. The focus is basically on the probability limit theorems for uniform and Plancherel random partitions, say, limit shapes and fluctuations. This is actually a variant of the classic laws of large numbers and central limit theorems in probability texts. The paper is organized as follows.

The rest of the Introduction introduces some necessary notations, for instance, partition, Young diagram, standard Young tableau. We also give the hook formula and the RSK correspondence.

Uniform and Plancherel measures are two of the most natural and important probability measures on the space of partitions. Their treatment involve apparently different techniques, but is in essence based on a common idea: conditioning device or poissonization and depoissonization method. The limiting distributions involved are Gumbel distribution, Tracy-Widom distribution and Gauss law, three of the most commonly used in random phenomena. The limit shapes of random Young diagrams under uniform and Plancherel measures are described by two distinct curve equations, one of which has infinite length while the other is finite. These two measures and their properties shall be discussed in Section 2 and 3 respectively. For clarity and simplicity, many complicated and technical analysis are not given in the contexts. The interested reader is referred to the original papers in References for more details.

Besides uniform and Plancherel measures, there are many other types of measures on partitions, which are either well studied or worthwhile further work. For sake of survey, we mention in Section 4 three interesting measures: multiplicative measures, Schur measure and 3-dimensional partitions. They are more general, possibly more difficult, than uniform and Plancherel measures. In Section 4 we also remark that it is very interesting and challenging to use the Stein-Chen method to understand the Tracy-Widom distribution.

Acknowledgement: This paper is dedicated to my teacher, Professor Zhengyan Lin, for his teaching me probability limit theory. I am indebted to him for encouragement and support over the years. This paper was completed when I was visiting the University of Leeds with the K. Wong Education Fellowship. Many thanks also go to Dr. Bogachev for numerous conversation during the visit.

1.1 Partitions

A partition of a positive integer n is a finite non-increasing sequence of positive integers $\lambda_1 \geqslant \lambda_2 \geqslant \cdots \geqslant \lambda_l > 0$ such that $\sum_{i=1}^{l} \lambda_i = n$. The λ_i are called the parts of the partition, the number l of parts is the length of λ.

Many times the partition $(\lambda_1, \lambda_2, \cdots, \lambda_l)$ will be denoted by λ, and we shall write $\lambda \vdash n$ to denote "λ is a partition of n". Sometimes it is convenient to use a notation which indicates the number of times each integer occurs as a part: $\lambda = (1^{r_1}, 2^{r_2}, \cdots, i^{r_i}, \cdots)$ where exactly r_i of the parts of λ are equal to i. The number $r_i = r_i(\lambda) = |\{j : \lambda_j = i\}|$ is called the multiplicity of i in λ. Note $\sum_{i\geqslant 1} i r_i = n$ and $\sum_{i\geqslant 1} r_i = l$.

The set of all partitions of n is denoted by $\mathcal{P}_n$, and the set of all partitions by $\mathcal{P}$, i.e., $\mathcal{P} = \cup_{n=0}^{\infty} \mathcal{P}_n$. Here by convention the empty sequence forms the only partition of zero. Among the most important and fundamental is the question of enumerating various set of partitions. Let $p(n)$ be the partition function, i.e., the number of partitions of n. Trivially, $p(0) = 0$; $p(n)$ increases quite rapidly with n. For example, $p(10) = 42$, $p(20) = 627$, $p(50) = 204226$, $p(100) = 190569292$, $p(200) = 3972999029388$.

One of the most elemental tools for treating partitions is the infinite product generating functions. Euler started the analytic theory of partitions by providing the explicit generating functions

$$\mathcal{F}(q) =: \sum_{n=0}^{\infty} p(n) q^n = \prod_{k=1}^{\infty} \frac{1}{1-q^k} \tag{1.1}$$

and a good deal more (see Ch.1 of Andrews [2]). We remark that on the one hand, for many problems it suffices to consider $\mathcal{F}(q)$ as a formal power series in q; on the other hand, much asymptotic work requires that the generating functions be analytic functions of the complex variable q. In actual fact, both approaches have their special merits.

The asymptotic theory starts 150 years after Euler, with the first letters of Ramanujan to Hardy in 1913. In a celebrated series of memoirs published in 1917 and 1918, Hardy and Ramanujan found (and was perfected by Rademacher) very precise estimates for the partition number

$p(n)$. In particular,

$$p(n) = \frac{1}{4n\sqrt{3}} e^{\pi\sqrt{\frac{2}{3}n}} (1 + O(\frac{1}{\sqrt{n}})). \tag{1.2}$$

See Ch.5 of Andrews [2] for more historic notes.

1.2 Young diagrams

Another effective elementary device for studying partitions is the graphical representation. To each partition λ is associated its Young diagram (shape), which can be formally defined as the set of points $(i, j) \in \mathbb{Z}^2$ such that $1 \leqslant j \leqslant \lambda_i$. In drawing such diagrams, by convention, the first coordinate i (the row index) increases as one goes downwards, and the second coordinate j (the column index) increases as one goes from the left to the right and these points are left justified. More often it is convenient to replace the nodes by unit squares (see Figure 1.1–left).

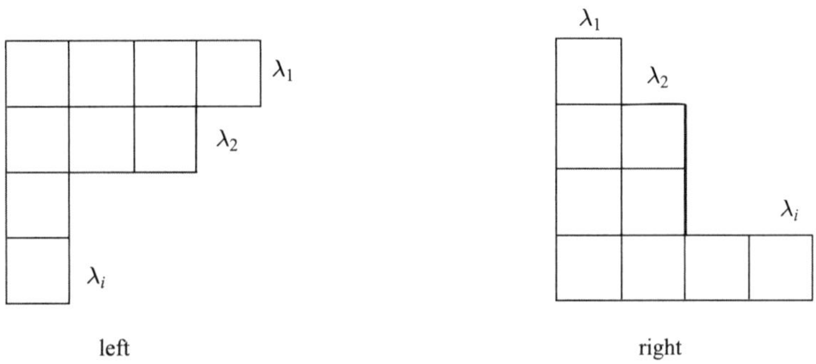

Figure 1.1: Young diarams

Such a representation is extremely useful when we consider applications of partitions to plane partitions or Young tableaux. Some authors (particularly in French) prefer the representation to be upside down in consistency with Descartes coordinate geometry (see Figure 1.1–right; see also MacDonald [29]); While other authors (for instance, the Russian scholars) like to flip the diagram and rotate it 135° for effective parameterizations (see Figure 1.2, see also Okounkov [31]).

We remark that these competing traditions of drawing a diagram are in essence equivalent, but have their own advantages in the study of distinct aspects of partitions. The diagram of a partition λ is still denoted by the same symbol λ.

The conjugate of a partition λ is the partition λ' whose diagram is the transpose of the diagram λ, i.e., the diagram obtained by reflection

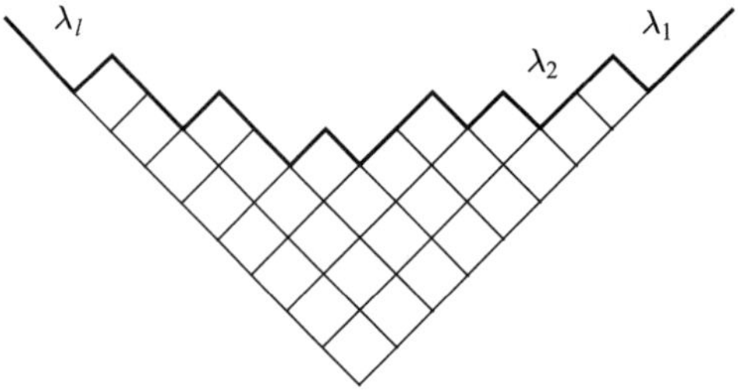

Figure 1.2: Young diarams

in the main diagonal. Hence the λ_i' is the number of squares in the i-th column of λ, or equivalently,

$$\lambda_i' = \sum_{k=i}^{\infty} r_k. \tag{1.3}$$

In particular, $\lambda_1' = l(\lambda)$ and $\lambda_1 = l(\lambda')$. Obviously, $\lambda'' = \lambda$.

The purpose of writing a Young diagram instead of just the partition, of course, is to put something in the squares. A standard Young tableau T with the shape $\lambda \vdash n$ is a one-to-one assignment of the numbers $1, 2, \cdots, n$ to the squares of λ in such a way that the numbers increase along the rows and down the columns. Let d_λ denote the total number of standard Young tableaux associated with a given shape λ. There are some remarkable closed formulas for d_λ in terms of "hook length". For a Young diagram λ, each square determines a hook, which consists of that square and all squares in its row to the right of the square or in its column below the square. The hook length of a square is the number of squares in its hook; denote by $h(i,j)$ the hook length of the square in the i-th row and j-th column. Then $h(i,j) = \lambda_i - j + \lambda_j' - i + 1$. Frame, Robinson and Thrall [15] discovered a so-called hook formula:

$$d_\lambda = \frac{n!}{H_\lambda}, \tag{1.4}$$

where H_λ is the product of hook lengths of λ, i.e., $H_\lambda = \prod_{(i,j)\in\lambda} h(i,j)$.

Greene, Nijenhuis and Wilf have given a short and elegant probabilistic interpretation of (1.4). An equivalent form of the hook formula is due to Frobenius (see p.10, MacDonald [29])

$$d_\lambda = n! \frac{\prod_{1\leqslant i<j\leqslant l}(\lambda_i - \lambda_j + j - i)}{\prod_{i=1}^{l}(\lambda_i + l - i)!}. \tag{1.5}$$

There is an intimate connection between $\mathcal{P}_n$ and the symmetric group S_n of permutations of $1, 2, \cdots, n$. Most direct is a bijection between the partitions of n and the set of conjugacy classes of S_n. Considerably deeper is the observation that the set of irreducible representations of S_n can be parameterized by $\lambda \in \mathcal{P}_n$ and d_λ is the degree (dimension) of the irreducible representation indexed by λ. Still, Robinson, Schensted and Knuth (see Sagan [35]) developed an algorithm for obtaining Young tableaux with the help of permutations. Let $\mathcal{T}_n$ be the set of standard Young tableaux with n squares. According to this algorithm, for any n there is a bijection, the so-called RSK correspondence, between S_n and pairs $T, T' \in \mathcal{T}_n$ with the same shape:

$$S_n \ni \pi \overset{\text{RSK}}{\longleftrightarrow} (T(\pi), T'(\pi)) \in \mathcal{T}_n \times \mathcal{T}_n. \tag{1.6}$$

The RSK correspondence is very intricate and has no obvious algebraic meaning at all, but it is very deep and allows us to understand many things. In particular, it gives an explicit proof of the Burnside identity:

$$\sum_{\lambda \vdash n} d_\lambda^2 = n!. \tag{1.7}$$

Moreover, the RSK correspondence provides us with a viable algorithm for computing the length of the longest increasing subsequence. For a given $\pi = (\pi_1, \pi_2, \cdots, \pi_n) \in S_n$ and $1 \leqslant i_1 < i_2 < \cdots < i_k \leqslant n$, we say $\pi_{i_1}, \pi_{i_2}, \cdots, \pi_{i_k}$ is an increasing subsequence of π if $\pi_{i_1} < \pi_{i_2} < \cdots < \pi_{i_k}$. Let $l_n(\pi)$ be the length of the longest increasing subsequence of π. The celebrated Erdös and Szekeres theorem states every π of S_n contains an increasing and/or decreasing subsequence of length at least $\sqrt{n}$ (see Steele [38, 39]). This can be proved by the pigeon-hole principle, but also follows from the RSK correspondence using the observation that a Young diagram of size n must have either width or height at least $\sqrt{n}$. More interestingly, $l_n(\pi)$ is exactly the number of squares in the first row of $T(\pi)$ or $T'(\pi)$, i.e., $l_n(\pi) = \lambda_1(T(\pi))$.

2 Random uniform partitions

We have defined the set $\mathcal{P}_n$ of partitions of n and known how to count its size $p(n)$. Now equip a probability measure on this set. As we will see below, this set bears many various natural measures. Certainly, the first natural measure is uniform, i.e., choose at random a partition with equal probability. Let P_n be the uniform measure on $\mathcal{P}_n$ defined by

$$P_n(\lambda) = \frac{1}{p(n)}, \quad \lambda \in \mathcal{P}_n. \tag{2.1}$$

It seems interesting to study the behaviors of the typical partitions (Young diagrams). For instance, what are the asymptotics of the maximal summand? Erdös' school first treated these questions and obtained very deep and interesting asymptotics. The most remarkable Erdös-Lehner theorem (see [13])is as follows. For all $x \in R$

$$\lim_{n\to\infty} P_n(\lambda \in \mathcal{P}_n : \frac{c}{\sqrt{n}}\lambda_1 - \log\frac{\sqrt{n}}{c} \leqslant x) = e^{-e^{-x}}, \tag{2.2}$$

where $c = \frac{\pi}{\sqrt{6}}$. Note that the limit is the very famous Gumbel distribution.

To prove (2.2), Erdös and Lehner used the Hardy-Ramanujan asymptotic formula for $p(n)$ (see (1.2)) and some finer combinatorial estimates for numbers of partitions with certain properties to find the limiting distribution of $l(\lambda)$ and thus — by the duality relation $l(\lambda) = \lambda_1'$ — the distribution of λ_1.

As known to us, the probabilistic argument of many combinatorial problems dates back to Erdös and his colleagues, but they usually investigate a single functional. Instead of a single functional, Vershik posed the problem of the limit shape in the early 1970s. Roughly speaking, they are concerned with the following problem: suppose we have some probability measures on the space $\mathcal{P}_n$ of partitions for all n, say μ_n, how do we scale the space $\mathcal{P}_n$ in order to obtain the true nontrivial limit distribution of the measures μ_n? Obviously, one can obtain rich information about asymptotics from the properties of the limit shape. See Vershik [44, 45, 46] for history and recent development.

2.1 Grand canonical ensembles

Let $q \in (0, 1)$. Define a probability measure P_q on $\mathcal{P}$ as follows:

$$P_q(\lambda) = \frac{q^{|\lambda|}}{\mathcal{F}(q)}, \quad \lambda \in \mathcal{P}, \tag{2.3}$$

where $|\lambda|$ is the number being partitioned by λ.

This P_q is particularly useful for the study of P_n because it is easily understood and many limit theorems as $q \to 1$ can be converted with some work to the corresponding statements for P_n as $n \to \infty$. We call the family $(\mathcal{P}_n, P_n)$ small canonical ensemble of partitions and the $(\mathcal{P}, P_q)$ grand canonical ensemble of partitions in view of similarities with statistical physics (see Vershik and Yakubovich [48]). The following lemma, due to Fristedt [16], summarizes some primary properties concerning the probabilities P_n and P_q.

Lemma 2.1. *For any $q \in (0, 1)$ and $n \geqslant 0$, we have*

(1)

$$P_q|_{\mathcal{P}_n} = P_n, \tag{2.4}$$

i.e., P_n is the conditional probability measure induced on $\mathcal{P}_n$ by P_q;

(2)

$$P_q = \frac{1}{\mathcal{F}(q)} \sum_{n=0}^{\infty} p(n) q^n P_n, \tag{2.5}$$

i.e., P_q is a convex combination of measures P_n;

(3) *Under $(\mathcal{P}, P_q)$, the random variables r_k are independent geometrically distributed with mean $q^k/(1-q^k)$.*

In this subsection we restrict ourselves to the grand canonical ensemble and obtain a first insight on the asymptotics of random Young diagrams. Let us start with the extremal distribution.

Since $\lambda_1 < m$ if and only if $r_j = 0$ for all $j \geqslant m$, then, by virtue of (3) of Lemma 2.1, direct calculations can be made to show the limiting distribution of λ_1 is the Gumbel distribution (see 2.2)). More generally, analogous extremal distributions hold for the first d largest summands (see Fristedt [16], Vershik and Yakubovich [48]).

Theorem 2.2. *Let $q = e^{-h}$. Define for $i \geqslant 1$*

$$W_i = h\lambda_i - |\log h|, \tag{2.6}$$

Then we have as $h \to 0$

$$(W_1, W_2, \cdots) \xrightarrow{d} (Y_1, Y_2, \cdots), \tag{2.7}$$

where $Y_1, Y_2, \cdots$, is a Markov chain such that the density of Y_1 is exp $(-x - e^{-x})$ *and the density of Y_i conditioned on $Y_{i-1} = x$ is* $\exp(-y - e^{-y} + e^{-x})$, $y \leqslant x$.

Note $\xrightarrow{d}$ in (2.7) means that $(W_1, W_2, \cdots, W_m)$ converges in distribution to $(Y_1, Y_2, \cdots, Y_m)$ for $m \geqslant 1$. Next let us look at the asymptotics in the deep bulk of the spectrum of Young diagrams.

To a partition $\lambda \in \mathcal{P}$, we assign a function φ_λ on $(0, \infty)$ by the following rule:

$$\varphi_\lambda(t) = \sum_{k=[t]}^{\infty} r_k, \qquad 0 < t < \infty, \tag{2.8}$$

where $[t]$ denotes the integer part of t. Clearly, by definition, $\varphi_\lambda(\cdot)$ is a monotone decreasing, piecewise constant function of t, and that $\int_0^\infty \varphi_\lambda(t)dt = |\lambda|$. This function describes actually the boundary of the Young diagram λ used by French mathematicians.

Again, using (3) of Lemma 2.1, one easily has for each $t > 0$

$$E_q(\varphi_\lambda(t)) = \sum_{k=[t]}^{\infty} \frac{q^k}{1-q^k}, \quad Var_q(\varphi_\lambda(t)) = \sum_{k=[t]}^{\infty} \frac{q^k}{(1-q^k)^2}. \tag{2.9}$$

Obviously, the Markov inequality can be used to show that $\varphi_\lambda(t)$ concentrates around its mean. In fact, one has a stronger result: the limit shape exists (see Vershik [45]).

Theorem 2.3. *Let $q = e^{-h}$. Consider the scaled function*

$$\tilde{\varphi}_h(t) = h\varphi_\lambda(\frac{t}{h}), \quad t > 0. \tag{2.10}$$

Then, as $h \to 0$, we have $\tilde{\varphi}_h \to \Psi$ in the sense of uniform convergence on compact sets, where Ψ is the function defined by

$$\Psi(t) = \int_t^{\infty} \frac{e^{-u}}{1-e^{-u}} du, \quad t > 0. \tag{2.11}$$

More precisely, for any $\varepsilon > 0$ and $0 < a < b < \infty$, there exists an $h_0 > 0$ such that for $0 < h < h_0$ we have

$$P_q(\lambda \in \mathcal{P} : \sup_{a \leqslant t \leqslant b} |\tilde{\varphi}_h(t) - \Psi(t)| > \varepsilon) < \varepsilon. \tag{2.12}$$

In other words, the image measure of P_q under $\tilde{\varphi}_h : \mathcal{P} \mapsto \mathcal{D}(0, \infty)$ (the space of right continuous function with left limits) has a limit which is singular measure concentrated on some continuous curve Ψ (see Figure 2.1).

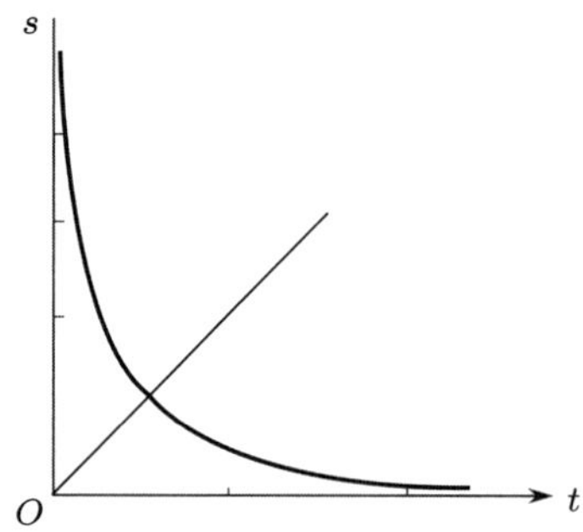

Figure 2.1: Temperley-Vershik curve

Note that Ψ in (2.11) is computable: let $s = \Psi(t)$, then

$$e^{-s} + e^{-t} = 1. \tag{2.13}$$

This curve was firstly obtained via a heuristic argument by the physist Temperley [42]; now it is called Temperley-Vershik's equation. On the other hand, we remark that 0 is a point of singularity of integrand in (2.11). In fact, the scaling constants at the edge and in the bulk are completely different.

For a probabilist, the law of large numbers is only the beginning and the questions about central limit theorem correction to the limit shape and local statistics of the Young diagram in various regions of the limit shape follow immediately. Conjecturally, the limit shape controls the answers to all these questions.

Theorem 2.4. *Let $q = e^{-h}$. Define*

$$V_h(t) = \frac{1}{h^{1/2}}(\tilde{\varphi}_h(t) - \Psi(t)), \quad t > 0. \tag{2.14}$$

Then under $(\mathcal{P}, P_q)$, we have as $h \to 0$

(1) *For each $t > 0$,*

$$V_h(t) \xrightarrow{d} N(0, \sigma^2(t)), \tag{2.15}$$

where

$$\sigma^2(t) = \int_t^\infty \frac{e^{-u}}{(1 - e^{-u})^2} du. \tag{2.16}$$

(2) *In $D(0, \infty)$,*

$$V_h \Rightarrow G \tag{2.17}$$

where G is a continuous Gaussian process with independent increments.

Note the classic theory of weak convergence is applicable to the normalized partial sums $V_h(t)$ of independent geometrically distributed random variables r_k (see Billingsley [5]). Indeed, (2.15) can be proved by a direct computation of characteristic functions, and finite dimensional distributions hold similarly; while the uniform tightness for (2.17) can be checked by using the fourth moment.

As the reader might notice, $|\lambda|$ is itself random under $(\mathcal{P}, P_q)$ and is actually an infinite sum of the kr_k's, so the mean is

$$\begin{aligned} E_q|\lambda| &= \sum_{k=1}^{\infty} \frac{ke^{-kh}}{1 - e^{-kh}} \\ &\sim \frac{1}{h^2} \int_0^\infty \frac{ue^{-u}}{1 - e^{-u}} du = \frac{\pi^2}{6h^2}, \end{aligned} \tag{2.18}$$

and $|\lambda|$ fluctuates around this mean like a Gaussian random variable (see Bloch and Okounkov [6], Pittel [33]).

2.2 Small canonical ensembles

In this subsection we turn to the second issue: how do we convert the results in the independent setting to the original partitions? By Lemma 2.1, for each n and q, the probability measure P_n is equal to the conditional probability measure of P_q obtained by conditioning P_q on the event $\{\lambda : |\lambda| = n\}$ (just $\mathcal{P}_n$). That is, for any set of partitions,

$$\begin{aligned} P_n(A) = P_q(A||\lambda| = n) &= \frac{P_q(A, |\lambda| = n)}{P_q(|\lambda| = n)} \\ &\leqslant \frac{P_q(A)}{P_q(|\lambda| = n)}. \end{aligned} \tag{2.19}$$

For this bound to be most effective, one needs to determine q for which $P_q(|\lambda| = n)$ is asymptotically the largest possible. It turns out that $q = e^{-h_n} =: e^{-\frac{c}{\sqrt{n}}}$ where $c = \frac{\pi}{\sqrt{6}}$ (see (2.2)) meets this requirement and that for this q,

$$P_q(|\lambda| = n) = (1 + o(1))\frac{1}{(96n^3)^{1/4}}. \tag{2.20}$$

This indeed follows from the local limit theorem. On the other hand, the choice of $h_n = \frac{c}{\sqrt{n}}$ is most natural if one notices that the equation $E_q|\lambda| = n$ has a unique solution, which is asymptotically equivalent to e^{-h_n} by (2.18).

We have the following weak equivalence of ensembles (see Vershik [45]).

Lemma 2.5. (1) *If the probability measures P_n have weak limit in $\mathcal{D}(0, \infty)$ as $n \to \infty$, then the measures P_q have the same limit as $q \to 1$;*

(2) *Conversely, if the measures P_q have weak limit as $q \to 1$ and the limit measure is singular, then the measures P_n have the same limit as $n \to \infty$.*

Statement (1) is obvious since P_q is a convex combination of the measures P_n. The proof of (2) is substantially based on the fact that the measures P_q are multiplicative; in fact, it is a Tauberian-type theorem. The problem of Tauberian-type theorem in general setting (for non-singular case and for general multiplicative measures) is of great interest and, seemingly, has not been studied yet. Now Lemma 2.5 together with Theorem 2.3 implies

Theorem 2.6. *Let h_n be as above. Consider the scaled function*

$$\tilde{\varphi}_n(t) = h_n \varphi_\lambda(\frac{t}{h_n}), \quad t > 0. \tag{2.21}$$

Then for any $\varepsilon > 0$ *and* $0 < a < b < \infty$, *there exist* n_0 *such that for* $n > n_0$

$$P_n(\lambda \in \mathcal{P}_n : \sup_{a \leqslant t \leqslant b} |\tilde{\varphi}_n(t) - \Psi(t)| > \varepsilon) < \varepsilon. \tag{2.22}$$

We say that the grand and small canonical ensembles are strongly equivalent for some functional **f** on partitions if the distributions of the functional **f** are asymptotically the same with respect to P_n and P_q with $q = e^{-h_n}$. According to Vershik and Yakubovich [48], the ensembles are equivalent for the functional of scaled largest parts in partition. Thus Theorem 2.2 yields the following results, which particularly recovers the Erdös and Lehner extremal distribution (see 2.2).

Theorem 2.7. *Let* h_n *be as above. Define for* $i \geqslant 1$

$$\hat{W}_i = h_n \lambda_i - |\log h_n|. \tag{2.23}$$

Then as $n \to \infty$,

$$(\hat{W}_1, \hat{W}_2, \cdots) \xrightarrow{d} (Y_1, Y_2, \cdots), \tag{2.24}$$

where Y_i*'s are given by Theorem 2.2.*

Note that there is no guarantee about the equivalence of distributions for a certain functional (see Vershik [44]). To extend the central limit theorem like Theorem 2.4 to the case of $(\mathcal{P}_n, P_n)$, we need some additional work.

By Lemma 2.1, the probability generating function of λ'_k under P_q is

$$E_q e^{x\lambda'_k} = \frac{1}{\mathcal{F}(q)} \sum_{n=0}^{\infty} p(n) q^n E_n e^{x\lambda'_k}. \tag{2.25}$$

On the other hand, $E_q e^{x\lambda'_k}$ is easily written as

$$E_q e^{x\lambda'_k} = \prod_{j=k}^{\infty} E_q e^{x r_j} = \prod_{j=k}^{\infty} \frac{1 - q^j}{1 - e^x q^j}. \tag{2.26}$$

Since this function is analytic in q in the disk $\{q \in \mathcal{C} : |q| < 1\}$, by the Cauchy integral formula, we have for any $\gamma > 0$

$$E_n e^{x\lambda'_k} = \frac{1}{2\pi p(n)\gamma^n} \int_{-\pi}^{\pi} e^{-in\theta} \mathcal{F}(\gamma e^{i\theta}) \prod_{j=k}^{\infty} \frac{1 - (\gamma e^{i\theta})^j}{1 - e^x (\gamma e^{i\theta})^j} d\theta. \tag{2.27}$$

It remains to choose a proper $\gamma > 0$ and to give a precise estimate for the contour integral. Note (1.2) and the following simple formula:

$$\mathcal{F}(e^{-z}) = \exp(\frac{\pi^2}{6z} + \frac{1}{2} \log \frac{z}{2\pi} + O(|z|)) \tag{2.28}$$

uniformly for $z \to 0$ within a corner $\{z : \operatorname{Im} z \leqslant \varepsilon \operatorname{Re} z, \operatorname{Re} z > 0\}$, $\varepsilon > 0$ being fixed. One can now prove for any real u and some $\tau^2(t)$

$$E_n \exp\left(\frac{u\tilde{\varphi}_n(t)}{h_n^{1/2}}\right) = \exp\left(\frac{u\Psi(t)}{h_n^{1/2}} + \frac{u^2}{2}\tau^2(t) + o(1)\right), \tag{2.29}$$

which obviously implies that $\tilde{\varphi}_n(t)$ after suitably scaled is approximately normal with variance $\tau^2(t)$ for each $t > 0$. Analogous results hold for multidimensional vectors. These are summarized as follows (see Pittel [33]).

Theorem 2.8. *Let h_n be as above. Define for $t > 0$*

$$V_n(t) = \frac{1}{h_n^{1/2}}(\tilde{\varphi}_n(t) - \Psi(t)). \tag{2.30}$$

Then for any $m \geqslant 1$ and $0 < t_1 < t_2 < \cdots < t_m$

$$(V_n(t_1), V_n(t_2), \cdots, V_n(t_m)) \xrightarrow{d} (V(t_1), V(t_2), \cdots, V(t_m)), \tag{2.31}$$

where $V(t), t > 0$ is a Gaussian process with covariance structure: for $s \leqslant t$

$$Cov(V(s), V(t)) = \frac{c^2}{(1-e^{-s})(1-e^{-t})}\left(\frac{e^{-t}}{1-e^{-s}} - \frac{l(s)l(t)}{2c^2}\right), \tag{2.32}$$

where

$$l(s) = se^{-s} + (1-e^{-s})\log(1-e^{-s}). \tag{2.33}$$

It is worth mentioning that the weak convergence of the process V_n is not proven yet. We don't know whether the uniform tightness is true for V_n. However, a slightly weaker condition can be proved, i.e., V_n is stochastically equi-continuous, uniformly for $t > 0$:

$$\lim_{\delta\to 0} \limsup_{n\to\infty} \sup_{t,s:|t-s|\leqslant\delta} P_n(|V_n(t) - V_n(s)| \geqslant \varepsilon) = 0. \tag{2.34}$$

And, fortunately, this stochastic equi-continuity can be used to prove convergence in distribution of $\mathbf{f}(V_n)$ for a broad class $\mathcal{G}$ of the integral functionals $\mathbf{f}$ of a form

$$\mathbf{f}(x) = \int_0^\infty g(t, x(t))dt, \tag{2.35}$$

i.e., for $\mathbf{f} \in \mathcal{G}$

$$\mathbf{f}(V_n) \xrightarrow{d} \mathbf{f}(V). \tag{2.36}$$

Since λ is uniformly distributed on $\mathcal{P}_n$, then so λ'. Equivalently,

$$(\lambda_1, \lambda_2, \cdots, \lambda_l) \stackrel{d}{=} (\lambda'_1, \lambda'_2, \cdots, \lambda'_{l'}). \tag{2.37}$$

Therefore, we can reformulate Theorems 2.6 and 2.8 in terms of λ; while Theorem 2.7 can be restated in terms of λ'. In fact, this symmetry can be directly seen from the limit curve equation (see (2.13)).

So far, we have seen some important limit theorems concerning the limit shape and the fluctuations of random uniform partitions. Frankly speaking, the probabilistic structure of a uniform partition is quite complicated. According to Theorem 2.7, for a fixed $k \geqslant 1$, the λ_k suitably scaled has exponential-type limit distribution. A natural question is what we can say about the k-th largest part λ_k if $k = k(n)$ varies as n, say, $k(n) \to \infty$ sufficiently slowly that $k(n) = o(n^{1/4})$. Likewise, Theorem 2.8 only describes the asymptotic distributional result for the intermediate parts of size $O(n^{1/2})$ (equivalently in the middle bulk); one naturally asks for the limiting distribution for the first $o(n^{1/2})$ parts and/or parts near to the other end. The reader is referred to Fristedt [16] and Pittel [33] for related answers.

2.3 Normal approximation of d_λ

In conclusion to this section, we review some work on the d_λ — the total number of standard Young tableaux associated with the partition λ. As we have seen, the d_λ is of great interest in its own right. In a seminal sequence of three papers, Szalay and Turán [41] undertook an asymptotic study of the random uniform partitions. A highlight of their work is a pioneering analysis of the limiting distribution of d_λ. Using the classic formula (1.5), Szalay and Turán showed that

$$\log d_\lambda = \frac{1}{2}\log n! - An = O_p(n^{7/8}\log^4 n), \tag{2.38}$$

where

$$A = \frac{1}{2}(1 - \log\frac{\pi^2}{6}) + \frac{6}{\pi^2}\int_0^\infty \frac{t\log t}{e^t - 1}dt. \tag{2.39}$$

Pittel [33] used the distribution identity (2.37) and refined estimates for the λ'_k to improve the error term to $O_p(n^{3/4}\log^{3/2} n)$. The power $3/4$ is indeed optimal. The following result implies that $\log d_\lambda/\sqrt{n!}$ is asymptotically normal, with mean $-An$, and standard deviation of order $n^{3/4}$ (see Pittel [34]).

Theorem 2.9. *Let $K(s,t)$ be a symmetric function on $[0,1]\times[0,1]$ defined by*

$$K(s,t)=\frac{1}{c}(s(1-t)-\frac{1}{2}L(s)L(t)),\quad 0\leqslant s\leqslant t\leqslant 1$$

where $L(t)=\frac{1}{c}(t\log t-(1-t)\log(1-t))$(remind $c=\frac{\pi}{\sqrt{6}}$). Introduce

$$\sigma^2=\frac{1}{c^2}\int_0^1\int_0^1 K(s,t)\phi(s)\phi(t)dsdt$$

where

$$\phi(t)=\int_0^\infty\frac{\log|\log u|}{(1-t+tu)^2}du$$

Then under P_n

$$\frac{1}{n^{3/4}}\left(\log\frac{d_\lambda}{\sqrt{n!}}+An\right)\xrightarrow{d}N(0,\sigma^2). \tag{2.40}$$

The proof of (2.40) consists of two steps: (1) show that, for almost all diagrams λ, $\log d_\lambda$ is well approximated by a linear function of the dual diagram λ', namely, for every $\varepsilon>0$,

$$\log\frac{d_\lambda}{\sqrt{n!}}=-An+\sum_{k=1}^{\lambda_1}\nu(k)(\lambda'_k-E(k))+O_p(n^{2/3+\varepsilon}) \tag{2.41}$$

for some deterministic functions $\nu(\cdot)$, $E(\cdot)$; (2) apply the functional central limit theorem (2.36) to the linear sums in (2.41).

3 Random Plancherel partitions

In this section we consider the Plancherel measure $P_{p,n}$ on the set $\mathcal{P}_n$ of partitions of n defined by

$$P_{p,n}(\lambda)=\frac{d_\lambda^2}{n!},\quad \lambda\in\mathcal{P}_n, \tag{3.1}$$

where p in the subscript stands for Plancherel. This measure arises naturally in representation-theoretic, combinatorial, and probabilistic problems. It is called Plancherel because the Fourier transform

$$L^2(S_n,\mu_s)\xrightarrow{\text{Fourier}}L^2(\hat{S}_n,P_{p,n}) \tag{3.2}$$

is an isometry just like in the classical Plancherel theorem, where μ_s is the uniform measure on S_n and $\hat{S}_n$ is the set of irreducible representations of S_n.

Recall $l_n(\pi)$ denotes the length of the longest increasing subsequence of π. Then by the RSK correspondence

$$\begin{aligned}\mu_s(\pi \in S_n : l_n(\pi) = k) &= \frac{|\{\pi \in S_n : l_n(\pi) = k\}|}{n!} \\ &= \sum_{\lambda:\lambda_1 = k} \frac{d_\lambda^2}{n!} \\ &= P_{p,n}(\lambda \in \mathcal{P}_n : \lambda_1 = k). \end{aligned} \tag{3.3}$$

In words, the Plancherel measure $P_{p,n}$ on $\mathcal{P}_n$ is the push-froward of the uniform measure μ_s on S_n. Thus the analysis of l_n is equivalent to a statistical problem in the "geometry" of the Young diagram.

On the other hand, we can understand the Plancherel measure in the following way. Consider the following chain of partitions:

$$\emptyset = \lambda^{(-n)} \nearrow \cdots \nearrow \lambda^{(-1)} \nearrow \lambda^{(0)} \searrow \lambda^{(1)} \searrow \cdots \searrow \lambda^{(n)} = \emptyset, \tag{3.4}$$

where $\mu \nearrow \nu$ means that ν is obtained from μ by adding a square, while $\mu \searrow \nu$ means that ν is obtained from μ by removing a square.

It is clear that the number of increasing chains is identical to that of decreasing chains. Also, the increasing chain may be viewed as a path in the Young graph which produces a standard tableau. Thus the Plancherel measure is the distribution of the central slice $\lambda^{(0)}$ of this chain induced by the uniform measure on all chains of the form (3.4).

3.1 Limit shapes

In this subsection we will give the limit shape of Young diagrams under the Plancherel measure in the spirit of Theorem 2.6. But let us first consider a basic problem on increasing subsequences, that is to determine the asymptotics of l_n as $n \to \infty$. Again, the study of l_n goes back to Erdös. It follows from the Erdös and Szekeres theorem that the expected value $E_s l_n$ satisfies

$$E_s l_n \geqslant \frac{1}{2}\sqrt{n}. \tag{3.5}$$

In 1960s Ulam, who had a long and enduring friendship with Erdös, computed $E_s l_n$ for n lying in the range $1 \leqslant n \leqslant 10$ using Monte Carlo techniques and conjectured that

$$\lim_{n\to\infty} \frac{E_s l_n}{\sqrt{n}} = \kappa \tag{3.6}$$

(see Deift [11], Steele [38]) . What became known as Ulam's problem was to prove that the limit in (3.6) indeed exists and compute the constant

κ. The first analytic result was due to Hammersley, who introduced a certain Poisson point process in the first quadrant with the express purpose of studying l_n. Then he proved the existence of the limit κ and

$$\frac{l_n}{\sqrt{n}} \xrightarrow{P} \kappa \tag{3.7}$$

as an application of subadditive ergodic theory (H. Kesten strengthened (3.7) to almost sure convergence). There has been significant interest in the determination of the exact value κ. see Aldous and Diaconis [1], Deift[11] for various associated results and some of the history.

Next we derive the value of κ making use of the distribution identity (3.3). To emphasize the dependence of λ upon the size n, we use $\lambda^{(n)}$ in place of λ in the following argument. Clearly, we can and do assume $\lambda^{(n)}$ is generated by adding a square to $\lambda^{(n-1)}$. Note $\lambda_1^{(n)} = \lambda_1^{(n-1)} + 1$ if and only if $\lambda^{(n)}$ is generated from $\lambda^{(n-1)}$ by adding an extra square at the first row of an extra column. Therefore,

$$\begin{aligned} E_{p,n}\lambda_1^{(n)} &= \sum_{k=1}^{n} E_{p,k}(\lambda_1^{(k)} - \lambda_1^{(k-1)}) \\ &= \sum_{k=1}^{n} P_{p,k}(\lambda_1^{(k)} - \lambda_1^{(k-1)} = 1). \end{aligned} \tag{3.8}$$

In turn,

$$P_{p,k}(\lambda_1^{(k)} - \lambda_1^{(k-1)} = 1) = \frac{d_{\lambda^{(k-1)}} d_{\lambda^{(k)*}}}{k!}, \tag{3.9}$$

where $\lambda^{(k)*}$ is the Young diagram with an extra square in the first row of $\lambda^{(k-1)}$. Now it easily follows

$$E_{p,n}\lambda_1^{(n)} \leqslant 2\sqrt{n}. \tag{3.10}$$

For the lower bound of κ, define the graph function $\lambda(x)$ and $\bar{\lambda}(x)$ associated with λ by

$$\lambda(0) = \lambda(1), \qquad \lambda(x) = \lambda_{\lceil x \rceil}, \quad x > 0 \tag{3.11}$$

and

$$\bar{\lambda}(x) = \frac{1}{\sqrt{n}}\lambda(\sqrt{n}x), \quad x \geqslant 0. \tag{3.12}$$

Obviously, $\int_0^\infty \bar{\lambda}(x)dx = 1$. Let $\mathcal{H}$ be the class of nonnegative nonincreasing functions on $[0, \infty)$ of integral unity. For $f \in \mathcal{H}$, define

$$I(f) = \int_0^\infty \int_0^{f(x)} \log(f(x) - y + f^{-1}(y) - x)dxdy. \tag{3.13}$$

Then by the hook formula (1.3)

$$P_{p,n}(\lambda) = \frac{n!}{H_\lambda^2}$$
$$= \exp(-(1+o(1))2n(I(\bar{\lambda}) + \frac{1}{2})), \tag{3.14}$$

where $o(1)$ is uniform over all λ . It immediately follows for any $\varepsilon > 0$

$$P_{p,n}(\lambda : \lambda_1 \leqslant (2-\varepsilon)\sqrt{n})$$
$$= \sum_{\lambda_1 \leqslant (2-\varepsilon)\sqrt{n}} P_{p,n}(\lambda)$$
$$\leqslant p(n) \exp(-(1+o(1))2n(\inf_{f \in \mathcal{H}_0(\varepsilon)} I(f) + \frac{1}{2})), \tag{3.15}$$

where $\mathcal{H}_0(\varepsilon) = \{f \in \mathcal{H} : f(0) = 2 - \varepsilon\}$.

Note $p(n)$ is at most $e^{\sqrt{6n}/3}$. It suffices to prove the infimum is strictly larger than $-1/2$ for $\varepsilon > 0$. This analysis leads to a calculus of variations minimization problem. The problem of minimizing the functional $I(f)$ over $\mathcal{H}$ was posed by Stanley, and Logan and Shepp [28] showed that the minimum is unique and has the value $-1/2$. Moreover, the minimizing function $y = \omega(x)$ such that $I(\omega) = -1/2$ is given parametrically by

$$x = \frac{2}{\pi}(\sin\theta - \theta\cos\theta), \quad y = x + 2\cos\theta, \quad 0 \leqslant \theta \leqslant \pi. \tag{3.16}$$

They also found the minimum of $I(f)$ subject to the constraints $f(0) \leqslant a$ and $f^{-1}(0) \leqslant b$; in particular

$$\inf_{f \in \mathcal{H}_0(\varepsilon)} I(f) = -1 + \frac{(2-\varepsilon)^2}{8} + \log\frac{2-\varepsilon}{2}$$
$$-(1 + \frac{(2-\varepsilon)^2}{4}) \log\frac{2(2-\varepsilon)^2}{4 + (2-\varepsilon)^2}, \tag{3.17}$$

which is a strictly convex, monotone decreasing function with minimum $-\frac{1}{2}$ at $\varepsilon = 0$. Thus, it follows for any $\varepsilon > 0$

$$P_{p,n}(\lambda : \lambda_1 \leqslant (2-\varepsilon)\sqrt{n}) \to 0. \tag{3.18}$$

Combining (3.10) and (3.18), we have proved

Theorem 3.1.

$$\kappa = 2. \tag{3.19}$$

It is natural to ask how the $\lambda_2, \lambda_3, \cdots$ behave asymptotically. Instead of a specific λ_k, Vershik and Kerov [47] investigated the limit shape of the Young diagram under the Plancherel measure. It is convenient to use the rotated coordinate system

$$u = y - x, \quad v = x + y. \tag{3.20}$$

In the (u, v) plane, the function λ in (3.11)($\bar{\lambda}$ in (resp. 3.12)) transforms into Λ (resp. $\bar{\Lambda}$). Note that $\Lambda'(u) = \pm 1$, $\Lambda(u) \geqslant |u|$ and $\Lambda(u) = |u|$ for sufficiently large values of u.

Let $\hat{\mathcal{H}}$ be the class of piece smooth functions $\hat{f}$ with $\hat{f}(u) \geqslant u$. Set for each piece smooth function $\hat{f} \in \hat{\mathcal{H}}$

$$\hat{I}(\hat{f}) = \frac{1}{4} \int\int_{u<v} [\log(v-u)](1 - \hat{f}'(v))(1 + \hat{f}'(u))dudv. \tag{3.21}$$

$\hat{I}(\hat{f})$ can be expressed in the form

$$\begin{aligned} 1 + 2\hat{I}(\hat{f}) &= \frac{1}{2} \int_{-\infty}^{\infty} \int_{-\infty}^{\infty} \left(\frac{\hat{f}(u) - \hat{f}(v)}{u - v}\right)^2 dudv + 2 \int_{|u|>2} \hat{f}(u) arch|\frac{u}{2}|du \\ &=: \frac{1}{2}||\hat{f}||_\theta^2 + 2 \int_{|u|>2} \hat{f}(u) arch|\frac{u}{2}|du. \end{aligned} \tag{3.22}$$

Note $||\hat{f}||_\theta$ is the Sobolev norm. Define the function $\Omega(u)$ by

$$\Omega(u) = \begin{cases} \frac{2}{\pi}(u \arcsin \frac{u}{2} + \sqrt{4 - u^2}), & |u| \leqslant 2, \\ |u|, & |u| > 2. \end{cases} \tag{3.23}$$

The following lemma, due to Vershik and Kerov [47], plays an important role in finding the limit shape (see Figure 3.1 and Theorem 3.3 below).

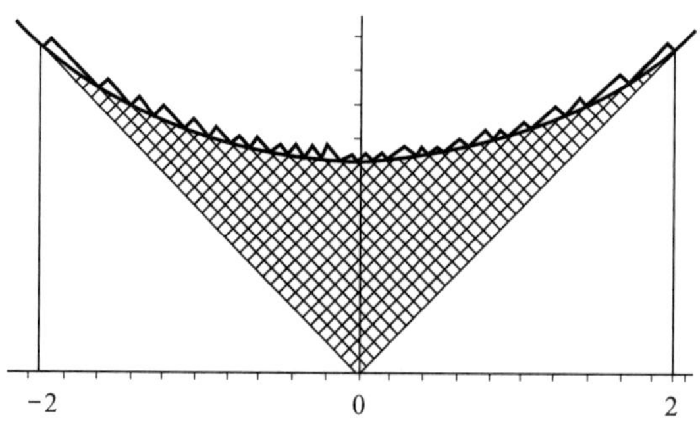

Figure 3.1: The limit shape $\Omega(u)$ under Plancherel measure

Lemma 3.2. (1) *Ω is the unique solution of the variational problem minimizing the functional $\hat{I}$ over $\hat{\mathcal{H}}$;*

(2) $\hat{I}(\Omega) = -\frac{1}{2}$.

It is easy to see

$$\Omega'(u) = \frac{2}{\pi}\arcsin\frac{u}{2}, \quad \Omega''(u) = \frac{2}{\pi\sqrt{4-u^2}}.$$

So, $\Omega(u)$ and $\Omega'(u)$ are continuous on the whole real line, while $\Omega''(u)$ is not.

Theorem 3.3. *Under $(\mathcal{P}_n, P_{p,n})$ we have as $n \to \infty$*

$$\sup_{-\infty<u<\infty} |\bar{\Lambda}(u) - \Omega(u)| \xrightarrow{P} 0. \tag{3.24}$$

To prove (3.24), observe an important estimate: there exists a constant C such that for any 1-Lipschtiz function $\hat{f}$ with compact support

$$\sup_{-\infty<u<\infty} |\hat{f}(u)| \leqslant C||\hat{f}||_\theta^{2/3}. \tag{3.25}$$

Thus it suffices to show

$$||\bar{\Lambda} - \Omega||_\theta \xrightarrow{P} 0. \tag{3.26}$$

Since $\bar{\Lambda}(u)-\Omega(u) \xrightarrow{P} 0$ for each $|u| > 2$, then $\int_{|u|>2}[\bar{\Lambda}(u)-\Omega(u)]\text{arch}|\frac{u}{2}|du$ is negligible. On the other hand,

$$1 + 2\hat{I}(\bar{\Lambda}) = \frac{1}{2}||\bar{\Lambda} - \Omega||_\theta^2 + 2\int_{|u|>2} [\bar{\Lambda}(u) - \Omega(u)]\text{arch}|\frac{u}{2}|du. \tag{3.27}$$

Thus for (3.26), it is sufficient to prove

$$\hat{I}(\bar{\Lambda}) \xrightarrow{P} -\frac{1}{2}. \tag{3.28}$$

Observe $I(\lambda) = \hat{I}(\Lambda)$ for all Young diagrams $\lambda \in \mathcal{P}_n$. Then a similar argument to (3.14) and (3.15) concludes the proof. An alternative proof can be found in Ivanov and Olshanski [21].

As a consequence of Theorem 3.3, we have for each $k \geqslant 1$

$$\frac{\lambda_k}{\sqrt{n}} \xrightarrow{P} 2 \tag{3.29}$$

and

$$\lambda_k - \lambda_{k+1} \xrightarrow{P} \infty. \tag{3.30}$$

3.2 Tracy-Widom distribution at the edge

In Logan and Shepp [28] the question about the fluctuations of $\bar{\lambda}$ around the curve ω and/or $\bar{\Lambda}$ around the curve Ω was posed (see p. 211, [28] for their original guess).

As in the uniform partitions in Section 2, one needs to consider separately two different cases: at the edge and in the bulk. The present subsection is devoted to the study of fluctuation at the edge. The appearance of Tracy-Widom distribution is definitely a miracle.

Let us start with the notion of the determinantal point process. A random point process $\mathcal{X}$ on the real line R is called determinantal with respect to a reference measure μ (the Lebesgue measure or the count measure) if the m-point correlation function ($m \geqslant 1$) is given by

$$P(\{x_1, \cdots, x_m\} \subset \mathcal{X}) = \det(K(x_i, x_j))_{1 \leqslant i,j \leqslant m}, \tag{3.31}$$

where $K(x, y)$ is a kernel of an integral operator $K : L^2(R, \mu) \to L^2(R, \mu)$, non-negative and locally trace class.

A Poisson point process on $\mathbb{R}$ with density $\rho(x)$ can be viewed as a somewhat degenerate determinantal process with kernel (see Johansson [24]),

$$K_{ext}(x, y) = \begin{cases} 0, & x \neq y, \\ \rho(x), & x = y. \end{cases} \tag{3.32}$$

Other important examples are sine, Bessel and Airy point processes. The Airy kernel is defined by

$$\mathbf{A}(x, y) =: \frac{A(x)A'(y) - A(y)A'(x)}{x - y}, \tag{3.33}$$

where $d\mu = dx$ and $A(x)$ is the Airy function satisfying the equation $A''(x) = xA(x)$. Equivalently,

$$\mathbf{A}(x, y) = \int_0^\infty A(x + t)A(y + t)dt. \tag{3.34}$$

Let $\zeta = (\zeta_1 > \zeta_2 > \cdots)$ be the corresponding random configuration. The following result describes the fluctuations of the Young diagram at the edge.

Theorem 3.4. *Under the Plancherel measure we have*

$$\left(\frac{\lambda_k - 2\sqrt{n}}{n^{1/6}}, k = 1, 2, \cdots\right) \xrightarrow{d} (\zeta_k, k = 1, 2, \cdots). \tag{3.35}$$

The distributions of ζ_k can be explicitly calculated. In fact, it is easy to see for any $t \in \mathbb{R}$

$$\begin{aligned} F_1(t) =: P(\zeta_1 \leqslant t) &= E \prod_{j=1}^{\infty} (1 - I_{(\zeta_j > t)}) \\ &= \sum_{k=0}^{\infty} \frac{(-1)^k}{k!} \int_{[0,t]^k} \det(\mathbf{A}(x_i, x_j))_{1 \leqslant i,j \leqslant k} dx_1 \cdots dx_k \\ &= \det(1 - \mathbf{A})_{L^2((t,\infty),dx)}, \end{aligned} \tag{3.36}$$

where the last determinant is called Fredholm determinant of the operator $\mathbf{A}$ on the space $L^2((t,\infty), dx)$.

It were Tracy and Widom [43] who discovered the Fredholm determinant could be governed by the Painlevé II equation. Let u be the global positive solution of $u''(x) = xu(x) + 2u^3(x)$ with boundary asymptotics

$$u(x) = \begin{cases} -(-\frac{x}{2})^{1/2}(1 + O(\frac{1}{x^2})), \; x \to -\infty, \\ \\ -A(x) + O(\frac{e^{-4x^{3/2}/3}}{x^{1/4}}), \; x \to \infty. \end{cases} \tag{3.37}$$

Then for any $t \in \mathbb{R}$

$$F_1(t) = \exp\left(- \int_t^{\infty} (x - t) u^2(x) dx \right). \tag{3.38}$$

Note that the right hand side of (3.38) is indeed a distribution function — the so called Tracy-Widom distribution. Interestingly, the Tracy-Widom distribution has occurred in many distinct areas, say, increasing subsequence, random matrices, random growth models, queueing theory.

It is worth mentioning that F_1 is asymmetric with mean ~ -1.7711 and variance ~ 0.8132, and tail behaviors (see Figure 3.2): for some positive constant C,

$$C^{-1} e^{-C|t|^3} \leqslant F_1(t) \leqslant C e^{-|t|^3/C}, \qquad t \to -\infty$$

and

$$C^{-1} e^{-Ct^{3/2}} \leqslant 1 - F_1(t) \leqslant C e^{-t^{3/2}/C}, \qquad t \to \infty.$$

For $k \geqslant 2$, one can prove the distribution F_k of ζ_k

$$F_k(t) = F_{k-1}(t) + \frac{1}{(k-1)!} \left(-\frac{\partial}{\partial \kappa}\right)^{(k-1)} |_{\kappa=1} F(t, \kappa), \tag{3.39}$$

where $F(t, \kappa)$ is defined in a similar way to (3.37) with $u(x) \sim -\sqrt{\kappa} A(x)$ when $x \to \infty$.

Now Theorem 3.4 is restated as follows.

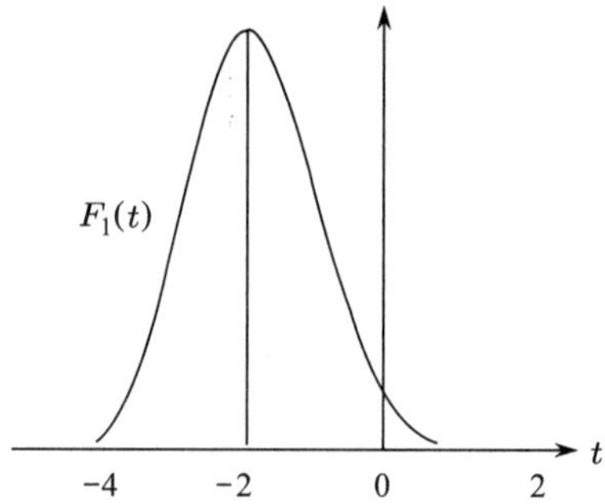

Figure 3.2: Tracy-Widom density

Theorem 3.5. *Under the Plancherel measure, we have for each $k \geqslant 1$*

$$\frac{\lambda_k - 2\sqrt{n}}{n^{1/6}} \xrightarrow{d} F_k. \tag{3.40}$$

The discovery of Tracy-Widom distribution is one of the most spectacular achievement in probability theory and mathematical physics in the 1990s. Baik, Deift and Johansson [3] first proved that F_1 in (3.38) is the limiting distribution of l_n (equivalently λ_1) suitably scaled; then [4] gave a proof of analog for λ_2 and conjectured the analogs are also true for general $k \geqslant 3$. Over the year 1999, almost within six months, three different groups had independently proved Theorem 3.4. The proof in Borodin, Okounkov and Olshanski [8] is representation theoretic, while the proof in Johannson[23] involves ideas form statistical mechanis together with the asymptotics of polynomials orthogonal with respect to a certain distinguished discrete measure. The proof in Okounkov [30], however, provides a picture relating random matrices and random permutations via the equivalence of two points of view on topological surfaces.

For the purpose of this survey, we sketch quickly the proof of [8]. For $\theta > 0$, consider the poissonization of the Plancherel measure

$$P_p^\theta(\lambda) = e^{-\theta} \sum_{n=0}^{\infty} \frac{\theta^n}{n!} P_{p,n}(\lambda). \tag{3.41}$$

This is a probability measure on $\mathcal{P}$. The fact of the matter is that one can compute the correlation functions of P_p^θ.

Set $\mathcal{D}(\lambda) = \{\lambda_i - i\} \subset \mathbb{Z}$ and define the modified Frobenius coordinates $Fr(\lambda)$ of a partition λ by

$$\begin{aligned} Fr(\lambda) &=: (\mathcal{D}(\lambda)+\frac{1}{2}) \triangle (\mathbb{Z}_{\leqslant 0}-\frac{1}{2}) \\ &= \{p_1+\frac{1}{2}, \cdots, p_d+\frac{1}{2}, -q_1-\frac{1}{2}, \cdots, -q_d-\frac{1}{2}\}, \end{aligned} \tag{3.42}$$

where $\triangle$ stands for the symmetric difference of two sets, d is the number of squares on the diagonal of λ. We rewrite the hook formula (1.4) to get

$$\frac{d_\lambda}{|\lambda|!} = \det\left(\frac{1}{(p_i+q_j+1)p_i!q_i!}\right)_{1\leqslant i,j\leqslant d}. \tag{3.43}$$

Thus for a partition λ

$$\begin{aligned} P_p^\theta(\lambda) &= e^{-\theta}\theta^{|\lambda|}(\frac{d_\lambda}{|\lambda|!})^2 \\ &= e^{-\theta}\det(\mathbf{L}(x_i, x_j;\theta))_{1\leqslant i,j\leqslant 2d}, \end{aligned} \tag{3.44}$$

where $Fr(\lambda) = \{x_1, \cdots, x_{2d}\}$ and

$$\mathbf{L}(x,y;\theta) = \begin{cases} 0, & xy > 0, \\ \frac{1}{x-y}\frac{\theta^{(|x|+|y|)/2}}{\Gamma(|x|+\frac{1}{2})\Gamma(|y|+\frac{1}{2})}, & xy < 0. \end{cases} \tag{3.45}$$

From this it follows

Lemma 3.6. *For any $X = \{x_1, x_2, \cdots, x_m\} \subset \mathbb{Z}$ we have*

$$P_p^\theta(\lambda \in \mathcal{P} : X \subset \mathcal{D}(\lambda)) = \det(\mathbf{J}(x_i, x_j;\theta))_{1\leqslant i,j\leqslant m}. \tag{3.46}$$

Here the kernel $\mathbf{J}$ *is given by the following formula*

$$\mathbf{J}(x,y;\theta) = \sqrt{\theta}\frac{J_x J_{y+1} - J_y J_{x+1}}{x-y}, \tag{3.47}$$

where $J_x =: J_x(2\sqrt{\theta})$ is the Bessel function of order x and argument $2\sqrt{\theta}$.

This lemma can be used to show Theorem 3.4 holds under the poissonized Plancherel measure P_p^θ, i.e,

$$(\frac{\lambda_k - 2\sqrt{\theta}}{\theta^{1/6}}, k=1,2,\cdots) \xrightarrow{d} (\zeta_k, k=1,2,\cdots), \quad \theta\to\infty. \tag{3.48}$$

To illustrate the basic idea for (3.48), we look only at the case $k = 1$:

$$P_p^\theta\Big(\lambda \in \mathcal{P} : \frac{\lambda_1 - 2\sqrt{\theta}}{\theta^{1/6}} < t\Big) = P_p^\theta(\lambda \in \mathcal{P} : \lambda_1 - 1 < 2\sqrt{\theta} + \theta^{1/6} t)$$
$$= \det(1 - \mathbf{J})_{l^2(m,m+1,\cdots)}, \tag{3.49}$$

where $m = \lceil 2\sqrt{\theta} + \theta^{1/6} t \rceil$. Since for all $t \in \mathbb{R}$,

$$\det(1 - \mathbf{J})_{l^2(m,m+1,\cdots)} \to \det(1 - \mathbf{A})_{L^2((t,\infty),dx)}, \tag{3.50}$$

then

$$\frac{\lambda_1 - 2\sqrt{\theta}}{\theta^{1/6}} \to \zeta_1, \quad \theta \to \infty. \tag{3.51}$$

We now convert to the original measures using the depoissonization techniques, which is slightly easier than the conditioning argument in Section 2. Given a sequence $b_n, n \geqslant 0$, its poissonization is by definition, the function

$$B(\theta) = e^{-\theta} \sum_{k=0}^\infty \frac{\theta^k}{k!} b_k = E b_N, \tag{3.52}$$

where N is a Poisson random variable with mean θ. One expects that $B(n) \sim b_n$ as $n \to \infty$ provided the variations of b_k for $|k - n| \leqslant \text{const}\sqrt{n}$ are small. One possible regularity condition on b_n which implies (3.52) is monotonicity. In the setting of Young diagrams, this monotonicity was established by Johansson (see also [8]): if we define

$$F_n(x_1, \cdots, x_m) =: P_{p,n}(\lambda \in \mathcal{P}_n : \lambda_i < x_i, 1 \leqslant i \leqslant m), \tag{3.53}$$

then for $(x_1, \cdots, x_m) \in \mathbb{R}^m$

$$F_{n+1}(x_1, \cdots, x_m) \leqslant F_n(x_1, \cdots, x_m). \tag{3.54}$$

This concludes Theorem 3.4. We remark that convergence of the moments also holds, but it requires pretty delicate tail probability estimates. See Baik, Deift and Johansson [3], Ledoux [26] and Su [40] for large deviations and related results.

3.3 CLT in the bulk

In this subsection, we investigate the fluctuations in the bulk of the spectrum of the Young diagrams. Just recently, Bogachev and Su [7] proved the classic central limit theorems hold like in the case of random uniform partitions. Define for $0 < x < 2$

$$\Delta_n(x) = \frac{\lambda(\sqrt{n}x) - \sqrt{n}\omega(x)}{\frac{1}{2\pi}\sqrt{\log n}}. \tag{3.55}$$

Theorem 3.7. *Under the Plancherel measure, we have, as $n \to \infty$, for $m \geqslant 1$ and $0 < x_1 < x_2 < \cdots < x_m < 2$*

$$(\Delta_n(x_1), \Delta_n(x_2), \cdots, \Delta_n(x_m)) \xrightarrow{d} (Z_1, Z_2, \cdots, Z_m), \tag{3.56}$$

where $Z_1, Z_2, \cdots, Z_m$ are independent random variables and each Z_i is normally distributed with mean zero and variance $\varrho^{-2}(x_i)$ where $\varrho(x) = \frac{1}{\pi}\arccos\frac{\omega(x)-x}{2}$.

One can reformulate Theorem 3.7 in coordinates u, v. Define for $-2 < u < 2$

$$\hat{\Delta}_n(u) = \frac{\Lambda(\sqrt{n}u) - \sqrt{n}\Omega(u)}{\frac{1}{\pi}\sqrt{\log n}}. \tag{3.57}$$

An elementary geometric argument shows

$$\hat{\Delta}_n(u) = \frac{1}{2}\varrho(x)\Delta_n(x)(1+\eta_n), \tag{3.58}$$

where $x_n \xrightarrow{P} x$ and $\eta_n \xrightarrow{P} 0$.

Theorem 3.8. *Under the Plancherel measure, we have as $n \to \infty$ for $m \geqslant 1$ and $0 < u_1 < u_2 < \cdots < u_m < 2$*

$$(\hat{\Delta}_n(u_1), \hat{\Delta}_n(u_2), \cdots, \hat{\Delta}_n(u_m)) \xrightarrow{d} (\hat{Z}_1, \hat{Z}_2, \cdots, \hat{Z}_m), \tag{3.59}$$

where $\hat{Z}_1, \hat{Z}_2, \cdots, \hat{Z}_m$ are independent standard normal random variables.

That the normalization constant is the same for all points is not surprising if one observes a Young diagram looks more symmetric and more balanced in coordinates u, v. The proof of Theorem 3.8 is again based on the Poissonization and depoissonization technique. We show only the following 1-dimensional case: under the Poissonized Plancherel measure, for $0 < x < 2$

$$\frac{\lambda(\sqrt{\theta}x) - \sqrt{\theta}\omega(x)}{\frac{1}{2\pi}\sqrt{\log\theta}} \xrightarrow{d} N(0, \varrho^{-2}(x)), \quad \theta \to \infty \tag{3.60}$$

Equivalently, for each $y \in \mathbb{R}$,

$$P_p^\theta(\lambda \in \mathcal{P} : \lambda(\sqrt{\theta}x) - \lceil\sqrt{\theta}x\rceil \leqslant a_\theta(x,y)) \to \Phi_x(y), \tag{3.61}$$

where $a_\theta(x,y) = \sqrt{\theta}(\omega(x) - x) + \frac{y}{2\pi}\sqrt{\log\theta}$ and $\Phi_x(y)$ denotes the distribution function of $N(0, \varrho^{-2}(x))$.

Let I_θ be the half infinite interval $[a_\theta(x,y),\infty)$, and let $|I_\theta|$ denote the number of $\lambda_i - i$ it contains. Thus, (3.61) can be rewritten as

$$P_p^\theta\left(\lambda\in\mathcal{P}: \frac{|I_\theta|-E_p^\theta(|I_\theta|)}{\sqrt{Var_p^\theta(|I_\theta|)}} \leqslant \frac{\lceil\sqrt{\theta}x\rceil - E_p^\theta(|I_\theta|)}{\sqrt{Var_p^\theta(|I_\theta|)}}\right) \to \Phi_x(y). \quad (3.62)$$

At this point, we need the following two lemmas.

Lemma 3.9. *Assume that $\mathcal{X}_t$, $t\in\mathbb{R}_+$ is a family of determinantal point process on the real line with kernel $K_t(x,y)$. Let I_t, $t\in\mathbb{R}_+$ be a set of intervals and $|I_t|$ the number of points of $\mathcal{X}_t$ in I_t. Assume, in addition,*

(1) the integral operator $A_t =: K - t\cdot\chi_{I_t}$ is a family of trace class operators;

(2) $Var(|I_t|) = tr(A_t - A_t^2)\to\infty$ as $t\to\infty$.

Then as $t\to\infty$

$$\frac{|I_t| - E|I_t|}{\sqrt{Var(|I_t|)}} \xrightarrow{d} N(0,1). \quad (3.63)$$

This lemma was first proved by Costin and Lebowitz [10] when $K_t(x,y)\equiv\frac{\sin\pi(x-y)}{\pi(x-y)}$. Soshinikov [36] extended it to a general determinantal random fields and studied at length the Airy, Bessel and sine kernels. Hough et al [20] gave a simple and elegant probabilistic interpretation.

Note that (1) of Lemma 3.9 is satisfied for many kernels, in particular $\mathbf{J}$ in Lemma 3.6. So it remains to verify (2) and to find out what the expectation $E|I_t|$ and variance $Var(|I_t|)$ behave asymptotically like.

Lemma 3.10. *Fix $0<x<2$ and $y\in\mathbb{R}$. We have as $\theta\to\infty$*

$$E_p^\theta(|I_\theta|) = \sqrt{\theta}x - y\varrho(x)\frac{1}{2\pi}\sqrt{\log\theta} + O(1) \quad (3.64)$$

and

$$Var_p^\theta(|I_\theta|) = \frac{1}{4\pi^2}\log\theta(1+o(1)). \quad (3.65)$$

The proof of Lemma 3.10 is based on a direct asymptotic analysis of the expressions for the expectation and variance. In so doing, the calculations are quite laborious and heavily use the asymptotics of the Bessel function $J_m(2\sqrt{\theta})$ in various regions of variation of the parameters (see Bogachev and Su [7]).

The multidimensional normal convergence follows similar ideas using Soshinikov's central limit theorem for linear statistics of the form $\sum_i \alpha_i|I_{t_i}|$ (see Soshinikov [37]). A direct calculation shows that the covariance is 0 between Z_i and Z_j, so that Z_i and Z_j are independent of each other. By independence, we have for any $\varepsilon > 0$ and two distinct $u_1, u_2 \in (-2, 2)$

$$\lim_{n\to\infty} P_{p,n}(|\hat{\Delta}_n(u_1) - \hat{\Delta}_n(u_2)| > \varepsilon) = P(|\hat{Z}_1 - \hat{Z}_2| > \varepsilon) > 0. \quad (3.66)$$

Hence, the weak convergence of process is not true at least under the natural choice of the space of continuous functions $C[-2,2]$ since the necessary condition of tightness breaks down. Analogous remark applies to the process $\Delta_n(u), u \in [0,2]$, considered in the space $D[0,2]$ of right continuous function with left limits.

3.4 Global fluctuations

Kerov obtained and announced a global central limit theorem in his short note [25] in 1993. There also contained the scheme of the proof and a number of fruitful ideas. Around 1999 Kerov found a simpler derivation of the theorem and started writing a detailed paper on this subject but had time only to finish the preliminary section. Ivanov and Olshanski [21] reconstructed the proof from Kerov's unpublished notes.

Consider the modified Chebyschev polynomials

$$u_k(u) =: U_k(\frac{u}{2}) = \sum_{j=0}^{[k/2]}(-1)^j\binom{k-j}{j}u^{k-2j}, \quad k = 0, 1, 2, \cdots.$$

Note that

$$u_k(2\cos\theta) = \frac{\sin(k+1)\theta}{\sin\theta}$$

and

$$\frac{1}{2\pi}\int_{-2}^{2} u_k(u)u_l(u)\sqrt{4-u^2}du = \begin{cases} 1, \ k = l, \\ 0, \ k \neq l. \end{cases}$$

Theorem 3.11. *Let for $k \geqslant 1$*

$$u_k^{(n)} = \int_{-\infty}^{\infty} u_k(u)(\Lambda(u) - \sqrt{n}\Omega(u))du, \quad (3.67)$$

and let $\xi_2, \xi_3, \cdots$ stand for a sequence of independent standard normal random variables. Then under the Plancherel measure we have

$$(u_k^{(n)}, k = 1, 2, \cdots) \xrightarrow{d} (\frac{2\xi_{k+1}}{\sqrt{k+1}}, k = 1, 2, \cdots). \quad (3.68)$$

The proof used essentially the moment method (see [21]). Consider the random series

$$\Xi(u)=\sum_{k=1}^{\infty}\frac{\xi_{k+1}u_k(u)\sqrt{4-u^2}}{\pi\sqrt{k+1}}. \tag{3.69}$$

This series correctly defines a generalized Gaussian process on the space of distributions with support on $[-2,2]$. Its trajectories are not ordinary functions but generalized functions, i.e., the functions in the space $(C^\infty(\mathbb{R}))'$ of compactly supported distributions on the real line. Informally, Theorem 3.11 claims that for the Plancherel diagram $\lambda\in\mathcal{P}_n$,

$$\hat{\Lambda}(u)\sim\Omega(u)+\frac{\Xi(u)}{\sqrt{n}}. \tag{3.70}$$

As we have seen, the fluctuations at the edge are large (of order $n^{1/6}$) and so might present a danger in the integral (3.67). But Kerov's theorem shows the edge of the spectrum in fact does not give any considerable contribution into the integral fluctuations.

To conclude this section, we remark that there is a striking similarity between random matrices and random partitions. The Tracy-Widom distribution was first discovered in the study of the eigenvalues of random matrices in the Gaussian Unitary Ensemble. A similar result about convergence to a generalized Gaussian process for eigenvalues of random matrices was obtained by Johansson [22]; while Gustavsson [19] established the central limit theorem in the bulk of the spectrum of eigenvalues of random matrices. However, random matrices and random partitions are linked only from the view of point of limiting distributions. It would be interesting to establish a finite identity relation between the eigenvalues of random matrices and the parts of random partitions.

4 Miscellaneous

In the last section we will briefly introduce other natural measures, among which are multiplicative measures and Schur measures. There have been many works on them in the literatures, see Okounkov [31], Vershik [44, 45] and references therein. So far we have only touched the so-called linear partitions; a natural generalization is higher dimensional partitions. We will give the definition of 3-dimensional (plane) partitions in this section. This is a very interesting and new field, which has close link with both multiplicative measures and Schur measures. At the end we will mention applications of the well-known Stein-Chen method to Plancherel measures, a real challenge.

4.1 Multiplicative measures

Consider a sequence of functions $f_k(z), k \geqslant 1$, analytic in the open disk $D = \{z \in \mathcal{C} : |z| < \gamma\}$, $\gamma = 1$ or $\gamma = \infty$, such that $f_k(0) = 1$ and assume that (i) the Taylor series

$$f_k(z) = \sum_{j=0}^{\infty} s_k(j) z^j \tag{4.1}$$

have all coefficients $s_k(j) \geqslant 0$ and (ii) the infinite product

$$\mathcal{F}(z) = \prod_{k=1}^{\infty} f_k(z^k) \tag{4.2}$$

converges in D. Then we can define a family of probability measures $P_q, q \in (0, \gamma)$, on $\mathcal{P}$ in the following way: put

$$P_q(\lambda \in \mathcal{P} : r_k(\lambda) = j) = \frac{s_k(j) q^{kj}}{f_k(q^k)}, \quad j \geqslant 0, \quad k \geqslant 1 \tag{4.3}$$

and assume that different r_k are independent. Thus

$$P_q(\lambda) = \frac{\prod_{k=1}^{\infty} s_k(r_k)}{\mathcal{F}(q)} q^{|\lambda|}, \quad \lambda \in \mathcal{P}. \tag{4.4}$$

We call P_q a multiplicative measure with parameter q on $\mathcal{P}$. The generating function $\mathcal{F}(z)$, along with its decomposition $\mathcal{F}(z) = \prod_{k=1}^{\infty} f_k(z^k)$, completely determines such a family. The class of multiplicative measures contain many important examples as discussed in Vershik [45]. Now define the measure P_n on $\mathcal{P}_n$ as follows:

$$P_n(\lambda \in \mathcal{P}_n : r_k(\lambda) = j) = \frac{s_k(j)}{Q_n}, \quad j \geqslant 0, \quad k \geqslant 1 \tag{4.5}$$

and

$$P_n(\lambda) = \frac{\prod_{k=1}^{\infty} s_k(r_k)}{Q_n}, \quad \lambda \in \mathcal{P}_n, \tag{4.6}$$

where

$$Q_n = \sum_{\lambda \in \mathcal{P}_n} \prod_{k=1}^{\infty} s_k(r_k). \tag{4.7}$$

Similarly to Lemma 2.1, we have, for any $q \in (0, \gamma)$ and $n \geqslant 0$, (1) P_n is the conditional probability measure induced on $\mathcal{P}_n$ by P_q; (2) P_q is a convex combination of measures P_n.

On the other hand, P_n is no longer uniform and λ is not identical in distribution to λ'. It would be interesting to investigate the limit shape and fluctuations under $(P_n, \mathcal{P}_n)$. See Vershik [45], Vershik and Yakubovich [48] for related works.

4.2 Schur measures

A generalization of the Plancherel measure is defined as follows. Introduce the following function of λ

$$P^x_{sch}(\lambda) = \frac{1}{\mathcal{Z}} s_\lambda(x) s_\lambda(\bar{x}), \tag{4.8}$$

where $\bar{x}$ is complex conjugate of x, s_λ are the Schur fucntion in auxiliary variables $x_1, x_2, \cdots$ and $\mathcal{Z}$ is the sum in the Cauchy identity for the Schur functions

$$\mathcal{Z} = \sum_{\lambda \in \mathcal{P}} s_\lambda(x) s_\lambda(\bar{x}) = \prod_{i,j} \frac{1}{1 - x_i \bar{x}_j} < \infty. \tag{4.9}$$

It is clear P^x_{sch} is a probability measure on $\mathcal{P}$ which we call the Schur measure. These measure (depending on countably many parameters) are in fact objects of fundamental importance, with profound connections to many central themes of mathematics and physics, including integrable system.

It is convenient to introduce another parameters for the Schur measure

$$t_k = \frac{1}{k} \sum_i x_i^k, \qquad \bar{t}_k = \frac{1}{k} \sum_i \bar{x}_i^k, \quad k = 1, 2, \cdots \tag{4.10}$$

In particular,

$$s_\lambda(x) = \sum_\rho \chi_\lambda(\rho) \prod_k \frac{t_k^{r_k(\rho)}}{(r_k(\rho))!}, \tag{4.11}$$

where $\chi_\lambda(\rho)$ is the character of a permutation ρ in the representation labeled by λ.

If we set $t = \bar{t} = (\sqrt{\theta}, 0, 0, \cdots)$ then P^x_{sch} specializes to the poissonized Plancherel measure P^θ_p. The fact of the matter is that the correlation function of the Schur measure has still a determinantal structure: for any finite set $X \subset \mathbb{Z}$

$$P^x_{sch}(\lambda \in \mathcal{P} : X \subset \mathcal{D}(\lambda)) = \det(K(x_i, x_j))_{x_i, x_j \in X} \tag{4.12}$$

for a certain kernel K, which has a nice generating function, and thus a contour integral representation, in terms of the parameters. From this one can use the steepest descent method to analyze the asymptotics of the correlation function and limit shapes. See Okounkov [31] for further discussions.

4.3 Plane partitions

A plane partition is by definition a 2-dimensional array of nonnegative numbers

$$\pi = (\pi_{ij}), \quad i,j = 1,2,\cdots \tag{4.13}$$

that are nonincreasing as a function of both i and j and such that its size

$$|\pi| =: \sum_{i,j} \pi_{ij} \tag{4.14}$$

is finite. For example,

$$\pi = \begin{pmatrix} 5\,3\,2\,1 \\ 4\,3\,1\,1 \\ 3\,2\,1 \\ 2\,1 \end{pmatrix}$$

is a plane partition of size 29, where the entries that are not shown are zeros. The plot of the function $(x,y) \mapsto \pi_{\lceil x\rceil,\lceil y\rceil}$, $x, y > 0$ is a 3-dimensional Young diagram with the volume $|\pi|$.

Let $\mathcal{P}_{3,n}$ be the set of all plane partitions of n, and let $\mathcal{P}_3$ be the set of all plane partitions, i.e., $\mathcal{P}_3 = \cup_{n=0}^{\infty}\mathcal{P}_{3,n}$. Denote by $p_3(n)$ the number of all plane partitions of n, then the generating function is

$$\sum_{n=0}^{\infty} p_3(n)q^n = \prod_{k=1}^{\infty} \frac{1}{(1-q^k)^k}. \tag{4.15}$$

This coincides $\mathcal{F}(q)$ in (4.2) with $f_k(q) = \frac{1}{(1-q)^k}$, $k \geqslant 1$.

Define a probability measure P_q on $\mathcal{P}_3$ by

$$P_q(\pi) = \prod_{k=1}^{\infty}(1-q^k)^k q^{|\pi|}. \tag{4.16}$$

The restriction $P_{3,n}$ of P_q to $\mathcal{P}_{3,n}$ is a uniform measure on $\mathcal{P}_{3,n}$. On the other hand, to a plane partition π, assign a sequence $\{\lambda(t)\}$ of its diagonal slices (linear partitions). In this way one can use the Schur process and correlation functions to study the limit shape and fluctuations of 3-dimensional diagrams. This is a very interesting probability model; only a few of papers devoted to this object. See Cerf and Kenyon [9], Ferrai and Spohn [14], Okounkov and Reshetikhin [32], Vershik and Yakubovich [48].

Up to now, there is no idea what happens in other dimensions. MacMahon conjectured that the generating function for the 4-dimensional partitions has the form

$$\prod_{k=1}^{\infty} \frac{1}{(1-z^k)^{\binom{k}{2}}}.$$

However, this is wrong.

4.4 Applications of the Stein-Chen method

One of the most important developments in probability theory in the last decades was the Stein-Chen method. The Stein-Chen method is a highly original technique and has been useful in proving normal and Poisson approximation theorems in probability problems with limited information such as the knowledge of only a few moments of the random variables. Fulman [18] initiated the study of the Plancherel measure by the Stein-Chen method. But as Fulman remarked in [18], it is very interesting and challenging to use the Stein-Chen method to understand the Theorem 3.4, giving explicit bounds on the convergence to the Tracy-Widom type distributions.

References

[1] D. Aldous, P. Diaconis. Longest increasing subsequences: from patience sorting to the Baik-Deift-Johansson theorem. *Bull. Am. Math. Soc.* **36** (1999), 413–432.

[2] G. E. Andrews, *The Theory of Partitions, Encyclopedia of Mathematics and Its Applications*, **2**, Addison–Wesley, Reading, MA, 1976.

[3] J. Baik, P. Deift, K. Johansson. On the distribution of the length of the longest increasing subsequence in a random permutation. *J. Am. Math. Soc.* **12** (1999), 1119–1178.

[4] J. Baik, P. Deift, K. Johansson (2000). On the distribution of the length of the second rowof a Young diagram under Plancherel measure. *Geomet. Funct. Anal.* **10** (2000), 702–731.

[5] P. Billingsley, Weak Convergence of Measures, 1968.

[6] S. Bloch, A. Okounkov, The Character of the infinite wedge representation. *Adv. Math.*, **149** (2000), 1–60.

[7] L. Bogachev, Z. G. Su, Central limit theorem for random partitions under the Plancherel measure. to appear in *Soviet Math. Dokl.* (2006).

[8] A. Borodin, A. Okounkov, G. Olshanski, Asymptotics of Plancherel measures for symmetric groups. *J. Am. Math. Soc.* **13** (2000), 481–515.

[9] R. Cerf, R. Kenyon, The low-temperature expansion of the Wulff crystal in the 3D Ising model. *Comm. Math. Phys.* **222**(2001), 147–179.

[10] O. Costin, J. Lebowitz, Gaussian fluctuations in random matrices. *Phys. Rev. Lett.* , **75** (1995), 69–72.

[11] P. Deift, Orthogonal polynomials and random matrices: A Riemann-Hilbert approach. *Courant Lecture Notes in Mathematics*, **3**, Courant Institute of Mathematical Science, New York, 1999.

[12] N. G. de Bruijn, *Asymptotic Methods in Analysis*, North-Holland Publishing Co.-Amsterdam, 1958.

[13] P. Erdös, J. Lehner, The distribution of the number of summands in the partition of a positive integer. *Duke Math. J.* Vol. **8** (1941), 335–345.

[14] P. Ferrari, H. Sphon, Step fluctuations for a faceted crystal. *J. Stat. Phys.* **113** (2003), 1–46.

[15] J. S. Frame, G. de B. Robinson, R. M. Thrall, The hook graphs of the symmetric groups. *Can. J. Math.* **6** (1954) 316–324.

[16] B. Fristedt, The structure of random partitions of large integers. *Trans. Am. Math. Soc.* Vol. **337** (1993), 703–735.

[17] G. Frobenius, Uber die character der symmetrischen gruppe. *Sitzungsber. Preuss. Akad. Berlin* (1903) 328–358.

[18] J. Fulman, Stein's method and Plancherel measure of the symmetric group. *Trans. Am. Math. Soc.* **357** (2004), 555–570.

[19] J. Gustavsson, Gaussian fluctuations of eigenvalues in the GUE. *Ann. Inst. H. Poincaré-Probabilités et Statistques*, **41** (2005), 151–178.

[20] J. Ben Hough, M. Krishnapur, Y. Peres and B. Virág, Determinantal processes and independence. *Probability Surveys.* **2** (2006), 206–229.

[21] V. Ivanov and G. Olshanski, Kerov's central limit theorem for the Plancherel measure on Young diagrams. *Symmetric functions 2001: Surveys of development and perspectives* , Kluwer Academic Publishers, Dordrecht, 2002.

[22] K. Johansson, On fluctuations of eigenvalues of random Hermitian matrices. *Duke Math. J.* **91**(1998), 151–204.

[23] K. Johansson, Discrete orthogonal polynomials ensembles and the Plancherel measure *Ann. math.* **153** (2001), 259–296.

[24] K. Johansson, From Gumbel to Tracy-Widom. *arXiv: Math. PR/ 0510181*

[25] S. Kerov, Gaussian limit for the Plancherel measure of the symmetric group *C. R. Acad. Sci. Paris*, **316** (1993), 303–308.

[26] M. Ledoux, Deviation inequalities on largest eigenvalues. *Concentration between probability and geometric functional analysis. GAFA Seminar Notes* (2005).

[27] M. Löwe, F. Merkl, S. Rolles. Moderate deviations for longest increasing subsequences: the lower tail. *J. Theoret. Probab.* **15** (2002), 1031–1047.

[28] B. F. Logan, L. A. Shepp, A variational problem for random Young tableaux. *Adv. Math.* **26** (1977), 206–222.

[29] I. G. MacDonald, *Symmetric functions and Hall polynomials.* Second edition, Oxford Science Publications, 1995.

[30] A. Okounkov, Random matrices and random permutations. *Internat. Math. Res. Notices* **20** (2000), 1043–1095.

[31] A. Okounkov, The use of random partitions. *arXiv: math-ph/ 0309015v1*

[32] A. Okounkov, N. Reshetikhin, Correlation function of Schur process with application to local geometry of a random 3-dimensional Young diagram. *J. Am. Math. Soc.* **16** (2003), 581–603.

[33] B. Pittel, On a likely shape of the random Ferrers digram. *Adv. Appl. Math.*, Vol. **18** (1997), 432–488.

[34] B. Pittel, On the distribution of the number of YOung tableaux for a uniformly random diagram. *Adv. Appl. Math.*, **29** (2002), 184–214.

[35] B. Sagan, *The Symmetric Group: Representations, Combinatorial Algorithms and Symmetric Functions.* Wadsworth &Books/Cole, Pacific Grove, Calif. 1991.

[36] A. Soshinikov, Gaussian fluctuations in Airy, Bessel, sin and other determinantal random point fields. *J. Stat. Phys.* , (2000) **100** 3/4, 491–522.

[37] A. Soshnikov, Gaussian limit for determinantal random point fields. *Ann. Probab.* **30** (2002), 171–187 .

[38] J. M. Steele, Variations on the Monotone Subsequence Problem of Erds and Szekeres, *In Discrete Probability and Algorithms* (Aldous, Diaconis, and Steele, eds.), 111–132, Springer Publishers, New York, 1995.

[39] J. M. Steele, Probability Theory and Combinatorial Optimization. *CBMS-NSF Regional Conference Series in Applied Mathematics* **69**. SIAM, 1997.

[40] Z. G. Su, Precise asymptotics for random matrices and random growth models. *Submitted* (2006).

[41] M. Szalay, P. Turan, On some problems of the statistical theory of partitions with applications to characters of the symmetric group, *Acta Math. Acad. Sci. Hungar.* I, **29** (1977), 361–379; II, 381–392; III **32** (1978) 129–155.

[42] H. Temperley, Statistical mechanics and the partition of numbers. The form of the crystal surfaces. *Proc. Cambridge Philos. Soc.* **48** (1952), 683–697.

[43] C. A. Tracy, H. Widom, Level-spacing distribution and the Airy kernel. *Comm. Math. Phys.* **159** (1994), 151–174.

[44] A. M. Vershik, Asymptotic combinatorics and algebraic analysis. *Proc. Inter. Cong. Math. Zürich 1994* **2** (1995) 1384–1394.

[45] A. M. Vershik, Statistical mechanics of combinatorial partitions, and their limit shapes. *Funct. Anal. Appl.* , Vol. **30** (1996), 90–105.

[46] A. M. Vershik, Limit shapes of typical geometric configurations and their applications. *J. Math. Sci.*, Vol. **119** (2004), 165–177.

[47] A. M. Vershik and S. Kerov, Asymptotics of the Plancherel measure of the symmetric group and the limit form of Young tableaux. *Soviet Math. Dokl.* **18** (1977), 527–531.

[48] A. M. Vershik, Yu. Yakubovich, Fluctuations of the maximal particle energy of the quantum ideal gas and random partitions. *Comm. Math. Phys.* Vol. **261** (2006), 795–769.

[49] G. N. Watson, *A treatise on the theory of Bessel functions.* Cambridge university Press, 1944.

[50] E. Wigner, Characteristic vectors of bordered matrices with infinite dimensions. *Ann. Math.* **62** (1955), 548–564.

Limit Theorems on Adaptive Designs in Clinical Trials *

Li-Xin Zhang

Department of Mathematics, Zhejiang University
Hangzhou 310027, China
E-mail: stazlx@zju.edu.cn

Abstract

Various adaptive designs (adaptive randomization procedures) have been proposed to clinical trials. This paper discusses several broad families of designs, such as the play-the-winner rule and Markov chain model, randomized play-the-winner rule and urn models, drop-the-loser rule, doubly biased coin adaptive design. Asymptotic theories are presented with several pivotal proofs. The effect of delayed responses, the power and variability comparison of these designs are also discussed.

1 Introduction

Adaptive designs, an important subdivision of experimental designs nowadays, are allocation rules in which the probability a treatment assigned to the coming patient depends upon the results of the previous patients in the study. The basic goal is to skew allocation probabilities to favor better treatment performance. Early important work on adaptive designs was carried out by Thompson (1933) and Robbins (1952). Since then, a steady stream of research (Zelen (1969), Wei and Durham (1978), Wei (1979), Eisele and Woodroofe (1995), etc) in this area has generated various treatment allocation schemes for clinical trials. This paper provides the recent theories of several broad families of designs. In Section 2, we state some limit results on martingale, which are the basic tools to derive the asymptotic properties of adaptive designs. In Section 3, we consider Zelen's play-the-winner (PW) rule and its generations by Lin, Bai, Chen and Hu (2003). In Section 4, we derive the asymptotic properties of the adaptive designs based on urn models, a large family of randomization procedures. In Section 5, the drop-the-loser rule is introduced. In Section 6, an important family of target-driven designs,

*Project supported by the National Natural Science Foundation of China (No. 10471126).

the doubly adaptive biased coin designs, are discussed. In Section 7, the effect of the delay of treatment results is discussed. In Section 8, we compare the variabilities of different type of adaptive designs. The lower bound of the asymptotic variability for a pre-specified allocation proportion is established, and asymptotic best adaptive designs are provided. Finally, further discussion and future topics are mentioned in Section 9. For the convenience of reading, we also give several pivotal proofs. The principle ideas of deriving asymptotic properties can be found from these proofs. Those who are not interested in the theoretical results can skip these proofs and go quickly to the last two sections.

The following notations and definitions are introduced to describe the randomized treatment allocation schemes. Given a clinical trial with K treatments. Let $\boldsymbol{X}_1, \boldsymbol{X}_2, \cdots$ be the sequence of random treatment assignments. For the m-th subject, $\boldsymbol{X}_m = (X_{m,1}, \cdots, X_{m,K})$ represents the assignment of treatment such that if the m-th subject is allocated to treatment k, then all elements in $\boldsymbol{X}_m$ are 0 except for the k-th component, $X_{m,k}$, which is 1. Let $N_{n,k}$ be the number of subjects assigned to treatment k in the first n assignments and write $\boldsymbol{N}_n = (N_{n,1}, \cdots, N_{n,K})$. Then $\boldsymbol{N}_n = \sum_{m=1}^n \boldsymbol{X}_m$. We are interested in the statistical behavior of proportions $N_{n,k}/n$, $k = 1, \cdots, K$.

2 Preliminaries, limit theorems on martingales

The martingale approach is the basic tool to investigate the asymptotic properties of adaptive designs. In this section, we state some limit theorems on martingales. For more results, one can refer to Hall and Heyde (1980) and Stout (1974) or other text books. Let $\{M_n, \mathcal{F}_n; n \geqslant 1\}$ be a real martingale sequence with $\Delta M_n = M_n - M_{n-1}$ being it difference, where $\{\mathcal{F}_n\}$ is a filter of σ-fields.

Theorem 2.1. (LLN) *Let $\eta_n > 0$ be a sequence of random variables such η_n is $\mathcal{F}_{n-1}$ measurable and is non-decreasing almost surely, and p is a real number in $(0, 2]$.*

(a) *Then with probability one, $M_n/\eta_n \to 0$ on the event $\{\eta_n \to \infty, \sum_{m=1}^\infty \mathsf{E}[|\Delta M_m|^p|\mathcal{F}_{m-1}]/\eta_m^p < \infty\}$.*

(b) *If $\sum_{m=1}^n \mathsf{E}[(\Delta M_m)^2|\mathcal{F}_{m-1}] \leqslant \eta_n$ a.s., then with probability one $M_n/\eta_n \to 0$ on the event $\{\eta_n \to \infty\}$.*

Theorem 2.2. (LIL) *Suppose $\sup_m \mathsf{E}[|\Delta M_m|^p] < \infty$ for some $p > 2$, then $M_n = O(\sqrt{n \log\log n})$ a.s..*

Theorem 2.3. (CLT) *Suppose there is a constant $\sigma \geqslant 0$ such that $\sum_{m=1}^n \mathsf{E}[(\Delta M_m)^2|\mathcal{F}_{m-1}]/n \xrightarrow{P} \sigma^2$. Here and in the sequel $\xrightarrow{P}$ denotes the *convergence in probability. Further assume that the conditional Lindberg condition*

$$\frac{1}{n}\sum_{m=1}^{n}\mathsf{E}\big[(\Delta M_m)^2 I\{|\Delta M_m|\geqslant \epsilon\sqrt{n}\}|\mathcal{F}_{m-1}\big]\xrightarrow{P}0,\quad \forall\epsilon>0$$

is satisfied. Then $M_n/\sqrt{n}\xrightarrow{d}N(0,\sigma^2)$. *Further,* $M_{[nt]}/\sqrt{n}\xrightarrow{d}\sigma W(t)$ *in* $D[0,\infty)$, *where* $W(t)$ *is a standard Brownian motion. Here and in the sequel* $\xrightarrow{d}$ *denotes the convergence in distribution.*

Theorem 2.4. (Skorokhod embedding theorem) *In a possibly enlarged probability space in which there is a standard motion* $B(t)$, *we can redefine the martingale sequence* $\{M_n,\mathcal{F}_n\}$ *without changing its distribution and define a non-decreasing sequence of random variables* $\{\tau_n\}$ *such that* τ_n *is* $\mathcal{F}_n$ *measurable,* $M_n=B(\tau_n)$ *and* $\mathsf{E}[\Delta\tau_m|\mathcal{F}_{m-1}]=\mathsf{E}[(\Delta M_m)^2|\mathcal{F}_{m-1}]$ *a.s.,* $\mathsf{E}[|\Delta\tau_m|^p|\mathcal{F}_{m-1}]\leqslant C_p\mathsf{E}[|\Delta M_m|^{2p}|\mathcal{F}_{m-1}]$ *a.s. for* $p\geqslant 1$.

Theorem 2.5. (Strong approximation) *Let* $\{\boldsymbol{M}_n,\mathcal{F}_n;\geqslant 1\}$ *be a martingale sequence in* $\mathbb{R}^d$ *space,* $\sum_n=\sum_{m=1}^n\mathsf{E}[(\Delta\boldsymbol{M}_m)'\Delta\boldsymbol{M}_m|\mathcal{F}_{m-1}]$. *Suppose that there exists a constant* $0<\epsilon<1$ *such that*

$$\sum_{n=1}^{\infty}\mathsf{E}\big[\|\Delta\boldsymbol{M}_n\|^2 I\{\|\Delta\boldsymbol{M}_n\|^2\geqslant n^{1-\epsilon}\}\big|\mathcal{F}_{n-1}\big]/n^{1-\epsilon}<\infty\quad a.s.,\qquad(2.1)$$

and that $\boldsymbol{T}$ *is a covariance matrix which is measurable with respect to* $\mathcal{F}_k$ *for some* $k\geqslant 0$. *Then for any* $\delta>0$, *(possibly in an enlarged probability space with* $\{\boldsymbol{M}_n\}$ *being redefined) there exist* $\kappa>0$ *and a d-dimensional standard Brownin* $\boldsymbol{W}(t)$, *independent of* $\boldsymbol{T}$, *such that*

$$\boldsymbol{S}_n-\boldsymbol{W}(n)\boldsymbol{T}^{1/2}=O(n^{1/2-\kappa})+O(\alpha_n^{1/2+\delta})\ a.s..$$

Here $\alpha_n=\max_{m\leqslant n}\|\boldsymbol{\Sigma}_m-m\boldsymbol{T}\|$. *Here and in the sequel* $\|\cdot\|$ *denotes the norm of a vector or a matrix.*

If $\sup_n\mathsf{E}\|\Delta\boldsymbol{M}_n\|^{2+\delta_0}<\infty$ for some $\delta_0>0$, then the condition (2.1) is satisfied. The proofs of Theorem 2.1 (a), Theorem 2.3 and Theorem 2.4 can be found in Hall and Heyde (1980). The proof of Theorem 2.1 (b) can be found in Stout (1974). Theorem 2.2 can be proved by Theorem 2.4 and the LIL of a Brownian motion. The proof of Theorem 2.5 is given in Zhang (2004). Also, Theorems 2.1–2.3 remain true for martingales in a $\mathbb{R}^d$ space with some necessary notations changed.

3 Play-the-winner rule and Markov chain adaptive design

Consider a two-arm clinical trial: two treatments (1 and 2) with dichotomous response (success and failure). Patients (subjects) are recruited into the clinical trial sequentially and respond immediately to treatments. Zelen (1969) proposed the following design, which is well known as the PW rule: *A success on a particular treatment generates a future trial on the same treatment with a new patient. A failure on a treatment generates a future trial on the alternate treatment.* Let p_i be the success probability of a patient on the treatment i, $q_i = 1 - p_i$ be the failure probability , $i = 1, 2$. Then

$$\frac{N_{n,1}}{n} \xrightarrow{P} \frac{q_2}{q_1 + q_2} \quad \text{and} \quad \sqrt{n}(\frac{N_{n,1}}{n} - \frac{q_2}{q_1 + q_2}) \xrightarrow{d} N(0, \sigma_{PW}^2),$$

where $\sigma_{PW}^2 = q_1 q_2 (p_1 + p_2)/(q_1 + q_2)^3$. Notice that $q_2/(q_1 + q_2) > 1/2$ if $p_1 > p_2$. So, if treatment 1 is "doing better", the PW rule favors treatment 1.

Lin, Bai, Chen and Hu (2003) extended the PW rule to a general Markov chain adaptive design. Suppose that at the stage m, the treatment 1 is assigned to the m-th patient. Then the $(m+1)$-th patient will be assigned either treatment 1 or treatment 2 according certain probabilities, which depend on the response of the m-th patient. Let α_s be the probability of assigning the $(m+1)$-th patient to treatment 1, when the response of the m-th patient to treatment 1 is "success", and let α_f be the probability of assigning the $(m+1)$-th patient to treatment 1, when the response of the m-th patient to treatment 1 is "failure". Similarly define β_s and β_f with treatment 2 instead of treatment 1 in the definitions of α_s and α_f. When $\alpha_s = 1$, $\alpha_f = 0$, $\beta_s = 1$, $\beta_f = 0$, we get Zelen's PW rule. A clinical application disadvantage of the PW rule is that it is fully deterministic, i.e., when the previous results are known, the assignment of the next subject is fully determined. The Markov chain adaptive design is not fully deterministic except when the parameter α_s, α_f, β_s and β_f take extreme values 0 and 1. When taking $\alpha_s = \alpha_f = \beta_s = \beta_f = 1/2$, we get the fully randomization procedure which allocates patients to each treatment with a probability 1/2. The more are the parameters near extreme values, the more is the procedure being deterministic. When $\alpha_s < \alpha_f$ and $\beta_s < \beta_f$, the Markov chain adaptive design is less ethical than the balanced allocation (c.f., Equation (3.1)). The parameters α_s, α_f, β_s and β_f can be chosen to reflect the trade-off between the degree of randomness and ethic.

Let $p_1(m) = P(\{\text{success } | X_{m,1} = 1\})$ and $p_2(m) = P(\{\text{success } | X_{m,1} = 0\})$ and $\alpha_m = p_1(m)\alpha_s + (1 - p_1(m))\alpha_f$, $\beta_m = p_2(m)\beta_s + (1 - p_2(m))\beta_f$. Then $\{X_{m,1}\}$ is a Markov chain with the transition probability matrix

$$\boldsymbol{P}_m = \begin{pmatrix} \alpha_m & 1 - \alpha_m \\ 1 - \beta_m & \beta_m \end{pmatrix}.$$

When $p_1(m) = p_1$ and $p_2(m) = p_2$ for all m, $\alpha_n \equiv \alpha$, $\beta_n \equiv \beta$, and $\{X_{m,1}\}$ is a homogeneous Markov chain with a stationary distribution

$(\mu, 1-\mu)$, where $\mu = (1-\beta)/(2-\alpha-\beta)$. Following from the central limit theorem for Markov chains we have that

$$\frac{N_{n,1}}{n} \xrightarrow{p} \mu \text{ and } \sqrt{n}(\frac{N_{n,1}}{n} - \mu) \xrightarrow{d} N(0, \sigma^2), \tag{3.1}$$

where

$$\sigma^2 = \mathsf{Var}_\mu\{X_{1,1}\} + 2\sum_{j=2}^{\infty} \mathsf{Cov}_\mu\{X_{1,1}, X_{j,1}\} = \frac{(1-\alpha)(1-\beta)(\alpha+\beta)}{(2-\alpha-\beta)^3}.$$

For non-homogeneous case, Lin, Bai, Chen and Hu (2003) proved (3.1) under the condition that

$$\sum_{m=1}^{\infty} \frac{|p_1(m) - p_1| + |p_2(m) - p_2|}{\sqrt{m}} < \infty.$$

Lin, Zhang, Chueng and Chan (2005) established the strong approximation for $N_{n,1}$, from which (3.1) follows immediately.

Theorem 3.1. *In a possibly enlarged probability space, we can redefine the sequence $\{N_{n,1}\}$ without changing its distribution, such that*

$$N_{n,1} - n\mu - \sigma W(n) = O((n \log\log n)^{1/4} (\log n)^{1/2}) + O(\Delta_n) \quad \text{a.s.},$$

where $\{W(t)\}$ is a standard Brownian motion and $\Delta_n = \sum_{m=1}^{n}(|\alpha_m - \alpha| + |\beta_m - \beta|)$.

Proof. Write $\Delta m_m = X_{m,1} - \mathsf{E}[X_{m,1}|\mathcal{F}_{m-1}]$, where $\mathcal{F}_{m-1}$ is the history σ-field generated by $X_{m,1}, \cdots, X_{1,1}$. Then $X_m = \Delta m_m + \mathsf{E}[X_{m,1}|\mathcal{F}_{m-1}] = \Delta m_m + 1 - \beta_m - (1-\alpha_m-\beta_m)X_{m-1,1}$. It follows that $N_{n,1} = m_n + 1 - \beta - (1-\alpha-\beta)N_{n-1,1} + O(\Delta_n) + O(1)$. So,

$$N_{n,1} = n\mu + m_n/(2-\alpha-\beta) + O(\Delta_n).$$

For the martingale m_n, we have $\mathsf{E}[(\Delta m_m)^2|\mathcal{F}_{m-1}] = (1-\beta_m)\beta_m + (\alpha_m - \beta_m)(1-\alpha_m-\beta_m)X_{m-1,1}$ and $|\Delta m_m| \leqslant 1$. It follows that

$$\begin{aligned}
&\sum_{m=1}^{n} \mathsf{E}[(\Delta m_m)^2|\mathcal{F}_{m-1}] \\
=&n(1-\beta)\beta + (\alpha-\beta)(1-\alpha-\beta)N_{n-1,1} + O(\Delta_n) \\
=&n(1-\beta)\beta + n(\alpha-\beta)(1-\alpha-\beta)\mu + \frac{(\alpha-\beta)(1-\alpha-\beta)}{2-\alpha-\beta} m_n + O(\Delta_n) \\
=&n\frac{(1-\alpha)(1-\beta)(\alpha+\beta)}{2-\alpha-\beta} + O(\Delta_n) + O(\sqrt{n\log\log n}) \\
=:&n\sigma_M^2 + O(\Delta_n) + O(\sqrt{n\log\log n}) \quad \text{a.s.}
\end{aligned}$$

due to the LIL (Theorem 2.1). Applying the Skorokhod embedding theorem (Theorem 2.4), we can write $m_m = B(\tau_m)$. Notice $\sum_{m=1}^n \big(\Delta\tau_m - \mathsf{E}[\Delta\tau_m|\mathcal{F}_{m-1}]\big)$ is also a martingale sequence. By the LIL, we conclude that

$$\begin{aligned}\tau_n &= \sum_{m=1}^n \mathsf{E}[\Delta\tau_m|\mathcal{F}_{m-1}] + O(\sqrt{n\log\log n})\\ &= \sum_{m=1}^n \mathsf{E}[(\Delta m_m)^2|\mathcal{F}_{m-1}] + O(\sqrt{n\log\log n})\\ &= n\sigma_M^2 + O(\Delta_n) + O(\sqrt{n\log\log n}) \text{ a.s..}\end{aligned}$$

It follows that

$$\begin{aligned}m_n &= B(\tau_n) = B(n\sigma_M^2) + O\Big(\sqrt{(\Delta_n + \sqrt{n\log\log n})\log n}\Big)\\ &= B(n\sigma_M^2) + O(\Delta_n) + O((n\log\log n)^{1/4}(\log n)^{1/2}) \text{ a.s.}\end{aligned}$$

by the sample properties of a Brownian motion (c.f., Csörgő and Révśz (1980)). The proof is now completed by letting $W(t) = B(t\sigma_M^2)/\sigma_M$ and noticing that $\sigma_M^2/(2-\alpha-\beta)^2 = \sigma^2$. □

For the multi-treatment case, we let $p_i(m) = \mathsf{P}(\{\text{success } | X_{m,i} = 1\})$. Assume the transition probability matrix of the Markov chain $\{\boldsymbol{X}_n\}$ is $\boldsymbol{H}_n = \{H_{ij}(n)\}$ which is a function of $p_i(n)$, $i = 1, \cdots, K$, i.e, $\mathsf{E}[\boldsymbol{X}_{n+1}|\boldsymbol{X}_n] = \boldsymbol{X}_n\boldsymbol{H}_n$. By using Theorem 2.5 instead of the Skorokhod embedding theorem, Zhang (2004) showed that $\boldsymbol{N}_n$ can be approximated by a multi-dimensional Browian motion:

$$\boldsymbol{N}_n - n\boldsymbol{v} - \boldsymbol{W}(n)\boldsymbol{\Sigma}^{1/2} = o(n^{1/2-\kappa}) + O\Big(\sum_{k=1}^n \|\boldsymbol{H}_k - \boldsymbol{H}\|\Big) \quad \text{a.s.,}$$

where $\kappa > 0$, $\boldsymbol{v} = (v_1, \cdots, v_K)$ is the left eigenvector corresponding to the largest eigenvalue of $\boldsymbol{H}$ with $v_1 + \cdots + v_K = 1$, $\{\boldsymbol{W}(t)\}$ is a K-dimensional standard Brownian motion, and

$$\boldsymbol{\Sigma} = (\boldsymbol{I} - \boldsymbol{H}' + \boldsymbol{v}'\boldsymbol{1})^{-1}(\text{diag}(\boldsymbol{v}) - \boldsymbol{H}'\text{diag}(\boldsymbol{v})\boldsymbol{H})(\boldsymbol{I} - \boldsymbol{H} + \boldsymbol{v}'\boldsymbol{1})^{-1}.$$

In particular, we have asymptotic normality, if

$$\sum_{k=1}^n \|\boldsymbol{H}_k - \boldsymbol{H}\| = o(n^{1/2}). \tag{3.2}$$

Zhang (2006) studied a kind of non-humongous Markov chain designs, in which $\boldsymbol{H}_n$ is a function of an estimated unknown parameters $\widehat{\boldsymbol{\theta}}$. In this case the condition (3.2) is not satisfied.

4 Randomized play-the-winner rule and urn models

To overcome the drawback that the PW rule is fully deterministic, Wei and Durham (1978) introduced the following randomized play-the-winner (RPW) rule: *We start with* (α, β) *balls* (*type* 1 *and* 2 *respectively*) *in the urn. If the type* k *ball is drawn, a patient is assigned to the treatment* k, $k = 1, 2$. *The ball is replaced and the patient's response is observed. A success on the treatment* 1 *or a failure on the treatment* 2 *generates a type* 1 *ball in the urn; a success on the treatment* 2 *or a failure on the treatment* 1 *generates a type* 2 *ball in the urn.* Let $Y_{n,1}$ ($Y_{n,2}$) be number of balls of type 1 (2) after n stage. From the results of Athreya and Karlin (1968), we have

$$\frac{Y_{n,1}}{Y_{n,1}+Y_{n,2}} \to \frac{q_2}{q_1+q_2} \quad \text{a.s.} \quad \text{and} \quad \frac{N_{n,1}}{n} \to \frac{q_2}{q_1+q_2} \quad \text{a.s..}$$

Here p_i (q_i) is the success (failure) probability of a patient on the treatment i, as usual. The limiting proportion is the same as that of the PW rule. We refer to it as urn proportion. When $p_1 + p_2 < 1.5$ (or $q_1 + q_2 > 0.5$), we have the following asymptotic normality:

$$\sqrt{n}\Big(\frac{Y_{n,1}}{Y_{n,1}+Y_{n,2}} - \frac{q_2}{q_1+q_2}\Big) \xrightarrow{d} N\Big(0, \frac{q_1 q_2}{(2(q_1+q_2)-1)(q_1+q_2)^2}\Big)$$

and $\sqrt{n}(N_{n,1}/n - q_2/(q_1+q_2)) \xrightarrow{d} N(0, \sigma^2_{\text{RPW}})$, where

$$\sigma^2_{\text{RPW}} = \frac{q_1 q_2[5 - 2(q_1+q_2)]}{[2(q_1+q_2)-1](q_1+q_2)^2}. \tag{4.1}$$

The asymptotic normality was first given in Smythe and Rosenberger (1995). When $q_1 + q_2 < 0.5$, the limiting distributions of both the urn composition and the allocation proportion are unknown. The RPW rule has the same limiting allocation proportion as the PW rule. But the asymptotic variability is much larger.

As multi-treatment extensions of the RPW rule, one large family of randomized adaptive designs can be developed from the generalized Polya urn (GPU) model. Urn models have also long been recognized as valuable mathematical apparatus in many areas including physical science, biological science, engineering, information science, the study of economic behaviors, etc.

Consider an urn containing balls of K types. Initially, the urn contains $\boldsymbol{Y}_0 = (Y_{0,1}, \cdots, Y_{0,k}, \cdots, Y_{0,K})$ balls, where $Y_{0,k}$ denotes the number of balls of type k, $k = 1, \cdots, K$. A ball is drawn at random. Its type is observed and the ball is then replaced. At the m-th stage, following a type k drawn, $D_{kj}(m)$ ($\geqslant 0$) balls of type j, for $j = 1, \cdots, K$, are added to the urn. $D_{ij}(m)$ is a random function of the response $\xi_{m,k}$ of the m-th subject on treatment k. The expectation of the total numbers of balls added in each stage is assumed to be the same (say γ), so

$$\sum_{j=1}^{K} \mathsf{E}[D_{kj}(m)|\mathcal{F}_{m-1}] = \gamma, \quad k, j = 1, \cdots, K, m = 1, 2, \cdots,$$

where $\mathcal{F}_{m-1}$ is the history σ-field. Without loss generality, we can assume $\gamma = 1$. Let $\boldsymbol{H}_m$ be the matrix comprising element $\{h_{kj}(m) = \mathsf{E}[D_{kj}(m)|\mathcal{F}_{m-1}]\}$ and $\boldsymbol{D}_m = \{D_{kj}(m)\}$. We refer to $\boldsymbol{D}_m$ as the adding rules and $\boldsymbol{H}_m$ as the design matrices. If $\boldsymbol{D}_m$, $m = 1, 2, \cdots$, are i.i.d., then $\boldsymbol{H}_m = \boldsymbol{H}$ for all n. In this case the model is said to be homogenous. In general, it is assumed that $\boldsymbol{H}_m \to \boldsymbol{H}$. Let $\boldsymbol{Y}_n = (Y_{n,1}, \cdots, Y_{n,K})$, where $Y_{n,k}$ represents the number of balls in the urn of type k after n-th stage. And let $\boldsymbol{v} = (v_1, \cdots, v_K)$ be the left eigenvector corresponding to the largest eigenvalue of $\boldsymbol{H}$ with $v_1 + \cdots + v_K = 1$. Then v_k is just the limiting proportion of both the patients assigned to treatment k and the type k balls in the urn. Bai, Hu and Zhang (2002), Hu and Zhang (2001) obtained the asymptotic properties via the strong approximation.

Theorem 4.1. *Suppose that $\{\boldsymbol{D}_m\}$ is sequence of i.i.d. random matrices with $\sup_m \mathsf{E}\|\boldsymbol{D}_m\|^{2+\delta} < \infty$. Let $\lambda_1 = \gamma = 1, \lambda_2, \cdots, \lambda_K$ be the eigenvalues of $\boldsymbol{H}$, and $\lambda = \max\{\mathrm{Re}(\lambda_2), \cdots, \mathrm{Re}(\lambda_K)\}$. If $\lambda < 1$, then*

$$\frac{N_{n,i}}{n} \to v_i \quad \text{a.s. and} \quad \frac{Y_{n,i}}{\sum_{j=1}^{K} Y_{nj}} \to v_i \quad \text{a.s..} \tag{4.2}$$

If $\lambda < 1/2$, then

$$\sqrt{n}(\frac{\boldsymbol{Y}_n}{n} - \boldsymbol{v}) \xrightarrow{d} N(0, \boldsymbol{\Sigma}) \text{ and } \sqrt{n}(\frac{\boldsymbol{N}_n}{n} - \boldsymbol{v}) \xrightarrow{d} N(0, \boldsymbol{\Sigma}^*). \tag{4.3}$$

Proof. Write $\boldsymbol{1} = (1, \cdots, 1)$, $|\boldsymbol{Y}_n| = \sum_{k=1}^{K} Y_{nk}$, $\widetilde{\boldsymbol{H}} = \boldsymbol{H} - \boldsymbol{1}'\boldsymbol{v}$, $\boldsymbol{M}_n = \sum_{m=1}^{N} \boldsymbol{X}_m(\boldsymbol{D}_m - \boldsymbol{H})$, $\boldsymbol{m}_m = \sum_{m=1}^{n}(\boldsymbol{X}_m - \mathsf{E}[\boldsymbol{X}_m|\mathcal{F}_{m-1}])$. Then $\boldsymbol{H}\boldsymbol{1}' = \boldsymbol{1}'$, $\boldsymbol{m}_n\boldsymbol{1}' = \boldsymbol{0}$, and the eigenvalues of $\widetilde{\boldsymbol{H}}$ are $0, \lambda_2, \cdots, \lambda_K$. We have the following lemma on matrices, the proof of which can be founded in Hu and Zhang (2004a). □

Lemma 4.2. *If $\Delta \boldsymbol{Q}_n = \Delta \boldsymbol{P}_n + \boldsymbol{Q}_{n-1}\widetilde{\boldsymbol{H}}/(n-1)$, $n \geqslant 2$, then*

$$\|\boldsymbol{Q}_n\| = O(\|\boldsymbol{P}_n\|) + \sum_{m=1}^{n} \frac{O(\|\boldsymbol{P}_m\|)}{m}(n/m)^{\lambda} \log^{\nu-1}(n/m),$$

where ν is the degree of the second largest eigenvalue of $\boldsymbol{H}$.

We prove (4.3) only. Notice

$$\begin{aligned}\Delta(\boldsymbol{Y}_n - n\boldsymbol{v}) &= \boldsymbol{X}_n\boldsymbol{D}_n - \boldsymbol{v} = \Delta\boldsymbol{M}_n + \Delta\boldsymbol{m}_n\boldsymbol{H} + \frac{\boldsymbol{Y}_{n-1}}{|\boldsymbol{Y}_{n-1}|}\boldsymbol{H} - \boldsymbol{v}\\ &= \frac{\boldsymbol{Y}_{n-1} - (n-1)\boldsymbol{v}}{n-1}\widetilde{\boldsymbol{H}} + \Delta\boldsymbol{M}_n + \Delta\boldsymbol{m}_n\widetilde{\boldsymbol{H}}\\ &\quad + \left(1 - \frac{|\boldsymbol{Y}_{n-1}|}{n-1}\right)\left(\frac{\boldsymbol{Y}_{n-1}}{|\boldsymbol{Y}_{n-1}|} - \boldsymbol{v}\right)\widetilde{\boldsymbol{H}}. \end{aligned} \tag{4.4}$$

Multiplying $\mathbf{1}'$ yields $|\boldsymbol{Y}_n| - n = \Delta\boldsymbol{M}_n\mathbf{1}' + |\boldsymbol{Y}_0| = O(\sqrt{n\log\log n})$ a.s. due to the LIL (Theorem 2.2). It follows that

$$\boldsymbol{M}_n + \boldsymbol{m}_n\boldsymbol{H} + \sum_{m=1}^{n-1}\left(1 - \frac{|\boldsymbol{Y}_m|}{m}\right)\left(\frac{\boldsymbol{Y}_m}{|\boldsymbol{Y}_m|} - \boldsymbol{v}\right)\widetilde{\boldsymbol{H}} = O(\sqrt{n\log\log n}) \text{ a.s..}$$

By applying Lemma 4.2 and noticing $\lambda < 1/2$, we obtain $\boldsymbol{Y}_n - n\boldsymbol{v} = O(\sqrt{n\log\log n})$ a.s.. Write $\boldsymbol{D}_1^{(k)} = (D_{k1}, \cdots, D_{kK})$, $\boldsymbol{\Sigma}_1 = \text{diag}(\boldsymbol{v}) - \boldsymbol{v}'\boldsymbol{v}$ and $\boldsymbol{\Sigma}_2 = \sum_{k=1}^K v_k \mathsf{Var}\{\boldsymbol{D}_1^{(k)}\}$. For the martingale $(\boldsymbol{M}_n, \boldsymbol{m}_n)$, we have $\sup_n \mathsf{E}\|\Delta\boldsymbol{M}_n\|^{2+\delta} < \infty$, $\|\Delta\boldsymbol{m}_n\| \leqslant K$, $\mathsf{E}[(\Delta\boldsymbol{M}_n)'\Delta\boldsymbol{m}_n|\mathcal{F}_{m-1}] = \mathbf{0}$ and

$$\begin{aligned}\sum_{m=1}^n \mathsf{E}\Big[(\Delta\boldsymbol{m}_m)'\Delta\boldsymbol{m}_m|\mathcal{F}_{m-1}\Big] &= \sum_{m=1}^n[\text{diag}(\frac{\boldsymbol{Y}_m}{|\boldsymbol{Y}_m|}) - (\frac{\boldsymbol{Y}_m}{|\boldsymbol{Y}_m|})'\frac{\boldsymbol{Y}_m}{|\boldsymbol{Y}_m|}]\\ &= n\boldsymbol{\Sigma}_1 + O(\sqrt{n\log\log n}) \text{ a.s.,}\\ \sum_{m=1}^n \mathsf{E}\Big[(\Delta\boldsymbol{M}_m)'\Delta\boldsymbol{M}_m|\mathcal{F}_{m-1}\Big] &= \sum_{m=1}^n\sum_{k=1}^K \mathsf{Var}\{\boldsymbol{D}_1^{(k)}\}Y_{mk}/|\boldsymbol{Y}_m|\\ &= n\boldsymbol{\Sigma}_2 + O(\sqrt{n\log\log n}) \text{ a.s..}\end{aligned}$$

By the strong approximation (Theorem 2.5), there are two independent d-dimensional standard Brownian motions $\boldsymbol{W}_1(t)$ and $\boldsymbol{W}_2(t)$ such that for some $\kappa > 0$,

$$\boldsymbol{m}_n = \boldsymbol{W}_1(n)\boldsymbol{\Sigma}_1^{1/2} + o(n^{1/2-\kappa}) \text{ a.s.}, \ \boldsymbol{M}_n = \boldsymbol{W}_2(n)\boldsymbol{\Sigma}_2^{1/2} + o(n^{1/2-\kappa}) \text{ a.s..} \tag{4.5}$$

Without loss generality, we assume $1/2 - \kappa > \lambda$ and $1/2 - \kappa > 1/4$. Let $\boldsymbol{G}_i(t)$ be the solution of the equation

$$\boldsymbol{G}_i(t) = \int_0^t \frac{\boldsymbol{G}_i(x)}{x}\widetilde{\boldsymbol{H}}dx + \boldsymbol{W}_i(t)\boldsymbol{\Sigma}_i^{1/2}, \ \ \boldsymbol{G}_i(0) = 0, \tag{4.6}$$

$i = 1, 2$. Notice $\{[\boldsymbol{W}_i(T(\cdot + s)) - \boldsymbol{W}_i(T\cdot)]/\sqrt{T}\} \stackrel{d}{=} \{\boldsymbol{W}_i(\cdot)\}$. It can be checked that $\boldsymbol{G}_i(t)$ is a Gaussian process with stationary increments and

$\mathsf{Var}\{\boldsymbol{G}_i(t)\} = t\mathsf{Var}\{\boldsymbol{G}_i(1)\}$, and

$$\int_0^n \frac{\boldsymbol{G}_i(x)}{x} dx = \sum_{m=1}^{n-1} \frac{\boldsymbol{G}_i(m)}{m} + O((\log n)^{3/2}) \text{ a.s.}. \tag{4.7}$$

Combing (4.4)–(4.7) yields

$$\boldsymbol{Y}_n - n\boldsymbol{v} - \boldsymbol{G}_2(n) - \boldsymbol{G}_1(n)\widetilde{\boldsymbol{H}} = \sum_{m=1}^{n-1} \frac{\boldsymbol{Y}_m - m\boldsymbol{v} - \boldsymbol{G}_2(m) - \boldsymbol{G}_1(m)\widetilde{\boldsymbol{H}}}{m} \widetilde{\boldsymbol{H}} + \boldsymbol{P}_n,$$

where

$$\begin{aligned} \boldsymbol{P}_n &= O((\log n)^{3/2}) + o(n^{1/2-\kappa}) + \sum_{m=1}^{n-1} \left(1 - \frac{|\boldsymbol{Y}_m|}{m}\right) \left(\frac{\boldsymbol{Y}_m}{|\boldsymbol{Y}_m|} - \boldsymbol{v}\right) \widetilde{\boldsymbol{H}} \\ &= O((\log n)^{3/2}) + o(n^{1/2-\kappa}) + \sum_{m=1}^{n-1} \left(\sqrt{(\log\log m)/m}\right)^2 = o(n^{1/2-\kappa}) \text{ a.s.}. \end{aligned}$$

Now, by applying Lemma 4.2 we conclude that

$$\boldsymbol{Y}_n - n\boldsymbol{v} = \boldsymbol{G}_2(n) + \boldsymbol{G}_1(n)\widetilde{\boldsymbol{H}} + o(n^{1/2-\kappa}) \text{ a.s.}.$$

Finally,

$$\begin{aligned} \boldsymbol{N}_n - n\boldsymbol{v} &= \boldsymbol{m}_n + \sum_{m=0}^{n-1} \frac{\boldsymbol{Y}_m}{|\boldsymbol{Y}_m|} - n\boldsymbol{v} = \boldsymbol{m}_n + \sum_{m=0}^{n-1} \left(\frac{\boldsymbol{Y}_m}{|\boldsymbol{Y}_m|} - \boldsymbol{v}\right) (\boldsymbol{I} - \boldsymbol{1}'\boldsymbol{v}) \\ &= \boldsymbol{m}_n + \sum_{m=1}^{n-1} \frac{\boldsymbol{Y}_m - m\boldsymbol{v}}{m} (\boldsymbol{I} - \boldsymbol{1}'\boldsymbol{v}) \\ &\quad + \sum_{m=1}^{n-1} \left(1 - \frac{|\boldsymbol{Y}_m|}{m}\right) \left(\frac{\boldsymbol{Y}_m}{|\boldsymbol{Y}_m|} - \boldsymbol{v}\right) (\boldsymbol{I} - \boldsymbol{1}'\boldsymbol{v}) + O(1) \\ &= \boldsymbol{W}_1(n) + \sum_{m=1}^{n-1} \frac{\boldsymbol{G}_2(m) + \boldsymbol{G}_1(m)\widetilde{\boldsymbol{H}}}{m} (\boldsymbol{I} - \boldsymbol{1}'\boldsymbol{v}) + o(n^{1/2-\kappa}) \\ &= \boldsymbol{W}_1(n) + \int_0^n \frac{\boldsymbol{G}_2(x) + \boldsymbol{G}_1(x)\widetilde{\boldsymbol{H}}}{x} dx (\boldsymbol{I} - \boldsymbol{1}'\boldsymbol{v}) + o(n^{1/2-\kappa}) \\ &= \boldsymbol{G}_1(n) + \int_0^n \frac{\boldsymbol{G}_2(x)}{x} dx (\boldsymbol{I} - \boldsymbol{1}'\boldsymbol{v}) + o(n^{1/2-\kappa}) \text{ a.s.}. \end{aligned}$$

(4.3) is now proved, where $\boldsymbol{\Sigma}$ and $\boldsymbol{\Sigma}^*$ are the variance-covariance matrices of normal random variables $\boldsymbol{G}_2(1) + \boldsymbol{G}_1(1)\widetilde{\boldsymbol{H}}$ and $\boldsymbol{G}_1(1) + \int_0^1 \boldsymbol{G}_2(x)/x\, dx$ $(\boldsymbol{I} - \boldsymbol{1}'\boldsymbol{v})$, respectively. For the details of specifying and estimating the variance-covariance matrices, one can refer to Bai and Hu (2005), Hu and Zhang (2004b), Zhang, Hu and Cheung (2006). □

The interested one can check that for the RPW rule, the equation (4.6) reduces to

$$G_{i,1}(t) = \lambda \int_0^t \frac{G_{i,1}(x)}{x} dx + \sigma_i W_i(t),\ G_{i,1}(0) = 0,$$

$G_{i,2}(t) = -G_{i,1}(t)$, and the solution is $G_{i,1} = \sigma_i t^{\lambda} \int_0^t x^{-\lambda} dW_i(x)$, $i = 1, 2$, where $\lambda = 1 - q_1 - q_2$, $\sigma_1^2 = v_1 v_2 = q_1 q_2/(q_1 + q_2)^2$ and $\sigma_2^2 = v_1 p_1 q_1 + v_2 p_2 q_2 = q_1 q_2 (p_1 + p_2)/(q_1 + q_2)$.

When only the second moment of $\boldsymbol{D}_m$ is assumed to be finite, we can show (4.3) with a similar argument by applying the weak convergence of martingales (Theorem 2.3) instead of the strong approximation. For details one can refer to Hu and Zhang (2001). Janson (2004) studied the urn models by embedding them to continuous branching processes and established the asymptotic normality in a different way.

Example 4.1. As a multi-treatment extension of the RPW rule, Wei (1979) proposed a GUP to allocate subjects, in which the urn is updated in the following way: at the n-th stage, if a subject is assigned to treatment k and cured, then a type k ball is added to the urn, otherwise, if treatment k for a subject fails, then $\frac{1}{K-1}$ balls are added to the urn for each of the other $K - 1$ treatments. In this urn model, $\boldsymbol{H} = \{h_{kj}, k, j = 1, \cdots, K\}$, where $h_{kk} = p_k$ and $h_{kj} = q_k/(K - 1)$ $(j \neq k)$, and p_k is the successful probability of treatment k, $q_k = 1 - p_k$. So $v_k = (1/q_K)/\sum_{j=1}^{K}(1/q_j)$.

For the non-homogenous case, Bai and Hu (2005) obtained (4.3) under the condition that

$$\sum_{m=1}^{\infty} \frac{\|\boldsymbol{H}_m - \boldsymbol{H}\|}{\sqrt{m}} < \infty \quad \text{a.s..} \tag{4.8}$$

This condition can be weakened to $\sum_{m=1}^{n} \|\boldsymbol{H}_m - \boldsymbol{H}\| = o(\sqrt{n})$ a.s. by use the argument in the above proof. An applicable class of non-homogenous urn models is the sequential estimation-adjusted urn (SEU) model, in which the urn is updated according to the current response and the current estimate of an unknown parameter, and so $\boldsymbol{H}_m = \boldsymbol{H}(\widehat{\boldsymbol{\theta}}_m)$ is a function of the estimator. In this case, the fastest convergence rate of $\boldsymbol{H}_m$ is $O_P(\sqrt{m})$ and the condition (4.8) is not satisfied. Zhang, Hu and Cheung (2006) established the asymptotic properties of SEU models.

Example 4.2. Bai, Hu and Shen (2002) proposed a GUP to allocate subjects, in which the urn is updated in the following way: at the m-th stage, if a subject is assigned to treatment k and cured, then a type k ball is added to the urn. If a failure, then $\frac{\widehat{p}_{m-1,j}}{\sum_{i\neq k}\widehat{p}_{m-1,i}}$ balls of each type $j \neq k$ are added, where $\widehat{p}_{m-1,j} = (S_{m-1,i}+1)/(N_{m-1,j}+1)$, and $S_{m-1,j}$ is the number of successes of treatment j in previous $m-1$ stages. This model is a SEU model with $\boldsymbol{H}(\boldsymbol{x}) = \{h_{ij}(\boldsymbol{x}); i,j = 1,\cdots,K\}$, where $h_{kk}(\boldsymbol{x}) = p_k$ and $h_{kj}(\boldsymbol{x}) = q_k x_j / \sum_{i\neq k} x_i$ $(j \neq k)$.

More examples and applications of SEU models can be found in Zhang, Hu and Chueng (2006), in which how to defined a SEU model by using the information of distribution parameters to target a pre-specified limiting allocation proportion is discussed in details.

5 Drop-the-loser rule

The asymptotic normality for the urn models can be obtained only when the condition $\lambda \leqslant 1/2$ is satisfied. This is a very strict condition. Even in the case of $K = 3$, it is hard to be satisfied and to check it is not a easy work. Also, when λ is close or exceeds $1/2$, the variability of an urn model is extremely high. Ivanova (2003) proposed a drop-the-loser (DL) rule which has the same limiting proportion as Wei (1979)'s rule (See Example 4.1) but has much smaller variability. Consider an urn containing balls of $K+1$ types, type $0, 1, \cdots, K$, when comparing K treatments. A ball is drawn at random. If it is type k, $k = 1, \cdots, K$, the corresponding treatment is assigned and the subject's response is observed. If the response is a success, the ball is replaced and the urn remains unchanged. If a failure, the ball is not replaced. When a type 0 ball is drawn, no subject is treated, and the ball is return to the urn together with one ball of each type k, $k = 1, \cdots, K$. Ivanova (2003, 2006) established the asymptotic normality after embedding the urn process to a death-and-immigration process. Here we give the strong approximation.

Theorem 5.1. *There is a K-dimensional standard Brownian motion $\boldsymbol{W}(t)$ such that*

$$\boldsymbol{N}_n - n\boldsymbol{v} = \boldsymbol{W}(n)\mathrm{diag}\Big(\sqrt{\frac{v_1 p_1}{q_1}}, \cdots, \sqrt{\frac{v_K p_K}{q_K}}\Big)(\boldsymbol{I} - \boldsymbol{1}'\boldsymbol{v}) + o(n^{1/2-\kappa}) \text{ a.s.},$$

for some $\kappa > 0$, where $v_k = (1/q_k)/\sum_{j=1}^{k}(1/q_j)$, $k = 1, \cdots, K$. Hence

$$\sqrt{n}(\boldsymbol{N}_n/n - \boldsymbol{v}) \xrightarrow{d} N\Big(\boldsymbol{0}, (\boldsymbol{I} - \boldsymbol{v}\boldsymbol{1}')\mathrm{diag}(\frac{v_1 p_1}{q_1}, \cdots, \frac{v_K p_K}{q_K})(\boldsymbol{I} - \boldsymbol{1}'\boldsymbol{v})\Big).$$

In particular, in the two-treatment case, $\sqrt{n}(\boldsymbol{N}_n/n - \boldsymbol{v}) \xrightarrow{d} N(0, \sigma_{DL}^2)$ with $\sigma_{DL}^2 = q_1 q_2 (p_1 + p_2)/(q_1 + q_2)^3$, the same as the σ_{PW}^2. For the generalizations of the DL rule and their applications, one can refer to Zhang, Chan, Cheung and Hu (2007), Sun, Cheung and Zhang (2007).

Proof of the theorem. Let $\boldsymbol{Z}_m = (Z_{m,0}, \cdots, Z_{m,K})$ be the urn compositions after the m-the assignment. And let μ_m be the number of draws of type 0 balls between the $(m-1)$-th assigment and the m-th assignment. Remember that when a type 0 ball is drawn, we add one ball of each treatment type, and when a treatment type ball is drawn, it is replaced only when the response is a success. So

$$Z_{m,k} - Z_{m-1,k} = \mu_m + X_{m,k}(\xi_{m,k} - 1) = \mu_m - X_{m,k}q_k + X_{m,k}(\xi_{m,k} - p_k),$$

where $\xi_{m,k} = 1$ if the response of the m-th subject on treatment k is a success, and 0 if failure. Let $M_{n,k} = \sum_{m=1}^n X_{m,k}(\xi_{m,k} - p_k)$, $\boldsymbol{M}_n = (M_{n,1}, \cdots, M_{n,K})$, and $\mathcal{A}_m$ be the σ-field generated by $\xi_{1,k}, \cdots, \xi_{m,k}$, $k = 1, \cdots, K$, and $\boldsymbol{X}_1, \cdots, \boldsymbol{X}_m, \boldsymbol{X}_{m+1}$. Then $\{\boldsymbol{M}_n, \mathcal{A}_m\}$ is a martingale. It follows that $Z_{n,k} - Z_{0,0} = \sum_{m=1}^n \mu_m - N_{n,k}q_k + M_{m,k}$. We can prove that $Z_{n,k} = o(n^{1/2-\delta_0})$ a.s. for some $\delta_0 > 0$. For details of the proof, we refer to Zhang, Hu, Chueng and Chan (2006b), Sun, Cheung and Zhang (2007). Hence

$$N_{n,k} = \sum_{m=1}^n \mu_m/q_k + M_{m,k}/q_k + o(n^{1/2-\delta_0}) \text{ a.s., } k = 1, \cdots, K,$$

which, together with the fact $N_{n,1} + \cdots + N_{n,K} = n$, implies

$$\boldsymbol{N}_n - n\boldsymbol{v} = \boldsymbol{M}_n \text{diag}(1/q_1, \cdots, 1/q_K)(\boldsymbol{I} - \boldsymbol{1}'\boldsymbol{v}) + o(n^{1/2-\delta_0}) \text{ a.s..}$$

So, $\boldsymbol{N}_n - n\boldsymbol{v} = O(\sqrt{n \log\log n})$ a.s. by the LIL (Theorem 2.2). On the other hand, for the martingale $\{\boldsymbol{M}_n\}$ we have

$$\begin{aligned}\sum_{m=1}^n \mathsf{E}[(\Delta\boldsymbol{M}_n)'\Delta\boldsymbol{M}_n|\mathcal{F}_{m-1}] &= \text{diag}(N_{n,1}p_1q_1, \cdots, N_{n,K}p_Kq_K) \\ &= n \,\text{diag}(v_1p_1q_1, \cdots, v_Kp_Kq_K) + O(\sqrt{n \log\log n}) \text{ a.s..}\end{aligned}$$

By applying the strong approximation (Theorem 2.5), we can define a K-dimensional Brownian motion $\boldsymbol{W}(t)$ such that

$$\boldsymbol{M}_n = \boldsymbol{W}(n)\text{diag}(\sqrt{v_1p_1q_1}, \cdots, \sqrt{v_Kp_Kq_K}) + o(n^{1/2-\kappa}) \text{ a.s..}$$

The proof is now completed. □

6 Doubly adaptive biased coin designs

The PW rule and urn model designs are a kind of design-driven adaptive designs, which are constructed with intuitive motivation. However, clinical trials are usual complex experiments on humans with multiple, often competing, objectives, including maximizing power to detect clinically relevant differences in treatment outcomes, maximizing the individual patient's personal experience while treated in the trial, and minimizing the total monetary cost of trial. These and other objectives can be defined in terms of optimization of function of the trial's parameters, the optimal allocation proportion is often a function of unknown parameters. Take a binary response clinical trail with two treatments 1 and 2 as an example. The well known Neyman proportion is

$$\rho(p_1, p_2) =: \frac{n_1}{n_1 + n_2} = \frac{\sqrt{p_1 q_1}}{\sqrt{p_1 q_1} + \sqrt{p_2 q_2}},$$

where p_k (q_k) is the probability of success (failure) of a trial treatment k, n_k is the number of subjects assigned to treatment k, $k = 1, 2$, The Neyman proportion maximizes the power of a test of the simple difference $p_1 - p_2$ for fixed sample size n. But if we implement Neyman allocation, when $p_1 + p_2 > 1$, we will assign more subjects to the inferior treatment, which will compromise the ethical objective. Rosenberger, et al (2001) discussed another important optimization criteria that minimize the expected number of treatment failures, $n_1 q_1 + n_2 q_2$, for fixed the variance, $p_1 q_1 / n_1 + p_2 q_2 / n_2$, of the statistic $\widehat{p}_1 - \widehat{p}_2$ under an alternative hypothesis $p_1 \neq p_2$. This leads the optimal proportion as follows.

$$\rho(p_1, p_2) = \frac{\sqrt{p_1}}{\sqrt{p_1} + \sqrt{p_2}}.$$

Many simulation studies have validated that a adaptive design with this proportion as its target performs very satisfactorily for both the consideration of ethic and the test of power. For other optimization criteria one can refer to Jennison and Turnbull (2000) and Rosenberger, et al. (2001). Both Neyman allocation and the allocation of Rosenberger,et al (2001) cannot be implemented directly in a clinical trail, because we do not know the values of p_1 and p_2. In this section, we introduce an important class of adaptive designs, doubly adaptive biased coin designs (DBCD), which is first proposed by Eisele (1994) and Eisele and Woodroofe (1995), to target a pre-specified allocation proportion.

Consider a clinical with K treatments. The outcome of a subject on treatment k has a distribution $f_k(\cdot|\theta_k)$. Write $\boldsymbol{\theta} = (\theta_1, \cdots, \theta_K)$. The pre-specified allocation proportion is $\boldsymbol{v} = \boldsymbol{\rho}(\boldsymbol{\theta}) = (\rho_1(\boldsymbol{\theta}), \cdots, \rho_m(\boldsymbol{\theta}))$. Here $\boldsymbol{\rho}(\boldsymbol{y})$ is assumed to be continuous function on the parameter space, taking the values on $(0,1)^{\otimes k}$ and twice differentiable at the true value of the parameter $\boldsymbol{\theta}$. A multi-treatment DBCD proposed by Hu and Zhang (2004a) is defined as follows.

To start, allocate M subjects to each treatment. At stage m, suppose $m-1 (\geqslant MK)$ subjects are allocated and the outcomes, $N_{m-1,k}$ outcomes of treatment k, $k=1,\cdots,K$, are observed. Let $\widehat{\theta}_{m-1,k}$ be the MLE of the parameter θ_k, $k=1,\cdots,K$. Write $\widehat{\boldsymbol{\theta}}_{m-1}=(\widehat{\theta}_{m-1,1},\cdots,\widehat{\theta}_{m-1,K})$, and let $\widehat{\boldsymbol{\rho}}_{m-1}=\rho(\widehat{\boldsymbol{\theta}}_{m-1})$ be the current estimate of the target allocation proportion. Now, the m-th subject is allocated to treatment k with a probability:

$$P_{m,k}=:\mathsf{P}(X_{m,k}=1|\mathcal{F}_{m-1})=g_k\big(\frac{\boldsymbol{N}_{m-1}}{m-1},\widehat{\rho}_{m-1}\big), \tag{6.1}$$

$k=1,\cdots,K$. Here $\boldsymbol{g}(\boldsymbol{x},\boldsymbol{y})=(g_1(\boldsymbol{x},\boldsymbol{y}),\cdots,g_K(\boldsymbol{x},\boldsymbol{y})):(0,1)^{\otimes 2k}\to(0,1)^{\otimes k}$ is the allocation function. Write $\boldsymbol{P}_m=(P_{m,1},\cdots,P_{m,K})$.

Theorem 6.1. *Suppose the distributions $f_1(\cdot|\theta_1),\cdots,f_K(\cdot|\theta_K)$ follow an exponential family. Let $\boldsymbol{g}(\boldsymbol{x},\boldsymbol{y})$ be defined as*

$$g_k(\boldsymbol{x},\boldsymbol{y})=\frac{y_k\left(\frac{y_k}{x_k}\right)^{\gamma}}{\sum_{j=1}^K y_j\left(\frac{y_j}{x_j}\right)^{\gamma}},\quad k=1,\cdots,K;\quad \gamma\geqslant 0. \tag{6.2}$$

Then

$$\boldsymbol{N}_n-n\boldsymbol{v}=O(\sqrt{n\log\log n})\text{ a.s.}\quad\text{and}\quad \sqrt{n}(\boldsymbol{N}_n/n-\boldsymbol{v})\xrightarrow{d}N(\boldsymbol{0},\boldsymbol{\Sigma}),$$

where

$$\boldsymbol{\Sigma}=\boldsymbol{\Sigma}_{\boldsymbol{\rho}}+\frac{1}{1+2\gamma}(\mathrm{diag}(\boldsymbol{v})-\boldsymbol{v}'\boldsymbol{v}+\boldsymbol{\Sigma}_{\boldsymbol{\rho}}),$$

$$\boldsymbol{\Sigma}_{\boldsymbol{\rho}}=\left(\frac{\partial\boldsymbol{\rho}}{\partial\boldsymbol{\theta}}\right)'\mathrm{diag}\left((v_1I_1(\theta_1))^{-1},\cdots,v_KI_1(\theta_K))^{-1}\right)\frac{\partial\boldsymbol{\rho}}{\partial\boldsymbol{\theta}}$$

and $I_k(\theta_k)$ is the Fisher information function for a single observation on treatment k.

In particular, for the two-treatment case, suppose the targeted allocation proportion of treatment 1 is $\rho=\rho(\theta_1,\theta_2)$. Then $\sqrt{n}(N_{n,1}/n-\rho)\xrightarrow{d}N(0,\sigma^2_{DBCD})$ where

$$\sigma^2_{DBCD}=\sigma^2_\rho+\frac{1}{1+2\gamma}\{\rho(1-\rho)+\sigma^2_\rho\}$$

and $\sigma^2_\rho=\big(I_1(\theta_1)\rho\big)^{-1}(\partial\rho/\partial\theta_1)^2+\big(I_2(\theta_2)(1-\rho)\big)^{-1}(\partial\rho/\partial\theta_2)^2$.

Example 6.1. Consider the binary response clinical trail with two treatments. For the urn proportion $\rho=q_2/(q_1+q_2)$,

$$\sigma^2_{DBCD}=\frac{q_1q_2(p_1+p_2)}{(q_1+q_2)^3}+\frac{2q_1q_2}{(1+2\gamma)(q_1+q_2)^3}.$$

For the Neyman proportion $\rho = \sqrt{p_1q_1}/(\sqrt{p_1q_1}+\sqrt{p_2q_2})$, $\sigma^2_{DBCD} =$

$$\frac{\sqrt{p_1q_1p_2q_2}}{(1+2\gamma)(\sqrt{p_1q_1}+\sqrt{p_2q_2})^2}+ \frac{1+\gamma}{2(1+2\gamma)(\sqrt{p_1q_1}+\sqrt{p_2q_2})^3}\left(\frac{p_2q_2(q_1-p_1)^2}{\sqrt{p_1q_1}}+\frac{p_1q_1(q_2-p_2)^2}{\sqrt{p_2q_2}}\right).$$

For Rosenberger, et al's proportion $\rho = \sqrt{p_1}/(\sqrt{p_1}+\sqrt{p_2})$, $\sigma^2_{DBCD} =$

$$\frac{\sqrt{p_1p_2}}{(1+2\gamma)(\sqrt{p_1}+\sqrt{p_2})^2}+\frac{1+\gamma}{2(1+2\gamma)(\sqrt{p_1}+\sqrt{p_2})^3}\left(\frac{p_2q_1}{\sqrt{p_1}}+\frac{p_1q_2}{\sqrt{p_2}}\right).$$

From Theorem 6.1, we find that the asymptotic variability is a decreasing function of parameter γ. However, the degree of randomness of the design decreases when γ increases, because, as the value of γ becomes larger, the allocation probabilities shift faster to extreme values 0 and 1 if there is a bias between the current sample allocation and the estimated target. When $\gamma = \infty$, the variability of the procedure is minimized, but the procedure is completely predictable. The parameter γ can be chosen to reflect the trade-off between the degree of randomness and the variability.

The allocation function defined in (6.2) is very special though it has fine properties. For results for general allocation function $\boldsymbol{g}(\cdot,\cdot)$, one can refer to Hu and Zhang (2004a).

Proof the Theorem. Let $\xi_{m,k}$, $m=1,2,\cdots$, be i.i.d. random variables, which represent the outcomes on treatment k, $k=1,\cdots,K$. In clinical trial, only $X_{m,k}\xi_{m,k}$s are observed. Write $\boldsymbol{\xi}_m = (\xi_{m,1},\cdots,\xi_{m,k})$. For simplifying the proof, we assume that $\theta_k = \mathsf{E}\xi_{m,k}$ is the mean of the outcomes, and so we use the sample mean as it estimate:

$$\widehat{\theta}_{m,k} = \frac{\sum_{i=1}^m X_{m,k}\xi_{j,k}}{N_{m,k}},\ k=1,\cdots,K.$$

In practices, if necessary, we can add $\alpha>0$ in the numerator and $\beta>0$ in the denominator to avoid the nonsense case of $0/0$, or to use prior information to estimate the parameters. Assume $\mathsf{E}|\xi_{m,k}|^{2+\delta}<\infty$ and write $\sigma_k^2 = \mathsf{Var}\{\xi_{m,k}\}$. Write $Q_{m,k} = \sum_{i=1}^m X_{m,k}(\xi_{m,k}-\theta_k)$, then $Q_{m,k}$ is a martingale and $\widehat{\theta}_{m,k}-\theta_k = Q_{m,k}/N_{m,k}$. By the LIL (Theorem 2.2), we have

$$Q_{m,k} = O(\sqrt{n\log\log n})\ \text{ a.s..} \tag{6.3}$$

We first show the consistency of $\boldsymbol{N}_n/n$. If let $\mathcal{A}_m = \sigma(\boldsymbol{X}_1,\cdots,\boldsymbol{X}_{m+1}, \boldsymbol{\xi}_1,\cdots,\boldsymbol{\xi}_m)$, then $\sum_{i=1}^m \mathsf{E}[(X_{i,k}(\xi_{i,k}-\theta_k))^2|\mathcal{A}_{i-1}] = \sigma_k^2 N_{m,k}$. By Theorem 2.1 (b), it follows that

$$\widehat{\theta}_{m,k}\to\theta_k\ \text{ a.s. on the event }\ \{N_{m,k}\to\infty\}. \tag{6.4}$$

On the event $\{N_{m,k} < \infty\}$, $\widehat{\theta}_{m,k}$ will fix to a value eventually. In either case, $\widehat{\theta}_{m,k}$ has a limit $\widetilde{\theta}_k$ in the parameter space, $k = 1, \cdots, K$. By the continuity of $\boldsymbol{\rho}(\cdot)$, $\widehat{\boldsymbol{\rho}}_{m-1} \to \boldsymbol{\rho}(\widetilde{\theta}_1, \cdots, \widetilde{\theta}_K) := \widetilde{\boldsymbol{v}} \in (0,1)^{\otimes K}$ a.s.. Notice that the minimum of $\sum_j y_j(y_j/x_j)^\gamma$ over $\sum_j x_j = 1$ and $x_j \geqslant 0$ is $\sum_j y_j$. It is easily seen that $g_k(\boldsymbol{x}, \boldsymbol{y}) \leqslant y_k(y_k/x_k)^\gamma < y_k$ if $x_k > y_k$ and $\sum_j x_j = \sum_j y_j = 1$. So, $P_{m,k} \leqslant \widehat{\rho}_{m-1,k}$ if $N_{m-1,k}/(m-1) > \widehat{\rho}_{m-1,k}$. Denote $M_{n,k} = \sum_{m=1}^n (X_{m,k} - \mathsf{E}[X_{m,k}|\mathcal{F}_{m-1}])$. Let $S_n = \max\{m \geqslant MK + 1 : N_{m-1,k} \leqslant (m-1)\widehat{\rho}_{m-1,k}\}$ and $\max\{\emptyset\} = MK$. Then

$$\begin{aligned} N_{n,k} &= N_{S_n,k} + \sum_{m=S_n+1}^{n} (X_{m,k} - \mathsf{E}[X_{m,k}|\mathcal{F}_{m-1}]) + \sum_{m=S_n+1}^{n} P_{m,k} \\ &\leqslant 1 + N_{S_n-1,k} + M_{n,k} - M_{S_n,k} + \sum_{m=S_n}^{n-1} \widehat{\rho}_{m,k} \\ &\leqslant 1 + N_{MK-1,k} + (M_{n,k} - M_{S_n,k}) + (S_n - 1)\widehat{\rho}_{S_n-1,k} + \sum_{m=S_n}^{n-1} \widehat{\rho}_{m,k}. \end{aligned} \tag{6.5}$$

Notice that $|M_{n,k} - M_{S_n,k}| \leqslant \max_{m \leqslant n} |M_{m,k}| = O(\sqrt{n \log\log n})$ a.s. by the LIL, and $\widehat{\rho}_{m,k} \to \widetilde{v}_k$. We conclude that $\limsup_{n\to\infty} N_{n,k}/n \leqslant \widetilde{v}_k$ a.s., $k = 1, \cdots, K$. From the fact that $\sum_{k=1}^n N_{n,k}/n = \sum_{k=1}^n \widetilde{v}_k = 1$, we conclude that $N_{n,k}/n \to \widetilde{v}_k \in (0,1)$ a.s., which implies that $N_{n,k} \to \infty$, $k = 1, \cdots, K$. Hence $\widetilde{\theta}_k$ and θ_k ($\widetilde{\boldsymbol{v}}$ and $\boldsymbol{v}$) must be identical by (6.4). We have proved the consistency of $\boldsymbol{N}_n/n$. Further, according to (6.3), we have $\widehat{\boldsymbol{\theta}}_m - \boldsymbol{\theta} = O\big(\sqrt{\log\log m}/\sqrt{m}\big)$ a.s., and then $\widehat{\boldsymbol{\rho}}_m - \boldsymbol{v} = O\big(\sqrt{\log\log m}/\sqrt{m}\big)$ a.s., which together with (6.5), yields

$$\begin{aligned} N_{n,k} - nv_k &\leqslant O(\sqrt{n \log\log n}) + O(\sqrt{S_n \log\log S_n}) \\ &\quad + \sum_{m=S_n}^{n-1} O\big(\sqrt{\log\log m}/\sqrt{m}\big) = O(\sqrt{n\log\log n}) \ \text{ a.s..} \end{aligned}$$

By the fact that $\sum_{k=1}^n N_{n,k}/n = \sum_{k=1}^n \widetilde{v}_k = 1$ again, we conclude that

$$\boldsymbol{N}_n - n\boldsymbol{v} = O(\sqrt{n\log\log n}) \ \text{ a.s..} \tag{6.6}$$

Now, we begin the proof of the asymptotic normality. Write $\boldsymbol{M}_n = (M_{n,1}, \cdots, M_{n,K})$ and $\boldsymbol{Q}_n = (Q_{n,1}, \cdots, Q_{n,K})$. Then by (6.3) and (6.6),

$$\widehat{\theta}_{m,k} - \theta_k = \frac{Q_{m,k}}{mv_k} + O\left(\frac{\log\log m}{m}\right) \ \text{ a.s..}$$

It is easily seen that $\partial \boldsymbol{g}/\partial \boldsymbol{x}|_{\boldsymbol{x}=\boldsymbol{y}} = -\gamma(\boldsymbol{I}-\mathbf{1}'\boldsymbol{v})$ and $\partial \boldsymbol{g}/\partial \boldsymbol{y}|_{\boldsymbol{x}=\boldsymbol{y}} = (\gamma+1)(\boldsymbol{I}-\mathbf{1}'\boldsymbol{v})$. By the Taylor formula, we have

$$
\begin{aligned}
\boldsymbol{P}_m - \boldsymbol{v} &= \boldsymbol{\rho}\big(\frac{\boldsymbol{N}_{m-1}}{m-1}, \widehat{\boldsymbol{\rho}}_{m-1}\big) - \boldsymbol{\rho}(\boldsymbol{v},\boldsymbol{v}) \\
&= -\gamma\big(\frac{\boldsymbol{N}_{m-1}}{m-1} - \boldsymbol{v}\big)(\boldsymbol{I}-\mathbf{1}'\boldsymbol{v}) + (\gamma+1)\big(\widehat{\boldsymbol{\theta}}_{m-1}-\boldsymbol{\theta}\big)\frac{\partial\boldsymbol{\rho}}{\partial\boldsymbol{\theta}}(\boldsymbol{I}-\mathbf{1}'\boldsymbol{v}) \\
&\quad + O(\big\|\frac{\boldsymbol{N}_{m-1}}{m-1}-\boldsymbol{v}\big\|^2) + (\big\|\widehat{\boldsymbol{\theta}}_{m-1}-\boldsymbol{\theta}\big\|^2) \\
&= -\gamma\big(\frac{\boldsymbol{N}_{m-1}}{m-1} - \boldsymbol{v}\big) + (\gamma+1)\big(\widehat{\boldsymbol{\theta}}_{m-1}-\boldsymbol{\theta}\big)\frac{\partial\boldsymbol{\rho}}{\partial\boldsymbol{\theta}} + O\Big(\frac{\log\log m}{m}\Big) \\
&= -\gamma\big(\frac{\boldsymbol{N}_{m-1}}{m-1} - \boldsymbol{v}\big) + (\gamma+1)\frac{\boldsymbol{Q}_{m-1}}{(m-1)}\mathrm{diag}\big(\frac{1}{\boldsymbol{v}}\big)\frac{\partial\boldsymbol{\rho}}{\partial\boldsymbol{\theta}} + O\Big(\frac{\log\log m}{m}\Big) \\
&= O\Big(\sqrt{\frac{\log\log m}{m}}\Big).
\end{aligned}
$$

Hence

$$
\begin{aligned}
\boldsymbol{N}_n - n\boldsymbol{v} &= \sum_{m=1}^{n}(\boldsymbol{X}_m - \mathsf{E}[\boldsymbol{X}_m|\mathcal{F}_{m-1}]) + \sum_{m=1}^{n}(\boldsymbol{P}_m - \boldsymbol{v}) \\
&= \boldsymbol{M}_n - \gamma\sum_{m=1}^{n-1}\frac{\boldsymbol{N}_m - m\boldsymbol{v}}{m} + (\gamma+1)\sum_{m=1}^{n-1}\frac{\boldsymbol{Q}_m}{m}\mathrm{diag}\big(\frac{1}{\boldsymbol{v}}\big)\frac{\partial\boldsymbol{\rho}}{\partial\boldsymbol{\theta}} + o((\log n)^2) \ \text{ a.s..}
\end{aligned}
$$

On the other hand, it is easily checked that, for the martingale $(\boldsymbol{M}_n, \boldsymbol{Q}_n)$ we have

$$
\begin{aligned}
\sum_{m=1}^{n}\mathsf{E}[(\Delta\boldsymbol{M}_m)'(\Delta\boldsymbol{M}_m)|\mathcal{F}_{m-1}] &= \sum_{m=1}^{n}(\mathrm{diag}(\boldsymbol{P}_m) - \boldsymbol{P}_m'\boldsymbol{P}_m) \\
&= n(\mathrm{diag}(\boldsymbol{v}) - \boldsymbol{v}'\boldsymbol{v}) + O(\sqrt{n\log\log n}) \ \text{ a.s.,} \\
\sum_{m=1}^{n}\mathsf{E}[(\Delta Q_{m,k})^2|\mathcal{F}_{m-1}] = \sum_{m=1}^{n}P_{m,k}\sigma_k^2 &= nv_k\sigma_k^2 + O(\sqrt{n\log\log n}) \text{ a.s.,}
\end{aligned}
$$

$\mathsf{E}[(\Delta\boldsymbol{M}_m)'(\Delta\boldsymbol{Q}_m)|\mathcal{F}_{m-1}] = \mathbf{0}$ and $\mathsf{E}[(\Delta Q_{m,k}Q_{m,j}|\mathcal{F}_{m-1}] = 0$, $k \neq j$. So, by the strong approximation (Theorem 2.5), there are two independent K-dimensional standard Brownian motions $\boldsymbol{W}(t)$ and $\boldsymbol{B}(t)$ such that for some $\kappa > 0$, $\boldsymbol{M}_n = \boldsymbol{W}(n)\boldsymbol{\Sigma}_1^{1/2} + o(n^{1/2-\kappa})$ a.s. and $Q_{n,k} = B_k(n)\sigma_k\sqrt{v_k} + o(n^{1/2-\kappa})$ a.s., $k = 1, \cdots, K$, where $\boldsymbol{\Sigma}_1 = \mathrm{diag}(\boldsymbol{v}) - \boldsymbol{v}'\boldsymbol{v}$. It follows that

$$
\begin{aligned}
\boldsymbol{N}_n - n\boldsymbol{v} = &\boldsymbol{W}(n)\boldsymbol{\Sigma}_1^{1/2} - \gamma\sum_{m=1}^{n-1}\frac{\boldsymbol{N}_m - m\boldsymbol{v}}{m} + (\gamma+1)\sum_{m=1}^{n-1}\frac{\boldsymbol{B}(m)}{m}\boldsymbol{\Sigma}_{\boldsymbol{\rho}}^{1/2} \\
&+ o(n^{1/2-\kappa}) \ \text{ a.s..}
\end{aligned}
$$

Hence, $\boldsymbol{N}_n - n\boldsymbol{v} = \boldsymbol{G}(n) + +o(n^{1/2-\kappa})$ a.s., where

$$\boldsymbol{G}(t) = t^{-\gamma}\int_0^t x^{\gamma} d\boldsymbol{W}(x)\boldsymbol{\Sigma}_1^{1/2} + (\gamma+1)t^{-\gamma}\int_0^t x^{\gamma-1}\boldsymbol{B}(x)dx\boldsymbol{\Sigma}_{\boldsymbol{\rho}}^{1/2}$$

is the solution of the equation

$$\boldsymbol{G}(t) = \boldsymbol{W}(t)\boldsymbol{\Sigma}_1^{1/2} - \gamma\int_0^t \frac{\boldsymbol{G}(x)}{x}dx + (\gamma+1)\int_0^t \frac{\boldsymbol{B}(x)}{x}dx\boldsymbol{\Sigma}_{\boldsymbol{\rho}}^{1/2}.$$

It follows that $\sqrt{n}(\boldsymbol{N}_n/n - \boldsymbol{v}) \xrightarrow{d} N(\boldsymbol{0}, \boldsymbol{\Sigma})$ with

$$\begin{aligned}\boldsymbol{\Sigma} =& \mathsf{Var}\{\boldsymbol{G}(1)\}\\ =&\boldsymbol{\Sigma}_1\int_0^1 x^{2\gamma}dx + (\gamma+1)^2\boldsymbol{\Sigma}_{\boldsymbol{\rho}}\int_0^1\int_0^1 x^{\gamma-1}y^{\gamma-1}(x\wedge y)dxdy\\ =&\frac{1}{2\gamma+1}\boldsymbol{\Sigma}_1 + \frac{2(\gamma+1)}{2\gamma+1}\boldsymbol{\Sigma}_{\boldsymbol{\rho}}.\end{aligned}$$

The proof is now completed. □

7 Delayed responses

In practices, the outcomes in clinical trials are not available immediately prior to the treatment allocation of the next subject. The estimating of the parameters, and the updating of the urn when using urn models, can only be processed according to observed responses. The effect of the delay of treatment results is finst studied in theory by Bai, Hu and Rosenberger (2002) for the urn compositions in an urn model design with discrete responses. After that, Hu and Zhang (2004b), Zhang, Chan, Chueng and Hu (2007), Sun, Cheung and Zhang (2007) and Zhang, Hu, Chueng and Chan (2006a) have shown that the delay machine does not effect the asymptotic properties of the sample allocation proportions for many adaptive designs if the delay degree decays with a power rate. The basic reason is that the total delayed responses is a high order of square root of the sample size when the delay degree decays with a power rate.

To describe the delay machine, we let t_m be the entry time of the m-th subject, where t_m is an increasing sequence of random variables. Assume that $\{t_{m+1}-t_m\}$ is a sequence of independent random variables. The response time of the m-th subject on treatment k is denoted by $r_m(k)$. Suppose $\{r_m(k); m \geqslant 1\}$ are sequences of independent random variables, $k = 1, \cdots, K$. Further, assume that $\{t_{m+1} - t_m, r_m(k); k = 1, \cdots, K, m \geqslant n\}$ is independent of the assignments $\boldsymbol{X}_1, \cdots, \boldsymbol{X}_n$.

Assumption 7.1. *Let $\delta_k(m,l) = I\{r_m(k) \geqslant t_{m+l} - t_m\}$ be an indicator function that takes the value 1 if the outcome of the m-th subject on treatment k occurs after at least another l subjects arrive, and 0 otherwise. Suppose for some constants $C > 0$ and $\gamma \geqslant 2$,*

$$\mu_k(m,l) = P(\delta_k(m,n) = 1) \leqslant Cl^{-\gamma}, \quad m,l = 1,2,\cdots, k = 1,\cdots,K.$$

This assumption is easily satisfied. A practical approach is to assume that the entry mechanism generates a Poisson process and the delay time has an exponential distribution in which both $\{r_m(k)\}$ and $\{t_{m+1} - t_m\}$ are sequences of i.i.d. exponential random variables with means $\lambda_k > 0$ and $\lambda_0 > 0$, respectively. This approach is common in clinical studies and the probability $\mu_k(m,l)$ is $\big(\lambda_k/(\lambda_0 + \lambda_k)\big)^l$.

Let $S^{\text{obs}}_{m,k}$ (resp. $N^{\text{obs}}_{m,k}$) be the sum (resp. the number) of the outcomes on treatment k observed prior to the $(m+1)$-th assignment, and $S_{m,k}$ (resp. $N_{m,k}$) be the sum (resp. the number) of all the outcomes of those being assigned to treatment k in the first m subjects, $k = 1,\cdots,K$.

Theorem 7.2. *Suppose Assumptions 7.1 is satisfied, and the responses on each treatment are i.i.d. random variables having finite $(2+\delta)$-th moments. Then for some $0 < \delta_0 < \frac{1}{2} - \frac{1}{2+\delta}$, we have*

$$R_{n,k} =: S_{m,k} - S^{\text{obs}}_{m,k} = o(n^{1/2-\delta_0}) \quad \text{a.s.} \tag{7.1}$$

and $N_{n,k} - N^{\text{obs}}_{n,k} = o(n^{1/2-\delta_0})$ a.s., $k = 1,\cdots,K$.

Proof. Let $I_k(m,l)$ be the indicator function, which takes value 1 if the outcome $\xi_{m,k}$ on treatment k of the m-th subject occurs after the $(m+l)$-th assignment and before the $(m+l+1)$-th assignment, $k = 1,\cdots,K$. For given m and l, if $I_k(m,l) = 1$, we observe a response $X_{m,k}\xi_{m,k}$. Hence, between the m-th assignment and the $(m+1)$-th assignment, the observed outcomes are $I_k(m,0)X_{m,k}\xi_{m,k}$, $I_k(m-1,1)X_{m-1,k}\xi_{m-1,k}$, $\cdots$, $I_k(1,m-1)X_{1,k}\xi_{1,k}$, $k = 1,\cdots,K$, and the sum of those on treatment k is $\sum_{l=0}^{m-1} I_k(m-l,l)X_{m-l,k}\xi_{m-l,k} = \sum_{l=1}^{m} I_k(l,m-l)X_{l,k}\xi_{l,k}$. Hence,

$$S^{\text{obs}}_{m,k} = \sum_{m=1}^{n}\sum_{l=1}^{m} I_k(l,m-l)X_{l,k}\xi_{l,k} = \sum_{m=1}^{n}\sum_{j=m}^{n} X_{m,k}\xi_{m,k}I_k(m,j-m).$$

On the other hand, it is obvious that $S_{m,k} = \sum_{m=1}^{n} X_{m,k}\xi_{m,k} = \sum_{m=1}^{n}\sum_{j=m}^{\infty} X_{m,k}\xi_{m,k}I_k(m,j-m)$. It follows that

$$R_{n,k} = \sum_{m=1}^{n}\sum_{j=n+1}^{\infty} X_{m,k}\xi_{m,k}I_k(m,j-m).$$

Let $0<\phi<1/2$ be a number whose value will be specified later. Write $l_n=[n^\phi]$. Then

$$\begin{aligned}
|R_{n,k}|=&\Big|\Big(\sum_{m=n-l_n+1}^{n}+\sum_{m=1}^{n-l_n}\Big)\sum_{j=n-m+1}^{\infty}I_k(m,j)X_{m,k}\xi_{m,k}\Big|\\
\leqslant&\sum_{m=n-l_n+1}^{n}X_{m,k}|\xi_{m,k}|+\sum_{m=1}^{n}\delta_k(m,l_m)X_{m,k}|\xi_{m,k}|\\
\leqslant&Cn^\phi+\sum_{m=n-l_n+1}^{n}\big(|\xi_{m,k}|-\mathsf{E}[|\xi_{m,k}|]\big)+\sum_{m=1}^{n}X_{m,k}\mathsf{E}[\delta_k(m,l_m)|\xi_{m,k}|]\\
&+\sum_{m=1}^{n}X_{m,k}\big(\delta_k(m,l_m)|\xi_{m,k}|-\mathsf{E}[\delta_k(m,l_m)|\xi_{m,k}|]\big).
\end{aligned}\tag{7.2}$$

For $\sum_{m=1}^n\big(|\xi_{m,k}|-\mathsf{E}[|\xi_{m,k}|]\big)$, due to Theorems 1.2.1 and 2.6.6 of Csörgő and Révész (1981),

$$\begin{aligned}
\sum_{m=n-l_n+1}^{n}\big(|\xi_{m,k}|-\mathsf{E}[|\xi_{m,k}|]\big)=&O(\sqrt{l_n\log n})+o(n^{\frac{1}{2+\delta}})\\
=&o(n^\phi)+o(n^{\frac{1}{2+\delta}})\ \text{ a.s..}
\end{aligned}$$

For the martingale $\sum_{m=1}^n X_{m,k}\big(\delta_k(m,l_m)|\xi_{m,k}|-\mathsf{E}[\delta_k(m,l_m)|\xi_{m,k}|]\big)$, notice

$$\begin{aligned}
&\mathsf{E}\Big(X_{m,k}\big(\delta_k(m,l_m)|\xi_{m,k}|-\mathsf{E}[\delta_k(m,l_m)|\xi_{m,k}|]\big)\Big)^2\\
\leqslant&\mathsf{E}\big(\delta_k(m,l_m)|\xi_{m,k}|\big)^2\leqslant\big(\mathsf{E}\delta_k(m,l_m)\big)^{\frac{\delta}{2+\delta}}\big(\mathsf{E}|\xi_{m,k}|^{2+\delta}\big)^{\frac{2}{2+\delta}}\leqslant m^{-\frac{\phi\gamma\delta}{2+\delta}}.
\end{aligned}$$

due to Assumptions 7.1 and the assumption of finite $(2+\delta)$-th moments. So, by the LLN (Theorem 2.1 (a)), $\sum_{m=1}^n X_{m,k}\big(\delta_k(m,l_m)|\xi_{m,k}|-\mathsf{E}[\delta_k(m,l_m)|\xi_{m,k}|]\big)=o(n^{\frac12-\frac{\phi\gamma\delta}{2(2+\delta)}}\log n)$ a.s. Also

$$\begin{aligned}
\sum_{m=1}^{n}X_{m,k}\mathsf{E}[\delta_k(m,l_m)|D_{m,k}|]\leqslant&\sum_{m=1}^{n}\big(\mathsf{E}\delta_k(m,l_m)\big)^{\frac{1+\delta}{2+\delta}}\big(\mathsf{E}|D_{m,k}|^{2+\delta}\big)^{\frac{1}{2+\delta}}\\
\leqslant&C\sum_{m=1}^{n}m^{-\frac{\phi\gamma(1+\delta)}{2+\delta}}\leqslant Cn^{1-\frac{\phi\gamma(1+\delta)}{2+\delta}}.
\end{aligned}$$

Combining the above arguments yields

$$R_{n,k}=O(n^\phi)+O(n^{1-\frac{\phi\gamma(1+\delta)}{2+\delta}})+o(n^{\frac{1}{2+\delta}})+o(n^{\frac12-\frac{\phi\gamma\delta}{2(2+\delta)}}\log n)\ \text{ a.s..}$$

Choosing $\phi=\frac{2+\delta}{1+(1+\delta)\gamma}$ and $0<\delta_0<\min\{\frac12-\phi,\frac12-\frac{1}{2+\delta},\frac{\phi\gamma\delta}{2(2+\delta)}\}$ yields (7.1). The proof of $N_{n,k}-N_{n,k}^{\text{obs}}=o(n^{1/2-\delta_0})$ a.s. is similar. □

8 Variability, power and asymptotic best adaptive designs

The variability of the sample proportion $N_{n,k}/n$ is an important quantity, which measures the distance between $N_{n,k}/n$ and its limit v_k. The smaller is the variability, the smaller is the probability that there is large bias between $N_{n,k}/n$ and v_k. When using a adaptive design with high variability, a clinical might result in assigning more subjects to the inferior treatment making the allocation even less ethical than equal allocation. Also, a trial with high variability might result in assigning only a few subjects to one of the the treatments decreasing the efficiency in the test or the estimation of parameters. According to (4.1), the asymptotic variability of the RPW rule is very high unless both treatments have low success rates. It is extremely high when $q_1 + q_2$ is close to $1/2$, and it is showed that $n\mathsf{Var}\{N_{n,1}/n\} \to \infty$ when $q_1 + q_2 < 1/2$. The RPW rule used in ECMO, 1985, trial assigned only one patient to the less successful control therapy (see Royall, 1991, for discussion). The relationship among the power, the target allocation and the variability of the designs is fist revealed by Hu and Rosenberger (2003) in theory, though simulation studies had indicated there is strong relationship among these quantities. Hu and Rosenberger (2003) proved that the average power of a statistical test of the difference of distribution parameters is a decreasing function of the variability of the designs. In Section 6, we have found that the asymptotic variability of DBCD is a decreasing function of the parameter γ. When $\gamma \to \infty$, the variability tends to its minimums

$$\boldsymbol{\Sigma}_\rho = \left(\frac{\partial \boldsymbol{\rho}}{\partial \boldsymbol{\theta}}\right)' \operatorname{diag}\left((v_1 I_1(\theta_1))^{-1}, \cdots, (v_K I_1(\theta_K))^{-1}\right) \frac{\partial \boldsymbol{\rho}}{\partial \boldsymbol{\theta}}.$$

Hu, Rosenberger and Zhang (2006) proved that this limit is the lower bound of the asymptotic variability of adaptive designs among all adaptive designs which have the same limiting proportion.

Assume the following regularity conditions:

1. The parameter space $\boldsymbol{\Theta}_k$ is an open subset in $\mathbb{R}^d$, $k = 1, \cdots, K$;
2. The distributions of outcomes $f_1(\cdot|\boldsymbol{\theta}_1)$, $\cdots$, $f_K(\cdot|\boldsymbol{\theta}_K)$ follow an exponential family;
3. For the limiting allocation proportion $\boldsymbol{\rho}(\boldsymbol{\theta}) = (\rho_1(\boldsymbol{\theta}), \cdots, \rho_K(\boldsymbol{\theta})) \in (0,1)^{\otimes K}$,
$$\frac{\boldsymbol{N}_{n,j}}{n} \to \rho_j(\boldsymbol{\theta}) \text{ a.s. } j = 1, \cdots, K;$$
4. For a positive definite matrix $\boldsymbol{V}(\boldsymbol{\theta})$,
$$\sqrt{n}\left(\frac{\boldsymbol{N}_n}{n} - \boldsymbol{\rho}(\boldsymbol{\theta})\right) \xrightarrow{d} N(\boldsymbol{0}, \boldsymbol{V}(\boldsymbol{\theta})).$$

Theorem 8.1. *Under regularity conditions 1-4, there exists a* $\boldsymbol{\Theta}_0 \subset \boldsymbol{\Theta} = \boldsymbol{\Theta}_1 \otimes \cdots \otimes \boldsymbol{\Theta}_K$ *with Lebesgue measure* 0 *such that for every* $\boldsymbol{\theta} \in \boldsymbol{\Theta} - \boldsymbol{\Theta}_0$,

$$\boldsymbol{V}(\boldsymbol{\theta}) \geqslant \left(\frac{\partial \rho(\boldsymbol{\theta})}{\partial \boldsymbol{\theta}}\right)' \boldsymbol{I}^{-1}(\boldsymbol{\theta}) \frac{\partial \rho(\boldsymbol{\theta})}{\partial \boldsymbol{\theta}} := \boldsymbol{\Sigma}_{LB},$$

where $\boldsymbol{I}(\boldsymbol{\theta}) = \text{diag}\big(\rho_1(\boldsymbol{\theta})\boldsymbol{I}_1(\boldsymbol{\theta}_1), \cdots, \rho_K(\boldsymbol{\theta})\boldsymbol{I}_K(\boldsymbol{\theta}_K)\big)$ *and* $\boldsymbol{I}_k(\boldsymbol{\theta}_k)$ *is the Fisher information for a single observation on treatment* $k = 1, \cdots, K$.

We refer to a adaptive design that attains the lower bound as *asymptotic best* for that particular allocation $\boldsymbol{\rho}(\boldsymbol{\theta})$. Table 8 gives the lower bounds of the asymptotic variabilitites (σ^2_{LB}) for urn proportion (UP), Roenberger, et al's proportion (RP) and Neyman proportion (NP) in a binary response clinical trial for two-treatments. It is interesting to note that the Zelen (1969)'s deterministic design (PW rule) is asymptotic best among all procedures with limiting allocation proportion $q_2/(q_1+q_2)$, the urn proportion. The RPW rule has the same limiting proportion, but it is not asymptotic best because $\sigma^2_{\text{RPW}} > \sigma^2_{LB}$. The DL rule proposed by Ivanova (2003) is a random procedure having the same urn proportion and the asymptotic variability $\sigma^2_{DL} = q_1 q_2 (p_1 + p_2)/(q_1 + q_2)^3$. So, the DL rule is asymptotic best. However, both the RPW rule and the DL rule can only target this particular proportion, which is not optimal in any formal sense, and can only be used for binary responses (c.f., Hu and Rosenberger, 2003).

Table 1: Lower bounds of the asymptotic variabilities for urn proportion, Rosenberger et al's proportion and Neyman proportion.

	ρ	σ^2_{LB}
UP	$\frac{q_2}{q_1+q_2}$	$\frac{q_1 q_2 (p_1+p_2)}{(q_1+q_2)^3}$
RP	$\frac{\sqrt{p_1}}{\sqrt{p_1}+\sqrt{p_2}}$	$\frac{1}{4(\sqrt{p_1}+\sqrt{p_2})^3}\left(\frac{p_2 q_1}{\sqrt{p_1}} + \frac{p_1 q_2}{\sqrt{p_2}}\right)$
NP	$\frac{\sqrt{p_1 q_1}}{\sqrt{p_1 q_1}+\sqrt{p_2 q_2}}$	$\frac{1}{4(\sqrt{p_1 q_1}+\sqrt{p_2 q_2})^3}\left(\frac{p_2 q_2 (1-2p_1)^2}{\sqrt{p_1 q_1}} + \frac{p_1 q_1 (1-2p_2)^2}{\sqrt{p_2 q_2}}\right)$

DBCD is also not asymptotic best (except $\gamma = \infty$). But, it can target any desired allocation and can be used for general responses, for example, continuous response. Zhang, Hu and Cheung (2006) proposed an urn model, the SEU model, which can target any pre-specified allocation proportion and can be used for general responses. The drop-the-loser rule has also been generalized to GDL model, a kind of urn model with immigration, by Zhang, Chan, Cheung and Hu (2007) and Sun, Chueng and Zhang (2007), by using the estimators of unknown distribution parameters, such that it can target any pre-specified allocation proportion and can be used for general responses. For a general pre-specified allocation proportion $\boldsymbol{\rho} = \boldsymbol{\rho}(\boldsymbol{\theta})$ in a K-treatment trial, the asymptotic variance-covariance matrices of SEU, GDL and DBCD are given in Table 8. Among these models, the DBCD can approach the lower bound for large values of γ. However, the procedure becomes more deterministic as γ becomes larger, and hence careful tuning of γ must be done to counter the trade-off between the randomness and variability. The use of DBCD with $\gamma = 2$ was strongly recommended in Hu and Rosenberer (2003) for binary response trials with two treatments, according to the simulation study.

Table 2: The asymptotic variability of SEU, GDL and DBCD for a same limiting allocation

Model	SEU	GDL	DBCD
Variability	$\text{diag}(\boldsymbol{\rho}) - \boldsymbol{\rho}'\boldsymbol{\rho} + 6\boldsymbol{\Sigma}_{LB}$	$2\boldsymbol{\Sigma}_{LB}$	$\frac{\text{diag}(\boldsymbol{\rho})-\boldsymbol{\rho}'\boldsymbol{\rho}}{1+2\gamma} + \frac{2+2\gamma}{1+2\gamma}\boldsymbol{\Sigma}_{LB}$

Very recently, we have found a randomized biased coin design (RBCD) for two-treatment clinical trails, a kind of DBCD, which preserves randomization, attains the lower bound, and can target any allocation. Whether or not an urn model can be defined to have these properties is still an open problem. In the RBCD, instead of using a continuous allocation function $g(x, y)$, we use a discrete function: $g(x, y) = \alpha y$ if $x > y$, y if $x = y$ and $1 - \alpha(1 - y)$ if $x < y$, where $0 < \alpha < 1$. That is, the m-th subject is allocated to treatment 1 with a probability

$$\begin{aligned}\mathsf{P}(X_{m,1} = 1|\mathcal{F}_{m-1}) &= P_{m,1} \\ &= \begin{cases} \alpha\rho(\widehat{\boldsymbol{\theta}}_{m-1}), & \text{if } N_{m-1,1}/(m-1) > \rho(\widehat{\boldsymbol{\theta}}_{m-1}), \\ \rho(\widehat{\boldsymbol{\theta}}_{m-1}), & \text{if } N_{m-1,1}/(m-1) = \rho(\widehat{\boldsymbol{\theta}}_{m-1}), \\ 1 - \alpha(1 - \rho(\widehat{\boldsymbol{\theta}}_{m-1})), & \text{if } N_{m-1,1}/(m-1) < \rho(\widehat{\boldsymbol{\theta}}_{m-1}). \end{cases}\end{aligned}$$

When $\rho(\boldsymbol{\theta}) \equiv 1/2$, the RBCD is just the Efron (1971)'s biased coin design. The following theorem gives the asymptotic results for the RBCD, the proof of which will not be presented here.

Theorem 8.2. *Suppose $\rho(\cdot, \cdot)$ is a continuous function on the parameter space and twice differentiable at the true value of the parameter $\boldsymbol{\theta} = (\theta_1, \theta_2)$, the distributions $f_1(\cdot|\theta_1)$ and $f_2(\cdot|\theta_2)$ of the responses follow an exponential family. Then $N_{n,1}/n - \rho(\boldsymbol{\theta}) = O(\sqrt{\log\log n/n})$ a.s. and $\sqrt{n}\,(N_{n,1}/n - \rho(\boldsymbol{\theta})) \xrightarrow{d} N(0, \sigma_\rho^2)$, where*

$$\sigma_\rho^2 = (\partial\rho(\boldsymbol{\theta})/\partial\theta_1)^2\,(\rho(\boldsymbol{\theta})I_1(\theta_1))^{-1} + (\partial\rho(\boldsymbol{\theta})/\partial\theta_2)^2\,((1-\rho(\boldsymbol{\theta}))I_2(\theta_2))^{-1}.$$

Further, $\max_{m\leqslant n}|N_{m1} - m\rho(\boldsymbol{\theta}) - \sigma_\rho W(m)| = o(\sqrt{n})$ in probability, where $W(t)$ is a standard Brownian motion.

9 Discussion

In this paper, we have discussed several classes of adaptive designs. The PW rule is the simplest procedure and has small variability. But it is too deterministic to be used in clinical trials. The RPW rule and urn models are random procedures. But their variabilities are very high. Theoretical results, simulation studies and a real example in ECMO trial all indicate that the statistical test in using a adaptive design with high variability is not powerful. However, besides in adaptive designs, urn models have wide applications in many areas including biological science, random algorithm and sampling, information science, etc. And the urn models have strong relationship with multi-type branching processes. The study of urn models has been of interest in a long history. The DL rule is fully randomized and has the smallest variability among all the adaptive designs with limiting allocation proportion $q_2/(q_1 + q_2)$. But it can only target this particular proportion and can only be used for binary responses. When it is generalized to be able to target any pre-specified allocation, the variability is no longer the smallest (c.f. Table 8). Among the adaptive designs mentioned in this paper, the DBCD and RBCD are the only procedures that preserves randomization, attains or can approach the lower bound of the variability, can target any allocation and can be used for general discrete or continuous responses.

The examples considered in this paper are binary response clinical trials. In practices, the responses of clinical trials appear in various types. For more examples and discussion, we refer to a new book of Hu and Rosneberger (2006). In many clinical trials, covariate information is available that has a strong influence on the responses of patients. For instance, the efficacy of a hypertensive drug is related to a patient's initial blood pressure and cholesterol level, whereas the effectiveness of a cancer treatment may depend on whether the patient is a smoker or a non-smoker. The theory of an adaptive design in using covariate information is much more complicated than those without covarites. A limit success in deriving the asymptotic properties of covariate-adjusted adaptive designs has been achieved by Zhang, Hu Cheung and Chan (2007). The power study and evaluation of the covariate-adjusted adaptive designs are our future studies. Also, in many clinical trials, the observed responses are usual survival data. Though it has been shown that the delay of treatment results does not effect the asymptotic properties in many adaptive designs, it is assume that the delayed responses are finally observed if the time is long enough. It is an interesting topic of studying the properties of the adaptive designs with missed or censored data.

Finally, in this paper, we only present the asymptotic results. It is important to check the accuracy of the asymptotic approximations when using the theoretical results to evaluate or compare designs. Simulations have indicated that in most cases these designs closely approximate asymptotic results for a moderate sample size of $n = 100$.

Acknowledgment

The author thanks an anonymous referee for helpful comments. Special thanks go to Professor Lin Zhengyan, the author's Ph. D. supervisor, for his help and encouragement with enthusiasm constantly.

References

[1] Athreya, K. B. and Karlin, S. (1968). Embedding of urn schemes into continuous time branching processes and related limit theorems. *Ann. Math. Stat.* **39** 1801–1817.

[2] Bai, Z. D. and Hu, F. (1999). Asymptotic theorem for urn models with nonhomogeneous generating matrices. *Stochastic Process. Appl.* **80** 87–101.

[3] Bai, Z. D. and Hu, F. (2005). Strong consistency and asymptotic normality for urn models. *Ann. Appl. Probab.* **15** 914–940.

[4] Bai, Z.D., Hu, F. and Rosenberger, W.F. (2002). Asymptotic properties of adaptive designs for clinical trials with delayed response. *Ann. Stat.* **30** 122–139.

[5] Bai, Z. D., Hu, F. and Shen, L. (2002). An adaptive design for multi-arm clinical trials. *J. Multi. Anal.* **81** 1–18.

[6] Bai, Z. D., Hu, F. and Zhang, L.-X. (2002). The Gaussian approximation theorems for urn models and their applications. *Ann. Appl. Probab.* **12** 1149–1173.

[7] Efron, B. (1971). Forcing a sequential experiment to be balanced. *Biometrika* **62** 347–352.

[8] Eisele, J. (1994). The doubly adaptive biased coin design for sequential clinical trials. *J. Stat. Plann. Infer.* **38** 249–262.

[9] Eisele, J. and Woodroofe, M. (1995). Central limit theorems for doubly adaptive biased coin designs. *Ann. Stat.* **23** 234–254.

[10] Hall, P. and Heyde, C. C. (1980). *Martingale Limit Theory and its Applications.* Academic Press, London.

[11] Hu, F. and Rosenberger, W. F. (2003). Optimality, variability, power evaluating response-adaptive randomization procedures for treatment comparisons. *J. Am. Stat. Assoc.* **98** 671–678.

[12] Hu, F. and Rosenberger, W. F. (2006). *The Theory of Response-Adaptive Randomization in Clinical Trials.* John Wiley and Sons. New York.

[13] Hu, F., Rosenberger, W. F. and Zhang, L.-X. (2006). Asymptotically best response-adaptive randomization procedures. *J. Stat. Plann. Infer.* **136** 1911–1922.

[14] Hu, F. and Zhang, L.-X. (2001). The weak and strong invariance for generalized Friedman's urn model. Manuscript. http://www.math.zju.edu.cn/zlx/mypapers2/URNHDIM3.pdf

[15] Hu, F. and Zhang, L.-X. (2004a). Asymptotic properties of doubly adaptive biased coin designs for multi-treatment clinical trials. *Ann. Stat.* **32** 268–301.

[16] Hu, F. and Zhang, L.-X. (2004b). The asymptotic normality of urn models for clinical trials with delayed response. *Bernoulli* **10** 447–463.

[17] Ivanova, A. V. (2003). A play-the-winner type urn model with reduced variability. *Metrika* **58** 1–13.

[18] Ivanova, A. V. (2006). Urn designs with immigration: Useful connection with continuous time stochastic processes. *J. Stat. Plann. Infer.* **136** 1836–1844.

[19] Ivanova, A. V. and Rosenberger, W. F. (2003). A comparison of urn designs for randomized clinical trials of $K > 2$ treatments. *J. Biopharmaceutial Statistics* **10** 93–107.

[20] Janson, S. (2004). Functional limit theorems for multitype branching processes and generalized Pólya urns. *Stochastic Process. Appl.* **110** 177–245.

[21] Jennison, C. and Turnbull, B. W. (2000). *Group Sequential Methods with Applications to Clinical Trials.* Chapman and Hall/CRC, Boca Raton, FL.

[22] Lin, Z. Y., Bai, Z. D., Chen, Y. M. and Hu, F. (2003). Adaptive designs based on Markov chains (I) (II). *J. Biomath.* **18** 597–609 (in Chinese).

[23] Lin, Z. Y., Zhang, L.-X., Chueng, S. H. and Chan, W. S. (2005). Strong approximation of a Markov chain. *Chinese Ann. Math.* **26A** 241–250 (in Chinese); *Chinese J. Contemporary Math.* **26** (2006) 283–290 (in English).

[24] Robbins, H. (1952). Some aspects of the sequential design of experiments. *Bull. Am. Math. Soc.* **58** 527–535.

[25] Rosenberger, W. F. (1999). Randomized play-the-winner clinical trials: review and recommendation. *Controlled Clinical Trials* **2** 328–342.

[26] Rosenberger, W. F., Flournoy, N. and Durham, S. D. (1997). Asymptotic normality of maximum likelihood estimators from multiparameter response-driven design. *J. Stat. Plann. Infer.* **60** 69–76.

[27] Rosenberger, W. F. and Hu, F. (2004). Maximizing power and minimizing treatment failures. *Clinical Trials* **1** 141–147.

[28] Rosenberger, W. F. and Lachin, J. M. (1995). *Randomization in Clinical Trials: Theory and Practice.* John Wiley and Sons, New York.

[29] Rosenberger, W. F., Stallard, N., Ivanova, A., Harper, C. N. and Ricks, M. L. (2001). Optimal adaptive designs for binary response trials. *Biometrics* **57** 909–913.

[30] Rosenberger, W. F., Vidyashankar, A. N. and Agarwal, D. K. (2001). Covariate-adjusted response-adaptive designs for binary response. *J. Biopharmaceutial Statist.* **11** 227–236.

[31] Royall, R. M. (1991). Ethics and statistics in randomized clinical trials (Disc:p63–88). *Statistical Science* **6** 52–62.

[32] Smythe, R. T. and Rosenberger, W. F. (1995). Play-the-winner, generalized Polya urns, and Markov branching processes. In *Adaptive Designs* (Flournoy, N. and Rosenberger, W. F., eds.). Hayward: Institute of Mathematical Statistics, 13–22.

[33] Stout, W. F. (1974). *Almost sure convergence.* Academic Press, New York.

[34] Sun, R., Cheung, S. H. and Zhang, L.-X. (2007). A generalized drop-the-loser rule for multi-treatment clinical trials. *J. Stat. Plann. Infer.* **137** 2011–2023.

[35] Thompson, W. R. (1933). On the likelihood that one unknown probability exceeds another in view of the evidence of the two samples. *Biometrika* **25** 275–294.

[36] Wei, L. J. (1979). The generalized Polya's urn design for sequential medical trials. *Ann. Stat.* **7** 291–296.

[37] Wei, L. J. and Durham, S. (1978). The randomized pay-the-winner rule in medical trials. *J. Am. Stat. Assoc.* **73** 840–843.

[38] Yao, Q. and Wei, L. J. (1996). Play the winner for phase II/III clinical trials. *Statistics in Medicine* **15** 2413–2423.

[39] Zelen, M. (1969). Play-the-winner rule and the controlled clinical trial. *J. Am. Stat. Assoc.* **64** 131–146.

[40] Zhang, L.-X. (2004). Strong approximations of martingale vectors and their applications in Markov-chain adaptive designs. *Acta Math. Appl. Sinica, English Series* **20**(2) 337–352.

[41] Zhang, L.-X. (2006). Asymptotic results on a class of adaptive multi-treatment designs. *J. Multi. Anal.* **97** 586–605.

[42] Zhang, L.-X., Chan, W. S., Cheung. S. H. and Hu, F. (2007). A A generalized drop-the-loser urn for clinical trials with delayed responses. *Statistica Sinica* **15** 387–409.

[43] Zhang, L.-X., Hu, F., and Cheung. S. H. (2006). Asymptotic theorems of sequential estimation-adjusted urn models. *Ann. Appl. Probab.* **16** 340–369.

[44] Zhang, L-.X., Hu, F., Cheung, S. H. and Chan, W. S. (2006a). Doubly adaptive biased coin designs with delayed responses. Manuscript.

[45] Zhang, L-.X., Hu, F., Cheung, S. H. and Chan, W. S. (2006b). Immigrated urn models– asymptotic properties and applications. Manuscript.

[46] Zhang, L-.X., Hu, F., Cheung, S. H. and Chan, W. S. (2007). Asymptotic properties of covariate-adjusted response-adaptive designs. *Ann. Stat.* **35** (3) to appear. http://arxiv.org/abs/math.ST /0610518.

Functional Limit Theorems for Gaussian Processes*

Wensheng Wang
Department of Statistics, East China Normal University
Shanghai 200062, China
and
Department of Mathematics, Hangzhou Normal University
Hangzhou 310036, China
E-mail: wswang@stat.ecnu.edu.cn

Abstract

In this article we present a survey of some developments on the studies of functional limit theorems for Gaussian processes.

1 Introduction

There have been a lot of developments on the studies of sample path properties of Gaussian processes since the publication of the remarkable book of Csörgő and Révész (1981). Most of these research results since then have been summarized and reviewed in the books of Lin and Lu (1992) and Lin, Lu and Zhang (2001). Recently, there have been some developments on the studies of functional limit theorems of Gaussian processes. In this article I present a survey of these developments on the studies of functional limit theorems for Gaussian processes (my own knowledge of which is quite limited). The basic aim here is to ease the reader into some of the more recent accounts of this subject.

Let $\{W(t), t \geqslant 0\}$ be Brownian motion (BM) in $\mathbb{R}^1$ and $C_0[0,1]$ be the space of continuous functions on the closed interval $[0,1]$ which vanish at the origin. Assume $H_\mu \subset C_0[0,1]$ is the Hilbert space of absolutely continuous functions on $[0,1]$ whose unit ball is the set

$$U = \Big\{f(t) = \int_0^t f'(s)ds, 0 \leqslant t \leqslant 1 : \int_0^1 |f'(s)|^2 ds \leqslant 1\Big\}. \tag{1.1}$$

*Research supported by National Natural Science Foundation of China (10401037) and Doctoral Program Foundation of the Ministry of Education of China (20060269016).

Here the inner product is given by

$$\|f\|_\mu = \Big(\int_0^1 |f'(s)|^2 ds \Big)^{1/2}, \quad f \in H_\mu, \tag{1.2}$$

and $\|f\|_\infty$ denotes the usual sup-norm on $C_0[0,1]$, i.e. $\|f\|_\infty = \sup_{0 \leqslant x \leqslant 1} |f(x)|$.

Let

$$f_n(x) = \frac{W(nx)}{\sqrt{2n \log\log n}}, \quad 0 \leqslant x \leqslant 1.$$

Here and throughout $\log x = \log_e \max(1, x)$ and $\log\log x = \log(\log x)$. Then $\{f_n(\cdot),\ n \geqslant 3\}$ is a sequence of stochastic processes with sample paths almost surely in $C_0[0,1]$. Strassen (1964) proved the following theorem.

Theorem 1.1 (functional law of the iterated logarithm). *With probability one, the sequence $\{f_n(x), 0 \leqslant x \leqslant 1\}_{n \geqslant 3}$ is relatively compact in $C_0[0,1]$, and the set of its limit points is U.*

An equivalent result is the following theorem.

Theorem 1.2. *We have*

$$\lim_{n\to\infty} \inf_{f \in U} \sup_{0 \leqslant x \leqslant 1} |f_n(x) - f(x)| = 0 \quad \text{a.s.}, \tag{1.3}$$

and for any $f \in U$

$$\liminf_{n\to\infty} \sup_{0 \leqslant x \leqslant 1} |f_n(x) - f(x)| = 0 \quad \text{a.s.}. \tag{1.4}$$

(1.3) and (1.4) imply the following well-known law of the iterated logarithm.

Corollary 1.3 (LIL). *We have*

$$\limsup_{n\to\infty} f_n(1) = \limsup_{n\to\infty} \sup_{0 \leqslant x \leqslant 1} |f_n(x)| = 1 \quad \text{a.s.}.$$

Bolthausen (1978) and Grill (1987, 1991) gave the exact convergence rates of (1.3) and (1.4). Oodaira (1972), Goodman and Kuelbs (1991), Kuelbs, Li and Talagrand (1994) and Monrad and Rootzén (1995) generalized the above theorem to some more general classes of Gaussian processes. (In fact, they gave the exact rates of convergence for functional LIL of Gaussian processes). Chan, Csörgő and Révész (1978),

Révész (1979) and Chen (1998) obtained functional Csörgő-Révész large increment theorem for Wiener processes, which generalized Strassen's functional LIL (see Theorem 1.2 above). Mueller (1981) and Chen and Csörgő (2001) studied functional modulus of continuity of Wiener processes (saparately, by using different proof methods). For more references, we refer the interested reader to the references cited therein. Define

$$M_{t,h}(x) = \frac{W(t+hx) - W(t)}{(2h\log(h^{-1}))^{1/2}}, \quad h \in (0,1),\ t \in [0, 1-h],\ 0 \leqslant x \leqslant 1.$$

Then the functional modulus of continuity, given by Mueller (1981) and Chen and Csörgő (2001) saparately, states the following theorem.

Theorem 1.4 (functional modulus of continuity). *We have*

$$\lim_{h\to 0} \sup_{0\leqslant t\leqslant 1-h} \inf_{f\in U} \|M_{t,h} - f\|_\infty = 0 \quad \text{a.s.}, \tag{1.5}$$

and for any $f \in U$

$$\lim_{h\to 0} \inf_{0\leqslant t\leqslant 1-h} \|M_{t,h} - f\|_\infty = 0 \quad \text{a.s..} \tag{1.6}$$

(1.5) and (1.6) imply the following well-known modulus of continuity of Wiener processes (see P. Lévy (1937, 1948)).

Corollary 1.5 (modulus of continuity). *We have*

$$\lim_{h\to 0} \sup_{0\leqslant t\leqslant 1-h} M_{t,h}(1) = \lim_{h\to 0} \sup_{0\leqslant t\leqslant 1-h} \sup_{0\leqslant x\leqslant 1} |M_{t,h}(x)| = 1 \quad \text{a.s..}$$

Remark 1.6. From Theorem 1.4 we can see that the important difference between Theorem 1.4 and classic Strassen-type theorems (see Theorem 1.2) is the fact that Theorem 1.4 implies "lim" results while classic Strassen-type theorems only imply "lim sup" results. The key reason for this is (1.6), which in classic Strassen-type theorems has a weaker replacement as follows:

$$\liminf_{h\to 0} \inf_{0\leqslant t\leqslant 1-h} \sup_{0\leqslant x\leqslant 1} |M_{t,h}(x) - f(x)| = 0 \quad \text{a.s..}$$

We note in passing that, in the same spirit, via (1.6), Csörgő and Révész (1981) succeeded in replacing "lim sup" in the original P. Lévy moduli by "lim" as in Corollary 1.5.

Let a_T be a positive function of $T \geqslant 3$. Introduce the following notations:

$$\beta_T = (2a_T(\log(Ta_T^{-1}) + \log\log T))^{-\frac{1}{2}},$$

$$Y_{t,T}(x) = \beta_T(W(t + xa_T) - W(t)), \qquad 0 \leqslant x \leq 1, \;\; 0 \leqslant t \leqslant T - a_T.$$

Révész (1979) and Chen and Csörgő (2001) obtained the following functional Csörgő-Révész large increment theorem.

Theorem 1.7 (functional large increment theorem). *Assume that* $0 < a_T \leqslant T$ *satisfies the following conditions:*

(i) a_T *is non-decreasing*

(ii) Ta_T^{-1} *is non-decreasing.*

Then, with probability one, $\{Y_{t,T}(x) : \; 0 \leqslant x \leqslant 1, \;\; 0 \leqslant t \leqslant T - a_T, \;\; T \geqslant 3\}$ *(as* $T \to \infty$*) is relatively compact in* $C_0[0,1]$*, and the set of its limit points is* U*. If we also assume that*

(iii) $\displaystyle\lim_{T\to\infty} \frac{\log(Ta_T^{-1})}{\log\log T} = \infty,$

then we also have that

$$\lim_{T\to\infty} \sup_{0\leqslant t\leqslant T-a_T} \inf_{f\in U} \sup_{0\leqslant x\leqslant 1} |Y_{t,T}(x) - f(x)| = 0 \quad \text{a.s.,} \tag{1.7}$$

and for any $f \in U$

$$\lim_{T\to\infty} \inf_{0\leqslant t\leqslant T-a_T} \sup_{0\leqslant x\leqslant 1} |Y_{t,T}(x) - f(x)| = 0 \quad \text{a.s..} \tag{1.8}$$

Theorem 1.7 implies the following well-known Csörgő-Révész large increment theorem (Csörgő and Révész (1979, 1981)).

Corollary 1.8 (Csörgő-Révész large increment theorem). *Assume that* $0 < a_T \leqslant T$ *satisfies conditions* (i) *and* (ii). *Then we have*

$$\limsup_{T\to\infty} \sup_{0\leqslant t\leqslant T-a_T} Y_{t,T}(1) = \limsup_{T\to\infty} \sup_{0\leqslant t\leqslant T-a_T} \sup_{0\leqslant x\leqslant 1} |Y_{t,T}(x)| = 1 \quad \text{a.s..}$$

If we also assume that (iii) *is satisfied, then we also have*

$$\lim_{T\to\infty} \sup_{0\leqslant t\leqslant T-a_T} Y_{t,T}(1) = \lim_{T\to\infty} \sup_{0\leqslant t\leqslant T-a_T} \sup_{0\leqslant x\leqslant 1} |Y_{t,T}(x)| = 1 \quad \text{a.s..}$$

We will conclude this section by giving the proofs of Corollaries 1.3, 1.5 and 1.8. It is sufficient to show Corollary 1.8, the proofs of the others are similar. The proof of Corollary 1.8 given below, is not new (see Chp. 3 in Chen (1998)).

Proof of Corollary 1.8. The first part results of Theorem 1.7 imply that (1.7) and the following are true: for any $f \in U$

$$\liminf_{T\to\infty} \inf_{0\leqslant t\leqslant T-a_T} \sup_{0\leqslant x\leqslant 1} |Y_{t,T}(x) - f(x)| = 0 \quad \text{a.s..} \tag{1.9}$$

By (1.7), we have

$$\begin{aligned}
&\limsup_{T\to\infty} \sup_{0\leqslant t\leqslant T-a_T} \sup_{0\leqslant x\leqslant 1} |Y_{t,T}(x)| \\
&\quad \leqslant \limsup_{T\to\infty} \sup_{0\leqslant t\leqslant T-a_T} \inf_{f\in U} \sup_{0\leqslant x\leqslant 1} |Y_{t,T}(x) - f(x)| + \sup_{f\in U} \sup_{0\leqslant x\leqslant 1} |f(x)| \\
&\quad = \sup_{f\in U} \sup_{0\leqslant x\leqslant 1} |f(x)| = 1 \quad \text{a.s..}
\end{aligned}$$

Also, via (1.9), for any $f \in U$

$$\begin{aligned}
&\limsup_{T\to\infty} \sup_{0\leqslant t\leqslant T-a_T} Y_{t,T}(1) \\
&\quad \geqslant f(1) - \liminf_{T\to\infty} \inf_{0\leqslant t\leqslant T-a_T} \sup_{0\leqslant x\leqslant 1} |Y_{t,T}(x) - f(x)| \\
&\quad = f(1) \quad \text{a.s..}
\end{aligned}$$

Hence

$$\limsup_{T\to\infty} \sup_{0\leqslant t\leqslant T-a_T} Y_{t,T}(1) \geqslant \sup_{f\in U} f(1) = 1 \quad \text{a.s..}$$

Finally, if condition (iii) is also satisfied, then (1.8) is true. Via (1.8), for any $f \in U$

$$\begin{aligned}
&\liminf_{T\to\infty} \sup_{0\leqslant t\leqslant T-a_T} Y_{t,T}(1) \\
&\quad \geqslant f(1) - \lim_{T\to\infty} \inf_{0\leqslant t\leqslant T-a_T} \sup_{0\leqslant x\leqslant 1} |Y_{t,T}(x) - f(x)| \\
&\quad = f(1) \quad \text{a.s..}
\end{aligned}$$

Hence

$$\liminf_{T\to\infty} \sup_{0\leqslant t\leqslant T-a_T} Y_{t,T}(1) \geqslant \sup_{f\in U} f(1) = 1 \quad \text{a.s..}$$

Therefore, the proof of Corollary 1.8 is now complete.

Wang (2002 a,b) studied the exact convergence rates of Theorems 1.4 and 1.7 respectively. Wei (1999) studied functional modulus of continuity and functional large increment theorems for l^p-valued Wiener processes under Hölder norm. Hwang, Lin and Wang (2006) studied functional limit theorems for Gaussian processes under Hölder norm.

Recently, there have been also some developments on the studies of functional limit theorems for multi-parameter Gaussian processes. Chen (1998) first studied functional laws of the iterated logarithm for two-parameter Wiener processes (or Brownian sheet). Choi, Lin and Wang (2004) extended Chen (1998)'s results and obtained functional laws of the iterated logarithm for fractional Brownian sheet. Wang (2004) studied functional Csörgő-Révész large increment theorem for two-parameter Wiener processes.

Throughout the remainder of the paper we will concentrate on obtaining information regarding functional limit theorems for increments of general Gaussian processes, which generalize Theorems 1.4 and 1.7. In Section 2, we recall some basic facts about Gaussian random vectors, and in Section 3 we state a functional limit theorem for increments of Gaussian processes with stationary increments. Theorems relating to functional modulus of continuity and functional large increment theorem for fractional Brownian motion, which generalize the related results for Wiener processes mentioned above, are stated in Section 4. The proofs will be given in Sections 5 and 6.

2 Gaussian random vectors

Let B^* be the topological dual of B with norm $\|\cdot\|$ and X be a centered B-valued Gaussian random vector with law $\mu = \mathcal{L}(X)$. It is well known that there is a unique Hilbert space $H_\mu \subset B$ (also call the reproducing Hilbert space (RHS) generated by μ) such that μ is determined by considering the pair (B, H_μ) as an abstract Wiener space (see Gross (1970)). For example, if $B = C_0[0,1]$ and μ is Wiener measure, then the unique Hilbert space H_μ has unit ball as in (1.1) with inner product norm given by (1.2).

The Hilbert space H_μ can be described as the completion of the range of the mapping $S : B^* \to B$ defined by the Bochner integral

$$Sf = \int_B xf(x)d\mu(x), \quad f \in B^*,$$

where the completion is in the inner product norm

$$< Sf, Sg >_\mu = \int_B f(x)g(x)d\mu(x), \quad f, g \in B^*.$$

We use $\|\cdot\|_\mu$ to denote the inner product norm induced on H_μ, and for well known properties and various relationships between μ, H_μ and B, see Lemma 2.1 in Kuelbs (1976). In particular, if $\{\alpha_k, k \geqslant 1\}$ is a sequence in B^*, orthonormal in $L^2(\mu)$ such that $\{\Delta\alpha_k, k \geqslant 1\}$ is a CONS in $H_\mu \subset B$, then the operators defined for $d \geqslant 1$ by

$$\Pi_d(x) = \sum_{k=1}^{d} \alpha_k(x)\Delta\alpha_k \quad \text{and} \quad Q_d(x) = x - \Pi_d(x) \tag{2.1}$$

are continuous mappings from B to B. Furthermore, when restricted to H_μ, Π_d and Q_d are orthogonal projections onto their ranges, it is also known that for μ centered Gaussian, $\lim_{d\to\infty} \|Q_d(x)\| = 0$ with μ-probability one, and

$$\mathrm{E}\|Q_d(X)\| \downarrow 0, \quad \text{as} \quad d \uparrow \infty, \tag{2.2}$$

and that for $f \in H_\mu$ we can define the stochastic inner product for μ almost all x in B by

$$< x, f >^{\sim} = \lim_{d\to\infty} \sum_{k=1}^{d} \alpha_k(x) < f, \Delta\alpha_k(x) >_\mu . \tag{2.3}$$

Because $< f, \Delta\alpha_k(x) >_\mu = \alpha_k(f)$, it is easy to see that $< \cdot, f >^{\sim}$ is $N(0, \sigma^2)$ with $\sigma^2 = < f, f >_\mu$. The Cameron-Martin translation formula for Gaussian measures states that

$$\mu(E + f) = \int_E \exp\{-\frac{1}{2}\|f\|_\mu^2 - < x, f >^{\sim}\} d\mu(x) \tag{2.4}$$

for Borel subsets E of B. This is well known, but a particularly nice proof is contained in Proposition 2.1 of de Acosta (1983).

The following theorem is the large deviation principle for a Gaussian probability measure in a separable Banach space, which is well-known, and one can see its proof in Azencott (1980) or Jain (1984), e.g..

Theorem 2.1 (large deviation principle). *Let μ be a centered Gaussian probability measure on $(B, \mathcal{B})$, where B is a separable Banach space with norm $\|\cdot\|$ and $\mathcal{B}$ its Borel sets, and for $A \in \mathcal{B}$ let $\mu_\varepsilon(A) = \mu\{x : \sqrt{\varepsilon}x \in A\}$. Then $\{\mu_\varepsilon, \varepsilon > 0\}$ satisfies the large deviation principle with rate function given by*

$$I(x) = \begin{cases} \frac{1}{2}\|x\|_\mu^2, & \text{if } x \in H_\mu, \\ \infty, & \text{otherwise,} \end{cases}$$

where H_μ and $\|\cdot\|_\mu$ are defined as above. That is, the family of probability measure $\{\mu_\varepsilon, \varepsilon > 0\}$ and the function I on $(B, \mathcal{B})$ satisfy the following four conditions:

(i) *$I : B \to [0, \infty]$ is lower semi-continuous, $I \not\equiv \infty$.*

(ii) *Given $\ell < \infty$, the set $\{x \in B : I(x) \leqslant \ell\}$ is compact.*
(iii) *For a closed set $F \subset B$*

$$\limsup_{\varepsilon \to 0} \varepsilon \log \mu_\varepsilon(F) \leqslant - \inf_{x \in F} I(x).$$

(iv) *For an open set $G \subset B$*

$$\liminf_{\varepsilon \to 0} \varepsilon \log \mu_\varepsilon(G) \geqslant - \inf_{x \in G} I(x).$$

The following proposition is an immediate consequence of (iii) of Theorem 2.1.

Proposition 2.2. *Let X be a centered Gaussian random vector with values in B, where B is a separable Banach space with norm $\|\cdot\|$. Let $\alpha > 0$ and let $U_\alpha = \{f \in H_\mu : \|f\|_\mu \leqslant \alpha\}$, where H_μ is the* RHS *with $\mu = \mathcal{L}(X)$. Then, for any $\delta > 0$ and any $0 < \tau < 1$ there exists $\lambda_0 = \lambda_0(\tau, \delta) > 0$ such that for any $\lambda \geqslant \lambda_0$*

$$P\Big(\inf_{f \in U_\alpha} \Big\|\frac{X}{\lambda} - f\Big\| \geqslant \delta\Big) \leqslant \exp\Big\{-(1-\tau)\lambda^2 \inf_{f \in ((U_\alpha)^\delta)^c} I(f)\Big\}, \tag{2.5}$$

where the following notations are used:

$$F^\delta = \{g \in B : \inf_{f \in F} \|f - g\| < \delta\} \tag{2.6}$$

and F^c is the complement of F.

Schilder (1966) first proved the above theorem for Wiener measure in $C_0[0,1]$. Here we state Schilder (1966)'s theorem as a corollary of Theorem 2.1. Let $B = C([0,T]; \mathbb{R}^d)$, the space of continuous $\mathbb{R}^d$-valued functions on $[0,T]$ and $\mathcal{B}$ its Borel sets. As usual, $\|x\| = \sup_{0 \leqslant t \leqslant T} |x(t)| = \sup_{0 \leqslant t \leqslant T} (x_1(t)^2 + \ldots + x_d(t)^2)^{1/2}$. Let μ be the standard Wiener measure on B. Then

$$H_\mu = \{x \in B : x \text{ absolutely continuous}, \int_0^T |\dot{x}(t)|^2 dt < \infty\},$$

$$\langle x, y \rangle = \int_0^T \langle \dot{x}(t), \dot{y}(t) \rangle_{\mathbb{R}^d} dt.$$

Let

$$I^T(x) = \begin{cases} \frac{1}{2}\|x\|_\mu^2 = \frac{1}{2}\int_0^T |\dot{x}(t)|^2 dt, & \text{if } x \in H_\mu, \\ \infty, & \text{otherwise.} \end{cases}$$

Theorem 2.3 (Schilder). *If μ is standard Wiener measure in $C([0,T];\mathbb{R}^d)$ and $\mu_\varepsilon(A) = \mu\{x : \sqrt{\varepsilon}x \in A\}$ for $A \in \mathcal{B}$, then $\{\mu_\varepsilon, \varepsilon > 0\}$ satisfies the large deviation principle with rate function I^T.*

The following large deviation principle (Proposition 2.4 below) is an extension of (2.5), by applying it we can consider the convergence rate for clustering of Gaussian processes, see Goodman and Kuelbs (1991). See Proposition 2.1 and Remark 2.1 of Wang (2001) or Theorem 2.1 of Goodman and Kuelbs (1991) for its proof.

Proposition 2.4. *Let X be a centered Gaussian random vector with values in B. Let $Q_d (d \geqslant 1)$ be the linear operator of (2.1) and $U_\alpha = \{f \in H_\mu : \|f\|_\mu \leqslant \alpha\}$, where $\alpha > 0$ and H_μ is the RHS with $\mu = \mathcal{L}(X)$. Let d_λ be an integer such that*

$$d_\lambda \geqslant \inf\{m \geqslant 1 : \mathrm{E}\|Q_m X\|/m \leqslant 2\tau \log\lambda/\lambda\}, \quad (\lambda \geqslant 1), \tag{2.7}$$

and $\delta_\lambda = \gamma d_\lambda \log\lambda/\lambda^2$ with $\gamma > 3\tau$, where $\tau = \sup_{x\in U_\alpha} \|x\|_\infty$. Then, for any $\delta \geqslant \delta_\lambda$ (where δ may dependent on λ), it holds that

$$\begin{aligned} &P\Big(\inf_{f\in U_\alpha} \Big\|\frac{X}{\lambda} - f\Big\| \geqslant \delta\Big) \\ &\leqslant \frac{1}{\sqrt{2\pi(d_\lambda+1)}} \exp\Big\{-\frac{(\alpha\lambda(1+\delta))^2}{2} + \frac{d_\lambda - 1}{2}\log\frac{(\alpha\lambda(1+\delta))^2 e}{d_\lambda - 1}\Big\} \end{aligned} \tag{2.8}$$

for any $\lambda \geqslant \lambda_0$ with some $\lambda_0 > 0$.

To end this section we quote the following Fernique type inequality due to Csáki, Csörgő and Shao (1992) (see Lemma 2.1 therein), which is very useful to study sample path properties of Gaussian processes.

Lemma 2.5 (Fernique type inequality). *Let B be a separable Banach space with norm $\|\cdot\|$ and let $\{\Gamma(t), -\infty < t < \infty\}$ be a stochastic process with values in B. Let P be the probability measure generated by $\Gamma(\cdot)$. Assume that $\Gamma(\cdot)$ is P-almost surely continuous with respect to $\|\cdot\|$ and that for $|t| \leqslant t_0 \leqslant \infty$, $0 < x^* \leqslant x$ and $0 < h \leqslant h_0 \leqslant \infty$ there exist non-negative monotone non-decreasing functions $\sigma_1(h)$ and $\sigma_2(h)$ such that*

$$P(\|\Gamma(t+h) - \Gamma(t)\| \geqslant x\sigma_1(h) + \sigma_2(h)) \leqslant K\exp(-\gamma x^\beta)$$

with some $K, \gamma, \beta > 0$. *Then*

$$P\Big(\sup_{0\leqslant t\leqslant T}\sup_{0\leqslant s\leqslant a}\|\Gamma(t+s)-\Gamma(t)\| \geqslant x(\sigma_1(a)+\sigma_1(a,k)) + \sigma_1^*(a,k)+\sigma_2(a)+\sigma_2(a,k)\Big) \leqslant 4\Big(\frac{T}{a}+1\Big)K2^{2^{k+1}}\exp(-\gamma x^{\beta}) \tag{2.9}$$

for any $0 \leqslant T \leqslant t_0 \leqslant \infty$, $0 < a \leqslant h_0 \leqslant \infty$, $x \geqslant x^*$ *and any positive* $k \geqslant 3$, *where*

$$\sigma_1(a,k) = 2^{3+(1/\beta)}\int_{2^{k-3}}^{\infty}\frac{\sigma_1(ae^{-z})}{z}dz,$$
$$\sigma_2(a,k) = 6\int_{2^{k-3}}^{\infty}\frac{\sigma_2(ae^{-z})}{z}dz,$$
$$\sigma_1^*(a,k) = 4\Big(\frac{14}{\gamma}\Big)^{1/\beta}\beta\int_{2^{(k-2)/\beta}}^{\infty}\sigma_1(ae^{-z^{\beta}})dz.$$

For related further developments along these lines we refer to Csörgő and Shao (1993).

3 Functional limit theorems for Gaussian processes with stationary increments

We restrict our attention to Gaussian processes with stationary increments. For such processes we obtain the necessary independence by splitting up their spectral representation.

Throughout this section we let $X = \{X(t), t \geqslant 0\}$ be a real-valued, centered Gaussian process with almost surely continuous sample paths and assume that $X(0) = 0$ and that X has stationary increments and continuous covariance function

$$R(s,t) = \mathrm{E}[X(s)X(t)] = \int_{\mathbb{R}}\Big(e^{is\lambda}-1\Big)\Big(e^{-it\lambda}-1\Big)\Delta(d\lambda), \tag{3.1}$$

where the symmetric spectral measure Δ satisfies

$$\int_{\mathbb{R}}\frac{\lambda^2}{1+\lambda^2}\Delta(d\lambda) < \infty.$$

There exists a centered, complex-valued Gaussian random measure $W(d\lambda)$ such that

$$X(t) = \int_{\mathbb{R}}\Big(e^{it\lambda}-1\Big)W(d\lambda). \tag{3.2}$$

The measures W and Δ are related by the identity $\mathrm{E}[W(E)\overline{W(F)}] = \Delta(E \cap F)$ for all Borel sets E and F in $\mathbb{R}$. Furthermore, $W(-E) = \overline{W(E)}$. Put $\sigma^2(h) = \mathrm{E}[X(t+h) - X(t)]^2$.

Moreover, suppose that $\sigma(x)$ is a regularly varying function both at zero and at ∞ with a positive exponent β, namely

$$\sigma(x) = x^{\beta}\ell(x), \quad 0 < \beta < 1, \tag{3.3}$$

where $\ell(\cdot)$ is slowly varying both at zero and at ∞, that is, $\lim_x \ell(tx)/\ell(x) = 1$ for all $t > 0$; and suppose that X is self-similar in the sense that for any $h > 0$,

$$X(h\cdot) \overset{d}{=} \sigma(h)X(\cdot), \tag{3.4}$$

Let $U = U_1 = \{f \in H_\mu : \|f\|_\mu \leqslant 1\}$ with $\mu = \mathcal{L}(X)$, where $H_\mu \subset B$ is the RHS defined as in Section 2 with $B = C_0[0,1]$. The following result is a functional limit theorem for Gaussian processes. The proof of which given in Section 5 below, is not new (see Wang (2005)). Here I will give its proof for attempting to illustrate the basic technicalities to show functional limit theorems of Gaussian processes.

Theorem 3.1. *Let A_T and $0 < a_T \leqslant A_T$ be functions of T satisfying*

(i) *A_T, a_T and A_T/a_T are nondecreasing;*

(ii) *when A_T is bounded, a_T tends to zero; or $A_T \to \infty$ as $T \to \infty$.*

If

$$\Delta_{t,T}(x) = \frac{X(t + a_T x) - X(t)}{\sigma(a_T)(2(\log \frac{A_T - a_T}{a_T} + \log\log A_T))^{1/2}},$$

$0 \leqslant x \leqslant 1, 0 \leqslant t \leqslant A_T - a_T$, *then we have*

$$\lim_{T\to\infty} \sup_{0\leqslant t\leqslant A_T - a_T} \inf_{f\in U} \|\Delta_{t,T} - f\|_\infty = 0 \quad \text{a.s.} \tag{3.5}$$

and, for any $f \in U$,

$$\liminf_{T\to\infty} \inf_{0\leqslant t\leqslant A_T - a_T} \|\Delta_{t,T} - f\|_\infty = 0 \quad \text{a.s.}. \tag{3.6}$$

If we also assume that the following condition is satisfied:

(iii) $\lim_{T\to\infty} \dfrac{\log(A_T a_T^{-1})}{\log\log A_T} = r, \quad (0 \leqslant r \leqslant \infty),$

then we also have that

$$\liminf_{T\to\infty} \sup_{0\leqslant t\leqslant A_T - a_T} \inf_{f\in U_{\alpha(r)}} \|\Delta_{t,T} - f\|_\infty = 0 \quad \text{a.s.} \tag{3.7}$$

and, for any $f \in U_{\alpha(r)}$,

$$\lim_{T\to\infty} \inf_{0\leqslant t\leqslant A_T - a_T} \|\Delta_{t,T} - f\|_\infty = 0 \quad \text{a.s.}, \tag{3.8}$$

where

$$U_{\alpha(r)} = \{f \in C_0[0,1] : \|f\|_\mu \leqslant \alpha(r)\} \quad \text{with} \quad \alpha(r) = \begin{cases} \frac{r}{r+1}, & 0 \leqslant r < \infty, \\ 1, & r = \infty. \end{cases} \tag{3.9}$$

In closing this section, we prove a large deviation inequality for increments of general Gaussian processes, which is an extension of Propositions 2.2 and 2.4 with $B = C_0[0,1]$ endowed with the usual sup-norm, and will be used in the proof of Theorem 3.1 and may be of independent interest.

Proposition 3.2. *Let d_λ, δ_λ and U_α be defined as in Proposition 2.3 with $B = C_0[0,1]$. Then, for any $a > 0$, $0 < b \leqslant a$ and $1 > \delta \geqslant \delta_\lambda$ (where δ may dependent on λ), it holds that*

$$\begin{aligned} &P\Big(\sup_{0\leqslant t\leqslant a-b}\inf_{f\in U_\alpha}\Big\|\frac{X(t/b+\cdot)-X(t/b)}{\lambda}-f\Big\|_\infty \geqslant 3\delta\Big) \\ &\leqslant \frac{Ca}{b}\exp\Big\{-\frac{(\alpha\lambda(1+\delta))^2}{2}-\frac{2}{\beta}\log\delta+\frac{d_\lambda-1}{2}\log\frac{(\alpha\lambda(1+\delta))^2 e}{d_\lambda-1}\Big\} \end{aligned} \tag{3.10}$$

for any $\lambda \geqslant \lambda_0$ with some $\lambda_0 > 0$, where C is a positive constant.

Particularly, if δ is independent of λ, then, for any $\delta > 0$ and any $0 < \tau < 1$ there exists $\lambda_0 = \lambda_0(\tau,\delta) > 0$ such that for any $\lambda \geqslant \lambda_0$, $a > 0$, $0 < b \leqslant a$, we have

$$\begin{aligned} &P\Big(\sup_{0\leqslant t\leqslant a-b}\inf_{f\in U_\alpha}\Big\|\frac{X(t/b+\cdot)-X(t/b)}{\lambda}-f\Big\|_\infty \geqslant 2\delta\Big) \\ &\leqslant \frac{Ca}{b}\exp\Big\{-(1-\tau)\lambda^2\inf_{f\in((U_\alpha)^\delta)^c} I(f)\Big\}, \end{aligned} \tag{3.11}$$

where the following notations are used:

$$F^\delta = \{g \in B : \inf_{f\in F}\|f-g\| < \delta\}$$

and F^c is the complement of F, and C is a positive constant.

Proof. We first show (3.10). For any positive number t and integer $n > 0$, let $t_n = [t2^n]/2^n$. We write

$$\begin{aligned} &P\Big(\sup_{0\leqslant t\leqslant a-b}\inf_{f\in U_\alpha}\Big\|\frac{X(t/b+\cdot)-X(t/b)}{\lambda}-f\Big\|_\infty \geqslant 3\delta\Big) \\ &\leqslant P\Big(\sup_{0\leqslant t\leqslant a-b}\inf_{f\in U_\alpha}\Big\|\frac{X(t_n/b+\cdot)-X(t_n/b)}{\lambda}-f\Big\|_\infty \geqslant \delta\Big) \\ &\quad +2P\Big(\sup_{0\leqslant t\leqslant a-b}\frac{\|X(t/b+\cdot)-X(t_n/b+\cdot)\|_\infty}{\lambda} \geqslant \delta\Big) \\ &=: I_1 + I_2. \end{aligned} \tag{3.12}$$

Let $0 < \tau < \beta$. Take n, such that $2^{2(\beta-\tau)(n+1)} > C_0/(b^{2(\beta-\tau)}\delta^2) \geqslant 2^{2(\beta-\tau)n}$, where C_0 is a positive number and will be specified later. Since $\mathrm{Card}\{t_n, 0 \leqslant t < a\} \leqslant 2^n a$, and $X(\cdot)$ is stationary, by Proposition 2.4 we get

$$\begin{aligned} I_1 &\leqslant 2^n a P\Big(\inf_{f\in U_\alpha}\Big\|\frac{X}{\lambda} - f\Big\|_\infty \geqslant \delta\Big) \\ &\leqslant \frac{C_0^{1/2(\beta-\tau)}a}{b\delta^{1/\beta}}\exp\Big\{-\frac{(\alpha\lambda(1+\delta))^2}{2} + \frac{d_\lambda - 1}{2}\log\frac{(\alpha\lambda(1+\delta))^2 e}{d_\lambda - 1}\Big\} \\ &\leqslant \frac{C_0^{1/2(\beta-\tau)}a}{b}\exp\Big\{-\frac{(\alpha\lambda(1+\delta))^2}{2} - \frac{1}{\beta-\tau} \\ &\qquad + \frac{d_\lambda - 1}{2}\log\frac{(\alpha\lambda(1+\delta))^2 e}{d_\lambda - 1}\Big\}. \end{aligned} \tag{3.13}$$

By Lemma 2.4, we have that for any $\varepsilon > 0$, there exists a positive $K = K(\varepsilon)$ such that

$$P\Big(\sup_{0\leqslant t\leqslant T}\sup_{0\leqslant s\leqslant h}|X(t+s) - X(t)| \geqslant (1+\varepsilon)x\sigma(h)\Big) \leqslant \frac{KT}{h}\exp(-x^2/2) \tag{3.14}$$

for any $T > 0$, $0 < h \leqslant T$ and $x \geqslant x_0$ with some $x_0 > 0$.

Note that $\sup_{0\leqslant t\leqslant a}|t_{n+j+1} - t_{n+j}| \leqslant 2^{-n-j-1}$ and $\mathrm{Card}\{X(t_{n+j+1}) - X(t_{n+j}), 0 \leqslant t \leqslant a\} \leqslant 2^{n+j+1}(a+1)$. Put $D = \frac{1}{3}\sum_{j=0}^\infty\Big(\sqrt{\frac{j}{2^{2(\beta-\tau)j}}} + \frac{1}{\sqrt{2^{2(\beta-\tau)j}}}\Big)$ and $x_j = \frac{1}{D}\sqrt{\frac{j/\lambda^2+1}{2^{2(\beta-\tau)j}}}$. Then, by (3.14) and noting that $\sigma(2^{-n-j-1}b^{-1}) \leqslant 2^{-(\beta-\tau)(n+j+1)}b^{-(\beta-\tau)}$ for large C_0, (here we have used the facts that $2^{n(\beta-\tau)}b^{\beta-\tau} \geqslant C_0^{1/2}$ and that $x^\varepsilon\ell(x) \to 0$ as $\varepsilon \to 0$ for any $\varepsilon > 0$), we get

$$\begin{aligned} I_2 &\leqslant 2P\Big(\bigcup_{0\leqslant t\leqslant a-b}\Big\{\sum_{j=0}^\infty\|X\Big(\frac{t_{n+j+1}}{b}+\cdot\Big) \\ &\qquad - X\Big(\frac{t_{n+j}}{b}+\cdot\Big)\|_\infty \geqslant \delta\lambda\sum_{j=0}^\infty x_j\Big\}\Big) \\ &\leqslant 2\sum_{j=0}^\infty P\Big(\bigcup_{0\leqslant t\leqslant a-b}\Big\{\Big\|X\Big(\frac{t_{n+j+1}}{b}+\cdot\Big) - X\Big(\frac{t_{n+j}}{b}+\cdot\Big)\Big\|_\infty \geqslant \delta\lambda x_j\Big\}\Big) \\ &\leqslant Ca\sum_{j=0}^\infty 2^{n+j+1}P\Big(\Big\|X\Big(\frac{t_{n+j+1}}{b}+\cdot\Big) - X\Big(\frac{t_{n+j}}{b}+\cdot\Big)\Big\|_\infty \geqslant \delta\lambda x_j\Big) \\ &\leqslant Ca\sum_{j=0}^\infty 2^{2(n+j+1)}b\exp\{-C\delta^2\lambda^2 2^{2(\beta-\tau)(n+j+1)}b^{2(\beta-\tau)}x_j^2\} \end{aligned}$$

$$\leqslant \frac{CC_0^{1/(\beta-\tau)}a}{b\delta^{2/(\beta-\tau)}}\sum_{j=0}^{\infty}4^j\exp\Big\{-CC_0\lambda^2\frac{j/\lambda^2+1}{D^2}\Big\}$$

$$\leqslant \frac{CC_0^{1/(\beta-\tau)}a}{b}\exp\Big\{-CC_0\frac{\lambda^2}{D^2}-\frac{2}{(\beta-\tau)}\log\delta\Big\}G, \qquad (3.15)$$

where the almost sure continuity of $X(\cdot)$ is used and $G=\sum_{j=0}^{\infty}4^j\exp\{-\frac{CC_0}{D^2}j\}$. Here, and in the sequel, C stands for an unspecified positive and finite constant whose value may be different in each occurrence. Now, making $C_0>0$ large enough and letting $\tau\downarrow 0$, and combining (3.12), (3.13) and (3.15), we obtain (3.10).

Next we show (3.11). By applying Proposition 2.2, along the same lines of the proof (3.10), we get (3.11) immediately. This completes the proof.

4 Applications to fractional Brownian motion

Throughout this section we let $X=\{X(t),t\geqslant 0\}$ be fractional Brownian motion (FBM) of index α with $0<\alpha<1$, i.e., $\{X(t),t\geqslant 0\}$ is a real-valued Gaussian process with mean zero, stationary increments, $X(0)=0$ and covariance function

$$\mathrm{E}[X(s)X(t)]=\frac{1}{2}\{|s|^{2\alpha}+|t|^{2\alpha}-|s-t|^{2\alpha}\}.$$

It is known that when $\alpha=1/2$, X is Brownian motion (BM). Let U be defined as in Section 3 with $\mu=\mathcal{L}(X)$. Put

$$Y_{t,h}(x)=\frac{X(t+hx)-X(t)}{h^{\alpha}(2\log(h^{-1}))^{1/2}},\ 0\leqslant x\leqslant 1, 0\leqslant t\leqslant 1-h, 0<h<1,$$

and

$$Z_{t,T}(x)=\frac{X(t+a_Tx)-X(t)}{a_T^{\alpha}(2(\log(Ta_T^{-1})+\log\log T))^{1/2}},\ 0\leqslant x\leqslant 1, 0\leqslant t\leqslant T-a_T,$$

where a_T is a positive function of T.

There are a lot of papers in the literature to investigate limit behaviors of FBM. Oodaira (1972), Goodman and Kuelbs (1991) and Monrad and Rootzén (1995) studied functional LIL and its convergence rates for FBM. Ortega (1984) investigated the size of increments for FBM and obtained large increment theorem for FBM. Wang (2001) investigated functional limit theorem for increments of FBM and proved that with probability one

$$\{Y_{t,h}(x), 0 \leqslant x \leqslant 1, 0 < h < 1\} \subset C_0[0,1]$$

is relatively compact in $C_0[0,1]$, and the set of its limit points (as $h \to 0$) is contained in U. But Theorem 4.1 below is the functional modulus of continuity of FBM, which implies that the set of its limit points is U, and so it gives a further result. Theorem 4.2 below is functional large increment theorem for FBM. Theorems 4.1 and 4.2 below generalize the related results for BMs (see Corollaries 4.2 and 4.4 below).

Theorem 4.1. *We have*

$$\lim_{h\to 0} \sup_{0\leqslant t\leqslant 1-h} \inf_{f\in U} \|Y_{t,h} - f\|_\infty = 0 \quad \text{a.s.} \tag{4.1}$$

and, for any $f \in U$,

$$\lim_{h\to 0} \inf_{0\leqslant t\leqslant 1-h} \|Y_{t,h} - f\|_\infty = 0 \quad \text{a.s..} \tag{4.2}$$

Corollary 4.2. *Theorem 4.1 implies Theorem 1.4.*

Theorem 4.3. *Assume that $0 < a_T \leqslant T$ satisfies that both a_T and T/a_T are non-decreasing. Then we have*

$$\lim_{T\to\infty} \sup_{0\leqslant t\leqslant T-a_T} \inf_{f\in U} \|Z_{t,T} - f\|_\infty = 0 \quad \text{a.s.} \tag{4.3}$$

and, for any $f \in U$,

$$\liminf_{T\to\infty} \inf_{0\leqslant t\leqslant T-a_T} \|Z_{t,T} - f\|_\infty = 0 \quad \text{a.s..} \tag{4.4}$$

If we also assume that the following condition is satisfied:

(iii) $\lim_{T\to\infty} \dfrac{\log(T a_T^{-1})}{\log\log T} = r, \quad (0 \leqslant r \leqslant \infty),$

then we also have that

$$\liminf_{T\to\infty} \sup_{0\leqslant t\leqslant T-a_T} \inf_{f\in U_{\alpha(r)}} \|Z_{t,T} - f\|_\infty = 0 \quad \text{a.s.} \tag{4.5}$$

and, for any $f \in U_{\alpha(r)}$,

$$\lim_{T\to\infty} \inf_{0\leqslant t\leqslant T-a_T} \|Z_{t,T} - f\|_\infty = 0 \quad \text{a.s.,} \tag{4.6}$$

where $U_{\alpha(r)}$ is defined as in (3.9).

Corollary 4.4. *Theorem 4.3 implies Theorem 1.7.*

5 Proof of Theorem 3.1

It is enough to show that (3.5)–(3.8) hold. By condition (ii), A_T is either bounded or unbounded. In the bounded case, a_T tends to zero; in the unbounded case, the limit inferior of a_T is larger than zero. Next, we consider only the unbounded case. The bounded case can be dealt with in the same way.

Proof of (3.5). Let $1 < \theta < 2$. Take $T_n = \inf\{T : A_T \geqslant \theta^n\}$, $n \geqslant 1$. For all $T \geqslant 3$, there exists integer n such that $T_{n-1} \leqslant T \leqslant T_n$. Then, without loss of generality, considering $A_T - a_T \leqslant A_{T_n} - a_{T_n}$, (otherwise, we can consider $A_T - a_T \leqslant A_{T_n} - a_{T_{n-1}}$); and considering $a_T \to \infty$, otherwise, if $\lim_{T\to\infty} a_T = a_0$ then, by self-similarity, we need only consider

$$\limsup_{T\to\infty} \sup_{0\leqslant t\leqslant A_T - a_0} \inf_{f\in U} \sup_{0\leqslant x\leqslant 1} \left|\frac{X(t/a_0 + x) - X(t/a_0)}{\beta(T)} - f(x)\right| = 0 \quad \text{a.s.},$$

whose proof can be dealt with in the same way. Write

$$\begin{aligned}
&\limsup_{T\to\infty} \sup_{0\leqslant t\leqslant A_T - a_T} \inf_{f\in U} \|\Delta_{t,T} - f\|_\infty \\
&\leqslant \limsup_{n\to\infty} \sup_{T_{n-1}\leqslant T\leqslant T_n} \sup_{0\leqslant t\leqslant A_{T_n} - a_{T_n}} \inf_{f\in U} \|\Delta_{t,T} - f\|_\infty \\
&\leqslant \limsup_{n\to\infty} \sup_{0\leqslant t\leqslant A_{T_n} - a_{T_n}} \inf_{f\in U} \|\Delta_{t,T_n} - f\|_\infty \\
&\quad + \limsup_{n\to\infty} \sup_{T_{n-1}\leqslant T\leqslant T_n} \sup_{0\leqslant t\leqslant A_{T_n} - a_{T_n}} \|\Delta_{t,T} - \Delta_{t,T_n}\|_\infty \\
&=: J_1 + J_2.
\end{aligned} \tag{5.1}$$

Put

$$\beta(T) := \left(2\left(\log \frac{A_T - a_T}{a_T} + \log\log A_T\right)\right)^{1/2}$$

and for any $\varepsilon \in (0, 1)$, put

$$A_n = \left\{\sup_{0\leqslant t\leqslant A_{T_n} - a_{T_n}} \inf_{f\in U} \|\Delta_{t,T_n} - f\|_\infty \geqslant 2\varepsilon\right\},$$

$$B_n = \left\{\sup_{T_{n-1}\leqslant T\leqslant T_n} \sup_{0\leqslant t\leqslant A_{T_n} - a_{T_n}} \|\Delta_{t,T} - \Delta_{t,T_n}\|_\infty \geqslant \varepsilon\right\}.$$

By self-similarity (3.4) and by taking $\lambda = \beta(T_n)$, $b = a_{T_n}$, $a = A_{T_n} - a_{T_n}$, $\alpha = 1$ and $\delta = \varepsilon$ in (3.11) we get

$$\begin{aligned} P(A_n) &= P\left\{ \sup_{0 \leqslant t \leqslant A_{T_n} - a_{T_n}} \inf_{f \in U} \left\| \frac{X(t/a_{T_n} + \cdot) - X(t/a_{T_n})}{\beta(T_n)} - f \right\|_\infty \geqslant 2\varepsilon \right\} \\ &\leqslant C \frac{A_{T_n}}{a_{T_n}} \exp\left\{ -\eta \beta_{T_n}^2 \right\} \\ &\leqslant C n^{-2\eta}, \end{aligned} \tag{5.2}$$

where

$$\eta =: (1-\tau) \inf_{f \in ((U^\varepsilon)^c)} I(f), \quad 0 < \tau < 1.$$

By choosing $\tau > 0$ small enough we have

$$\eta > 1/2.$$

Hence,

$$\sum_{n=1}^{\infty} P(A_n) < \infty,$$

which by the Borel-Cantelli lemma implies that

$$P\left(\limsup_{n \to \infty} A_n \right) = 0. \tag{5.3}$$

Therefore we obtain

$$J_1 = 0 \quad \text{a.s..} \tag{5.4}$$

since $\varepsilon \in (0,1)$ is arbitrary.

To estimate J_2. Note that $1 \leqslant A_{T_{n+1}}/A_{T_n} \leqslant \theta^{n+2}/\theta^n = \theta^2$ and that $1 \leqslant a_{T_{n+1}}/a_{T_n} \leqslant A_{T_{n+1}}/A_{T_n} \leqslant \theta^2$. Then for $T_{n-1} \leqslant T \leqslant T_n$ and $n \geqslant n_0$ with some integer $n_0 > 0$, it is easy to see that $1 \leqslant \sigma(a_{T_n})/\sigma(a_T) \leqslant \theta^{3\beta}$ (since $a_T \to \infty$) and $0 \leqslant \beta^2(T_n) - \beta^2(T) \leqslant 3 \log \theta$. Thus

$$\frac{\sigma(a_{T_n})\beta(T_n)}{\sigma(a_T)\beta(T)} - 1 \leqslant \theta^{3\beta}\sqrt{1 + 3\log\theta} - 1 =: \eta_0.$$

Write

$$
\begin{aligned}
&B_n \\
&\subset \Big\{ \sup_{T_{n-1}\leqslant T\leqslant T_n} \sup_{0\leqslant t\leqslant A_{T_n}-a_{T_n}} \sup_{0\leqslant x\leqslant 1} \frac{|X(t+a_T x)-X(t+a_{T_n}x)|}{\sigma(a_{T_n})\beta_{T_n}} \geq \varepsilon/2 \Big\} \\
&\quad \bigcup \Big\{ \sup_{T_{n-1}\leqslant T\leqslant T_n} \sup_{0\leqslant t\leqslant A_{T_n}-a_{T_n}} \sup_{0\leqslant x\leqslant 1} \Big(\Big| \frac{\sigma(a_{T_n})\beta(T_n)}{\sigma(a_T)\beta(T)} - 1 \Big| \Big) \\
&\qquad\qquad \times \frac{|X(t+a_T x)-X(t)|}{\sigma(a_{T_n})\beta(T_n)} \geq \varepsilon/2 \Big\} \\
&\subset \Big\{ \sup_{0\leqslant t\leqslant A_{T_n}} \sup_{0\leqslant s\leqslant a_{T_n}-a_{T_{n-1}}} \frac{|X(t+s)-X(t)|}{\sigma(a_{T_n})\beta_{T_n}} \geq \varepsilon/2 \Big\} \\
&\quad \bigcup \Big\{ \sup_{0\leqslant t\leqslant A_{T_n}-a_{T_n}} \sup_{0\leqslant s\leqslant a_{T_n}} \frac{|X(t+a_T x)-X(t)|}{\sigma(a_{T_n})\beta(T_n)} \geq \varepsilon/(2\eta_0) \Big\} \\
&=: B_{n,1} \bigcup B_{n,2}.
\end{aligned}
\tag{5.5}
$$

By using self-similarity and Lemma 2.5, we get that for any $\mu \in (0,1)$

$$
\begin{aligned}
&P(B_{n,1}) \\
&= P\Big(\sup_{0\leqslant t\leqslant A_{T_n}/a_{T_n}} \sup_{0\leqslant s\leqslant (a_{T_n}-a_{T_{n-1}})/a_{T_n}} \frac{|X(t+s)-X(t)|}{\beta_{T_n}} \geq \varepsilon/2 \Big) \\
&\leqslant P\Big(\sup_{0\leqslant t\leqslant A_{T_n}/a_{T_n}} \sup_{0\leqslant s\leqslant 1-\theta^{-2}} \frac{|X(t+s)-X(t)|}{\beta_{T_n}} \geq \varepsilon/2 \Big) \\
&\leqslant C \frac{A_{T_n}}{a_{T_n}} \exp\Big\{ -\frac{(1-\mu)\varepsilon^2}{4\sigma^2(1-\theta^2)} \beta_{T_n}^2 \Big\} \\
&\leqslant C \Big(\frac{A_{T_n}}{a_{T_n}} \Big)^{1-2\eta_1} n^{-2\eta_1},
\end{aligned}
\tag{5.6}
$$

where

$$\eta_1 = \frac{(1-\mu)\varepsilon^2}{4\sigma^2(1-\theta^2)}.$$

Choosing $\theta > 1$ close one enough such that $\eta_1 > 1/2$, via (5.6) and Borel-Cantelli lemma we get

$$P\left(\limsup_{n\to\infty} B_{n,1} \right) = 0. \tag{5.7}$$

Similar to the proof of (5.7), we can get easily (by letting $\theta \downarrow 1$) that

$$P\left(\limsup_{n\to\infty} B_{n,2} \right) = 0. \tag{5.8}$$

Combining (5.5), (5.7) and (5.8) we obtain

$$J_2 = 0 \quad \text{a.s.} \tag{5.9}$$

since $\varepsilon \in (0,1)$ is arbitrary. Therefore, by (5.4), (5.9) and (5.1), (3.5) follows.

Proof of (3.6). For $n \geqslant 1$, let

$$T_n = \begin{cases} \inf\{T : A_T/a_T \geqslant n^n\}, & \text{if } \rho = \infty, \\ \inf\{T : a_T \geqslant n^n\}, & \text{if } \rho < \infty, \end{cases}$$

where $\lim A_T/a_T = \rho$. Then, it is sufficient to show that

$$\lim_{n\to\infty} \inf_{0\leqslant t\leqslant A_{T_n}-a_{T_n}} \|\Delta_{t,T_n} - f\|_\infty = 0 \quad \text{a.s..} \tag{5.10}$$

We only consider the case of $\rho = \infty$. The proof of the case of $\rho < 1$ is similar and so omitted. In case of $\rho = \infty$, we have $\dfrac{A_{T_{n-1}}}{a_{T_{n-1}}} / \dfrac{A_{T_n}}{a_{T_n}} \leqslant (n-2)^{(n-2)}/n^n \to 0$. Let $t_k = kD$, $k = 0, 1, \cdots, \rho_n := [A_{T_n}/(Da_{T_n})]$, where $D > 3$ is a sufficiently large number, which will be specified later, and let

$$V_k(x) = X(t_k + x) - X(t_k), \quad k \geqslant 0, 0 \leqslant x \leqslant 1.$$

Then, by (3.4), it is enough to prove that

$$\lim_{n\to\infty} \min_{\rho_{n-1}<k\leqslant\rho_n} \left\| \frac{V_k}{\beta(T_n)} - f \right\|_\infty = 0 \quad \text{a.s.,} \tag{5.11}$$

where $\beta(T)$ is defined as above.

It follows from (3.1) that for $0 < h < 1$,

$$\sigma^2(h) = 2\int_{\mathbb{R}} (1 - \cos(h\lambda))\Delta(d\lambda). \tag{5.12}$$

We have

$$\begin{aligned} \int_{|\lambda|\geqslant 1/h} \Delta(d\lambda) &\leqslant \frac{1}{1-\sin 1}\int_{|\lambda|\geq 1/h} \Big(1 - \frac{\sin(h\lambda)}{h\lambda}\Big)\Delta(d\lambda) \\ &= \frac{1}{(1-\sin 1)h}\int_{|\lambda|\geq 1/h}\int_0^h (1-\cos(u\lambda))du\Delta(d\lambda) \\ &\leqslant \frac{1}{(1-\sin 1)h}\int_0^h\int_{\mathbb{R}} (1-\cos(u\lambda))\Delta(d\lambda)du \\ &\leqslant 4\sigma^2(h). \end{aligned} \tag{5.13}$$

Similarly, by (5.12)

$$\begin{aligned} \int_{|\lambda|\leqslant 1/h} |\lambda|^2\Delta(d\lambda) &\leqslant 4h^{-2}\int_{|\lambda|\leqslant 1/h} (1-\cos(h\lambda))\Delta(d\lambda) \\ &\leqslant 4h^{-2}\sigma^2(h). \end{aligned} \tag{5.14}$$

For $0 < \varepsilon < 1$ and $0 \leqslant k \leqslant \rho_n$, put

$$s_k = s(k) = \exp(-((k+3)D)^{1+\varepsilon}/2),$$

$$d_k = \exp\{((k+3)D)^{1+\varepsilon} + ((k+3)D)^{\varepsilon}\}.$$

For $k = 0, 1, \cdots$ and $0 \leqslant t \leqslant 1$, define

$$X^{(k)}(t) = \int_{|\lambda| \in (d_{k-1}, d_k]} (e^{it\lambda} - 1) W(d\lambda),$$

$$\tilde{X}^{(k)}(t) = \int_{|\lambda| \notin (d_{k-1}, d_k]} (e^{it\lambda} - 1) W(d\lambda),$$

$$\tilde{Y}^{(k)}(h,t) = \int_{|\lambda| \notin (hd_{k-1}, hd_k]} (e^{iht\lambda} - 1) W(d\lambda), \quad \tilde{Y}^{(k)}(s_k, t) =: \tilde{Y}^{(k)}(t).$$

Clearly,

$$X^{(k)}(\cdot),\ k = 0, 1, \ldots \text{ are independent}, \tag{5.15}$$

$$X(t) = X^{(k)}(t) + \tilde{X}^{(k)}(t). \tag{5.16}$$

and, by (3.4), for any $h > 0$,

$$\sigma(h) \tilde{X}^{(k)}(\cdot) \stackrel{d}{=} \tilde{Y}^{(k)}(h, \cdot). \tag{5.17}$$

By standard Borel-Cantelli arguments. (5.11) follows if we prove that

$$\sum^{\infty} P\Big(\min_{\rho_{n-1} < k \leqslant \rho_n} \Big\| \frac{X^{(k)}(t_k + \cdot) - X^{(k)}(t_k)}{\beta(T_n)} - f \Big\|_{\infty} \leqslant \eta \Big) = \infty \tag{5.18}$$

and

$$\sum^{\infty} P\Big(\max_{0 \leqslant k \leqslant \rho_n} \Big\| \frac{\tilde{X}^{(k)}(t_k + \cdot) - \tilde{X}^{(k)}(t_k)}{\beta(T_n)} \Big\|_{\infty} \geqslant \eta \Big) < \infty \tag{5.19}$$

for any $\eta > 0$ since the events in probabilities in (5.18) are independent.

From (5.13) and (5.14), we obtain that for $k \geqslant 0$, $0 \leqslant t \leqslant 1$ and for some $0 < \mu < 1 - \beta$ and D big enough,

$$\begin{aligned} \operatorname{Var}(\tilde{Y}^{(k)}(t)) &= 2 \int_{|\lambda| \notin (s_k d_{k-1}, s_k d_k]} (1 - \cos(s_k t \lambda)) \Delta(d\lambda) \\ &\leqslant \int_{|\lambda| \leqslant s_k d_{k-1}} s_k^2 t^2 |\lambda|^2 \Delta(d\lambda) + 4 \int_{|\lambda| \geq s_k d_k} \Delta(d\lambda) \\ &\leqslant 4 s_k^4 d_{k-1}^2 \sigma^2(s_k/(s_k^2 d_{k-1})) + 16 \sigma^2(s_k/(s_k^2 d_k)) \\ &\leqslant 16 (s_k^2 d_{k-1})^{2\mu} \sigma^2(s_k) + 16 \sigma^2(s_k/(s_k^2 d_k)) \\ &\leqslant 32 \exp(-C(kD)^{\varepsilon}) \sigma^2(s_k) \\ &\leqslant 32 \exp(-C D^{\varepsilon}) \sigma^2(s_k). \end{aligned} \tag{5.20}$$

Therefore, for every $0 \leqslant s, t \leqslant 1$, $|s-t| \leqslant h \leqslant 1$,

$$\operatorname{Var}(\tilde{Y}^{(k)}(s) - \tilde{Y}^{(k)}(t)) \leqslant \tilde{\sigma}^{(k)}(h)^2,$$

where $\tilde{\sigma}^{(k)}(h)^2 = \min\left\{\sigma^2(h), 64\exp(-CD^{\varepsilon})\sigma^2(s_k)\right\}$.

By (5.17) and Lemma 2.5, we get easily

$$\begin{aligned}
&P\Big(\max_{0\leqslant k\leqslant \rho_n}\Big\|\frac{\tilde{X}^{(k)}(t_k+\cdot) - \tilde{X}^{(k)}(t_k)}{\beta(T_n)}\Big\|_\infty \geqslant \eta\Big) \\
&\leqslant \sum_{k=0}^{\rho_n} P\Big(\Big\|\frac{\tilde{Y}^{(k)}(t_k+\cdot) - \tilde{Y}^{(k)}(t_k)}{\sigma(s_k)\beta(T_n)}\Big\|_\infty \geqslant \eta\Big) \\
&\leqslant \frac{CA_{T_n}^2}{a_{T_n}^2}\exp\Big(-\frac{\eta^2\beta^2(T_n)}{128\exp(-CD^{\varepsilon})}\Big) \\
&\leqslant Cn^{-2}
\end{aligned} \tag{5.21}$$

by choosing D large enough. This proves (5.19).

By (2.4) and Jensen's inequality, we have

$$\begin{aligned}
&\rho_n P(\|X - f\beta(T_n)\|_\infty \leqslant \eta\beta(T_n)) \\
&= \rho_n \exp(-\tfrac{1}{2}\|f\|_\mu^2\beta^2(T_n)) \int_{\{x\in C_0[0,1]:\|x\|_\infty\leqslant \eta\beta(T_n)\}} \exp(-<x,f>^\sim)d\mu(x) \\
&\geqslant \rho_n \exp(-\tfrac{1}{2}\beta^2(T_n))P(\|X\|_\infty \leqslant \eta\beta(T_n)) \\
&\geqslant C(n\log n)^{-1}.
\end{aligned} \tag{5.22}$$

For notation simplicity only, we let

$$\begin{aligned}
\xi_1(n,k,\cdot) &= (X^{(k)}(t_k+\cdot) - X^{(k)}(t_k))/\beta(T_n), \\
\xi_2(n,k,\cdot) &= (\tilde{X}^{(k)}(t_k+\cdot) - \tilde{X}^{(k)}(t_k))/\beta(T_n).
\end{aligned} \tag{5.23}$$

Thus, by noting $\rho_n - \rho_{n-1} \geqslant \rho_n/2$ for large n, (5.22) and (5.21),

$$\begin{aligned}
&\sum_{k=\rho_{n-1}+1}^{\rho_n} P(\|\xi_1(n,k,\cdot) - f\|_\infty \leqslant \eta/2) \\
&\geqslant \sum_{k=\rho_{n-1}+1}^{\rho_n} P\Big(\Big\|\frac{X(t_k+\cdot) - X(t_k)}{\beta(T_n)} - f\Big\|_\infty \leqslant \eta/4\Big) \\
&\qquad - \sum_{k=\rho_{n-1}+1}^{\rho_n} P(\|\xi_2(n,k,\cdot)\|_\infty \geq \eta/4) \\
&= (\rho_n/2)P(\|X - f\beta(T_n)\|_\infty \leqslant \eta\beta(T_n)/4) \\
&\qquad - \sum_{k-\rho_{n-1}+1}^{\rho_n} P(\|\xi_2(n,k,\cdot)\|_\infty \geq \eta/4) \\
&\geqslant C(n\log n)^{-1}.
\end{aligned}$$

Note that (by (5.15)),

$$\begin{aligned}
&P\Big(\min_{\rho_{n-1}+1\leqslant k\leqslant\rho_n}\|\xi_1(n,k,\cdot)-f\|_\infty\leqslant\eta\Big)\\
&=1-\prod_{k=\rho_{n-1}+1}^{\rho_n}\Big(1-P(\|\xi_1(n,k,\cdot)-f\|_\infty\leqslant\eta/2)\Big)\\
&\geq 1-\prod_{k=\rho_{n-1}+1}^{\rho_n}\exp\Big(-P(\|\xi_1(n,k,\cdot)-f\|_\infty\leqslant\eta/2)\Big)\\
&=1-\exp\Big(-\sum_{k=\rho_{n-1}+1}^{\rho_n}P(\|\xi_1(n,k,\cdot)-f\|_\infty\leqslant\eta/2)\Big)\\
&\geqslant 1-\exp(C(n\log n)^{-1})\\
&\geqslant C(n\log n)^{-1},
\end{aligned}$$

where for obtaining the last inequality we have used the following inequality: $1-e^{-x}\geqslant x/2$ for any $0\leqslant x\leqslant 1/5$. Then, by (5.21),

$$\begin{aligned}
&P\Big(\min_{\rho_{n-1}+1\leqslant k\leqslant\rho_n}\Big\|\frac{X(t_k+\cdot)-X(t_k)}{\beta(T_n)}-f\Big\|_\infty\leqslant\eta\Big)\\
&\geqslant P\Big(\min_{\rho_{n-1}+1\leqslant k\leqslant\rho_n}\|\xi_1(n,k,\cdot)-f\|_\infty\leqslant\eta/2\Big)\\
&\quad -P\Big(\max_{\rho_{n-1}+1\leqslant k\leqslant\rho_n}\|\xi_2(n,k,\cdot)\|_\infty\geq\eta/2\Big)\\
&\geqslant C(n\log n)^{-1}-\sum_{k=\rho_{n-1}+1}^{\rho_n}P(\|\xi_2(n,k,\cdot)\|_\infty\geq\eta/2)\\
&\geqslant C(n\log n)^{-1}.
\end{aligned}$$

Hence, we obtain

$$\sum^\infty P\Big(\min_{\rho_{n-1}+1\leqslant k\leqslant\rho_n}\Big\|\frac{X(t_k+\cdot)-X(t_k)}{\beta(T_n)}-f\Big\|_\infty\leqslant\eta\Big)=\infty,$$

which, together with (5.16) and (5.19), implies that (5.18) holds. This proves (3.6).

Proof of (3.7). For $n\geqslant 1$, take $T_n=\inf\{T: A_T\geqslant e^{e^n}\}$. It is enough to show that for any $\varepsilon>0$

$$\limsup_{n\to\infty}\sup_{0\leqslant t\leqslant A_{T_n}-a_{T_n}}\inf_{f\in U_{\alpha(r)}}\|\Delta_{t,T_n}-f\|_\infty<\varepsilon\quad\text{a.s.}. \tag{5.24}$$

For any $\tau>0$, by Proposition 3.2, for n large enough we have

$$\begin{aligned}
&P\Big(\sup_{0\leqslant t\leqslant A_{T_n}-a_{T_n}}\inf_{f\in U_{\alpha(r)}}\|\Delta_{t,T_n}-f\|\geqslant 2\varepsilon\Big)\\
&\leqslant\frac{A_{T_n}}{a_{T_n}}\exp\Big\{-2(1-\tau)\inf_{f\in((U_{\alpha(r)})^\varepsilon)^c}I(f)\log\frac{A_{T_n}\log A_{T_n}}{a_{T_n}}\Big\}\\
&=\frac{A_{T_n}}{a_{T_n}}\Big(\frac{a_{T_n}}{A_{T_n}\log A_{T_n}}\Big)^{2(1-\tau)\inf_{f\in((U_{\alpha(r)})^\varepsilon)^c}I(f)}.
\end{aligned}$$

Taking τ small enough, there exists some $\delta > 0$ such that

$$2(1-\tau) \inf_{f\in((U_{\alpha(r)})^{\varepsilon})^c} I(f) > (1+\delta)\alpha(r).$$

Hence

$$\begin{aligned} &P\Big(\sup_{0\leqslant t\leqslant A_{T_n}-a_{T_n}} \inf_{f\in U_{\alpha(r)}} \|\Delta_{t,T_n} - f\|_\infty \geqslant \varepsilon\Big) \\ &\leqslant C\Big(\frac{A_{T_n}}{a_{T_n}}\Big)^{1-(1+\delta)\alpha(r)} \Big(\frac{1}{\log A_{T_n}}\Big)^{(1+\delta)\alpha(r)}. \end{aligned}$$

If $1-(1+\delta)\alpha(r) < 0$, then

$$\begin{aligned} &P\Big(\sup_{0\leqslant t\leqslant A_{T_n}-a_{T_n}} \inf_{f\in U_{\alpha(r)}} \|\Delta_{t,T_n} - f\|_\infty \geq \varepsilon\Big) \\ &\leqslant C\Big(\frac{1}{\log A_{T_n}}\Big)^{(1+\delta)\alpha(r)} \\ &\leqslant Ce^{-(1+\delta)\alpha(r)n}, \end{aligned}$$

otherwise, noting that for n large enough and $\mu > 0$

$$\frac{A_{T_n}}{a_{T_n}} \leqslant (\log A_{T_n})^{r+\mu}$$

from the condition (iii), we have

$$\begin{aligned} &P\Big(\sup_{0\leqslant t\leqslant A_{T_n}-a_{T_n}} \inf_{f\in U_{\alpha(r)}} \|\Delta_{t,T_n} - f\|_\infty \geq \varepsilon\Big) \\ &\leqslant C\Big(\frac{1}{\log A_{T_n}}\Big)^{-(r+\mu)+(r+\mu)(1+\delta)\alpha(r)+(1+\delta)\alpha(r)} \\ &\leqslant Ce^{(\mu-\delta r-\mu(1+\delta)\alpha(r))n}. \end{aligned}$$

Therefore, taking $\mu < \delta r$, we obtain

$$\sum^{\infty} P\Big(\sup_{0\leqslant t\leqslant A_{T_n}-a_{T_n}} \inf_{f\in U_{\alpha(r)}} \|\Delta_{t,T_n} - f\|_\infty \geq \varepsilon\Big) < \infty,$$

which, via Borel-Cantelli lemma, implies (5.24).

Proof of (3.8). For $\theta > 1$, take $T_n = \inf\{T : A_T \geqslant \theta^n\}$. For all $T \geqslant 3$, there exists integer n such that $T_n \leqslant T \leqslant T_{n+1}$. Then, without loss of generality, considering $A_{T_n} - a_{T_n} \leqslant A_T - a_T$, (otherwise, we can consider $A_{T_n} - a_{T_{n+1}} \leqslant A_T - a_T$), for all $f \in U_{\alpha(r)}$ we have

$$\begin{aligned} &\inf_{0\leqslant t\leqslant A_T-a_T} \|\Delta_{t,T} - f\|_\infty \\ &\leqslant \inf_{0\leqslant t\leqslant A_{T_n}-a_{T_n}} \|\Delta_{t,T_n} - f\|_\infty + \sup_{T_n\leqslant T\leqslant T_{n+1}} \sup_{0\leqslant t\leqslant A_{T_n}-a_{T_n}} \|\Delta_{t,T_n} - \Delta_{t,T}\|_\infty \\ &\leqslant \min_{0\leqslant k\leqslant \rho_n} \|\Delta_{t_k a_{T_n},T_n} - f\|_\infty + \sup_{T_n\leqslant T\leqslant T_{n+1}} \sup_{0\leqslant t\leqslant A_{T_n}-a_{T_n}} \|\Delta_{t,T_n} - \Delta_{t,T}\|_\infty \\ &=: J_n^{(1)} + J_n^{(2)}, \end{aligned} \tag{5.25}$$

where t_k, ρ_n be defined as in the proof of (3.6).

Note that $1 \leqslant A_{T_{n+1}}/A_{T_n} \leqslant \theta^{n+2}/\theta^n = \theta^2$ and that $1 \leqslant a_{T_{n+1}}/a_{T_n} \leqslant A_{T_{n+1}}/A_{T_n} \leqslant \theta^2$. By applying (3.14) and standard Borel-Cantelli arguments, it is easy to prove that (by letting $\theta \downarrow 1$)

$$\limsup_{n\to\infty} J_n^{(2)} = 0 \quad \text{a.s..}$$

Therefore, to conclude the desired result, it is sufficient to show that

$$\limsup_{n\to\infty} J_n^{(1)} = 0 \quad \text{a.s..} \tag{5.26}$$

Note that (by (3.4))

$$\begin{aligned}
&P\Big(J_n^{(1)} \geqslant \varepsilon\Big) \\
&\leqslant \prod_{k=0}^{\rho_n} P\Big(\|\xi_1(n,k,\cdot) - f\|_\infty \geq \varepsilon/2\Big) + P\Big(\max_{0\leqslant k\leqslant \rho_n} \|\xi_2(n,k,\cdot)\|_\infty \geq \varepsilon/2\Big) \\
&\leqslant \prod_{k=0}^{\rho_n} \Big\{P(\|X - f\beta(T_n)\|_\infty \geq \varepsilon\beta(T_n)/4) + P(\|\xi_2(n,k,\cdot)\|_\infty \geq \varepsilon/4)\Big\} \\
&\quad + P\Big(\max_{0\leqslant k\leqslant \rho_n} \|\xi_2(n,k,\cdot)\|_\infty \geq \varepsilon/2\Big) \\
&=: J_n^{(3)} + J_n^{(4)},
\end{aligned} \tag{5.27}$$

where $\xi_1(n,i,\cdot), \xi_2(n,i,\cdot)$ are defined as in (5.23).

From (5.21), we get

$$\sum^{\infty} J_n^{(4)} < \infty. \tag{5.28}$$

Similarly to the arguments in (5.21) and (5.22), we get that for $0 < \delta < \alpha(r)$

$$\begin{aligned}
&P(\|X - f\beta(T_n)\|_\infty \geqslant \eta\beta(T_n)) \\
&\leqslant \exp\{-P(\|X - f\beta(T_n)\|_\infty \leqslant \eta\beta(T_n))\} \\
&\leqslant \exp\{-\exp(-\tfrac{1}{2}(\alpha(r) - \delta)\beta^2(T_n))\}
\end{aligned}$$

and

$$P\Big(\Big\|\frac{\tilde{X}^{(k)}(t_k + \cdot) - \tilde{X}^{(k)}(t_k)}{\beta(T_n)}\Big\|_\infty \geq \varepsilon\Big) \leqslant \exp\{-C\exp(CD^\varepsilon)\beta^2(T_n)\}.$$

Since, by condition (iii), for n large enough and $\mu > 0$

$$\frac{A_{T_n}}{a_{T_n}} > (\log A_{T_n})^{r-\mu},$$

then, it follows that for n large enough

$$\begin{aligned} J_n^{(3)} &\leqslant C\exp\{-\rho_n\exp(-\tfrac{1}{2}(\alpha(r)-\delta)\beta^2(T_n))\} \\ &\leqslant C\exp\Big\{-\Big(\frac{A_{T_n}}{a_{T_n}}\Big)^{1-\alpha(r)+\delta}\Big(\frac{1}{\log A_{T_n}}\Big)^{\alpha(r)-\delta}\Big\} \\ &\leqslant C\exp\Big\{-\Big(\frac{1}{\log A_{T_n}}\Big)^{-(r-\mu)(1-\alpha(r)+\delta)+\alpha(r)-\delta}\Big\} \\ &\leqslant C\exp\Big\{-(n\log\theta)^{\delta(r+1)\alpha(r)+\delta-\mu(\delta+1-\alpha(r))}\Big\}. \end{aligned}$$

Hence, taking $\mu<\delta/(\delta+1)$, it is obtained that

$$\sum^{\infty} J_n^{(3)}<\infty. \tag{5.29}$$

Therefore, combining (5.27) – (5.29), we obtain

$$\sum^{\infty} P(J_n^{(1)}\geqslant\varepsilon)<\infty,$$

which, via Borel-Cantelli lemma, implies that (5.26) holds. This proves (3.8). Now, the proof of Theorem 3.1 is completed.

6 Proof of Theorems 4.1 and 4.3

It is easy to see that the condition of Theorem 3.1 is satisfied for FBM. Hence, Theorem 4.1 follows from Theorem 3.1 with $A_T=1$ and $a_T=h$, while Theorem 4.3 follows from Theorem 3.1 with $A_T=T$.

References

[1] R. Azencott, Grandes deviations et applications, Lecture Notes in Math. 774, 1–249, Springer-verlag, New York, 1980.

[2] E. Bolthausen, On the speed of convergence in Strassen's law of the iterated logarithm, Ann. Probab. 6(1978), 668–672.

[3] A. H. C. Chan, M. Csörgő and P. Révész, Strassen type limit points for moving averages of a Wiener process, Canad. J. Stat. 6(1978), 57–75.

[4] B. Chen, Ph. D. Dissertation, Carleton Univ., Ottawa, Canada, 1998.

[5] B. Chen and M. Csörgő, A functional modulus of continuity for a Wiener process, Stat. Probab. Lett. 51(2001), 215–223.

[6] Y. K. Choi, Z. Lin and W. Wang, Strassen-type laws of the iterated logarithm for a fractional Brownian sheet, Stoch. Analysis Appl. 24(2004).

[7] E. Csáki, M. Csörgő and Q. M. Shao, Fernique type inequalities and moduli of continuity for l^2-valued Ornstein-Uhlenbeck processes, Ann. Inst. Henri. Poincaré Probab. Statist. 28(1992), 479–517.

[8] M. Csörgő, and P. Révész, (1979), How big are the increments of a Wiener process, Ann. Probab. 7(1979), 731–737.

[9] M. Csörgő, and P. Révész, Strong Approxiamtions in Probability and Statistics, Academic Press, New York, 1981.

[10] M. Csörgő and Q. M. Shao, Strong limit theorems for large and small increments of l^p-valued Gaussian processes, Ann. Probab. 21(1993), 1958–1990.

[11] A. de Acosta, Small deviations in the functional central limit theorem with applications to functional laws of the iterated logarithm, Ann. Probab. 11(1983), 78–101.

[12] V. Goodman and J. Kuelbs, Rate of clustering for some Gaussian self-similar processes, Probab. Th. Rel. Fields 88(1991), 47–75.

[13] K. Grill, On the rate of convergence in Strassen's LIL, Probab. Th. Rel. Fields 74(1987), 583–589.

[14] K. Grill, A liminf result in Strassen's law of the iterated logarithm, Probab. Th. Rel. Fields 89(1991), 149–157.

[15] L. Gross, Lectures in modern analysis and applications II, Lecture Notes in Math. Vol 140, Springer, Berlin, 1970.

[16] N. C. Jain, An introduction to large deviations, Lecture Notes in Math. 1153, 273–296, 1984.

[17] J. Kuelbs, A strong convergence theorem for Banach space valued random variables, Ann. Probab. 4(1976), 744–771.

[18] J. Kuelbs, W. V. Li and M. Talagrand, Lim inf results for Gaussian samples and Chung's functional LIL, Ann. Probab. 22(1994), 1879–1903.

[19] K. S. Hang, Z. Lin and W. Wang, The fuctional limit theorems for fractional Brownian motion in Holder norm, Acta math. Sinica, English Series 22(2006), 1767–1780.

[20] Z. Lin and C. Lu, Strong Limit Theorems, Kluwer Academic Publishers and Science Presess, Dordrecht-Beijing, 1992.

[21] Z. Lin, C. Lu and L. X. Zhang, Path Properties of Gaussian Processes, Kluwer Academic Publishers and Science Presess, Dordrecht-Beijing, and Zhejiang Univ. Press, Hangzhou, 2001.

[22] D. Monrad and H. Rootzén, (1995). Small values of Gaussian processes and functional laws of the iterated logarithm, Probab. Th. Rel. Fields 101(1995), 173–192.

[23] C. Mueller, A unification of Strassen's law and Lévy's modulus of continuity, Z. Wahrsch. Verw. Geb. 56(1981), 163–179.

[24] H. Oodaira, On Strassen's version of the law of the iterated for Gaussian processes, Z. Wahrsch. Verw. Geb. 21(1972), 289–299.

[25] J. Ortega, On the size of the increments of non-stationary Gaussian processes, Stoch. Process. Appl. 18(1984), 47–56.

[26] P. Révész, A generalization of Strassen's functional law of the iterated logarithm, Z. Wahrsch. Verw. Geb. 50(1979), 257–264.

[27] M. Schilder, Some asymptotic formulae for Wiener integrals, Trans. Am. Math. Soc. 125(1966), 63–85.

[28] V. Strassen, An invariance principle for the law of the iterated logarithm, Z. Wahrsch. Verw. Geb. 3(1964), 211–226.

[29] W. Wang, On a functional limit result of increments of a fractional Brownian motion, Acta Math. Hug. 93(2001), 153–170.

[30] W. Wang, The exact rates of convergence of the functional Csörgő-Révész increment theorem of a Wiener process, Acta Math. Sinica, English Series 18(2002a), 727–736.

[31] W. Wang, The exact rates of convergence of the functional modulus of continuity of a Wiener process, J. Math. Res. Exposit. 22(2002b), 507–524.

[32] W. Wang, How big are the Csörgő-Révész increments of two-parameter Wiener processes, Science in China 34(2004), 499–512.

[33] W. Wang, Functional limit theorems for increments of Gaussian samples, J. Theo. Probab. 18(2005), 327–343.

[34] Q. C. Wei, Ph. D. Dissertation, Zhejiang Univ., Hangzhou, China, 2000.

Strong Local Nondeterminism and Sample Path Properties of Gaussian Random Fields*

Yimin Xiao

Department of Statistics and Probability, Michigan State University East Lansing, MI 48824, USA

E-mail: xiao@stt.msu.edu

Abstract

We provide sufficient conditions for a real-valued Gaussian random field $X = \{X(t), t \in \mathbb{R}^N\}$ to be strongly locally nondeterministic. As applications, we establish small ball probability estimates, Hausdorff measure of the sample paths, sharp Hölder conditions and tail probability estimates for the local times of a large class of Gaussian random fields.

1 Introduction and definitions of local nondeterminism

In recent years, many authors have applied Gaussian processes and fields with stationary increments as stochastic models in telecommunications, turbulence, image processing, geostatistics, hydrology and finance. See, for example, Addie et al. (1999), Anh et al. (1999), Mannersalo and Norros (2002), Mueller and Tribe (2002), Bonami and Estrade (2003), Cheridito (2004), Benson et al. (2006), Biermé et al. (2007). These applications have raised many interesting theoretical questions about Gaussian random fields in general.

One of the major difficulties in studying the probabilistic, analytic or statistical properties of Gaussian random fields is the complexity of their dependence structures. As a result, many of the existing tools from theories on Brownian motion, Markov processes and martingales fail for Gaussian random fields; and one often has to use general principles for Gaussian processes or to develop new tools. In this paper, we show that in many circumstances, the properties of *local nondeterminism* can help us to overcome this difficulty so that many elegant and deep results of Brownian motion (and Markov processes) can be extended to Gaussian random fields.

*Research partially supported by the NSF grant DMS-0404729.

This paper is a companion to Xiao (2006), where several results in this paper have been announced without proof. Further developments on other forms of local nondeterminism, especially for anisotropic Gaussian random fields, and their applications have been given in Khoshnevisan and Xiao (2004), Khoshnevisan, Wu and Xiao (2006), Wu and Xiao (2006), Ayache, Wu and Xiao (2006) and Xiao (2006b).

The concept of local nondeterminism (LND, in short) of a Gaussian process was first introduced by Berman (1973) to unify and extend his methods for studying the existence and joint continuity of local times of Gaussian processes. Let $X = \{X(t),\, t \in \mathbb{R}_+\}$ be a separable Gaussian process with mean 0 and let $J \subset \mathbb{R}_+$ be an interval. Assume that $\mathbb{E}[X(t)^2] > 0$ for all $t \in J$ and there exists $\delta > 0$ such that

$$\sigma^2(s,t) = \mathbb{E}\big[(X(s) - X(t))^2\big] > 0$$

for all $s,\ t \in J$ with $0 < |s - t| < \delta$. Recall from Berman (1973) that X is called *locally nondeterministic* on J if for every integer $m \geqslant 2$,

$$\lim_{\varepsilon \to 0} \inf_{t_m - t_1 \leqslant \varepsilon} V_m > 0, \tag{1.1}$$

where V_m is the relative prediction error:

$$V_m = \frac{\mathrm{Var}\big(X(t_m) - X(t_{m-1}) | X(t_1), \cdots, X(t_{m-1})\big)}{\mathrm{Var}\big(X(t_m) - X(t_{m-1})\big)}$$

and the infimum in (1.1) is taken over all ordered points $t_1 < t_2 < \cdots < t_m$ in J with $t_m - t_1 \leqslant \varepsilon$.

This definition of LND was extended by Cuzick (1978) who defined local ϕ-nondeterminism by replacing the variance $\sigma^2(t_m, t_{m-1})$ by $\phi(t_m - t_{m-1})$, where $\phi : \mathbb{R}_+ \to \mathbb{R}_+$ is an arbitrary function with $\phi(0) = 0$. It follows from Berman (1973, Lemma 2.3) that (1.1) is equivalent to the following property: For every integer $m \geqslant 2$, there exist positive constants c_m and ε (both may depend on m) such that

$$\mathrm{Var}\bigg(\sum_{k=1}^m u_k \left(X(t_k) - X(t_{k-1})\right)\bigg) \geqslant c_m \sum_{k=1}^m u_k^2\, \sigma^2(t_{k-1}, t_k) \tag{1.2}$$

for all ordered points $t_1 < t_2 < \cdots < t_m$ in J with $t_m - t_1 < \varepsilon$ and $u_k \in \mathbb{R}$ $(k = 1, \cdots, m)$. Pitt (1978) used (1.2) to define local nondeterminism of a Gaussian random field $X = \{X(t),\, t \in \mathbb{R}^N\}$ with values in $\mathbb{R}^d$ by introducing a partial order among $t_1, \cdots, t_m \in \mathbb{R}^N$.

Roughly speaking, (1.2) suggests that the increments of X are asymptotically independent so that many of the results on the local times of Brownian motion can be extended to general Gaussian random fields. For example, Berman (1972, 1973), Pitt (1978) have applied LND to prove the joint continuity and Hölder conditions of the local times of a large class of Gaussian processes. See the comprehensive survey of Geman and Horowitz (1980) and the references therein for further information. Moreover, local nondeterminism has also been applied by Cuzick (1978) to study the moments of the zero crossing number of a stationary Gaussian process; by Rosen (1984) and Berman (1991) to study the existence and regularity of intersection local times; by Kahane (1985) to study the geometric properties of the images and level sets of fractional Brownian motion. Because of its various applications, it has been an interesting question to determine when a Gaussian process is locally nondeterministic. Some sufficient conditions for Gaussian processes to be locally nondeterministic can be found in Berman (1973, 1988, 1991), Cuzick (1978) and Pitt (1978).

On the other hand, it is known that the property of local nondeterminism defined in (1.1) or (1.2) is not enough for establishing fine regularity properties such as the law of the iterated logarithm and the modulus of continuity for the local times of Gaussian processes. For studying these and many other problems on Gaussian processes, the concept of strong local nondeterminism (SLND) has proven to be more appropriate. See Monrad and Pitt (1987), Csörgő et al. (1995), Monrad and Rootzén (1995), Talagrand (1995, 1998), Xiao (1996, 1997a, b, c), Kasahara et al. (1999), Xiao and Zhang (2002), just to mention a few.

The following definition of the strong local ϕ-nondeterminism (SL ϕ ND) was essentially given by Cuzick and DuPreez (1982) for Gaussian processes (i.e., $N = 1$). For Gaussian random fields, Definition 1.1 is more general than the definition of *strong local α-nondeterministism* of Monrad and Pitt (1987).

Definition 1.1. *Let $X = \{X(t), t \in \mathbb{R}^N\}$ be a real-valued random field with $0 < \mathbb{E}[X(t)^2] < \infty$ for $t \in J$, where $J \subseteq \mathbb{R}^N$ is a hyper-rectangle. Let $\phi : \mathbb{R}_+ \to \mathbb{R}_+$ be a given function such that $\phi(0) = 0$ and $\phi(r) > 0$ for $r > 0$. Then X is said to be* strongly locally ϕ-nondeterministic *on J if there exist positive constants c_0 and r_0 such that for all $t \in J$ and all $0 < r \leqslant \min\{|t|,\ r_0\}$,*

$$\operatorname{Var}\big(X(t)|X(s) : s \in J,\ r \leqslant |s-t| \leqslant r_0\big) \geqslant c_0\,\phi(r). \tag{1.3}$$

For a stationary Gaussian process $X = \{X(t), t \in \mathbb{R}\}$, Cuzick and DuPreez (1982) have given a sufficient condition for X to be strongly locally ϕ-nondeterministic in terms of its spectral measure F. More precisely, they have proven that if the absolutely continuous part of $dF(\lambda)$ has the property that

$$\frac{dF(\lambda/r)}{\phi(r)} \geqslant h(\lambda)d\lambda, \qquad \forall 0 < r \leqslant r_0 \tag{1.4}$$

and

$$\int_0^\infty \frac{\log h(\lambda)}{1+\lambda^2}\, d\lambda > -\infty, \tag{1.5}$$

then X is SLϕND. Their proof uses the ideas from Cuzick (1977) and relies on the special properties of stationary Gaussian processes.

Cuzick and DuPreez (1982, p. 811) point out that it appears to be difficult to establish conditions under which general Gaussian processes possess the various forms of strong local nondeterminism. There have only been a few known examples of strongly locally nondeterministic Gaussian random fields, one of them is the fractional Brownian motion which has been under extensive investigations in the last decade. A (standard) fractional Brownian motion $B_\alpha = \{B_\alpha(t), t \in \mathbb{R}^N\}$ of index α $(0 < \alpha < 1)$ is a centered, real-valued Gaussian random field with covariance function

$$\mathbb{E}\big(B_\alpha(t)B_\alpha(s)\big) = \frac{1}{2}\big(|t|^{2\alpha} + |s|^{2\alpha} - |t-s|^{2\alpha}\big).$$

The strong local ϕ-nondeterminism of B_α with $\phi(r) = r^{2\alpha}$ follows from Lemma 7.1 of Pitt (1978), where the self-similarity of B_α has played an essential role. Note that when $N = 1$, the property of strong local $r^{2\alpha}$-nondeterminism of B_α can also be derived from the above result of Cuzick and DuPreez (1982) by using the Lamperti transformation.

In the studies of Gaussian processes $X = \{X(t), t \in \mathbb{R}\}$, due to the simple order structure of $\mathbb{R}$, it is sometimes enough to assume that X is *one-sided* strongly locally ϕ-nondeterministic, namely, for some constant $c_0 > 0$

$$\mathrm{Var}\big(X(t)|X(s) : s \in J,\ r \leqslant t-s \leqslant r_0\big) \geqslant c_0\, \phi(r), \tag{1.6}$$

see Cuzick (1978), Berman (1972, 1978), Monrad and Rootzén (1995). When $X = \{X(t), t \in \mathbb{R}\}$ is a Gaussian process with stationary increments, some sufficient conditions in terms of the variance function $\sigma^2(h) = \mathbb{E}\big[\big(X(t+h) - X(t)\big)^2\big]$ for the *one-sided* strong local nondeterminism have been obtained earlier. Marcus (1968) and Berman (1978) have proved that if $\sigma(h) \to 0$ as $h \to 0$ and $\sigma^2(h)$ is concave on $(0, \delta)$ for some $\delta > 0$, then X is *one-sided* strongly locally ϕ-nondeterministic for $\phi(r) = \sigma^2(r)$.

The main objective of this paper is to prove a sufficient condition for a Gaussian random field $X = \{X(t),\, t \in \mathbb{R}^N\}$ with stationary increments to be strongly locally ϕ-nondeterministic [cf. Theorem 2.1 below]. In particular, we show that a spectral condition similar to that of Berman (1988) for the ordinary LND of X actually implies that X is strongly locally ϕ-nondeterministic and, moreover, $\phi(r)$ is comparable to the variance function $\sigma^2(h)$ with $|h| = r$; see Theorem 2.5 for details.

Our results on SLϕND have many applications. In Section 3, we apply them to study the sample path properties of Gaussian processes with stationary increments. In particular, we extend the small ball probability estimates of Monrad and Rootzén (1995), Shao and Wang (1995) and Stoltz (1996), the results on the exact Hausdorff measure of Talagrand (1995) and Xiao (1996, 1997a, b), the local and uniform Hölder conditions and tail probability of the local times of Xiao (1997a) and Kasahara et al. (1999), to more general Gaussian random fields. More applications of SLϕND to other problems for Gaussian random fields can be found in Shieh and Xiao (2006), Tudor and Xiao (2006).

Throughout the rest of this paper, unspecified positive and finite constants will be denoted by K which may have different values from line to line. Specific constants in Section j will be denoted by $K_{j,1}, K_{j,2}, \cdots$. For two non-negative functions f and g on $\mathbb{R}^N$, we denote $f \asymp g$ if there exists a finite constant $K \geqslant 1$ such that $K^{-1} f(x) \leqslant g(x) \leqslant K\, f(x)$ for all x in some neighborhood of 0 or infinity. This will be clear from the context.

2 Spectral conditions for strong local nondeterminism

Let $X = \{X(t),\, t \in \mathbb{R}^N\}$ be a real-valued, centered Gaussian random field with $X(0) = 0$. We assume that X has stationary increments and continuous covariance function $R(s,t) = \mathbb{E}\big[X(s)X(t)\big]$. According to Yaglom (1957) [see also Dudley (1973)], $R(s,t)$ can be represented as

$$R(s,t) = \int_{\mathbb{R}^N} (e^{i\langle s,\lambda\rangle} - 1)(e^{-i\langle t,\lambda\rangle} - 1)\Delta(d\lambda) + \langle s, Qt\rangle, \tag{2.1}$$

where $\langle x, y\rangle$ is the ordinary scalar product in $\mathbb{R}^N$, Q is an $N \times N$ nonnegative definite matrix and $\Delta(d\lambda)$ is a nonnegative symmetric measure on $\mathbb{R}^N \backslash \{0\}$ satisfying

$$\int_{\mathbb{R}^N} \frac{|\lambda|^2}{1+|\lambda|^2}\, \Delta(d\lambda) < \infty. \tag{2.2}$$

The measure Δ is called the *spectral measure* of X.

It follows from (2.1) that X has the following stochastic integral representation:

$$X(t) = \int_{\mathbb{R}^N} (e^{i\langle t,\lambda\rangle} - 1) W(d\lambda) + \langle \mathbf{Y}, t\rangle, \tag{2.3}$$

where $\mathbf{Y}$ is an N-dimensional Gaussian random vector with mean 0 and $W(d\lambda)$ is a centered complex-valued Gaussian random measure which is independent of $\mathbf{Y}$ and satisfies

$$\mathbb{E}\left(W(A)\overline{W(B)}\right) = \Delta(A \cap B) \quad \text{and} \quad W(-A) = \overline{W(A)}$$

for all Borel sets $A,\ B \subseteq \mathbb{R}^N$. From now on, we will assume $\mathbf{Y} = 0$. This is equivalent to assuming $Q = 0$ in (2.1). Consequently, we have

$$\sigma^2(h) = \mathbb{E}\big[\big(X(t+h) - X(t)\big)^2\big] = 2\int_{\mathbb{R}^N} \big(1 - \cos\langle h, \lambda\rangle\big)\ \Delta(d\lambda). \quad (2.4)$$

It is important to note that $\sigma^2(h)$ is a negative definite function and can be viewed as the characteristic exponent of a symmetric infinitely divisible distribution; see Berg and Forst (1975) for more information on negative definite functions.

If the function $\sigma^2(h)$ depends only on $|h|$, then X is called an *isotropic* random field. In this paper, we only consider Gaussian random fields which are *approximately isotropic*, that is, $\sigma^2(h) \asymp \phi(|h|)$ in a neighborhood of $h = 0$ for some nondecreasing function ϕ. As shown by Ayache and Xiao (2005), Ayache, Wu and Xiao (2006), the properties of anisotropic Gaussian random fields can be very different from those considered in this paper. For a systematic studies of anisotropic Gaussian random fields, see Xiao (2007a).

Our main results of this section are Theorems 2.1 and 2.5 below. Their proofs rely on the ideas from Kahane (1985), Pitt (1975, 1978) and Berman (1988, 1991).

Theorem 2.1. *Let $X = \{X(t), t \in \mathbb{R}^N\}$ be a mean zero, real-valued Gaussian random field with stationary increments and $X(0) = 0$, and let f be the density function of the absolutely continuous part Δ_c of the spectral measure Δ of X. Assume that there exist two locally bounded functions $\phi(r) : \mathbb{R}_+ \to \mathbb{R}_+$ and $q(\lambda) : \mathbb{R}^N \to \mathbb{R}_+$ satisfying the following conditions: $\phi(0) = 0$ and $\phi(r) > 0$ for $r > 0$,*

$$\frac{f(\lambda/r)}{\phi(r)} \geqslant \frac{r^N}{q(\lambda)}, \qquad \forall\, r \in (0,1] \ \text{ and } \ \lambda \in \mathbb{R}^N \quad (2.5)$$

and there exists a positive and finite constant η such that

$$q(\lambda) \leqslant |\lambda|^\eta, \qquad \forall\ \lambda \in \mathbb{R}^N \text{ with } |\lambda| \text{ large enough.} \quad (2.6)$$

Then for every $T > 0$, there exists a positive constant $K_{2,1}$ such that for all $t \in [-T, T]^N \backslash \{0\}$ and all $0 < r \leqslant \min\{1, |t|\}$,

$$\mathrm{Var}\big(X(t) | X(s) :\ s \in I,\ |s-t| \geqslant r\big) \geqslant K_{2,1}\, \phi(r). \quad (2.7)$$

In particular, X is strongly locally ϕ-nondeterministic on the hypercube $[-T, T]^N$.

In order to prove Theorem 2.1, we will make use of the following lemma which implies that SLND of X is determined by the behavior of the spectral measure Δ at infinity.

Lemma 2.2. *Assume the density function f of Δ_c satisfies the conditions (2.5) and (2.6). Then for any fixed constants $T > 0$ and $K_{2,2} > 0$, there exists a positive and finite constant $K_{2,3}$ such that for all functions g of the form*

$$g(\lambda) = \sum_{k=1}^{n} a_k \big(e^{i\langle s^k, \lambda\rangle} - 1\big), \tag{2.8}$$

where $a_k \in \mathbb{R}$ and $s^k \in [-T, T]^N$, we have

$$\big|g(\lambda)\big| \leqslant K_{2,3}\, |\lambda| \left(\int_{\mathbb{R}^N} |g(\xi)|^2 f(\xi)\, d\xi \right)^{1/2} \tag{2.9}$$

for all $|\lambda| < K_{2,2}$.

Proof. It follows from (2.5) and (2.6) that there exists a positive constant K such that

$$f(\lambda) \geqslant \frac{K}{|\lambda|^{\eta}}, \qquad \text{for all } \lambda \in \mathbb{R}^N \ \text{ with } |\lambda| \ \text{ large.}$$

Hence Proposition 5 of Pitt (1975) implies that for every constant $T > 0$, the measure Δ_c is regular at $[-T, T]^N$. [Pitt (1975, p.304) gives the definition of regularity for finite measures only, an extension to any σ-finite measure is immediate]. Let $\mathcal{G}$ be the collection of the functions $g(z)$ defined by (2.8) with $a_k \in \mathbb{R}$, $s^k \in [-T, T]^N$ and $z \in \mathbb{C}^N$. Since each $g \in \mathcal{G}$ is an entire function, it follows from Proposition 1 of Pitt (1975) [see also Pitt (1978, p.326)] that

$$K_{2,3} = \sup_{z \in U(0, K_{2,2})} \left\{ \sup_{g \in \mathcal{G}} \left\{ |g(z)| : \int_{\mathbb{R}^N} |g(\lambda)|^2 f(\lambda)\, d\lambda \leqslant 1 \right\} \right\} < \infty,$$

where $U(0, K_{2,2}) = \{z \in \mathbb{C}^N : |z| < K_{2,2}\}$ is the open ball of radius $K_{2,2}$ in $\mathbb{C}^N$. Since $g(0) = 0$ and g is analytic in $U(0, K_{2,2})$, Schwartz's lemma implies

$$\big|g(z)\big| \leqslant K_{2,3}\, |z| \left(\int_{\mathbb{R}^N} |g(\xi)|^2 f(\xi) d\xi \right)^{1/2}$$

for all $z \in U(0, K_{2,2})$. This proves (2.9). □

Now we are ready to prove Theorem 2.1.

Proof of Theorem 2.1. Working in the Hilbert space setting, the conditional variance in (2.7) is the square of the $L^2(\mathbb{P})$-distance of $X(t)$ from the subspace generated by $\{X(s) : s \in I,\ |s - t| \geqslant r\}$. Hence it is sufficient to show that there exists a constant $K_{2,1}$ such that for every $t \in [-T, T]^N \backslash \{0\}$, $0 < r \leqslant \min\{1, |t|\}$, the inequality

$$\mathbb{E}\Big(X(t) - \sum_{k=1}^{n} a_k X(s^k)\Big)^2 \geqslant K_{2,1}\, \phi(r) \tag{2.10}$$

holds for all integers $n \geqslant 1$, all $a_k \in \mathbb{R}$ and $s^k \in [-T, T]^N$ satisfying $|s^k - t| \geqslant r$, $(k = 1, 2, \cdots, n)$.

It follows from (2.1) or (2.3) that

$$\begin{aligned} &\mathbb{E}\Big(X(t) - \sum_{k=1}^{n} a_k X(s^k)\Big)^2 \\ &= \int_{\mathbb{R}^N} \Big| e^{i\langle t,\lambda\rangle} - 1 - \sum_{k=1}^{n} a_k \big(e^{i\langle s^k,\lambda\rangle} - 1\big)\Big|^2 \Delta(d\lambda) \\ &\geqslant \int_{\mathbb{R}^N} \Big| e^{i\langle t,\lambda\rangle} - \sum_{k=0}^{n} a_k\, e^{i\langle s^k,\lambda\rangle}\Big|^2 f(\lambda) d\lambda, \end{aligned} \tag{2.11}$$

where $a_0 = 1 - \sum_{k=1}^{n} a_k$ and $s_0 = 0$. Now we choose a bump function $\delta(\cdot) \in C^\infty(\mathbb{R}^N)$ with values in $[0, 1]$ such that $\delta(0) = 1$ and it vanishes outside the open unit ball. Let $\widehat{\delta}$ be the Fourier transform of δ. It is known that $\widehat{\delta}(\lambda)$ is also in $C^\infty(\mathbb{R}^N)$ and decays rapidly as $\lambda \to \infty$. Let $\delta_r(t) = r^{-N}\delta(t/r)$, then the Fourier inversion formula gives

$$\delta_r(t) = (2\pi)^{-N} \int_{\mathbb{R}^N} e^{-i\langle t,\lambda\rangle}\, \widehat{\delta}(r\lambda)\, d\lambda\,.$$

Since $\min\{|s^k - t|, 0 \leqslant k \leqslant n\} \geqslant r$ we have $\delta_r(t - s^k) = 0$ for all $k = 0, 1, \cdots, n$. Hence

$$\begin{aligned} &\int_{\mathbb{R}^N} \Big(e^{i\langle t,\lambda\rangle} - \sum_{k=0}^{n} a_k\, e^{i\langle s^k,\lambda\rangle}\Big) e^{-i\langle t,\lambda\rangle}\, \widehat{\delta}(r\lambda)\, d\lambda \\ &= (2\pi)^N \Big(\delta_r(0) - \sum_{k=0}^{n} a_k\, \delta_r(t - s^k)\Big) \\ &= (2\pi)^N\, r^{-N}. \end{aligned} \tag{2.12}$$

Now we will make use of the conditions (2.5) and (2.6). We choose the constant $K_{2,2}$ in Lemma 2.2 such that (2.6) holds for all $|\lambda| \geqslant K_{2,2}$ and split the integral in (2.12) over $\{\lambda : |\lambda| < K_{2,2}\}$ and $\{\lambda : |\lambda| \geqslant K_{2,2}\}$. Denote the two integrals by I_1 and I_2, respectively. It follows from

Lemma 2.2 that

$$
\begin{aligned}
I_1 &\leqslant \int_{|\lambda|<K_{2,2}} \left| e^{i\langle t,\lambda\rangle} - \sum_{k=0}^{n} a_k\, e^{i\langle s^k,\lambda\rangle} \right| |\widehat{\delta}(r\lambda)|\, d\lambda \\
&\leqslant K_{2,3} \left[\int_{\mathbb{R}^N} \left| e^{i\langle t,\lambda\rangle} - \sum_{k=0}^{n} a_k\, e^{i\langle s^k,\lambda\rangle} \right|^2 f(\lambda)\, d\lambda \right]^{1/2} \\
&\qquad \times \int_{|\lambda|<K_{2,2}} |\lambda|\, |\widehat{\delta}(r\lambda)|\, d\lambda \\
&\leqslant K_{2,4} \left[\mathbb{E}\Big(X(t) - \sum_{k=1}^{n} a_k X(s^k) \Big)^2 \right]^{1/2},
\end{aligned}
\tag{2.13}
$$

where the last inequality follows from (2.11) and the boundedness of $\widehat{\delta}$, and where $K_{2,4} > 0$ is a finite constant depending on T and $K_{2,2}$.

On the other hand, by the Cauchy-Schwarz inequality and (2.11), we have

$$
\begin{aligned}
I_2^2 &\leqslant \int_{|\lambda|\geqslant K_{2,2}} \left| e^{i\langle t,\lambda\rangle} - \sum_{k=0}^{n} a_k\, e^{i\langle s^k,\lambda\rangle} \right|^2 f(\lambda)\, d\lambda \\
&\qquad \times \int_{|\lambda|\geqslant K_{2,2}} \frac{1}{f(\lambda)} \left|\widehat{\delta}(r\lambda)\right|^2 d\lambda \\
&\leqslant \mathbb{E}\left(X(t) - \sum_{k=1}^{n} a_k X(s^k) \right)^2 \\
&\qquad \times r^{-N} \int_{|\lambda|\geqslant K_{2,2} r} \frac{1}{f(\lambda/r)} \left|\widehat{\delta}(\lambda)\right|^2 d\lambda.
\end{aligned}
\tag{2.14}
$$

By using (2.5) and (2.6), we deduce

$$
\begin{aligned}
&\int_{|\lambda|\geqslant K_{2,2} r} \frac{1}{f(\lambda/r)} \left|\widehat{\delta}(\lambda)\right|^2 d\lambda \\
&\leqslant \phi(r)^{-1}\, r^{-N} \int_{|\lambda|\geqslant K_{2,2} r} q(\lambda) \left|\widehat{\delta}(\lambda)\right|^2 d\lambda \\
&\leqslant K\, \phi(r)^{-1}\, r^{-N} \left\{ \int_{|\lambda|<K_{2,2}} q(\lambda) \left|\widehat{\delta}(\lambda)\right|^2 d\lambda \right. \\
&\qquad \left. + \int_{|\lambda|\geqslant K_{2,2}} |\lambda|^{\eta} \left|\widehat{\delta}(\lambda)\right|^2 d\lambda \right\} \\
&= K\, \phi(r)^{-1}\, r^{-N}.
\end{aligned}
\tag{2.15}
$$

Combining (2.14) and (2.15) yields

$$I_2^2 \leqslant K\,\phi(r)^{-1}\,r^{-2N}\,\mathbb{E}\Big(X(t)-\sum_{k=1}^{n}a_kX(s^k)\Big)^2. \tag{2.16}$$

Finally, we square both sides of (2.12) and use (2.13) and (2.16) to obtain

$$(2\pi)^{2N}\,r^{-2N} \leqslant K_{2,5}\,\phi(r)^{-1}\,r^{-2N}\,\mathbb{E}\Big(X(t)-\sum_{k=1}^{n}a_kX(s^k)\Big)^2.$$

This implies (2.10) and hence the theorem is proven. □

In order to apply Theorem 2.1 to investigate the sample path properties of the Gaussian random field X, we need to study the relationship between $\phi(|h|)$ and the function $\sigma^2(h)$. In the following, we show that under a condition analogous to that of Berman (1988, 1991), there is a non-decreasing function ϕ such that X is SLϕND and the functions $\phi(|h|)$ and $\sigma^2(h)$ are comparable. More precisely, we assume that the spectral measure Δ is absolutely continuous and its density function $f(\lambda)$ satisfies the following condition [when $N=1$, this is due to Berman (1988)]:

$$\begin{aligned} 0<\underline{\alpha} &= \frac{1}{2}\liminf_{\lambda\to\infty}\frac{\beta_N\,|\lambda|^N f(\lambda)}{\Delta\{\xi:|\xi|\geqslant|\lambda|\}} \\ &\leqslant \frac{1}{2}\limsup_{\lambda\to\infty}\frac{\beta_N\,|\lambda|^N f(\lambda)}{\Delta\{\xi:|\xi|\geqslant|\lambda|\}} = \overline{\alpha}<1, \end{aligned} \tag{2.17}$$

where $\beta_1=2$ and for $N\geqslant 2$, $\beta_N=\mu(S^{N-1})$ is the area [i.e., the $(N-1)$-dimensional Lebesgue measure) of S^{N-1}. At the end of this section, we will give several examples of Gaussian random fields satisfying condition (2.17).

In the rest of this section, we define $\phi(r)=\Delta\{\xi:|\xi|\geqslant r^{-1}\}$ and $\phi(0)=0$. Then the function ϕ is non-decreasing and continuous on $[0,\infty)$. The following lemma lists some properties of ϕ which will be useful later.

Lemma 2.3. *Assume the condition (2.17) holds. Then for any $\varepsilon\in(0,2\min\{\underline{\alpha},1-\overline{\alpha}\})$, there exists a constant $r_0>0$ such that for all $0<x\leqslant y\leqslant r_0$,*

$$\Big(\frac{x}{y}\Big)^{2\overline{\alpha}+\varepsilon} \leqslant \frac{\phi(x)}{\phi(y)} \leqslant \Big(\frac{x}{y}\Big)^{2\underline{\alpha}-\varepsilon}. \tag{2.18}$$

Consequently, we have

(i). $\lim_{r\to 0}\phi(r)/r^2=\infty$.

(ii). *The function ϕ has the following doubling property: there exists a constant $K_{2,6} > 0$ such that for all $0 < r < r_0/2$,*

$$\phi(2r) \leqslant K_{2,6}\, \phi(r). \tag{2.19}$$

Proof. For $N = 1$, (2.18) was proved by Berman (1988). Extension to $N > 1$ is easy and a proof is included for completeness. Denote $G(r) = \Delta\{\xi : |\xi| \geqslant r\}$. Then we can write

$$G(r) = \int_r^\infty \rho^{N-1} \int_{S^{N-1}} f(\rho\,\theta)\mu(d\theta)\, d\rho, \tag{2.20}$$

where μ is the surface measure on the unit sphere S^{N-1}. It follows that

$$\frac{d}{dr}\big[\log G(r)\big] = -\frac{r^{N-1}\int_{S^{N-1}} f(r\theta)\mu(d\theta)}{G(r)}.$$

Thus we derive the identity

$$\frac{G(x)}{G(y)} = \exp\left(\int_x^y \frac{r^N \int_{S^{N-1}} f(r\theta)\mu(d\theta)}{G(r)}\,\frac{dr}{r}\right) \tag{2.21}$$

for all x, $y > 0$. Note that the condition (2.17) and Fatou's lemma imply that

$$\begin{aligned} 0 < \underline{\alpha} &\leqslant \frac{1}{2}\liminf_{r\to\infty} \frac{r^N \int_{S^{N-1}} f(r\theta)\mu(d\theta)}{G(r)} \\ &\leqslant \frac{1}{2}\limsup_{r\to\infty} \frac{r^N \int_{S^{N-1}} f(r\theta)\mu(d\theta)}{G(r)} \leqslant \overline{\alpha} < 1. \end{aligned} \tag{2.22}$$

Hence for any $\varepsilon \in (0, 2\min\{\underline{\alpha},\, 1-\overline{\alpha}\})$, there exists $r_0 > 0$ such that for all $r \geqslant r_0^{-1}$, we have

$$2\underline{\alpha} - \varepsilon < \frac{r^N \int_{S^{N-1}} f(r\theta)\mu(d\theta)}{G(r)} < 2\overline{\alpha} + \varepsilon. \tag{2.23}$$

Therefore, (2.18) follows from (2.21) and (2.23). □

Remark 2.4. *The equation (2.18) shows that, under the assumption that the spectral measure Δ has a density $f(\lambda)$, condition (2.17) is more general than assuming ϕ is regularly varying at 0. Using the terminology of Bingham et al. (1987, pp.65–67), (2.18) implies that ϕ is extended regularly varying at 0 with upper and lower Karamata indices $2\overline{\alpha}$ and $2\underline{\alpha}$, respectively. Under (2.17), a necessary and sufficient condition for $\phi(r)$ to be regularly varying at 0 of index 2α is that the limit*

$$\alpha = \frac{1}{2}\lim_{r\to\infty} \frac{r^N \int_{S^{N-1}} f(r\theta)\mu(d\theta)}{\Delta\{\xi : |\xi| \geqslant r\}}$$

exists; this follows from Theorem 2.1.1 in Bingham et al. (1987) and (2.23). □

The following theorem shows that the assumption (2.17) implies that X is SL ϕ ND and $\phi(r)$ is comparable with $\sigma^2(h)$ with $|h| = r$ near 0. In Section 3, we will show that it is often more convenient to use the function ϕ to characterize the probabilistic and geometric properties of X.

Theorem 2.5. *Let $X = \{X(t), t \in \mathbb{R}^N\}$ be a mean zero, real-valued Gaussian random field with stationary increments and $X(0) = 0$. Assume that the spectral measure Δ of X has a density function f that satisfies (2.17). Then*

$$0 < \liminf_{h \to 0} \frac{\sigma^2(h)}{\phi(|h|)} \leqslant \limsup_{h \to 0} \frac{\sigma^2(h)}{\phi(|h|)} < \infty. \tag{2.24}$$

Moreover, for every $T > 0$, X is strongly locally ϕ-nondeterministic on the hypercube $[-T, T]^N$.

Proof. The proof of (2.24) is based on the proof of Theorem 3.1 of Berman (1991) which deals with the case of $N = 1$. Let $T, \tau > 0$ be any constants with $T\tau < 1$. By (2.4), we can write $\sigma^2(h)$ as

$$\begin{aligned}\sigma^2(h) = 2 &\int_{|\lambda| \leqslant T} \big(1 - \cos\langle h, \lambda\rangle\big) f(\lambda) d\lambda \\ &+ 2 \int_{T < |\lambda| \leqslant 1/\tau} \big(1 - \cos\langle h, \lambda\rangle\big) f(\lambda) d\lambda \\ &+ 2 \int_{|\lambda| > 1/\tau} \big(1 - \cos\langle h, \lambda\rangle\big) f(\lambda) d\lambda \\ := &\ 2(J_1 + J_2 + J_3).\end{aligned} \tag{2.25}$$

First we prove the left inequality in (2.24). Let $0 < \varepsilon < 2 \min\{\underline{\alpha}, 1 - \overline{\alpha}\}$ be fixed. Condition (2.17) implies the existence of a $\tau_0 \in (0, r_0)$ [r_0 is given in Lemma 2.3] such that

$$2\underline{\alpha} - \varepsilon \leqslant \frac{\beta_N |\lambda|^N f(\lambda)}{\phi(1/|\lambda|)} \leqslant 2\overline{\alpha} + \varepsilon \tag{2.26}$$

for all $\lambda \in \mathbb{R}^N$ with $|\lambda| \geqslant 1/\tau_0$. It follows from (2.26) and Lemma 2.3

that for $\tau = |h| < \tau_0$ in (2.25),

$$\begin{aligned}
\frac{J_3}{\phi(|h|)} &\geqslant \frac{2\underline{\alpha} - \varepsilon}{\beta_N} \int_{|\lambda|>1/|h|} \big(1 - \cos\langle h, \lambda\rangle\big) \frac{\phi(1/|\lambda|)}{\phi(|h|)} \frac{d\lambda}{|\lambda|^N} \\
&\geqslant \frac{2\underline{\alpha} - \varepsilon}{\beta_N} \int_{|\lambda|>1/|h|} \big(1 - \cos\langle h, \lambda\rangle\big) \frac{1}{(|\lambda||h|)^{2\overline{\alpha}+\varepsilon}} \frac{d\lambda}{|\lambda|^N} \\
&= \frac{2\underline{\alpha} - \varepsilon}{\beta_N} \int_{|\xi|>1} \Big(1 - \cos\langle \frac{h}{|h|}, \xi\rangle\Big) \frac{d\xi}{|\xi|^{N+2\overline{\alpha}+\varepsilon}} \\
&\geqslant K_{2,7},
\end{aligned} \tag{2.27}$$

where $K_{2,7}$ is a positive constant. In the above, the equality follows from a change of variable and the last inequality follows from Lemma 3.3 in Xiao (2003). It is clear that (2.25) and (2.27) imply the left inequality in (2.24).

In order to prove the right inequality in (2.24), we estimate J_1, J_2 and J_3 separately. Since $1 - \cos\langle h, \lambda\rangle \leqslant |h|^2|\lambda|^2$, we have

$$\begin{aligned}
\frac{J_1}{\phi(|h|)} &\leqslant \int_{|\lambda|\leqslant T} |h|^2|\lambda|^2 \frac{f(\lambda)}{\phi(|h|)}\, d\lambda \\
&\leqslant K \frac{|h|^2}{\phi(|h|)} \to 0 \qquad \text{as } h \to 0,
\end{aligned} \tag{2.28}$$

by (2.2) and Lemma 2.3. Next, suppose we have chosen the constant $T > 1/\tau_0$ so that (2.26) holds for all $\lambda \in \mathbb{R}^N$ with $|\lambda| > T$. Thus by taking $\tau = |h| < \tau_0$ in (2.25), we derive

$$\begin{aligned}
\frac{J_2}{\phi(|h|)} &\leqslant \frac{2\overline{\alpha} + \varepsilon}{\beta_N} \int_{T<|\lambda|\leqslant 1/|h|} \big(1 - \cos\langle h, \lambda\rangle\big) \frac{\phi(1/|\lambda|)}{\phi(|h|)} \frac{d\lambda}{|\lambda|^N} \\
&\leqslant \frac{2\overline{\alpha} + \varepsilon}{\beta_N} \int_{T<|\lambda|\leqslant 1/|h|} \big(1 - \cos\langle h, \lambda\rangle\big) \frac{1}{\big(|h||\lambda|\big)^{2\overline{\alpha}+\varepsilon}} \frac{d\lambda}{|\lambda|^N} \\
&\leqslant \frac{2\overline{\alpha} + \varepsilon}{\beta_N} \int_{|\xi|\leqslant 1} \frac{d\xi}{|\xi|^{N-2(1-\overline{\alpha})+\varepsilon}} < \infty.
\end{aligned} \tag{2.29}$$

Similar to (2.29), we use (2.26), Lemma 2.3 and the inequality $1 - \cos\langle h, \lambda\rangle \leqslant 2$ to deduce

$$\begin{aligned}
\frac{J_3}{\phi(|h|)} &\leqslant \frac{2\overline{\alpha} + \varepsilon}{\beta_N} \int_{|\lambda|\geqslant 1/|h|} \big(1 - \cos\langle h, \lambda\rangle\big) \frac{\phi(1/|\lambda|)}{\phi(|h|)} \frac{d\lambda}{|\lambda|^N} \\
&\leqslant \frac{2\overline{\alpha} + \varepsilon}{\beta_N} \int_{|\lambda|\geqslant 1/|h|} \big(1 - \cos\langle h, \lambda\rangle\big) \frac{1}{\big(|h||\lambda|\big)^{2\underline{\alpha}-\varepsilon}} \frac{d\lambda}{|\lambda|^N} \\
&\leqslant \frac{2(2\overline{\alpha} + \varepsilon)}{\beta_N} \int_{|\xi|\geqslant 1} \frac{d\xi}{|\xi|^{N+2\underline{\alpha}-\varepsilon}} < \infty.
\end{aligned} \tag{2.30}$$

Therefore the right inequality in (2.24) follows from (2.28), (2.29) and (2.30). This finishes the proof of (2.24).

Finally, note that Condition (2.17), together with Lemma 2.3, implies that (2.5) and (2.6) hold with $q(r) = K_{2,8}\, r^{N+2\overline{\alpha}+\varepsilon}$. Therefore, for any $T > 0$, the strong local ϕ-nondeterminism of X on $I = [-T, T]^N$ follows from Theorem 2.1. □

Let $X = \{X(t), t \in \mathbb{R}^N\}$ be a stationary random field with mean 0, variance 1 and spectral measure Δ. Then X can be represented as

$$X(t) = \int_{\mathbb{R}^N} e^{i\langle t,\lambda\rangle} W(d\lambda), \quad \forall t \in \mathbb{R}^N. \tag{2.31}$$

Clearly, Theorems 2.1 and 2.5 are applicable to the Gaussian random field $Y = \{Y(t), t \in \mathbb{R}^N\}$ defined by $Y(t) = X(t) - X(0)$. Furthermore, we remark that the proofs of Theorems 2.1 and 2.5 remain effective for X itself. Either way we have the following partial extension of the result of Cuzick and DuPreez (1982) mentioned in the Introduction to $N > 1$. It is not known to me whether (2.6) can be replaced by the weaker condition (1.5).

Corollary 2.6. *Let $X = \{X(t), t \in \mathbb{R}^N\}$ be a stationary Gaussian random field with mean 0 and variance 1. The following statements hold:*

(i). *If the spectral measure Δ of X has an absolutely continuous part with density f satisfying (2.5) and (2.6), then for every $T > 0$, X is strongly locally ϕ-nondeterministic on the hypercube $[-T, T]^N$.*

(ii). *If the spectral density of X satisfies (2.17), then (2.24) holds and X is SLϕND on the hypercube $[-T, T]^N$.*

We end this section with some examples of Gaussian random fields whose SLND can be determined.

Example 2.7. Let $B_\alpha = \{B_\alpha(t),\, t \in \mathbb{R}^N\}$ be an N-parameter fractional Brownian motion in $\mathbb{R}$ with Hurst index $\alpha \in (0, 1)$, then its spectral density is given by

$$f_\alpha(\lambda) = c(\alpha, N)\frac{1}{|\lambda|^{2\alpha+N}},$$

where $c(\alpha, N) > 0$ is a normalizing constant such that $\sigma^2(h) = |h|^{2\alpha}$; see e.g., Kahane (1985). Clearly, the condition (2.17) holds with $\underline{\alpha} = \overline{\alpha} = \alpha$. As mentioned earlier, the strong local nondeterminism of B_α was first proved by Pitt (1978). □

Example 2.8. Consider the mean zero Gaussian random field $X = \{X(t),\, t \in \mathbb{R}^N\}$ with stationary increments and spectral density

$$f_{\gamma,\beta}(\lambda) = \frac{c(\gamma,\beta,N)}{|\lambda|^{2\gamma}(1+|\lambda|^2)^{\beta}},$$

where γ and β are constants satisfying

$$\beta+\gamma > \frac{N}{2}, \quad 0 < \gamma < 1+\frac{N}{2}$$

and $c(\gamma,\beta,N) > 0$ is a normalizing constant. Since the spectral density $f_{\gamma,\beta}$ involves both the Fourier transforms of the Riesz kernel and the Bessel kernel, Anh et al. (1999) call the corresponding Gaussian random field the fractional Riesz-Bessel motion with indices β and γ; and they have shown that these Gaussian random fields can be used for modelling simultaneously long range dependence and intermittency.

It is easy to check that Condition (2.17) is satisfied with $\underline{\alpha} = \overline{\alpha} = \gamma + \beta - \frac{N}{2}$. Moreover, since the spectral density $f_{\gamma,\beta}(x)$ is regularly varying at infinity of order $2(\beta+\gamma) > N$, by a result of Pitman (1968) we know that, if $\gamma+\beta-\frac{N}{2} < 1$, then $\sigma(h)$ is regularly varying at 0 of order $\gamma+\beta-N/2$ and

$$\sigma(h) \sim |h|^{\gamma+\beta-N/2} \qquad \text{as } h \to 0.$$

Theorem 2.5 implies that X is SLND with respect to $\sigma^2(h)$. Hence, many sample path properties of the d-dimensional fractional Riesz-Bessel motion X with indices β and γ can be can be derived from the results in Section 3. □

Example 2.9. Let $0 < \alpha < 1$ and $0 < c_1 < c_2$ be constants such that $(\alpha c_2)/c_1 < 1$. For any increasing sequence $\{b_n,\, n \geqslant 0\}$ of real numbers such that $b_0 = 0$ and $b_n \to \infty$, define the function f on $\mathbb{R}^N$ by

$$f(\lambda) = \begin{cases} c_1\,|\lambda|^{-(2\alpha+N)} & \text{if } |\lambda| \in (b_{2k}, b_{2k+1}], \\ c_2\,|\lambda|^{-(2\alpha+N)} & \text{if } |\lambda| \in (b_{2k+1}, b_{2k+2}]. \end{cases} \tag{2.32}$$

Some elementary calculation shows that, when $\lim_{n\to\infty} b_{n+1}/b_n = \infty$, Condition (2.17) is satisfied with $\underline{\alpha} = (\alpha c_1)/c_2 < \overline{\alpha} = (\alpha c_2)/c_1$. Note that in this case,

$$\frac{c_1}{c(\alpha,N)}\,|h|^{2\alpha} \leqslant \sigma^2(h) \leqslant \frac{c_2}{c(\alpha,N)}\,|h|^{2\alpha}, \qquad \forall\, h \in \mathbb{R}^N,$$

where $c(\alpha,N)$ is the constant in Example 2.7, and

$$\frac{c_1\,\beta_N}{2\alpha}\,r^{2\alpha} \leqslant \phi(r) \leqslant \frac{c_2\,\beta_N}{2\alpha}\,r^{2\alpha}, \qquad \forall\, r > 0.$$

However, both functions are not regularly varying at the origin. □

Next, we present a class of Gaussian random fields for which (2.17) does not hold, but Theorem 2.1 is still applicable.

Example 2.10. For any given constants $0 < \alpha_1 < \alpha_2 < 1$ and any increasing sequence $\{b_n,\ n \geqslant 0\}$ of real numbers such that $b_0 = 0$ and $b_n \to \infty$, define the function f on $\mathbb{R}^N$ by

$$f(\lambda) = \begin{cases} |\lambda|^{-(2\alpha_1+N)} & \text{if } |\lambda| \in (b_{2k}, b_{2k+1}], \\ |\lambda|^{-(2\alpha_2+N)} & \text{if } |\lambda| \in (b_{2k+1}, b_{2k+2}]. \end{cases} \tag{2.33}$$

Using such functions f as spectral densities, we obtain a quite large class of Gaussian random fields with stationary increments that are significantly different from the fractional Brownian motion. If X is such a Gaussian random field, then it follows from (2.4) and (2.20) that there exist positive constants $K_{2,9}$ and $K_{2,10} \geqslant 1$ such that

$$K_{2,9}^{-1}\,|h|^{2\alpha_2} \leqslant \sigma^2(h) \leqslant K_{2,9}\,|h|^{2\alpha_1} \tag{2.34}$$

for all $h \in \mathbb{R}^N$ with $|h| \leqslant 1$, and

$$K_{2,10}^{-1}\,r^{2\alpha_2} \leqslant \phi(r) \leqslant K_{2,10}\,r^{2\alpha_1}, \qquad \forall\, 0 < r \leqslant 1. \tag{2.35}$$

Now we choose a strictly increasing sequence $\{b_n\}$ such that for all $k \geqslant 1$,

$$b_{2k+1}^{2\alpha_2}\left(b_{2k}^{-2\alpha_1} - b_{2k+1}^{-2\alpha_1}\right) \leqslant 1, \tag{2.36}$$

$$b_{2k+1}^{2-2\alpha_1} - b_{2k}^{2-2\alpha_1} \leqslant b_{2k}^{2-2\alpha_2}\,\frac{1}{k^2} \tag{2.37}$$

and

$$\frac{b_{2k+2}}{b_{2k+1}} \geqslant (k+1)^{1/\alpha_2}. \tag{2.38}$$

This can be done inductively: choose b_{2k+1} close to b_{2k} so that both (2.36) and (2.37) hold; then choose b_{2k+2} so that (2.38) holds.

We claim that the following properties hold:

(i) $\phi(r) \asymp r^{2\alpha_2}$ for $r \in (0,1)$.

(ii) $\sigma^2(h) \asymp |h|^{2\alpha_2}$ for all $h \in \mathbb{R}^N$ with $|h| \leqslant 1$.

(iii) condition (2.17) is not satisfied.

(iv) the corresponding Gaussian random field X is SLϕND on all hypercubes $I = [-T, T]^N$.

In order to verify (i), by (2.35), we only need to show $\phi(r) \leqslant K_{2,11}\,r^{2\alpha_2}$ for some finite constant $K_{2,11}$. For any $r > 0$ small, there exists an integer

$k_0 > 0$ such that either $r^{-1} \in [b_{2k_0}, b_{2k_0+1})$ or $r^{-1} \in [b_{2k_0+1}, b_{2k_0+2})$. In the first case

$$\begin{aligned}\phi(r) &\leqslant \int_{b_{2k_0} \leqslant |\lambda| \leqslant b_{2k_0+1}} |\lambda|^{-(2\alpha_1+N)}\, d\lambda \\ &\qquad + \int_{b_{2k_0+1} \leqslant |\lambda| \leqslant b_{2k_0+2}} |\lambda|^{-(2\alpha_2+N)}\, d\lambda + \cdots \\ &= \frac{K}{2\alpha_1}\left[b_{2k_0}^{-2\alpha_1} - b_{2k_0+1}^{-2\alpha_1} + b_{2k_0+2}^{-2\alpha_1} - b_{2k_0+3}^{-2\alpha_1} + \cdots\right] \\ &\qquad + \frac{K}{2\alpha_2}\left[b_{2k_0+1}^{-2\alpha_2} - b_{2k_0+2}^{-2\alpha_2} + b_{2k_0+3}^{-2\alpha_2} - b_{2k_0+4}^{-2\alpha_2} + \cdots\right].\end{aligned} \tag{2.39}$$

Clearly, the second sum is bounded above by $K\, r^{2\alpha_2}$. It follows from (2.36) and (2.38) that the first sum is bounded above by $K\, r^{2\alpha_2}$ as well. This verifies (i) when $r^{-1} \in [b_{2k_0}, b_{2k_0+1})$.

In the second case when $r^{-1} \in [b_{2k_0+1}, b_{2k_0+2})$, we have

$$\begin{aligned}\phi(r) &\leqslant \int_{r^{-1} \leqslant |\lambda| \leqslant b_{2k_0+2}} |\lambda|^{-(2\alpha_2+N)}\, d\lambda \\ &\qquad + \int_{b_{2k_0+2} \leqslant |\lambda| \leqslant b_{2k_0+3}} |\lambda|^{-(2\alpha_1+N)}\, d\lambda + \cdots \\ &= \frac{K}{2\alpha_2}\left[r^{2\alpha_2} - b_{2k_0+2}^{-2\alpha_2} + b_{2k_0+3}^{-2\alpha_2} - b_{2k_0+4}^{-2\alpha_2} + \cdots\right] \\ &\qquad + \frac{K}{2\alpha_1}\left[b_{2k_0+2}^{-2\alpha_1} - b_{2k_0+3}^{-2\alpha_1} + b_{2k_0+4}^{-2\alpha_1} - b_{2k_0+5}^{-2\alpha_1} + \cdots\right] \\ &\leqslant K\, r^{2\alpha_2},\end{aligned} \tag{2.40}$$

where the last inequality follows from (2.36) and (2.38).

Next we verify (ii). Because of (2.34), we only need to show $\sigma^2(h) \leqslant K_{2,12}\, |h|^{2\alpha_2}$ for all $h \in \mathbb{R}^N$ with $|h| \leqslant 1$. Fix such an $h \in \mathbb{R}^N$, let k_1 be the integer such that $|h|^{-1} \in [b_{2k_1}, b_{2k_1+1})$ or $|h|^{-1} \in [b_{2k_1+1}, b_{2k_1+2})$. In both cases, (2.36) and (2.37) imply $|\langle h, \lambda\rangle| \leqslant |h|\,|\lambda| \leqslant 2$ for all $|\lambda| \in [b_{2k}, b_{2k+1})$ and all $k \leqslant k_1$. For such λ, $1 - \cos\langle h, \lambda\rangle \leqslant (|h|\,|\lambda|)^2$. It follows from (2.4) that

$$\begin{aligned}\sigma^2(h) &\leqslant 2 \int_{\mathbb{R}^N} (1 - \cos\langle h, \lambda\rangle)\, \frac{d\lambda}{|\lambda|^{2\alpha_2+N}} \\ &\qquad + 2\sum_{k=0}^{k_1} \int_{b_{2k} \leqslant |\lambda| \leqslant b_{2k+1}} (|h|\,|\lambda|)^2 \frac{d\lambda}{|\lambda|^{2\alpha_1+N}} \\ &\qquad + 2\sum_{k=k_1+1}^{\infty} \int_{b_{2k} \leqslant |\lambda| \leqslant b_{2k+1}} \frac{d\lambda}{|\lambda|^{2\alpha_1+N}}.\end{aligned} \tag{2.41}$$

By Example 2.7, the first integral equals $K\,|h|^{2\alpha_2}$. Moreover, a few lines of elementary calculation using (2.36) and (2.37) show that both sums in (2.41) are at most $K\,|h|^{2\alpha_2}$. This proves (ii).

It follows from (2.39) that

$$\limsup_{\lambda\to\infty} \frac{\beta_N\,|\lambda|^N f(\lambda)}{\Delta\{\xi : |\xi| \geqslant |\lambda|\}} = \infty.$$

Thus (2.17) is not satisfied.

Finally, (2.33) and (i) above imply that (2.5) and (2.6) hold with $q(\lambda) = |\lambda|^{N+2\alpha_2}$. Therefore, Theorem 2.2 implies that the Gaussian random field X with spectral density (2.33) is SLϕND. □

So far we have not considered SLϕND for Gaussian random fields with stationary increments and *discrete* spectral measures. A systematic treatment for such Gaussian random fields will be given in Xiao (2007b). Here we content ourselves with the following example: Let $\{X_n, Y_n, n \geqslant 0\}$ be a sequence of independent standard normal random variables. Then for each $t \in \mathbb{R}$, the random Fourier series

$$Y(t) = \frac{\sqrt{8}}{\pi} \sum_{n=0}^{\infty} \frac{1}{2n-1}\Big(X_n \cos((2n-1)t) + Y_n \sin((2n-1)t)\Big) \qquad (2.42)$$

converges almost surely [see Kahane (1985)], and $Y = \{Y(t),\ t \in \mathbb{R}\}$ is a centered, periodic and stationary Gaussian process with mean 0 and covariance function

$$R(s,t) = 1 - \frac{2}{\pi}|s-t| \quad \text{for} \quad -\pi \leqslant s-t \leqslant \pi. \qquad (2.43)$$

It can be verified that the spectrum measure Δ of Y is discrete with $\Delta(\{2n-1\}) = (2n-1)^{-2}$ for all $n \in \mathbb{R}$. Shieh and Xiao (2006) have proved that Y is strongly locally ϕ-nondeterministic with $\phi(r) = r$, and it satisfies (3.3) in Section 3.

3 Sample path properties of Gaussian random fields

In the studies of Gaussian random fields with stationary increments, the variance function $\sigma^2(h)$ has played a significant role and it is typically assumed to be regularly varying at 0 and/or monotone in $|h|$. See Csörgő et al. (1995), Kasahara et al. (1999), Monrad and Rootzén (1995), Talagrand (1995, 1998), Xiao (1996, 1997a, b, 2003) and the references therein. In this section, we show that the regularly varying assumption on $\sigma^2(h)$ can be significantly weakened and the monotonicity assumption can be removed.

We will consider the Gaussian random field $X = \{X(t),\, t \in \mathbb{R}^N\}$ in $\mathbb{R}^d$ defined by

$$X(t) = \big(X_1(t), \cdots, X_d(t)\big), \qquad \forall t \in \mathbb{R}^N, \tag{3.1}$$

where the coordinator processes $X_1, \cdots, X_d$ are independent copies of a real-valued, centered Gaussian random field $Y = \{Y(t), t \in \mathbb{R}^N\}$. We call Y the associated random field. In the rest of this paper, we will often assume that Y satisfies the following condition (C):

(C1) There exist positive constants δ_0, $K_{3,1}$, $K_{3,2}$ and a non-decreasing, right continuous function $\phi : [0, \delta_0) \to [0, \infty)$ such that $\phi(0) = 0$ and

$$\frac{\phi(2r)}{\phi(r)} \leqslant K_{3,1}, \quad \forall\, r \in [0, \delta_0/2) \tag{3.2}$$

and for all $t \in \mathbb{R}^N$ and $h \in \mathbb{R}^N$ with $|h| \leqslant \delta_0$,

$$K_{3,2}^{-1}\phi(|h|) \leqslant \mathbb{E}\big[\big(Y(t+h) - Y(t)\big)^2\big] \leqslant K_{3,2}\phi(|h|). \tag{3.3}$$

(C2) For any $T > 0$, the process Y is strongly locally ϕ-nondeterministic on $[-T, T]^N$.

It follows from Theorem 2.5 that for any Gaussian random field Y with stationary increments and spectral density satisfying (2.17), condition (C) is satisfied. We point out that the setting of this section is more general than that of Section 2. In particular, our results in this section are applicable to Gaussian random fields with stationary increments and discrete spectral measures, as long as they satisfy condition (C).

3.1 Small ball probability

In recent years, there has been much interest in studying the small ball probability of Gaussian processes. We refer to Li and Shao (2001) and Lifshits (1999) for extensive surveys on small ball probabilities, their applications and open problems.

Our next theorem gives estimates on the small ball probability of Gaussian random fields satisfying the condition (C). In particular, the upper bound in (3.4) confirms a conjecture of Shao and Wang (1995), under a much weaker condition.

Theorem 3.1. *Let $X = \{X(t), t \in \mathbb{R}^N\}$ be a Gaussian random field in $\mathbb{R}$ satisfying the condition (C). Then there exist positive constants $K_{3,3}$*

and $K_{3,4}$ *such that for all* $x \in (0,1)$,

$$\exp\left(-\frac{K_{3,3}}{[\phi^{-1}(x^2)]^N}\right) \leqslant \mathbb{P}\left\{\max_{t\in[0,1]^N} |X(t)| \leqslant x\right\} \leqslant \exp\left(-\frac{K_{3,4}}{[\phi^{-1}(x^2)]^N}\right), \tag{3.4}$$

where $\phi^{-1}(x) = \inf\{y : \phi(y) > x\}$ *is the right-continuous inverse function of* ϕ.

Proof. Equip $I = [0,1]^N$ with the canonical metric

$$d(s,t) = (\mathbb{E}|X(s) - X(t)|^2)^{1/2}, \quad s,\, t \in I$$

and denote by $N_d(I,\varepsilon)$ the smallest number of d-balls of radius $\varepsilon > 0$ needed to cover I. Then it is easy to see from (C1) that for all $\varepsilon \in (0,1)$,

$$N_d(I,\varepsilon) \leqslant K\left(\frac{1}{\phi^{-1}(\varepsilon^2)}\right)^N := \Psi(\varepsilon).$$

Moreover, it follows from condition (C1) that Ψ has the doubling property, i.e., $\Psi(\varepsilon) \leqslant \Psi(\varepsilon/2) \leqslant K\,\Psi(\varepsilon)$. Hence the lower bound in (3.4) follows from a result of Talagrand (1993); see also Ledoux (1996, p.257).

The proof of the upper bound in (3.4) is based on condition (C2) and an argument in Monrad and Rootzén (1995). For any integer $n \geqslant 2$, we choose n^N points $t_{n,\mathbf{i}} \in [0,1]^N$, where

$$t_{n,\mathbf{i}} = \left(\frac{i_1}{n}, \cdots, \frac{i_N}{n}\right), \qquad \mathbf{i} = (i_1, \cdots, i_N) \in \{1, \cdots, n\}^N,$$

and denote them [in any order] by $t_{n,k}$ $(k = 1, 2, \cdots, n^N)$. Then

$$\mathbb{P}\left\{\max_{t\in[0,1]^N} |X(t)| \leqslant x\right\} \leqslant \mathbb{P}\left\{\max_{1\leqslant k\leqslant n^N} |X(t_{n,k})| \leqslant x\right\}. \tag{3.5}$$

By Anderson's inequality for Gaussian measures and the SLϕND of X, we derive the following upper bound for the conditional probabilities

$$\mathbb{P}\Big\{|X(t_{n,k})| \leqslant x \big| X(t_{n,j}),\ 1 \leqslant j \leqslant k-1\Big\} \leqslant \Phi\left(\frac{K\,x}{\phi^{1/2}(n^{-1})}\right), \tag{3.6}$$

where $\Phi(x)$ is the distribution function of a standard normal random variable. It follows from (3.5) and (3.6) that

$$\mathbb{P}\left\{\max_{t\in[0,1]^N} |X(t)| \leqslant x\right\} \leqslant \left[\Phi\left(\frac{K\,x}{\phi^{1/2}(n^{-1})}\right)\right]^{n^N}. \tag{3.7}$$

By taking n to be the smallest integer $\geqslant \left[\phi^{-1}(x^2)\right]^{-1}$, we obtain the upper bound in (3.4). □

Combining Theorem 3.1 with Theorem 7.1 in Li and Shao (2001) yields the following Chung's law of the iterated logarithm. When σ is regularly varying, this is also obtained in Xiao (1997a).

Corollary 3.2. *Let $X = \{X(t), t \in \mathbb{R}^N\}$ be an (N, d)-Gaussian random field defined by (3.1). Suppose that the associated Gaussian random field Y has stationary increments and spectral measure Δ. If Y satisfies condition (C) and its spectral measure Δ satisfies*

$$\liminf_{\lambda \to \infty} |\lambda|^{N+2} \Delta\big(B(\lambda, r)\big) > 0, \tag{3.8}$$

where $B(\lambda, r) = \{x \in \mathbb{R}^N : |x - \lambda| \leqslant r\}$. Then there exists a positive and finite constant $K_{3,5}$ such that

$$\liminf_{r \to 0} \frac{\sup_{t \in [0,r]^N} |X(t)|}{\phi^{1/2}\big(r/(\log\log(1/r))^{1/N}\big)} = K_{3,5}, \qquad a.s.. \tag{3.9}$$

Proof. By applying Theorem 3.1 and slightly modifying the proof of Theorem 7.1 in Li and Shao (2001) to each component X_k $(k = 1, \cdots, d)$ of X, we derive that there exists a positive constant $K_{3,6} \geqslant 1$ such that

$$K_{3,6}^{-1} \leqslant \liminf_{r \to 0} \frac{\sup_{t \in [0,r]^N} |X(t)|}{\phi^{1/2}\big(r/(\log\log(1/r))^{1/N}\big)} \leqslant K_{3,6}, \qquad \text{a.s..} \tag{3.10}$$

Since the components of X are independent, (3.8) implies that the zero-one law of Pitt and Tran (1979) holds for X at $t = 0$. Hence (3.9) follows from this and (3.10). □

Remark 3.3. *When Δ has a density function which satisfies (2.17), (3.8) follow easily from (2.26) and Lemma 2.3. Hence (3.9) holds.* □

We can also consider the small ball probability of Gaussian random fields under the Hölder-type norm. Let $\kappa : \mathbb{R}_+ \to \mathbb{R}_+$ be a continuous and non-decreasing function such that $\kappa(r) > 0$ for all $r > 0$. For any function $y \in C_0([0,1]^N)$, we consider the functional

$$\|y\|_\kappa = \sup_{s,t \in [0,1]^N, s \neq t} \frac{|y(s) - y(t)|}{\kappa(|s - t|)}. \tag{3.11}$$

When $\kappa(r) = r^\alpha$, $\|\cdot\|_\kappa$ is the α-Hölder norm on $C_0([0,1]^N)$ and is denoted by $\|\cdot\|_\alpha$.

The following theorem uses SLϕND to improve the results of Stolz (1996). We mention that the conditions of Theorem 2.1 of Kuelbs, Li and Shao (1995) can be weakened in a similar way.

Theorem 3.4. *Let $X = \{X(t), t \in \mathbb{R}^N\}$ be a Gaussian random field in $\mathbb{R}$ satisfying condition (C). If for some constant $\beta > 0$,*

$$\frac{\phi^{1/2}(r)}{\kappa(r)} \asymp r^{\beta}, \quad \forall r \in (0,1). \tag{3.12}$$

Then there exist positive constants $K_{3,7}$ and $K_{3,8}$ such that for all $\varepsilon \in (0,1)$,

$$\exp\left(-\frac{K_{3,7}}{\varepsilon^{N/\beta}}\right) \leqslant \mathbb{P}\Big\{\|X\|_{\kappa} \leqslant \varepsilon\Big\} \leqslant \exp\left(-\frac{K_{3,8}}{\varepsilon^{N/\beta}}\right). \tag{3.13}$$

Proof. The lower bound in (3.13) follows directly from Theorem 1.1 of Stolz (1996). The proof of the upper bound in (3.13) is a modification of the proof of Theorem 1.3 of Stolz (1996), using (C2) in place of Lemma 7.1 of Pitt (1978). We leave it to the interested reader. □

3.2 Hausdorff dimension and Hausdorff measure of the range

In this section we consider the fractal properties of the range $X([0,1]^N)$ and graph $\mathrm{Gr}X([0,1]^N) = \{(t, X(t)) : t \in [0,1]^N\}$ of the Gaussian random field in $\mathbb{R}^d$ defined by (3.1). In particular, we will show that the Hausdorff dimension of $X([0,1]^N)$ and $\mathrm{Gr}X([0,1]^N)$ can be determined mainly by the upper index of ϕ at 0 defined by

$$\alpha^* = \inf\left\{\beta \geqslant 0 : \lim_{r\to 0} \frac{\phi(r)}{r^{2\beta}} = \infty\right\} \tag{3.14}$$

with the convention $\inf \emptyset = \infty$. Analogously, we can define the lower index of ϕ at 0 by

$$\alpha_* = \sup\left\{\beta \geqslant 0 : \lim_{r\to 0} \frac{\phi(r)}{r^{2\beta}} = 0\right\}. \tag{3.15}$$

Clearly, $0 \leqslant \alpha_* \leqslant \alpha^* \leqslant \infty$. When the real-valued Gaussian random field $Y = \{Y(t), t \in \mathbb{R}^N\}$ associated with (3.1) has stationary increments and a continuous covariance function. Then the above upper and lower indices α^* and α_* coincide with the upper and lower indices of $\sigma(h)$, where

$$\sigma^2(h) = \mathbb{E}\big[\big(Y(t+h) - Y(t)\big)^2\big], \qquad \forall\, h \in \mathbb{R}^N. \tag{3.16}$$

In this case, we also call α^* and α_* the upper and lower indices of Y. See Adler (1981) for more information.

The following example shows that it is possible to have $\alpha^* = \infty$.

Example 3.5. Let $N \geqslant 2$ and let Δ be a Borel measure on $\mathbb{R}^N$ with support in a linear subspace L of $\mathbb{R}^N$ and satisfying (2.2). If Y is a Gaussian random field with stationary increments and spectrum measure Δ, then for all h in the linear subspace of $\mathbb{R}^N$ that is orthogonal to L, we have $\sigma^2(h) = 0$. Thus $\alpha^* = \infty$. □

Lemma 3.6 below shows that under quite general conditions the inequality $\alpha^* \leqslant 1$ holds. In particular, this is true for all the Gaussian random fields considered in Section 2.

Lemma 3.6. *Let $Y = \{Y(t), t \in \mathbb{R}^N\}$ be a Gaussian random field in $\mathbb{R}$ with stationary increments and spectrum measure Δ. If $N = 1$, or $N \geqslant 2$ and Δ has an absolutely continuous part with density $f(\lambda)$. Then $\alpha^* \leqslant 1$.*

Proof. It follows from (2.4) that

$$\begin{aligned}\sigma^2(h) &\geqslant \int_{|\lambda| \leqslant |h|^{-1}} (1 - \cos\langle h, \lambda\rangle)\Delta(d\lambda) \\ &\geqslant K|h|^2 \int_{|\lambda| \leqslant |h|^{-1}} \langle \frac{h}{|h|}, \lambda\rangle^2 \, \Delta(d\lambda).\end{aligned} \tag{3.17}$$

It is clear that when $N = 1$, we have $\sigma^2(h) \geqslant K|h|^2$ for all $h \in \mathbb{R}$ with $|h|$ small enough. This implies $\alpha^* \leqslant 1$ whenever $N = 1$.

Now we assume that $N \geqslant 2$. It follows from (3.17) that

$$\begin{aligned}\sigma^2(h) &\geqslant K|h|^2 \int_{|\lambda| \leqslant |h|^{-1}} \langle \frac{h}{|h|}, \lambda\rangle^2 \, f(\lambda)\, d\lambda \\ &\geqslant K\,|h|^2 \int_{S^{N-1}} \langle \frac{h}{|h|}, \theta\rangle^2 \, \mu(d\theta) \int_0^{|h|^{-1}} \rho^{N+1} f(\rho\theta)\, d\rho.\end{aligned} \tag{3.18}$$

Since $f(\lambda) > 0$ on a set of positive N-dimensional Lebesgue measure, we see that for all $h \in \mathbb{R}^N$ with $|h|$ small enough,

$$\int_{S^{N-1}} \langle \frac{h}{|h|}, \theta\rangle^2 \, \mu(d\theta) \int_0^{|h|^{-1}} \rho^{N+1} f(\rho\theta)\, d\rho \geqslant K_{3,9} \tag{3.19}$$

for some constant $K_{3,9} > 0$. Hence we have $\alpha^* \leqslant 1$. □

Remark 3.7. *It follows from Lemma 2.3 that if the spectral measure Δ has a density function that satisfies the condition (2.17), then $\underline{\alpha} \leqslant \alpha^* \leqslant \alpha_* \leqslant \overline{\alpha}$. Example 2.9 shows that it is possible to have $\underline{\alpha} < \alpha^* = \alpha_* < \overline{\alpha}$.* □

The following result gives general formulas for the Hausdorff dimensions of the range $X([0,1]^N)$ and the graph $\mathrm{Gr}X([0,1]^N) = \{(t, X(t)) : t \in [0,1]^N\}$ in terms of the upper index α^*. When $\alpha_* = \alpha^* \in (0,1)$, (3.20) and (3.21) are essentially due to Adler (1981); otherwise, they seem to be new.

Theorem 3.8. *Let $X = \{X(t), t \in \mathbb{R}^N\}$ be the Gaussian random field in $\mathbb{R}^d$ defined by (3.1). If the associated random field Y satisfies condition (C1) and $0 < \alpha_* \leqslant \alpha^* < 1$, then*

$$\dim_{\mathrm{H}} X([0,1]^N) = \min\left\{d, \frac{N}{\alpha^*}\right\}, \quad \text{a.s.,} \tag{3.20}$$

$$\dim_{\mathrm{H}} \mathrm{Gr}X([0,1]^N) = \min\left\{N + (1-\alpha^*)d, \frac{N}{\alpha^*}\right\}, \quad \text{a.s.,} \tag{3.21}$$

where $\dim_{\mathrm{H}}$ denotes Hausdorff dimension.

Remark 3.9. (a). *We can allow the components $X_1, \cdots, X_d$ in (3.1) to have different distributions. If the upper index of X_i is α_i^*, then the formulae for $\dim_{\mathrm{H}} X([0,1]^N)$ and $\dim_{\mathrm{H}} \mathrm{Gr}X([0,1]^N)$ analogous to those in Theorem 2.1 in Xiao (1995) hold.*

(b). *The proof of Theorem 3.8 also gives explicit expressions for $\dim_{\mathrm{H}} X(E)$ and $\dim_{\mathrm{H}} \mathrm{Gr}X(E)$ for all Borel sets $E \subset \mathbb{R}^N$ with $\dim_{\mathrm{H}} E = \dim_{\mathrm{P}} E$, where $\dim_{\mathrm{P}}$ denotes packing dimension [see Falconer (1990)]. However, the question of determining $\dim_{\mathrm{H}} X(E)$ and $\dim_{\mathrm{H}} \mathrm{Gr}X(E)$ for an arbitrary Borel set $E \subset \mathbb{R}^N$ is more involved and remains to be open.*

(c). *Another open problem is to determine the packing dimensions of $X([0,1]^N)$ and $\mathrm{Gr}X([0,1]^N)$. We conjecture that, under the assumptions of Theorem 3.8, the packing dimension analog of (3.20) and (3.21) hold with α^* being replaced by α_*.* □

For the proof of Theorem 3.8 as well as the proofs of Theorems 3.11 and 3.14 below, we need the following lemma on the modulus of continuity of Y, which is reminiscent to Corollary 2.3 or Theorem 2.10 of Dudley (1973), and on the tail probability of the supremum of Y.

Lemma 3.10. *Assume that the Gaussian random field $Y = \{Y(t), t \in \mathbb{R}^N\}$ in $\mathbb{R}$ satisfies Conditions (C1) and $0 < \alpha_* \leqslant \alpha^* < 1$. Let*

$$\omega_Y(\delta) = \sup_{\substack{t,\ t+s \in [0,1]^N \\ |s| \leqslant \delta}} |Y(t+s) - Y(t)|$$

be the uniform modulus of continuity of $Y(t)$ on $[0,1]^N$. Then there exists a finite constant $K_{3,10} > 0$ such that

$$\limsup_{\delta \to 0} \frac{\omega_Y(\delta)}{\sqrt{\phi(\delta) \log \frac{1}{\delta}}} \leqslant K_{3,10}, \quad \text{a.s..} \tag{3.22}$$

If, in addition, there is a constant $K_{3,11} > 0$ such that

$$\int_1^\infty \Big(\frac{\phi(ae^{-u^2})}{\phi(a)}\Big)^{1/2} du \leqslant K_{3,11}, \quad \textit{for all } a \in [0, \delta_0). \tag{3.23}$$

Then there exist positive constants $K_{3,12}$ and $K_{3,13}$ such that for all $r > 0$ small enough and $u \geqslant K_{3,12}\, \phi^{1/2}(r)$, we have

$$\mathbb{P}\Big\{\sup_{|t|\leqslant r} |Y(t)| \geqslant u\Big\} \leqslant \exp\Big(-\frac{u^2}{K_{3,13}\, \phi(r)}\Big). \tag{3.24}$$

Proof. Because $0 < \alpha_* \leqslant \alpha^* < 1$, the first part, i.e., (3.22), follows from Corollary 2.3 in Dudley (1973). The proof of the second part is based on the Gaussian isopermetric inequality [cf. Talagrand (1995)] and is standard. We include it for completeness.

Let $r < \delta_0$ and $S = \{t : \ |t| \leqslant r\}$. It follows from (3.3) that $d(s,t) \leqslant K_{3,2}^{1/2}\, \phi^{1/2}(|t-s|)$, we have

$$D := \sup\{d(s,t); s,\ t \in S\} \leqslant K_{3,2}^{1/2}\, \phi^{1/2}(r)$$

and

$$N_d(S, \varepsilon) \leqslant K\Big(\frac{r}{\phi^{-1}(\varepsilon^2/K_{3,2})}\Big)^N,$$

where ϕ^{-1} the inverse function of ϕ defined as in Theorem 3.1. Since $\alpha_* > 0$, there exists $\eta > 0$ such that

$$\sigma(r) \leqslant r^\eta \quad \text{for all} \quad r \in [0, \delta_0). \tag{3.25}$$

Some simple calculations and (3.3) yield

$$\begin{aligned}
\int_0^D \sqrt{\log N_d(S,\varepsilon)}\, d\varepsilon &\leqslant K \int_0^{K_{3,2}^{1/2}\phi^{1/2}(r)} \sqrt{\log\Big(\frac{r}{\phi^{-1}(\varepsilon^2/K_{3,2})}\Big)}\, d\varepsilon \\
&\leqslant K \int_0^r \sqrt{\log(r/t)}\ \, d\phi^{1/2}(t) \\
&= K \int_0^1 \frac{1}{u\sqrt{\log(1/u)}}\, \phi^{1/2}(ur) du \\
&\leqslant K \Big(\phi^{1/2}(r) + \int_1^\infty \phi^{1/2}(re^{-u^2}) du\Big) \\
&\leqslant K_{3,14}\, \phi^{1/2}(r)\ ,
\end{aligned}$$

where the last inequality follows from (3.23). It follows from Lemma 2.1

in Talagrand (1995) that for all $u \geqslant K_{3,14}\, \phi^{1/2}(r)$,

$$\begin{aligned} &\mathbb{P}\Big\{ \sup_{|t|\leqslant r} |Y(t)| \geqslant 2\,u \Big\} \\ &\leqslant \mathbb{P}\Big\{ \sup_{|t|\leqslant r} |Y(t)| \geqslant u + \int_0^D \sqrt{\log N_d(S,\varepsilon)}\, d\varepsilon \Big\} \\ &\leqslant \exp\Big(- \frac{u^2}{K_{3,15}\phi(r)} \Big). \end{aligned} \tag{3.26}$$

This proves (3.24) and the lemma. □

Proof of Theorem 3.8. The proofs of the lower bounds in (3.20) and (3.21) using a standard capacity argument are the same as in Adler (1981) or Kahane (1985), which also complete the proof of Theorem 3.8 when $\alpha^* = 0$.

Now we prove the upper bound in (3.20). Since $\dim_{\rm H} X([0,1]^N) \leqslant d$ automatically, we only need to show $\dim_{\rm H} X([0,1]^N) \leqslant N/\alpha^*$ a.s.. Note that for any $\gamma' < \gamma < \alpha^*$, it follows from (3.14) that there exists a sequence $r_n \to 0$ such that $\phi(r_n) \leqslant r_n^{2\gamma}$. For each fixed $n \geqslant 1$, divide $[0,1]^N$ into r_n^{-N} subcubes $C_{n,i}$ $(i = 1, \cdots, r_n^{-N})$ of side-length r_n. It follows from (3.22) in Lemma 3.10 that a.s. for n large enough, each $X(C_{n,i})$ can be covered by a ball of radius $r_n^{\gamma'}$ in $\mathbb{R}^d$. This implies that $\dim_{\rm H} X([0,1]^N) \leqslant N/\gamma'$ a.s. Since $\gamma' < \alpha^*$ is arbitrary, we have $\dim_{\rm H} X([0,1]^N) \leqslant \min\{d, N/\alpha^*\}$ a.s. This proves (3.20). The proof of the upper bound in (3.21) is similar and hence omitted. □

Now we consider the exact Hausdofff measure of the image $X([0,1]^N)$.

Theorem 3.11. *Let $X = \{X(t), t \in \mathbb{R}^N\}$ be a Gaussian random field defined in (3.1) such that the associated real-valued Gaussian random field Y has stationary increments and satisfies condition (C). In addition, we assume that the function ϕ satisfies (3.23) and there exists a constant $K_{3,16} > 0$ such that*

$$\int_1^{\delta_0/a} \Big(\frac{\phi(a)}{\phi(ax)} \Big)^{d/2} x^{N-1} dx \leqslant K_{3,16} \tag{3.27}$$

for all $a \in (0, \delta_0)$, then

$$0 < \varphi_1\text{-}m\big(X([0,1]^N)\big) < \infty \quad \text{a.s.}, \tag{3.28}$$

where $\varphi_1(r) = \big[\phi^{-1}(r^2)\big]^N \log\log 1/r$.

Remark 3.12. *If the random field Y has stationary increments with spectral density f satisfying condition (2.17), then it follows from Lemma 2.4 that (3.23) always holds and, moreover, (3.27) holds whenever $N < \underline{\alpha}\, d$.* □

Proof. Since the proof of (3.28) is similar to that in Xiao (1996), we will only point out the places where modifications have to be made.

To prove the lower bound in (3.28), we follow the standard procedure of using the upper density theorem of Rogers and Taylor (1961). For notational convenience, we assume $\delta_0 = 1$ and define the sojourn time

$$T(r) = \int_{[0,1]^N} \mathbf{1}_{B(0,r)}(X(t))\, dt$$

of X in the closed ball $B(0,r)$.

In order to establish a law of the iterated logarithm for $T(r)$ [the upper half], it is sufficient to show the following estimate of the moments: There exists a positive and finite constant $K_{3,17}$ such that for all integers $n \geqslant 1$,

$$\mathbb{E}\big[T(r)^n\big] \leqslant K_{3,17}^n\, n!\, \big[\phi^{-1}(r^2)\big]^{Nn}. \tag{3.29}$$

This can be proved by using induction, which is where the property of strong local nondeterminism will be needed. The details are given in Xiao (1996) and we only check (3.29) for $n = 1$. Denote $\psi(r) = \phi^{-1}(r^2)$. Note that $\phi(\psi(r)) \geqslant r^2$. Hence by (3.3), a change of variables and (3.27), we derive

$$\begin{aligned}
\mathbb{E}\big[T(r)\big] &\leqslant \int_{[0,1]^N} \min\left\{1,\ K\left(\frac{r}{\phi^{1/2}(|t|)}\right)^d\right\} dt \\
&\leqslant K \int_0^1 \min\left\{1,\ K\left(\frac{r^2}{\phi(\rho)}\right)^{d/2}\right\} \rho^{N-1}\, d\rho \\
&\leqslant K \int_0^{\psi(r)} \rho^{N-1} d\rho + K \int_{\psi(r)}^1 \left(\frac{r^2}{\phi(\rho)}\right)^{d/2} \rho^{N-1} d\rho \\
&\leqslant K\, \psi(r)^N \left\{1 + \int_1^{1/\psi(r)} \left(\frac{\phi(\psi(r))}{\phi(\psi(r)x)}\right)^{d/2} x^{N-1} dx\right\} \\
&\leqslant K\, \psi(r)^N < \infty.
\end{aligned} \tag{3.30}$$

In order to prove the upper bound in (3.28), we need to construct a sequence of economical coverings $\{B_n, n \geqslant 1\}$ for $X([0,1]^N)$ such that $\mathrm{diam} B_n \to 0$ as $n \to \infty$ and almost surely $\sum_n \varphi_1(\mathrm{diam} B_n) < \infty$. This has been done by Talagrand (1995) for fractional Brownian motion and by Xiao (1996) for any Gaussian random field with stationary increments such that its variance function $\sigma^2(h)$ is regularly varying at 0. By examining carefully the proofs in Xiao (1996), we see that the key ingredient for the construction of the desired coverings of $X([0,1]^N)$, Proposition 3.1 in Xiao (1996), is still valid under the present conditions. The rest of the proof is the same as in Xiao (1996). Therefore (3.28) is proven. □

Remark 3.13. *In a similar way, the results on the exact Hausdorff measure of the graph set* $\mathrm{Gr}X([0,1]^N)$ *in Xiao (1997a, c) can be extended to Gaussian random fields in this paper.* □

3.3 Local times and level sets of Gaussian random fields

Let $X = \{X(t), t \in \mathbb{R}^N\}$ be a Gaussian random field with stationary increments in $\mathbb{R}^d$ defined by (3.1). Suppose the associated real-valued random field Y satisfies (3.3) and for some $\varepsilon > 0$,

$$\int_{[0,1]^N} \frac{dh}{\sigma^{d+\varepsilon}(h)} < \infty.$$

If Y is locally nondeterministic on a cube $I \subset \mathbb{R}^N$, say, $I = [0,1]^N$, then it follows from Theorem 26.1 in Geman and Horowitz (1980) [see also Berman (1973) and Pitt (1978)] that X has a jointly continuous local time $L(x,t) := L(x, [0,t])$ for $(x,t) \in \mathbb{R}^d \times I$ and satisfies certain Hölder conditions in the time and space variables, respectively.

Under the assumptions of strong local nondeterminism and regular variation of $\sigma^2(h)$, Xiao (1997a) has established sharp local and uniform Hölder conditions for the local time $L(x,t)$ in the time variable t. Besides interest in their own right, such results are also useful in studying the fractal properties of the sample paths of X. In the following, we show that the results in Xiao (1997a) and Kasahara et al. (1999) still hold under the more general condition (C). For simplicity, we will only consider the case $N = 1$.

Theorem 3.14. *Let* $X = \{X(t), t \in \mathbb{R}\}$ *be a mean zero Gaussian process in* $\mathbb{R}^d$ *defined by (3.1) satisfying condition (C). In addition, we assume that the function* ϕ *satisfies (3.23) and there exist constants* $\gamma_0 \in (0,1)$ *and* $K_{3,18} > 0$ *such that*

$$\int_0^1 \left(\frac{\phi(a)}{\phi(as)}\right)^{\frac{d}{2}+\gamma_0} ds \leqslant K_{3,18} \tag{3.31}$$

for all $a \in (0, \delta_0)$. *Then the following properties hold:*

(i) X *has a local time* $L(x,t)$ *that is jointly continuous in* (x,t) *almost surely.*

(ii) *For any $B \in \mathcal{B}(\mathbb{R})$ define $L^*(B) = \sup_{x\in\mathbb{R}^d} L(x,B)$ be the maximum local time. Then there exists a positive constant $K_{3,19}$ such that for all $t \in \mathbb{R}$,*

$$\limsup_{r\to 0} \frac{L^*(B(t,r))}{\varphi_2(r)} \leqslant K_{3,19}, \quad \text{a.s.}, \tag{3.32}$$

and for all intervals $I \subseteq \mathbb{R}$, there exists a positive finite constant $K_{3,20}$ such that

$$\limsup_{r\to 0} \sup_{t\in I} \frac{L^*(B(t,r))}{\varphi_3(r)} \leqslant K_{3,20}, \quad \text{a.s.}, \tag{3.33}$$

where $B(t,r) = (t-r, t+r)$,

$$\varphi_2(r) = \frac{r}{\phi(r(\log\log 1/r)^{-1})^{d/2}} \quad \text{and} \quad \varphi_3(r) = \frac{r}{\phi(r(\log 1/r)^{-1})^{d/2}}.$$

Remark 3.15. *If X has stationary increments and its spectral measure satisfies (2.17) then (3.23) always holds. Moreover, if $1 > \overline{\alpha}\, d$, then Lemma 2.3 implies that (3.31) is satisfied for any $\gamma_0 \in (0, (1 - \overline{\alpha}d)/(2\overline{\alpha}))$.* □

The following states that the local Hölder condition for the maximum local time is sharp.

Remark 3.16. *By the definition of local times, we have that for any interval $Q \subseteq \mathbb{R}$,*

$$\begin{aligned} |Q| &= \int_{\overline{X(Q)}} L(x,Q)\, dx \\ &\leqslant L^*(Q) \cdot \Big(\sup_{s,t\in Q} |X(s)-X(t)|\Big)^d. \end{aligned} \tag{3.34}$$

If X has stationary increments and satisfies the conditions of Theorem 3.14, then Theorem 3.1 and the proof of Theorem 7.1 in Li and Shao (2001) imply the existence of a constant $K_{3,21} \geqslant 1$ such that for every $t \in \mathbb{R}$,

$$K_{3,21}^{-1} \leqslant \liminf_{r\to 0} \frac{\sup_{s\in B(t,r)} |X(s)-X(t)|}{\phi^{1/2}\big(r/(\log\log(1/r))^{1/N}\big)} \leqslant K_{3,21}, \quad \text{a.s..} \tag{3.35}$$

By taking $Q = B(t,r)$ in (3.34) and using the upper bound in (3.35), we derive the lower bound in the following

$$K_{3,22} \leqslant \limsup_{r\to 0} \frac{L^*(B(t,r))}{\varphi_2(r)} \leqslant K_{3,19}, \quad \text{a.s.}, \tag{3.36}$$

where $K_{3,22} > 0$ *is a constant and the upper bound is given by (3.32). A similar lower bound for (3.33) could also be established by using (3.34), if one proves that for every interval* $I \subseteq \mathbb{R}$,

$$\liminf_{r\to 0} \inf_{t\in I} \sup_{s\in B(t,r)} \frac{|X(s)-X(t)|}{\phi^{1/2}(r/(\log 1/r)^{1/N})} \leqslant K_{3,23}, \qquad \text{a.s..} \tag{3.37}$$

This is left to the interested reader. □

The proof Theorem 3.14 is similar to Xiao (1997a) which is based on getting sharp moment estimates for $L(x, B)$ and $L(x+y, B)-L(x, B)$ and on a chaining argument. We will not reproduce all the details. Instead, we give a simplified proof of the following key estimates.

Lemma 3.17. *Under the conditions of Theorem 3.14, there exist positive constants* $K_{3,24}$ *and* $K_{3,25}$ *such that for all integers* $n \geqslant 1$, $r(0, \delta_0)$, $x \in \mathbb{R}^d$ *and* $0 < \gamma < \gamma_0$, *we have*

$$\mathbb{E}\big[L(x,r)^n\big] \leqslant \frac{K_{3,24}^n\, r^n}{\phi(r/n)^{nd/2}} \tag{3.38}$$

and

$$\mathbb{E}\big[L(x+y,r)-L(x,r)\big]^n \leqslant \frac{K_{3,25}^n\, |y|^{n\gamma}\, r^n}{\big[\phi(r/n)\big]^{(d+2\gamma)n/2}}\, (n!)^\gamma. \tag{3.39}$$

For the proof of Lemma 3.17, we will need several lemmas. Lemma 3.18 is essentially due to Cuzick and DuPreez (1982) and Lemma 3.19 extends Lemma 3 of Kasahara et al. (1999).

Lemma 3.18. *Let* $Z_1, \cdots, Z_n$ *be mean zero Gaussian variables which are linearly independent. Then for any measurable function* $g: \mathbb{R} \to \mathbb{R}_+$,

$$\begin{aligned} &\int_{\mathbb{R}^n} g(v_1) e^{-\frac{1}{2}\mathrm{Var}(\sum_{j=1}^n v_j Z_j)}\, dv_1 \cdots dv_n \\ &= \frac{(2\pi)^{n-1}}{(\mathrm{detCov}(Z_1,\cdots,Z_n))^{1/2}} \int_{\mathbb{R}} g\Big(\frac{v}{\sigma_1}\Big) e^{-v^2/2} dv, \end{aligned} \tag{3.40}$$

where $\mathrm{detCov}(Z_1,\cdots,Z_n)$ *denotes the determinant of the covariance matrix of* $(Z_1,\cdots,Z_n)$, *and* $\sigma_1^2 = \mathrm{Var}(Z_1|Z_2,\cdots,Z_n)$ *is the conditional variance of* Z_1 *given* $Z_2, \cdots, Z_n$.

Lemma 3.19. *Let* $U(x)$ *be a right continuous, non-decreasing function on* $\mathbb{R}_+$ *with* $U(0) = 0$. *If there exists a constant* $K_{3,26} > 0$ *such that* $U(2t) \leqslant K_{3,26} U(t)$ *for all* $t > 0$, *then as* $n \to \infty$,

$$\left[\int_{\mathbb{R}_+^n \cap \{0 < t_1 + t_2 + \cdots + t_n \leqslant 1\}} dU(t_1) \cdots dU(t_n)\right]^{1/n} \asymp U(1/n). \tag{3.41}$$

Proof. The lower bound follows easily from

$$\begin{aligned}&\int_{\mathbb{R}_+^n\cap\{0<t_1+t_2+\cdots+t_n\leqslant 1\}} dU(t_1)\cdots dU(t_n)\\ &\geqslant \int_0^{1/n}\cdots\int_0^{1/n} dU(t_1)\cdots dU(t_n)\\ &=\big[U(1/n)\big]^n.\end{aligned}\tag{3.42}$$

To prove the upper bound, we follow the argument of Kasahara et al. (1999) and define the distribution functions F_n on $\mathbb{R}_+$ by

$$F_n(t)=\int_{\mathbb{R}_+^n\cap\{0\leqslant t_1+t_2+\cdots+t_n\leqslant t\}} dU(t_1)\cdots dU(t_n).$$

Then the integral on the left-hand side of (3.42) is $F_n(1)$. Note that the Laplace transform of F_n can be written as

$$\int_0^\infty e^{-st}dF_n(t)=\Big(\int_0^\infty e^{-st}\,dU(t)\Big)^n.$$

Hence we have

$$\begin{aligned}F_n(1)&\leqslant e^n\int_0^1 e^{-nt}dF_n(t)\leqslant e^n\int_0^\infty e^{-nt}dF_n(t)\\ &\leqslant\Big(e\int_0^\infty e^{-nt}dU(t)\Big)^n.\end{aligned}\tag{3.43}$$

It follows that

$$\limsup_{n\to\infty}\frac{1}{U(1/n)}\big[F_n(1)\big]^{1/n}\leqslant\limsup_{n\to\infty}\frac{e}{U(1/n)}\int_0^\infty e^{-nt}dU(t).\tag{3.44}$$

Now we split the last integral in (3.44) over the intervals $[0,n^{-1})$ and $[n^{-1}2^{k-1},n^{-1}2^k)$ $(k\geqslant 1)$, which gives

$$\begin{aligned}\int_0^\infty e^{-nt}dU(t)&\leqslant U(1/n)+\sum_{k=1}^\infty e^{-2^{k-1}}U(2^k/n)\\ &\leqslant U(1/n)\Big[1+\sum_{k=1}^\infty e^{-2^{k-1}}K_{3,26}^k\Big]\\ &=K_{3,27}\,U(1/n),\end{aligned}\tag{3.45}$$

where in deriving the second inequality, we have made use of the doubling property of U. Therefore, the upper bound in (3.41) follows from (3.44) and (3.45). □

Proof of Lemma 3.17. It follows from (25.5) and (25.7) in Geman and Horowitz (1980) [see also Pitt (1978)] that for any $x, y \in \mathbb{R}^d$, $B \in \mathcal{B}(\mathbb{R})$ and any integer $n \geqslant 1$, we have

$$\begin{aligned}\mathbb{E}\big[L(x,B)\big]^n = (2\pi)^{-nd} \int_{B^n}\int_{\mathbb{R}^{nd}} \exp\Big(-i\sum_{j=1}^n \langle u_j, x\rangle\Big) \\ \times\ \mathbb{E}\exp\Big(i\sum_{j=1}^n \langle u_j, X(t_j)\rangle\Big) d\overline{u}\, d\overline{t}\end{aligned} \tag{3.46}$$

and for any even integer $n \geqslant 2$,

$$\begin{aligned}&\mathbb{E}\big[L(x+y,B) - L(x,B)\big]^n \\ &= (2\pi)^{-nd} \int_{B^n}\int_{\mathbb{R}^{nd}} \prod_{j=1}^n \Big(e^{-i\langle u_j, x+y\rangle} - e^{-i\langle u_j, x\rangle}\Big) \\ &\quad \times\ \mathbb{E}\exp\Big(i\sum_{j=1}^n \langle u_j, X(t_j)\rangle\Big) d\overline{u}\, d\overline{t},\end{aligned} \tag{3.47}$$

where $\overline{u} = (u_1, \cdots, u_n)$, $\overline{t} = (t_1, \cdots, t_n)$, and each $u_j \in \mathbb{R}^d$, $t_j \in \mathbb{R}$. In the coordinate notation we then write $u_j = (u_j^1, \cdots, u_j^d)$.

Take $B = [0, r]$. It follows from (3.46) and the proof of Lemma 2.5 in Xiao (1997a) that for all integers $n \geqslant 1$,

$$\mathbb{E}\big[L(x,r)^n\big] \leqslant (2\pi)^{-nd/2} \int_{[0,\,r]^n} \frac{1}{\big[\det C_n(t_1, \cdots, t_n)\big]^{d/2}}\, dt_1 \cdots dt_n, \tag{3.48}$$

where $C_n(t_1, \cdots, t_n)$ denotes the covariance matrix of the Gaussian variables $X_1(t_1), \cdots, X_1(t_n)$. It is well known that

$$\begin{aligned}&\det\big(C_n(t_1, \cdots, t_n)\big) \\ &= \mathrm{Var}\big(X_1(t_1)\big) \prod_{j=2}^n \mathrm{Var}\big(X_1(t_j) | X_1(t_1), \cdots, X_1(t_{j-1})\big).\end{aligned} \tag{3.49}$$

We apply (C2) to derive that for any $0 < t_1 < t_2 < \cdots < t_n$,

$$K^n \prod_{j=1}^n \phi(t_j - t_{j-1}) \leqslant \det\big(C_n(t_1, \cdots, t_n)\big) \leqslant \prod_{j=1}^n \phi(t_j - t_{j-1}), \tag{3.50}$$

where $t_0 = 0$. By (3.48) — (3.50) and a simple substitution, we deduce that

$$\mathbb{E}\big[L(x,r)^n\big] \leqslant K^n\, n! \int_{0<t_1<\cdots<t_n\leqslant r} \prod_{j=1}^{n} \frac{1}{\big(\phi(t_j-t_{j-1})\big)^{d/2}}\, dt_1\cdots dt_n$$
$$\leqslant K^n\, n!\, r^n \int_{0<s_1+\cdots+s_n\leqslant 1} dU_1(s_1)\cdots dU_1(s_n), \tag{3.51}$$

where the function $U_1(t)$ is defined by

$$U_1(t) = \int_0^{\min\{t,1\}} \frac{ds}{(\phi(rs))^{d/2}}, \quad \text{for all } \ t \geqslant 0.$$

Since ϕ is non-decreasing, we see that $U_1(2t) \leqslant 2U_1(t)$ for all $t \geqslant 0$. Hence it follows from Lemma 3.19 that

$$\mathbb{E}\big[L(x,r)^n\big] \leqslant K_{3,28}^n\, n!\, r^n \big[U_1(1/n)\big]^n. \tag{3.52}$$

On the other hand, by (3.31), we derive that

$$U_1(t) \leqslant K_{3,29} \frac{t}{\big(\phi(rt)\big)^{d/2}}, \qquad \text{for all } 0 \leqslant t \leqslant 1. \tag{3.53}$$

Therefore, (3.38) follows from (3.52), (3.53) and Stirling's formula.

Now we turn to the proof of (3.39). By (3.47) and the elementary inequality

$$|e^{iu}-1| \leqslant 2^{1-\gamma}|u|^\gamma, \qquad \text{for all } \ u\in\mathbb{R},\ 0<\gamma<1,$$

we see that for all even integers $n \geqslant 2$ and any $0<\gamma<1$,

$$\mathbb{E}\big[L(x+y,r)-L(x,r)\big]^n \leqslant (2\pi)^{-nd} 2^{(1-\gamma)n} |y|^{n\gamma} \int_{[0,\,r]^n}\int_{\mathbb{R}^{nd}} \prod_{j=1}^{n}|u_j|^\gamma$$
$$\times \exp\left(-\frac{1}{2}\mathrm{Var}\Big(\sum_{j=1}^{n}\langle u_j,\ X(t_j)\rangle\Big)\right) d\overline{u}\, d\overline{t}. \tag{3.54}$$

Since for any $0<\gamma<1$, $|a+b|^\gamma \leqslant |a|^\gamma + |b|^\gamma$, we have

$$\prod_{j=1}^{n}|u_j|^\gamma \leqslant \sum{}' \prod_{j=1}^{n} |u_j^{k_j}|^\gamma, \tag{3.55}$$

where the summation $\sum'$ is taken over all $(k_1,\cdots,k_n) \in \{1,\cdots,d\}^n$. Fix such a sequence $(k_1,\cdots,k_n)$ and fix n points $0<t_1<\cdots<t_n\leqslant r$, we consider the integral

$$I_3 := \int_{\mathbb{R}^{nd}} \prod_{j=1}^{n} |u_j^{k_j}|^\gamma \ \exp\left(-\frac{1}{2}\mathrm{Var}\Big(\sum_{j=1}^{n}\langle u_j,\ X(t_j)\rangle\Big)\right) d\overline{u}.$$

It follows from (C2) that the Gaussian random variables $X_l(t_j)$ ($l = 1, \cdots, d,\ j = 1, \cdots, n$) are linearly independent. Hence we use the generalized Hölder's inequality and Lemma 3.18 to deduce that I_3 is at most

$$\begin{aligned}
&\prod_{j=1}^{n} \left\{ \int_{\mathbb{R}^{nd}} |u_j^{k_j}|^{n\gamma} \exp\Big[-\frac{1}{2} \mathrm{Var}\Big(\sum_{j=1}^{n} \sum_{l=1}^{d} u_j^l X_l(t_j) \Big) \Big] \, d\overline{u} \right\}^{1/n} \\
&= \frac{(2\pi)^{nd-1}}{\big[\det \mathrm{Cov}(X_l(t_j),\ 1 \leqslant l \leqslant d,\ 1 \leqslant j \leqslant n)\big]^{1/2}} \\
&\qquad \times \int_{\mathbb{R}} |v|^{n\gamma} \exp\Big(-\frac{v^2}{2} \Big) dv \ \prod_{j=1}^{n} \frac{1}{\sigma_j^{\gamma}} \\
&\leqslant \frac{K^n \, (n!)^{\gamma}}{\big[\det \mathrm{Cov}(Y(t_1), \cdots, Y(t_n))\big]^{d/2}} \prod_{j=1}^{n} \frac{1}{\sigma_j^{\gamma}},
\end{aligned} \tag{3.56}$$

where σ_j^2 is the conditional variance of $X_{k_j}(t_j)$ given $X_l(t_i)$ ($l \neq k_j$ or $l = k_j,\ i \neq j$) and the last inequality follows from Stirling's formula.

It follows from the independence of the Gaussian random fields X_1, $\cdots, X_n$ and condition (C) that

$$\sigma_j^2 \geqslant K \, \min\big\{ \phi(t_j - t_{j-1}),\ \phi(t_{j+1} - t_j) \big\},$$

where $t_0 := 0$. Hence

$$\prod_{j=1}^{n} \frac{1}{\sigma_j^{\gamma}} \leqslant K^n \prod_{j=1}^{n} \frac{1}{\big[\phi(t_j - t_{j-1})\big]^{\gamma}}. \tag{3.57}$$

Combining (3.56), (3.49), (3.50) and (3.57), we obtain

$$I_3 \leqslant K^n \, (n!)^{\gamma} \prod_{j=1}^{n} \frac{1}{\big[\phi(t_j - t_{j-1})\big]^{(d+2\gamma)/2}}. \tag{3.58}$$

It follows from (3.54), (3.55) and (3.58) that

$$\begin{aligned}
&\mathbb{E}\big[L(x+y, r) - L(x, r)\big]^n \leqslant K^n, |y|^{n\gamma} (n!)^{1+\gamma} \\
&\qquad \times \int_{0 < t_1 < \cdots < t_n \leqslant r} \prod_{j=1}^{n} \frac{1}{\big[\phi(t_j - t_{j-1})\big]^{(d+2\gamma)/2}} \, dt_1 \cdots dt_n \\
&\qquad \leqslant K^n |y|^{n\gamma} (n!)^{1+\gamma} \, r^n \int_{0 < s_1 + \cdots + s_n \leqslant 1} dU_2(s_1) \cdots dU_2(s_n),
\end{aligned} \tag{3.59}$$

where the function $U_2(t)$ is defined by

$$U_2(t) = \int_0^{\min\{t,1\}} \frac{ds}{(\phi(rs))^{(d+2\gamma)/2}}, \qquad t \geqslant 0.$$

Again U_2 has the doubling property. Hence it follows from Lemma 3.19 that

$$\mathbb{E}\big[L(x+y,r)-L(x,r)\big]^n \leqslant K_{3,30}^n\,|y|^{n\gamma}(n!)^{1+\gamma}\,r^n\,\Big[U_2\big(\frac{1}{n}\big)\Big]^n. \tag{3.60}$$

Finally, (3.31) implies that

$$U_2(t) \leqslant K_{3,31}\,\frac{t}{\big[\phi(rt)\big]^{(d+2\gamma)/2}}, \qquad \text{for all } \ t\in[0,1]. \tag{3.61}$$

Therefore, (3.39) follows from (3.60), (3.61) and Stirling's formula. □

Theorem 3.14 can be applied to determine the Hausdorff dimension and Hausdorff measure of the level set $X^{-1}(x)=\{t\in\mathbb{R}: X(t)=x\}$, where $x\in\mathbb{R}^d$. See Berman (1970, 1972), Adler (1981), Monrad and Pitt (1987) and Xiao (1997a). In the following theorem we prove a uniform Hausdorff dimension result for the level sets of the Gaussian process X, extending the previous results of Berman (1972), Monrad and Pitt (1987).

Theorem 3.20. *Let $X=\{X(t), t\in\mathbb{R}\}$ be a Gaussian process in $\mathbb{R}^d$ defined by (3.1) satisfying the conditions of Theorem 3.14. Then with probability one,*

$$\dim_{\rm H} X^{-1}(x) = 1-\alpha^* d, \quad \text{for all } \ x\in\mathcal{O}, \tag{3.62}$$

where $\mathcal{O}$ is the random open set defined by

$$\mathcal{O} = \bigcup_{s,t\in\mathbb{Q};\, s<t}\Big\{x\in\mathbb{R}^d:\ L(x,[s,t])>0\Big\}.$$

Proof. Let Ω_0 be the event on which the modulus of continuity for X [cf. (3.22)] and Theorem 3.14 hold. Clearly, $\mathbb{P}(\Omega_0)=1$. Now we choose and fix an $\omega\in\Omega_0$, and prove our conclusion for the sample path $X(\cdot,\omega)$.

To prove the upper bound in (3.62), it is sufficient to show that almost surely,

$$\dim_{\rm H}\big(X^{-1}(x)\cap[0,1]\big) \leqslant 1-\alpha^* d, \quad \text{for all } \ x\in\mathbb{R}^d. \tag{3.63}$$

For any integer $n\geqslant 1$, we divide the interval $[0,1]$ into 2^n subintervals $I_{n,k}=[(k-1)2^{-n}, k2^{-n}]$ $(k=1,\cdots,2^n)$. For every $x\in\mathbb{R}^d$, denote by $N(n,x)$ the number of k's such that $x\in X(I_{n,k})$. The modulus of continuity of X in Lemma 3.10 implies that if $x\in X(I_{n,k})$ then $X(I_{n,k})\subseteq B(x,\rho_n)$, where $\rho_n=K\sqrt{\phi(2^{-n})\,\log 2^n}$. Since the local time $L(y,1)$ is a continuous in y, it is bounded on $B(x,\rho_n)$. Hence we have

$$N(n,x)\,2^{-n} \leqslant \int_{B(x,\rho_n)} L(y,1)dy \leqslant K_{3,32}\,\rho_n^d, \tag{3.64}$$

where $K_{3,32}$ depends on ω. This gives

$$N(n,x) \leqslant K_{3,32}\, 2^n\, \rho_n^d. \tag{3.65}$$

Hence for every $0 < \gamma < \alpha^*$, there exists a sequence $\{n_k\}$ of positive integers such that $N(n_k, x) \leqslant K_{3,32}\, 2^{n_k(1-\gamma d)}$. This implies (3.63).

To prove the lower bound in (3.62), we note that the jointly continuous local time $L(x,t)$ of X can be extended to become a random Borel measure, denoted by $L(x,\cdot)$, on $X^{-1}(x)$; see Adler (1981). Moreover, for every $x \in \mathcal{O}$, $L(x,\cdot)$ is a positive measure.

Now for any $\gamma > \alpha^*$, (3.33) of Theorem 3.14 implies that almost surely, $L(x, B(t,r)) \leqslant K\, r^{1-\gamma d}$ for all $x \in \mathbb{R}^d$, all $t \in [0,1]$ and $r > 0$ small. By the Frostman lemma [cf. Kahane (1985)], we have almost surely $\dim_{\rm H} X^{-1}(x) \geqslant 1 - \gamma d$ for all $x \in \mathcal{O}$. Since $\gamma > \alpha^*$ is arbitrary, this proves the lower bound in (3.62) and hence the theorem. □

Remark 3.21. *It is an interesting question to characterize the random open set $\mathcal{O}$. Monrad and Pitt (1987) have given a real-valued* periodic *stationary Gaussian process X for which $\mathcal{O}$ is a proper subset of $\mathbb{R}$ [because the range of X is a.s. bounded]. They have shown a sufficient condition in terms of the spectral measure of a stationary (N,d)-Gaussian random field X so that $\mathcal{O} = \mathbb{R}^d$ holds. Monrad and Pitt (1987) also point out that the self-similarity of an (N,d)-fractional Brownian motion B_α implies that if $N > \alpha d$ then $\mathcal{O} = \mathbb{R}^d$ almost surely. However, we do not know whether $\mathcal{O} = \mathbb{R}^d$ is true for the (N,d)-Gaussian random fields satisfying the conditions of Theorem 2.5.* □

The local time $L(0,1)$ [i.e., $L(x,1)$ at $x = 0$] of a Gaussian process X sometimes appears as a limit in some limit theorems on the occupation measure of X; see, for example, Kasahara and Ogawa (1999) and the references therein. Since there is little knowledge on the explicit distribution of $L(0,1)$, it is of interest in estimating the tail probability $\mathbb{P}\{L(0,1) > x\}$ as $x \to \infty$. This problem has been considered by Kasahara et al. (1999) under some extra conditions on the Gaussian process X. The next theorem is an extension of their main result.

Theorem 3.22. *Let $X = \{X(t) : t \in \mathbb{R}\}$ be a mean 0 Gaussian process in $\mathbb{R}^d$ defined by (3.1). We assume that the associated Gaussian process Y satisfies Condition (C) and the condition (3.31) with $\gamma_0 = 0$. Then for $x > 0$ large enough,*

$$-\log \mathbb{P}\big\{L(0,1) > x\big\} \asymp \frac{1}{\phi^{-1}(1/x^2)}, \tag{3.66}$$

where ϕ^{-1} is the inverse function of ϕ as defined in Theorem 3.1.

Theorem 3.22 follows easily from the moment estimates for $L(0,1)$ in Lemma 3.24 and the following lemma on the tail probability of non-negative random variables. When ψ is a power function or a regularly varying function, Lemma 3.23 is well known.

Lemma 3.23. *Let ξ be a non-negative random variable and let $\psi : \mathbb{R}_+ \to \mathbb{R}_+$ be a non-decreasing function having the doubling property. If there exist positive constants $K_{3,33}$ and $K_{3,34}$ such that*

$$K_{3,33}^n\, \psi(n)^n \leqslant \mathbb{E}(\xi^n) \leqslant K_{3,34}^n\, \psi(n)^n$$

for all n large enough, then there exist positive constants $K_{3,35} > K_{3,34}$, $K_{3,36}$ and $K_{3,37}$ such that for all $x > 0$ large enough,

$$e^{-K_{3,36} x} \leqslant \mathbb{P}\big\{\xi \geqslant K_{3,35}\, \psi(x)\big\} \leqslant e^{-K_{3,37} x}. \tag{3.67}$$

Proof. The upper bound in (3.67) follows easily from Chebyshev's inequality and a monotonicity argument. In order to prove the lower bound, we follow the elementary argument of Talagrand (1998). By applying the Paley-Zygmund inequality [cf. Kahane (1985), p.8] to ξ^n, we have

$$\begin{aligned}
\mathbb{P}\Big\{\xi \geqslant \frac{K_{3,33}}{2}\, \psi(n)\Big\} &\geqslant \mathbb{P}\Big\{\xi^n \geqslant \frac{1}{2}\,\mathbb{E}(\xi^n)\Big\} \\
&\geqslant \frac{1}{4}\, \frac{\big[\mathbb{E}(\xi^n)\big]^2}{\mathbb{E}(\xi^{2n})} \\
&\geqslant \frac{1}{4}\, \frac{K_{3,33}^{2n}\, \psi(n)^{2n}}{K_{3,34}^{2n}\, \psi(2n)^{2n}}.
\end{aligned}$$

Now it is clear that the lower bound in (3.67) follows from the doubling property of ψ and a standard monotonicity argument. □

Lemma 3.24. *There exist positive and finite constants $K_{3,38}$ and $K_{3,39}$ such that for all integers $n \geqslant 1$*

$$\frac{K_{3,38}^n}{\phi(1/n)^{nd/2}} \leqslant \mathbb{E}\big[L(0,1)^n\big] \leqslant \frac{K_{3,39}^n}{\phi(1/n)^{nd/2}}. \tag{3.68}$$

Proof. As in the proofs of Lemma 2.5 in Xiao (1997a) or Lemma 1 in Kasahara et al. (1999), we derive from (3.46) that for any integer $n \geqslant 1$,

$$\mathbb{E}\big[L(0,1)^n\big] = (2\pi)^{-nd/2} \int_{[0,\,1]^n} \frac{1}{\big[\det C_n(t_1, \cdots, t_n)\big]^{d/2}}\, dt_1 \cdots dt_n. \tag{3.69}$$

It follows from (3.49) and (3.50) that

$$\Big[\mathbb{E}\big(L(0,1)^n\big)\Big]^{1/n} \asymp \left[n! \int_{0<t_1<\cdots<t_n\leqslant 1} \prod_{j=1}^{n} \frac{1}{\sigma^d(t_j-t_{j-1})}\, dt_1\cdots dt_n\right]^{1/n}$$
$$\asymp \left[n! \int_{0<s_1+s_2+\cdots+s_n\leqslant 1} dU_3(s_1)\cdots dU_3(s_n)\right]^{1/n}, \tag{3.70}$$

where the function $U_3(t)$ is defined by

$$U_3(t) = \int_0^{\min\{t,1\}} \frac{ds}{(\phi(s))^{d/2}}, \quad \forall\, t\geqslant 0.$$

Since ϕ is non-decreasing, we see that $U_3(2t) \leqslant 2U_3(t)$ for all $t \geqslant 0$. Hence it follows from Lemma 3.19 that

$$\Big[\mathbb{E}\big(L(0,1)^n\big)\Big]^{1/n} \asymp (n!)^{1/n} U_3(1/n). \tag{3.71}$$

Therefore, (3.68) follows from (3.53) and Stirling's formula as in the proof of Lemma 3.17. □

We end this section with the following open problem.

Question 3.25. Let $X = \{X(t) : t\in\mathbb{R}\}$ be the Gaussian process in Theorem 3.22. Does the limit $\lim_{x\to\infty} \phi^{-1}(1/x^2)\log \mathbb{P}\big\{L(0,1) > x\big\}$ exist? □

References

[1] R. J. Adler (1981), *The Geometry of Random Fields.* Wiley, New York.

[2] R. Addie, P. Mannersalo and I. Norros (2002), Performance formulae for queues with Gaussian input. *European Trans. Telecommunications* **13**(3), 183–196.

[3] V. V. Anh, J. M. Angulo and M. D. Ruiz-Medina (1999), Possible long-range dependence in fractional random fields. *J. Stat. Plann. Infer.* **80**, 95–110.

[4] A. Ayache, D. Wu and Y. Xiao (2006), Joint continuity of the local times of fractional Brownian sheets. *Ann. Inst. H. Poincare Probab. Stat.*, to appear.

[5] A. Ayache and Y. Xiao (2005), Asymptotic properties and Hausdorff dimensions of fractional Brownian sheets. *J. Fourier Anal. Appl.* **11**, 407–439.

[6] D. A. Benson, M. M. Meerschaert and B. Baeumer (2006), Aquifer operator-scaling and the efferct on solute mixing and dispersion. *Water Resour. Res.* **42**, W01415.

[7] C. Berg and G. Forst (1975), *Potential Theory on Locally Compact Abelian Groups.* Springer-Verlag, New York-Heidelberg.

[8] S. M. Berman (1972), Gaussian sample function: uniform dimension and Hölder conditions nowhere. *Nagoya Math. J.* **46**, 63–86.

[9] S. M. Berman (1973), Local nondeterminism and local times of Gaussian processes. *Indiana Univ. Math. J.* **23**, 69–94.

[10] S. M. Berman (1978), Gaussian processes with biconvex covariances. *J. Multivar. Anal.* **8**, 30–44.

[11] S. M. Berman (1988), Spectral conditions for local nondeterminism. *Stoch. Process. Appl.* **27**, 73–84.

[12] S. M. Berman (1991), Self-intersections and local nondeterminism of Gaussian processes. *Ann. Probab.* **19**, 160–191.

[13] H. Biermé, M. M. Meerschaert and H.-P. Scheffler (2007), Operator scaling stable random fields. *Stoch. Process. Appl.* **117**, 312–332.

[14] N. H. Bingham, C. M. Goldie and J. L. Teugels (1987), *Regular Variation.* Cambridge University Press, Cambridge.

[15] A. Bonami and A. Estrade (2003), Anisotropic analysis of some Gaussian models. *J. Fourier Anal. Appl.* **9**, 215–236.

[16] P. Cheridito (2004), Gaussian moving averages, semimartingales and option pricing. *Stoch. Process. Appl.* **109**, 47–68.

[17] M. Csörgő, Z.-Y. Lin and Q.-M. Shao (1995), On moduli of continuity for local times of Gaussian processes. *Stochastic Process. Appl.* **58**, 1–21.

[18] J. Cuzick (1977), A lower bound for the prediction error of stationary Gaussian processes. *Indiana Univ. Math. J.* **26**, 577–584.

[19] J. Cuzick (1978), Local nondeterminism and the zeros of Gaussian processes. *Ann. Probab.* **6**, 72–84.

[20] J. Cuzick and J. DuPreez (1982), Joint continuity of Gaussian local times. *Ann. Probab.* **10**, 810–817.

[21] R. M. Dudley (1973), Sample functions of the Gaussian proceses. *Ann. Probab.* **3**, 66–103.

[22] K. J. Falconer (1990), *Fractal Geometry.* John Wiley & Sons Ltd., Chichester.

[23] D. Geman and J. Horowitz (1980), Occupation densities. *Ann. Probab.* **8**, 1–67.

[24] J.-P. Kahane (1985), *Some Random Series of Functions.* 2nd edition, Cambridge University Press, Cambridge.

[25] Y. Kasahara and N. Ogawa (1999), A note on the local time of fractional Brownian motion. *J. Theoret. Probab.* **12**, 207–216.

[26] Y. Kasahara, N. Kôno and T. Ogawa (1999), On tail probability of local times of Gaussian processes. *Stoch. Process. Appl.* **82**, 15–21.

[27] D. Khoshnevisan, D. Wu and Y. Xiao (2006), Sectorial local nondeterminism and the geometry of the Brownian sheet. *Electron. J. Probab.* **11**, 817–843.

[28] D. Khoshnevisan and Y. Xiao (2004), Images of the Brownian sheet. *Trans. Am. Math. Soc.* to appear.

[29] J. Kuelbs, W. V. Li and Q.-M. Shao (1995), Small ball probabilities for Gaussian processes with stationary increments under Hölder norms. *J. Theoret. Probab.* **8**, 361–386.

[30] M. Ledoux (1996), Isoperimetry and Gaussian analysis. *Lecture Notes in Math.* **1648**, 165–294, Springer-Verlag, Berlin.

[31] W. V. Li and Q.-M. Shao (2001), Gaussian processes: inequalities, small ball probabilities and applications. In *Stochastic Processes: Theory and Methods.* Handbook of Statistics, **19**, (C. R. Rao and D. Shanbhag, editors), pp. 533–597, North-Holland.

[32] M. A. Lifshits (1999), Asymptotic behavior of small ball probabilities. In: *Probab. Theory and Math. Stat., Proc. VII International Vilnius Conference (1998).* Vilnius, VSP/TEV, pp. 533–597.

[33] P. Mannersalo and I. Norros (2002), A most probable path approach to queueing systems with general Gaussian input. *Comp. Networks* **40** (3), 399–412.

[34] M. B. Marcus (1968), Gaussian processes with stationary increments possessing discontinuous sample paths. *Pac. J. Math.* **26**, 149–157.

[35] M. B. Marcus and J. Rosen (2006), *Markov Processes, Gaussian Processes, and Local Times.* Cambridge University Press, Cambridge.

[36] D. Monrad and L. D. Pitt (1987), Local nondeterminism and Hausdorff dimension. In: *Progress in Probability and Statistics. Seminar on Stochastic Processes 1986*, (E, Cinlar, K. L. Chung, R. K. Getoor, Editors), pp.163–189, Birkhauser, Boston.

[37] D. Monrad and H. Rootzén (1995), Small values of Gaussian processes and functional laws of the iterated logarithm. *Probab. Theory Relat. Fields* **101**, 173–192.

[38] C. Mueller and R. Tribe (2002), Hitting properties of a random string. *Electron. J. Probab.* **7**, no. 10, 29 pp.

[39] E. J. G. Pitman (1968), On the behavior of the characteristic function of a probability sidtribution in the neighbourhood of the origin. *J. Aust. Math. Soc. Series A* **8**, 422–443.

[40] L. D. Pitt (1975), Stationary Gaussian Markov fields on $\mathbb{R}^d$ with a deterministic component. *J. Multivar. Anal.* **5**, 300–311.

[41] L. D. Pitt (1978), Local times for Gaussian vector fields. *Indiana Univ. Math. J.* **27**, 309–330.

[42] L. D. Pitt and L. T. Tran (1979), Local sample path properties of Gaussian fields. *Ann. Probab.* **7**, 477–493.

[43] C. A. Rogers and S. J. Taylor (1961), Functions continuous and singular with respect to a Hausdorff measure. *Mathematika* **8**, 1–31.

[44] J. Rosen (1984), Self-intersections of random fields. *Ann. Probab.* **12**, 108–119.

[45] Q.-M. Shao and D. Wang (1995), Small ball probabilities of Gaussian fields. *Probab. Theory Relat. Fields* **102**, 511–517.

[46] N.-R. Shieh and Y. Xiao (2006), Images of Gaussian random fields: Salem sets and interior points. *Studia Math.* **176**, 37–60. .

[47] W. Stolz (1996), Some small ball probabilities for Gaussian processes under nonuniform norms. *J. Theoret. Probab.* **9**, 613–630.

[48] M. Talagrand (1993), New Gaussian estimates for enlarged balls. *Geomet Funt. Anal.* **3**, 502–526.

[49] M. Talagrand (1995), Hausdorff measure of trajectories of multiparameter fractional Brownian motion. *Ann. Probab.* **23**, 767–775.

[50] M. Talagrand (1998), Multiple points of trajectories of multiparameter fractional Brownian motion. *Probab. Theory Relat. Fields* **112**, 545–563.

[51] C. A. Tudor and Y. Xiao (2006), Sample path properties of bifractional Brownian motion. *Submitted.*

[52] D. Wu and Y. Xiao (2006), Geometric properties of the images of fractional Brownian sheets. *J. Fourier Annal. Appl.*, to appear.

[53] Y. Xiao (1995), Dimension results for Gaussian vector fields and index-α stable fields. *Ann. Probab.* **23**, 273 – 291.

[54] Y. Xiao (1996), Hausdorff measure of the sample paths of Gaussian random fields. *Osaka J. Math.* **33**, 895–913.

[55] Y. Xiao (1997a), Hölder conditions for the local times and the Hausdorff measure of the level sets of Gaussian random fields. *Probab. Theory Relat. Fields* **109**, 129–157.

[56] Y. Xiao (1997b), Weak variation of Gaussian processes. *J. Theoret. Probab.* **10**, 849–866.

[57] Y. Xiao (1997c), Hausdorff measure of the graph of fractional Brownian motion. *Math. Proc. Cambridge Philos. Soc.* **122**, 565–576.

[58] Y. Xiao (2003), The packing measure of the trajectories of multiparameter fractional Brownian motion. *Math. Proc. Cambridge Philos. Soc.* **135**, 349–375.

[59] Y. Xiao (2006), Properties of local nondeterminism of Gaussian and stable random fields and their applications. *Ann. Fac. Sci. Toulouse Math.* **XV**, 157–193.

[60] Y. Xiao (2007a), Sample path properties of anisotropic Gaussian random fields. *Submitted.*

[61] Y. Xiao (2007b), Spectral conditions for strong local nondeterminism of Gaussian random fields. *In Preparation.*

[62] Y. Xiao and T. Zhang (2002), Local times of fractional Brownian sheets. *Probab. Theory Relat. Fields* **124**, 204–226.

[63] A. M. Yaglom (1957), Some classes of random fields in n-dimensional space, related to stationary random processes. *Th. Probab. Appl.* **2**, 273–320.

Large Deviations for Two-Parameter Gaussian Processes Related to Change-Point Analysis *

Bin Chen and Miklós Csörgő
School of Mathematics and Statistics, Carleton University
1125 Colonel By Drive
Ottawa, Ontario, Canada K1S 5B6
E-mail: mcsorgo@math.carleton.ca

Abstract

We obtain large deviation results for a Brownian sheet, a Kiefer process and a "tied down" Kiefer process in the q-metric $d_q(f,g) := \sup_{0\leqslant x\leqslant 1}\sup_{0<t\leqslant 1}\{|f(x,t) - g(x,t)|/q(t)\}$, where q is an appropriate weight function. As a consequence, we conclude some tail probability estimations that can be made use of in change-point analysis.

1 Introduction

Let $F(x)$ be a non-degenerate distribution function. We say that the separable real valued process $K := \{K_F(x,t), x \in \mathbb{R}, 0 \leqslant t < \infty\}$ is a Kiefer process indexed by F, if K_F is a Gaussian process with zero mean and covariance

$$EK_F(x,t)K_F(y,s) = (F(x\wedge y) - F(x)F(y))(t\wedge s)$$

for all $x, y \in \mathbb{R}$ and $0 \leqslant t, s < \infty$, where $\wedge$ stands for minimum. We say that the separable real valued process $\Gamma := \{\Gamma_F(x,t), x \in \mathbb{R}, 0 \leqslant t \leqslant 1\}$ is a "tied down" Kiefer process indexed by F, if Γ_F is a Gaussian process with $E\Gamma_F(x,t) = 0$ and

$$E\Gamma_F(x,t)\Gamma_F(y,s) = \{F(x\wedge y) - F(x)F(y)\}(t\wedge s - ts)$$

for all $x, y \in \mathbb{R}$ and $0 \leqslant t, s \leqslant 1$. Let $X_1 \cdots X_n \cdots$ be independent random variables. A natural way to test for the possibility of having

*Research supported by an NSERC Canada grant of M. Csörgő at Carleton University, Ottawa.

a change in the distribution of such a sequence of chronologically observed data up to time n is to compare their empirical distributions before and after the possibly unknown time of change $1 \leqslant k < n$, via studying the asymptotic distribution of the sequence of statistics (cf. B. S. Darkhovsky, 1984; D. Picard,1985; J. Deshayes and D. Picard, 1986; B. Szyszkowicz, 1991, 1994; M. Csörgő and B. Szyszkowicz, 1994; M. Csörgő, L. Horváth and B. Szyszkowicz [CsHSz], 1997)

$$\sup_{1\leqslant k<n}\sup_{x\in\mathbb{R}}\sqrt{n}\left|\frac{1}{k}\sum_{i=1}^{k}I\{X_i\leqslant x\}-\frac{1}{n-k}\sum_{i=k+1}^{n}I\{X_i\leqslant x\}\right|$$

$$=\sup_{1\leqslant k<n}\sup_{x\in\mathbb{R}}\left|\sum_{i=1}^{k}I\{X_i\leqslant x\}-\frac{k}{n}\sum_{i=1}^{n}I\{X_i\leqslant x\}\right|\Big/\sqrt{n}(k/n(1-k/n)).$$

These statistics, however, converge in distribution to ∞ as $n\to\infty$, even if the null assumption of no change in distribution were true. This is due to their tail behaviour being blown up too much by the weight function $((k/n)(1-k/n))$ in the denominator. Thus, one is led to considering "tied down" two-parameter empirical processes with weights, which should emphasize the possibility of having a change in distribution on the tails, but in such a way that their convergence in distribution should be nondegenerate. The building blocks for these "tied down" two-parameter empirical processes, $\hat{\beta}_n(x,t)/\hat{q}(t)$ with weights $\hat{q}(t)$, are the weighted two-parameter empirical processes, $\beta_n(x,t)/q(t)$ with appropriate weights $q(t)$, where

$$\beta_n(x,t)=n^{-\frac{1}{2}}\sum_{i=1}^{[nt]}(I\{X_i\leqslant x\}-F(x))$$

and

$$\hat{\beta}_n(x,t)=\begin{cases} n^{-\frac{1}{2}}(\sum_{i=1}^{[(n+1)t]}(I\{X_i\leqslant x\} \\ \qquad -\frac{[(n+1)t]}{n}\sum_{i=1}^{n}I\{X_i\leqslant x\}),\ 0\leqslant t<1, & x\in\mathbb{R}, \\ 0\ , & t=1,\ x\in\mathbb{R}. \end{cases}$$

Let Q_0 be the class of functions $q:(0,1)\to(0,\infty)$ that are strictly positive and are nondecreasing near 0. Then (cf. CsHSz, 1997, B. Szyszkowicz 1998), with $q\in Q_0$, as $n\to\infty$

$$\sup_{0<t\leqslant 1}\sup_{x\in\mathbb{R}}|\beta_n(x,t)|/q(t)\xrightarrow{d}\sup_{0<t\leqslant 1}\sup_{x\in\mathbb{R}}|K_F(x,t)|/q(t), \tag{1.1}$$

if and only if $J_0(q,c)<\infty$ for some $c>0$, where

$$J_0(q,c):=\int_{0+}^{1}t^{-1}\exp(-ct^{-1}q^2(t))dt. \tag{1.2}$$

Let $Q_{0,1}$ be the class of functions $\hat{q} : (0,1) \to (0,\infty)$ that are strictly positive and are nondecreasing near 0 and nonincreasing near 1. Then, with $\hat{q} \in Q_{0,1}$, as $n \to \infty$ we have (cf. Theorem 3.1 of CsHSz, 1997)

$$\sup_{0<t<1} \sup_{x\in\mathbb{R}} |\hat{\beta}_n(x,t)|/\hat{q}(t) \xrightarrow{d} \sup_{0<t<1} \sup_{x\in\mathbb{R}} |\Gamma_F(x,t)|/\hat{q}(t), \tag{1.3}$$

if and only if $J_0(\hat{q}, c) < \infty$ for some $c > 0$, where

$$J_{0,1}(\hat{q}, c) := \int_{0+}^{1-} \frac{1}{t(1-t)} \exp\left(-c\frac{\hat{q}^2(t)}{t(1-t)}\right) dt. \tag{1.4}$$

The thus resulting respective distributions of the two random variables in (1.1) and (1.3) are not known even in the case of $q(t) = \hat{q} = 1$. Hence estimating their distributions is of practical, as well as theoretical interest. This leads us to studying the tail behaviour of the distributions of the random variables $\sup_{0<t\leqslant 1} \sup_{x\in\mathbb{R}} |K_F(x,t)|/q(t)$ and $\sup_{0<t<1} \sup_{x\in\mathbb{R}} |\Gamma_F(x,t)|/\hat{q}(t)$. Naturally enough, our work is based on sample path large deviations for appropriate two-parameter Gaussian processes in appropriate weighted metrics. The desired tail probability estimates will follow as a consequence of such investigations.

Further to the results in (1.1) and (1.3) we note that, mutatis mutandis, they continue to hold true in terms of multivariate observations as well (cf. Theorem 3.1 of CsHSz, 1997). Concerning the nature of the integral tests (1.2) and (1.4) vis-à-vis the random variables on the respective right hand sides of (1.1) and (1.3), let

$$\Delta_0(t) := \sup_{x\in\mathbb{R}} |K_F(x,t)|, \tag{1.5}$$

$$\Delta_{0,1}(t) := \sup_{x\in\mathbb{R}} |\Gamma_F(x,t)|. \tag{1.6}$$

Then we have (cf. Theorems 2.1 and 2.3 of CsHSz, 1997)

(i) With $q \in Q_0$, $J_0(q,c) < \infty$ for some $c > 0$ if and only if

$$\limsup_{t\downarrow 0} \Delta_0(t)/q(t) < \infty, \quad \text{a.s.}, \tag{1.7}$$

(ii) with $\hat{q} \in Q_{0,1}$, $J_{0,1}(\hat{q},c) < \infty$ for some $c > 0$, if and only if, a.s.,

$$\limsup_{t\downarrow 0} \Delta_{0,1}(t)/\hat{q}(t) < \infty \quad \text{and} \quad \limsup_{t\uparrow 1} \Delta_{0,1}(t)/\hat{q}(t) < \infty \tag{1.8}$$

We note in passing that, as a consequence of (1.7) and (1.8) respectively, the respective results of (1.1) and (1.3) cannot be deduced via first proving a "corresponding" weak convergence result on $D[0,1]$ in a weighted sup-norm topology (cf. (i) versus (ii) of Theorem 2.3 of CsHSz, 1997). We (cf. Lemma 3.1 of CsHSz, 1997) overcome this difficulty via using strong approximation methods. For further results along these lines we refer to the books M. Csörgő and L. Horváth (1993, 1997) and to the paper M. Csörgő, R. Norvaiša and B. Szyszkowicz (1999).

Remark 1.1. *Later on we will make use of noting here that for any fixed $x_0 \in \mathbb{R}$ the stochastic process*

$$\begin{aligned}\{B(t),\, 0 \leqslant t \leqslant 1\} &= \{B_{x_0}(t),\, 0 \leqslant t \leqslant 1\}\\ &:= \{(F(x_0)(1-F(x_0)))^{-1/2}\Gamma_F(x_0,t),\ 0 \leqslant t \leqslant 1\}\end{aligned} \tag{1.9}$$

is a Brownian bridge, i.e., a mean zero Gaussian process with covariance function $(s,t) \mapsto (s \wedge t) - st$. Moreover, by (1.8), with $\hat{q} \in Q_{0,1}$ we have also

$$\textit{lim sup}_{t\downarrow 0}\ |B(t)|/\hat{q}(t) < \infty \ \ \textit{and} \ \ \textit{lim sup}_{t\uparrow 1}\ |B(t)|/\hat{q}(t) < \infty, \ \ \text{a.s..} \tag{1.10}$$

Hence, on using Theorem 3.3 of M. Csörgő, S. Csörgő, L. Horváth and D. M. Mason [CsCsHM] (1986) in combination with (1.10), we obtain that, with $\hat{q} \in Q_{0,1}$, $J_{0,1}(\hat{q}, c) < \infty$ for some $c > 0$. Assuming the latter, i.e., the if part, to conclude (1.8) is more difficult (cf. Theorem 2.3 of CsHSz, 1997).

Now, with large deviations in mind, let E be a metric space. We say that a function $I : E \to [0,\infty]$ is a *rate function*, if I is lower semicontinuous. We say that a *rate function* $I : E \to [0,\infty]$ *is good*, if $\{x \in E : I(x) \leqslant a\}$ is compact for each $a \geqslant 0$. Let $\{\theta_\varepsilon : \varepsilon > 0\}$ be a family of probability measures on E. We say that $\{\theta_\varepsilon, \varepsilon > 0\}$ satisfies the *large deviation principle* (LDP) with a rate function I, if it satisfies the following conditions :

a) for any open set $G \subset E$,

$$\liminf_{\varepsilon\to 0} \varepsilon \log \theta_\varepsilon(G) \geqslant -\inf_{x\in G} I(x),$$

b) for any closed set $H \subset E$,

$$\limsup_{\varepsilon\to 0} \varepsilon \log \theta_\varepsilon(H) \leqslant -\inf_{x\in H} I(x).$$

In Section 2, we prove an LDP for a Brownian sheet in q-metric equipped $C_0[0,1]^2$, which is an extension of that in the usual sup-norm equipped $C_0[0,1]^2$. We extend the results of Section 2 to Kiefer and "tied down" Kiefer processes in Sections 3 and 4, respectively, and then obtain probability estimates for their tails in q-metric.

2 Large deviations for Brownian sheet in q-metric

Let $C_0[0,1]^2$ be the space of continuous functions on $[0,1]^2$ which vanish along the axes. Let $W := \{W(x,t), 0 \leqslant x, t \leqslant 1\}$ be a standard Brownian sheet, i.e., a Gaussian process on $[0,1]^2$ with zero mean and covariance function

$$EW(x_1,t_1)W(x_2,t_2) = (x_1 \wedge x_2)(t_1 \wedge t_2).$$

Throughout the rest of this paper, we assume that $q(t)$ is a monotone increasing function on $[0,1]$, which is strictly positive for $t > 0$, and satisfies the following condition:

$$J_0(q,c) < \infty, \text{ for some } c > 0\ ,$$

where $J_0(q,c)$ is the integral defined in (1.2).

Define the q-metric on $C_0[0,1]^2$ by

$$d_q(f,g) := \sup_{0\leqslant x\leqslant 1}\ \sup_{0<t\leqslant 1} \{|f(x,t)-g(x,t)|/q(t)\},$$

for $f, g \in C_0[0,1]^2$. For any $\varepsilon > 0$, let P_ε be a probability measure on $C_0[0,1]^2$ corresponding to $\sqrt{\varepsilon}W$.

Theorem 2.1. *The net of probability measures $\{P_\varepsilon, \varepsilon > 0\}$ satisfies the large deviation principle in q-metric equipped $C_0[0,1]^2$ with the good rate function $I : C_0[0,1]^2 \to [0,\infty]$ that is defined by*

$$I(f) = \begin{cases} \frac{1}{2}\int_0^1\int_0^1 \left(\frac{\partial^2 f(x,t)}{\partial x \partial t}\right)^2 dxdt, & \text{if } f \text{ is an absolutely continuous} \\ & \text{function with square integrable } \frac{\partial^2 f}{\partial x\partial t}, \\ \infty, & \text{otherwise.} \end{cases}$$

Remark 2.2. *Via Theorem 2.1 it is easy to obtain a similar result for a standard Brownian motion (one parameter), which is due to M. Beibel and H. R. Lerche(1994).*

Corollary 2.3. *With $\|W\|_q := \sup_{0\leqslant x\leqslant 1}\sup_{0\leqslant t\leqslant 1}|W(x,t)|/q(t)$, we have*

$$\lim_{u\to\infty} \frac{\log P(\|W\|_q \geqslant u)}{u^2} = -\inf_{1<t\leqslant 1}\frac{q^2(t)}{2t}.$$

Lemma 2.4. [W.E. Pruitt, S. Orey 1973] *Let $0 \leqslant x_1 < x_2$, $0 \leqslant y_1 < y_2$.*

Then, for any $u > 0$,

$$P\Big(\sup_{x_1\leqslant s_1<s_2\leqslant x_2,y_1\leqslant t_1<t_2\leqslant y_2}\big|W(s_2,t_2)-W(s_2,t_1)-W(s_1,t_2)+W(s_1,t_1)\big|\geqslant u\Big)$$
$$\leqslant 4P(|W(x_2,y_2)-W(x_2,y_1)-W(x_1,y_2)+W(x_1,y_1)|\geqslant u).$$

Lemma 2.5. *For any* $b\in(0,1]$ *and* a *large enough, we have*

$$P\left(\sup_{0<t\leqslant b,0\leqslant x\leqslant 1}\frac{|W(x,t)|}{q(t)}\geqslant a\right)\leqslant CJ_0(q,\frac{a^2}{8})\exp\left\{-\frac{a^2}{8}\inf_{0<t\leqslant b}\frac{q^2(t)}{2t}\right\}.$$

Proof. By Lemma 2.4, we have

$$\begin{aligned}
&P\left(\sup_{0<t\leqslant b,0\leqslant x\leqslant 1}\frac{|W(x,t)|}{q(t)}\geqslant a\right)\\
&\leqslant\sum_{k=1}^{\infty}P(\sup_{0<t\leqslant b2^{-k+1},0\leqslant x\leqslant 1}|W(x,t)|\geqslant aq(2^{-k}))\\
&\leqslant 4\sum_{k=1}^{\infty}P(|W(1,b2^{-k+1})|\geqslant aq(b2^{-k}))\\
&\leqslant C\sum_{k=1}^{\infty}\exp\{-\frac{a^2q^2(b2^{-k})}{8b2^{-k}}\}\\
&\leqslant C\exp\{-\frac{a^2}{8}\inf_{0<t\leqslant b}\frac{q^2(t)}{2t}\}J_0(q,\frac{a^2}{8}).
\end{aligned}$$

Hence, this lemma is proved. □

Lemma 2.6. *For all* $a\geqslant 0$, *the set* $\{f|I(f)\leqslant a\}$ *is compact with respect to the* q*–metric* d_q.

Proof. Firstly, it is easy to show that $\{f|I(f)\leqslant a\}$ is compact with respect to $\|\cdot\|_\infty=\sup_{(x,t)\in[0,1]^2}|\cdot|$ via the usual way. Since, for $f,g\in C_0[0,1]^2$ and $b\in(0,1]$, it holds that

$$d_q(f,g)\leqslant\frac{1}{q(b)}\|f-g\|_\infty+\{I^{\frac12}(f)+I^{\frac12}(g)\}\sup_{0<t\leqslant b}\frac{\sqrt{2t}}{q(t)},$$

on noting that $J_0(q,c)<\infty$ for some $c>0$ implies $\lim_{t\to 0}\frac{t}{q^2(t)}=0$ (cf. CsCsHM, 1986), we conclude that $\{f|I(f)\leqslant a\}$ is also compact with respect to d_q. □

Lemma 2.6 implies the next result.

Lemma 2.7. *For any closed set $H \subset C_0[0,1]^2$ and $\delta > 0$, let $H^\delta := \{f : \inf_{g\in H} \|f-g\|_q < \delta\}$. Then*

$$\lim_{\delta\downarrow 0} \inf_{f\in H^\delta} I(f) = \inf_{f\in H} I(f).$$

Proof of Theorem 2.1. Throughout this proof, C denotes some constant. Firstly, we want to show that, for any closed set $H \subset C_0[0,1]^2$,

$$\limsup_{\varepsilon\to 0} \varepsilon \log P_\varepsilon(H) \leqslant -\inf_{f\in H} I(f). \tag{2.1}$$

In order to prove this, for any positive integer n and $f \in C_0[0,1]^2$, we define $f_n \in C_0[0,1]^2$ by

$$f_n(x,t) = a_n(k,i)xt + b_n(k,i)x + c_n(k,i)t + d_n(k,i),$$

$$(x,t) \in I_n(k,i),\ 0 \leqslant k,i \leqslant n,$$

where,

$$a_n(k,i) = n^2\left[f(\frac{k+1}{n},\frac{i+1}{n}) - f(\frac{k+1}{n},\frac{i}{n}) - f(\frac{k}{n},\frac{i+1}{n}) + f(\frac{k}{n},\frac{i}{n})\right],$$

$$I_n(k,i) = \left[\frac{k}{n},\frac{k+1}{n}\right] \times \left[\frac{i}{n},\frac{i+1}{n}\right],$$

and $b_n(k,i), c_n(k,i)$ and $d_n(k,i)$ are constants such that

$$f(x,t) = f_n(x,t), (x,t) \in \left\{ \left(\frac{k}{n},\frac{i}{n}\right), \left(\frac{k+1}{n},\frac{i}{n}\right), \left(\frac{k}{n},\frac{i+1}{n}\right), \left(\frac{k+1}{n},\frac{i+1}{n}\right) \right\}, 0 \leqslant k,i \leqslant n.$$

For any $\delta > 0$, let $H^\delta := \{f : \inf_{g\in H}\|f-g\|_q < \delta\}$. Then

$$\begin{aligned} P_\varepsilon(H) &\leqslant P_\varepsilon(f|f_n \in H^\delta) + P_\varepsilon(f|\|f_n - f\|_q \geqslant \delta) \\ &:= A + B. \end{aligned}$$

Let $l_\delta := \inf_{f\in H^\delta} I(f)$. Since $\frac{1}{\varepsilon n^2}\sum_{k=1}^n\sum_{i=1}^n a_n^2(k,i)$ is a $\chi^2_{n^2}$ random variable under P_ε, hence

$$\begin{aligned} A &\leqslant P_\varepsilon(f|I(f_n) \geqslant l_\delta) \\ &\leqslant \frac{1}{\Gamma(\frac{n^2}{2})2^{\frac{n^2}{2}}} \int_{\frac{2l_\delta}{\varepsilon}}^\infty \exp\{-\frac{x}{2}\} x^{\frac{n^2}{2}-1} dx \\ &\leqslant C(n)\exp\{-\frac{(1-\mu)}{\varepsilon} l_\delta\}, \end{aligned}$$

where $C(n)$ is a constant that depends on n, and μ is any positive number. Therefore,

$$\limsup_{\varepsilon\to 0} \varepsilon \log A \leqslant -l_\delta + \mu l_\delta,$$

and letting $\mu \to 0$, we obtain

$$\limsup_{\varepsilon\to 0} \varepsilon \log A \leqslant -l_\delta. \tag{2.2}$$

On the other hand, we have

$$\begin{aligned}
B &= P_\varepsilon(f|\|f_n - f\|_q \geqslant \delta) \\
&= P\left(\sqrt{\varepsilon} \sup_{0\leqslant x\leqslant 1, 0<t\leqslant 1} |W(x,t) - W_n(x,t)|/q(t) \geqslant \delta\right) \\
&\leqslant P\Big(\sup_{0\leqslant x\leqslant 1, 0<t\leqslant \frac{1}{n}} |W(x,t)|/q(t) \geqslant \delta/\sqrt{\varepsilon}\Big) \\
&\quad + P\Big(\sup_{0\leqslant x\leqslant 1, 0<t\leqslant \frac{1}{n}} |W_n(x,t)|/q(t) \geqslant \delta/\sqrt{\varepsilon}\Big) \\
&\quad + P\Big(\sup_{0\leqslant x\leqslant 1, \frac{1}{n}\leqslant t\leqslant 1} |W(x,t) - W_n(x,t)|/q(t) \geqslant \delta/\sqrt{\varepsilon}\Big) \\
&:= B_1 + B_2 + B_3.
\end{aligned}$$

By Lemma 2.5, we have

$$\limsup_{\varepsilon\to 0} \varepsilon \log B_1 \leqslant -\frac{\delta^2}{8} \inf_{0<t\leqslant \frac{1}{n}} \frac{q^2(t)}{2t}.$$

Since

$$\sup_{0\leqslant x\leqslant 1, 0<t\leqslant \frac{1}{n}} \frac{|W_n(x,t)|}{q(t)} = \max_{1\leqslant i\leqslant n} |nW(\frac{i}{n}, \frac{1}{n})| \sup_{0<t\leqslant \frac{1}{n}} \frac{t}{q(t)},$$

we have

$$B_2 \leqslant Cn \exp\{-\frac{\delta^2}{\varepsilon} \inf_{0<t\leqslant \frac{1}{n}} \frac{q^2(t)}{2t}\},$$

i.e.,

$$\limsup_{\varepsilon\to 0} \varepsilon \log B_2 \leqslant -\delta^2 \inf_{0<t\leqslant \frac{1}{n}} \frac{q^2(t)}{2t}.$$

Via Lemma 2.4, we obtain

$$\begin{aligned}
B_3 &\leqslant P\left(\max_{1\leqslant k,i\leqslant n}\sup_{(x,t)\in I_n(k,i)}|W(x,t)-W_n(x,t)|\geqslant \frac{\delta}{\sqrt{\varepsilon}}q(\frac{1}{n})\right)\\
&\leqslant \sum_{1\leqslant k,i\leqslant n} P\left(\sup_{(x,t)\in I_n(k,i)}|W(x,t)-W(\frac{k}{n},\frac{i}{n})|\geqslant \frac{\delta}{2\sqrt{\varepsilon}}q(\frac{1}{n})\right)\\
&\leqslant 8n^2 P\left(|W(1,1)|\geqslant \frac{\delta\sqrt{n}}{6\sqrt{\varepsilon}}q(\frac{1}{n})\right)\\
&\leqslant Cn^2\exp\{-\frac{\delta^2}{72\varepsilon}nq^2(\frac{1}{n})\}.
\end{aligned}$$

Consequently, we arrive at

$$\limsup_{\varepsilon\to 0}\varepsilon\log B_3\leqslant -\frac{\delta^2}{72}nq^2(\frac{1}{n}).$$

Therefore, we get

$$\limsup_{\varepsilon\to 0}\varepsilon\log B\leqslant -\frac{\delta^2}{36}\inf_{0<t\leqslant\frac{1}{n}}\frac{q^2(t)}{2t}. \tag{2.3}$$

Noting that $\lim_{t\downarrow 0}\frac{q^2(t)}{2t}=\infty$, and combining (2.2) with (2.3), for n large enough, we have

$$\limsup_{\varepsilon\to 0}\varepsilon\log(A+B)\leqslant -l_\delta. \tag{2.4}$$

Since H is a closed set, by Lemma 2.7, on letting $\delta\downarrow 0$, we obtain (2.1) from (2.4).

For completing the proof, we have to show that for any open set $G\subset C_0[0,1]^2$,

$$\liminf_{\varepsilon\to 0}\varepsilon\log P_\varepsilon(G)\geqslant -\inf_{f\in G}I(f). \tag{2.5}$$

Let $f\in G$ with $I(f)<\infty$. Then there exists $\delta>0$ such that

$$B_q(f,\delta):=\{g:\|f-g\|_q<\delta\}\subset G.$$

Let $P_\varepsilon(f)$ be the probability measure on the $C_0[0,1]^2$ corresponding to $h(x,t)-f(x,t)$, where $h(x,t)$ is distributed according to P_ε. Then, via the Cameron-Martin formula (cf. R. F. Bass,1988; or J. Yeh, 1963) and

Jensen's inequality, we have

$$\begin{aligned}
&P_\varepsilon(B_q(f,g))\\
&= P_\varepsilon(f)(B_q(\delta))\\
&= \int_{B_q(\delta)} \frac{dP_\varepsilon(\delta)}{dP_\varepsilon}(h)P_\varepsilon(dh)\\
&= \int_{B_q(\delta)} \exp\{\frac{1}{\varepsilon}\int_{[0,1]^2}\frac{\partial^2 f}{\partial x\partial y}dh - \frac{1}{\varepsilon}I(f)\}P_\varepsilon(dh)\\
&= \exp\{-\frac{1}{\varepsilon}I(f)\}P_\varepsilon(B_q(\delta))\\
&\qquad \cdot\frac{1}{P_\varepsilon(B_q(\delta))}\int_{B_q(\delta)} \exp\{\frac{1}{\varepsilon}\int_{[0,1]^2}\frac{\partial^2 f}{\partial x\partial t}dh\}P_\varepsilon(dh)\\
&\geqslant \exp\{-\frac{1}{\varepsilon}I(f)\}P_\varepsilon(B_q(\delta)\exp\{\frac{1}{\varepsilon P_\varepsilon(B_q(\delta))}\int_{B_q(\delta)}\int_{[0,1]^2}\frac{\partial^2 f}{\partial x\partial t}dhP_\varepsilon(dh)\}\\
&= \exp\{-\frac{1}{\varepsilon}I(f)\}P_\varepsilon(B_q(\delta)).
\end{aligned}$$

Since, by Lemma 2.5, we have

$$\lim_{\varepsilon\to 0} P_\varepsilon(B_q(\delta)) = 1,$$

we obtain

$$\liminf_{\varepsilon\to 0} \varepsilon \log P_\varepsilon(B_q(f,\delta)) \geqslant -I(f).$$

Therefore (2.5) is true, and the proof of Theorem 2.1 is complete. □

Proof of Corollary 2.3. Let $H_1 = \{f \in C_0[0,1]^2 : \|f\|_q \geqslant 1\}$, and $G_1 = \{f \in C_0[0,1]^2 : \|f\|_q > 1\}$. Then we have to show that

$$\inf_{f\in H_1} I(f) \leqslant \inf_{0<t\leqslant 1}\frac{q^2(t)}{2t} \leqslant \inf_{f\in G_1} I(f). \tag{2.6}$$

Since, for $f \in H_1$ with $I(f) < \infty$,

$$\frac{|f(x,t)|^2}{q^2(t)} \leqslant I(f)\frac{2t}{q^2(t)},$$

for any $(x,t) \in [0,1]\times(0,1]$, hence we have

$$\inf_{f\in H_1} I(f) \geqslant \inf_{0<t\leqslant 1}\frac{q^2(t)}{2t}.$$

On the other hand, for any $\varepsilon > 0$ and $u \in (0,1]$, we can define $g_{u,\varepsilon} \in G_1$ by

$$g_{u,\varepsilon}(x,t) = (1+\varepsilon)x[\frac{q(u)}{u} t I_{[0,u]}(t) + q(u) I_{(u,1]}(t)],$$

for any $(x,t) \in [0,1]^2$. Thus, we obtain

$$\inf_{f \in G_1} I(f) \leqslant \inf_{\varepsilon > 0, u \in (0,1]} I(g_{u,\varepsilon}) = \inf_{0 < u \leqslant 1} \frac{q^2(u)}{2u}.$$

Therefore (2.6) is true. □

3 Large deviations for Kiefer processes

Let $K := \{K(x,t), 0 \leqslant x, t \leqslant 1\}$ be a standard Kiefer process on $[0,1]^2$, that is a two-parameter Gaussian process with zero mean and covariance function

$$EK(x,t)K(y,s) = (x \wedge y - xy)(t \wedge s).$$

It is well known (cf., e.g., M. Csörgő and P. Révész, 1981) that, without loss of generality, a standard Kiefer process $K(\cdot,\cdot)$ on $[0,1] \times [0,\infty)$ can be defined by letting

$$K(x,t) = W(x,t) - xW(1,t),$$

for $0 \leqslant x \leqslant 1$ and $0 \leqslant t < \infty$, where $\{W(x,t),\ 0 \leqslant x,t < \infty\}$ is a standard Brownian sheet. For $\varepsilon > 0$, let P_ε^K be a probability measure on $C_0[0,1]^2$ corresponding to $\sqrt{\varepsilon} K$, with $K(\cdot,\cdot)$ on $[0,1]^2$. Then we have

Theorem 3.1. *The net of probability measures $\{P_\varepsilon^K, \varepsilon > 0\}$ satisfies the large deviation principle in q-metric equipped $C_0[0,1]^2$ with a good rate function $I_K : C_0[0,1]^2 \to [0,\infty]$ that is defined for any $f \in C_0[0,1]^2$ by*

$$I_K(f) = \begin{cases} I(f), \text{ if } I(f) < \infty, \text{ and } \sup_{0 \leqslant t \leqslant 1} |f(1,t)| = 0, \\ \\ \infty, \quad \text{otherwise.} \end{cases}$$

Toward proving Theorem 3.1, we quote the following contraction principle.

Lemma 3.2. (2.1.4 Lemma, J. D. Deuschel and D. W. Stroock, 1989) *Let X and Y be any two metric spaces and $f : X \longrightarrow Y$ a continuous function. Assume that the family of probability measures $\{P_\varepsilon\}$ on X satisfies a large deviation principle with good rate function I. Then the family of probability measures $\{P_\varepsilon \circ f^{-1}\}$ on Y also satisfies a large deviation principle with good rate function I' defined by*

$$I'(y) = \inf\{I(x);\, x \in X,\, y = f(x)\}.$$

Proof of Theorem 3.1. Theorem 3.1 is a straight consequence of Theorem 2.1 and the contraction principle(cf. Deuschel and Stroock,1989). □

From the latter theorem, we conclude

Corollary 3.3. *With* $\|K\|_q := \sup_{0\leqslant x\leqslant 1}\sup_{0<t\leqslant 1}|K(x,t)|/q(t)$, *we have*

$$\lim_{u\to\infty}\frac{\log P(\|K\|_q\geqslant u)}{u^2}=-\inf_{0<t\leqslant 1}\frac{2q^2(t)}{t}.$$

Proof of Corollary 3.3. Let $f\in\{g: I_K(g)<\infty \text{and} \|g\|_q\geqslant 1\}$. Then for any $(x,t)\in(0,1)\times(0,1]$

$$\frac{f^2(x,t)}{q^2(t)}\leqslant\frac{1}{2}\int_0^x\int_0^t(\frac{\partial^2 f(u,v)}{\partial u\partial v})^2dudv\cdot\frac{2xt}{q^2(t)},$$

and

$$\frac{f^2(x,t)}{q^2(t)}\leqslant\frac{1}{2}\int_x^1\int_0^t(\frac{\partial^2 f(u,v)}{\partial u\partial v})^2dudv\cdot\frac{2(1-x)t}{q^2}.$$

Thus

$$I_K(f)\geqslant\frac{f^2(x,t)}{q^2(t)}\inf_{0<t\leqslant 1}\frac{2q^2(t)}{t},$$

for any $(x,t)\in[0,1]\times(0,1]$, i.e., we have

$$I_K(f)\geqslant\inf_{0<t\leqslant 1}\frac{2q^2(t)}{t}.$$

On the other hand, for $\varepsilon>0, u\in(0,1]$ let

$$f_{u,\varepsilon}(x,t)=2(1+\varepsilon)q(u)[\tfrac{xt}{u}I_{[0,\frac{1}{2}]\times[0,u]}(x,t)+\tfrac{(1-x)t}{u}I_{(\frac{1}{2},1]\times[0,u]}(x,t)$$

$$+xI_{[0,\frac{1}{2}]\times(u,1]}(x,t)+(x-1)I_{(\frac{1}{2},1]\times(u,1]}(x,t)],$$

for any $(x,t)\in[0,1]^2$. Then, we have

$$\inf_{\|f\|_q>1}I_K(f)\leqslant\inf_{\varepsilon>0,u\in(0,1]}I_K(f_{u,\varepsilon})=\inf_{0<u\leqslant 1}\frac{2q^2(u)}{u}.$$

Therefore Corollary 3.3 is proved. □

Corollary 3.4. *Let* $F(x)$ *be a continuous distribution function on* R *and let* K_F *be a Kiefer process indexed by* F. *Then*

$$\lim_{u\to\infty}\frac{\log P(\sup_{x\in\mathbb{R}}\sup_{0<t\leqslant 1}|K_F(x,t)|/q(t)\geqslant u)}{u^2}=-\inf_{0<t\leqslant 1}\frac{2q^2(t)}{t}.$$

Clearly, Corollary 3.4 follows from Corollary 3.3 immediately.

4 Large deviations for "tied down" Kiefer processes

Let $\Gamma := \{\Gamma(x,t) : 0 \leqslant x,t \leqslant 1\}$ be a "tied down" standard Kiefer process $K(\cdot,\cdot)$ on $[0,1]^2$, i.e., a Gaussian process with zero mean and covariance function

$$E\Gamma(x,t)\Gamma(y,s) = (x \wedge y - xy)(t \wedge s - ts).$$

We can realize a "tied down" standard Kiefer process $\Gamma(\cdot,\cdot)$ on $[0,1]^2$ by letting

$$\Gamma(x,t) = K(x,t) - tK(x,1)$$

for $(x,t) \in [0,1]^2$, where $\{K(x,t), 0 \leqslant x,t \leqslant 1\}$ is a standard Kiefer process on $[0,1]^2$. For each $\varepsilon > 0$, let P_ε^Γ be a probability measure on $C_0[0,1]^2$ corresponding to $\sqrt{\varepsilon}\Gamma$. Then we have

Theorem 4.1. *The net of probability measures $\{P_\varepsilon^\Gamma : \varepsilon > 0\}$ satisfies the large deviation principle in q-metric equipped $C_0[0,1]^2$ with the good rate function $I_\Gamma : C_0[0,1]^2 \to [0,\infty]$ that for $f \in C_0[0,1]^2$ is defined by*

$$I_\Gamma(f) = \begin{cases} I(f), & \text{if } I(f) < \infty, \sup_{0\leqslant x\leqslant 1} |f(x,1)| = 0 \\ & \text{and } \sup_{0<t\leqslant 1} |f(1,t)| = 0, \\ \infty, & \text{otherwise.} \end{cases}$$

Corollary 4.2. *With $\|\Gamma\|_q := \sup_{0\leqslant x\leqslant 1} \sup_{0<t\leqslant 1} |\Gamma(x,t)|/q(t)$, we have*

$$\lim_{u\to\infty} \frac{\log P(\|\Gamma\|_q \geqslant u)}{u^2} = -\inf_{0<t<1} \frac{2q^2(t)}{t(1-t)}.$$

By the latter corollary and the fact that $\{\Gamma(x,1-t) : 0 \leqslant x,t \leqslant 1\}$ is also a "tied down" Kiefer process, we have

Corollary 4.3.

$$\lim_{u\to\infty} \frac{1}{u^2} \log P(\sup_{0\leqslant x\leqslant 1} \sup_{0\leqslant t<1} \frac{|\Gamma(x,t)|}{q(1-t)} \geqslant u) = -\inf_{0<t<1} \frac{2q^2(t)}{t(1-t)}.$$

Corollary 4.4. *Let $\hat{q}(t)$ be a function, defined on $[0,1]$,that is strictly positive on $(0,1)$, and satisfies the following conditions:*

(i) *$\hat{q}(\cdot)$ is increasing on $[0,\frac{1}{2}]$, and symmetric with respect to $\frac{1}{2}$,*

(ii) *$J_{0,1}(\hat{q},c) < \infty$, for some $c > 0$, where $J_{0,1}(\hat{q},c)$ is the integral defined in* (1.4). *Then, we have*

$$\lim_{u\to\infty} \frac{1}{u^2} \log P\left(\sup_{0\leqslant x\leqslant 1} \sup_{0<t<1} \frac{|\Gamma(x,t)|}{\hat{q}(t)} \geqslant u\right) = -\inf_{0<t<1} \frac{2\hat{q}^2(t)}{t(1-t)}.$$

Clearly, we also have

Corollary 4.5. *Let F be a continuous distribution function on $\mathbb{R}$, and let Γ_F be a "tied down" Kiefer process indexed by F. Assume that $\hat{q}$ satisfies the same conditions as in* Corollary 4.4. *Then*

$$\lim_{u\to\infty}\frac{1}{u^2}\log P\left(\sup_{x\in\mathbb{R}}\sup_{0<t<1}\frac{|\Gamma_F(x,t)|}{\hat{q}(t)}\geqslant u\right)=-\inf_{0<t<1}\frac{2\hat{q}^2(t)}{t(1-t)}.$$

Remark 4.6. *Let $\{B(t), 0\leqslant t\leqslant 1\}$ be a Brownian bridge. Then, under the same assumptions as those of Corollary 4.4 for $\hat{q}$, we also have a similar result as follows*

$$\lim_{u\to\infty}\frac{1}{u^2}\log P\left(\sup_{0<t<1}\frac{|B(t)|}{\hat{q}(t)}\geqslant u\right)=-\inf_{0<t<1}\frac{\hat{q}^2(t)}{2t(1-t)}.$$

Proof of Theorem 4.1. Consider the mapping $\Pi : C_0[0,1]^2 \to C_0[0,1]^2$ defined for $f \in C_0[0,1]^2$ by

$$\Pi f(x,t) = f(x,t) - xf(1,t) - t(f(x,1) - xf(1,1)).$$

Clearly, Π is continuous with respect to the q-metric $g_q(\cdot,\cdot)$ on $C_0[0,1]^2$. According to contraction principle of Lemma 3.2 and Theorem 2.1, we have that $\{P_\varepsilon^\Gamma : \varepsilon > 0\}$ satisfies an LDP with the good rate function $\hat{I}_\Gamma : C_0[0,1]^2 \to [0,\infty]$ that for any $f \in C_0[0,1]^2$ is defined by

$$\hat{I}_\Gamma(f) = \begin{cases} \inf_{g\in\Pi^{-1}(f)} I(g), & \text{if } \Pi^{-1}(f)\neq\emptyset,\\ \infty, & \text{if } \Pi^{-1}(f)=\emptyset. \end{cases}$$

Since $\Pi^{-1}(f) \neq \emptyset$ implies $f \in \Pi^{-1}(f)$, and $I(g) < \infty$ implies $I(g) \geqslant I(\Pi g)$, it is easy to obtain $\hat{I}_\Gamma(f) = I_\Gamma(f)$ for any $f \in C_0[0,1]^2$. Therefore, the proof of Theorem 4.1 is complete. □

Proof of Corollary 4.2. By Theorem 4.1, it suffices to show that

$$\inf_{\|f\|_q\geqslant 1} I_\Gamma(f) \geqslant \inf_{0<t<1}\frac{2q^2(t)}{t(1-t)} \geqslant \inf_{\|f\|_q>1} I_\Gamma(f).$$

Let $f \in C_0[0,1]^2$ with $I_\Gamma(f) < \infty$ and $\|f\|_q \geqslant 1$. Then, for any $(x,t) \in (0,1)^2$,

$$I_\Gamma(f) \geqslant \frac{f^2(x,t)}{q^2(t)}\,\frac{q^2(t)}{2t(1-t)x(1-x)} \geqslant \frac{f^2(x,t)}{q^2(t)}\inf_{0<t<1}\frac{2q^2(t)}{t(1-t)}.$$

Hence, we have

$$\inf_{\|f\|_q\geqslant 1} I_\Gamma(f) \geqslant \inf_{0<t<1}\frac{2q^2(t)}{t(1-t)}.$$

On the other hand, for any $\varepsilon > 0, v \in (0,1)$, let

$$h_{v,\varepsilon}(x,t) = 2(1+\varepsilon)q(v)[\tfrac{xt}{v} I_{[0,\frac{1}{2}]\times[0,v]}(x,t) + \tfrac{(1-x)t}{v} I_{(\frac{1}{2},1]\times[0,v]}(x,t)$$

$$+\tfrac{x(1-t)}{1-v} I_{[0,\frac{1}{2}]\times(v,1]}(x,t) + \tfrac{(1-x)(1-t)}{1-v} I_{(\frac{1}{2},1]\times(v,1]}(x,t)],$$

for any $(x,t) \in [0,1]^2$. Then

$$h_{v,\varepsilon}(x,t) \in \{f \in C_0[0,1]^2 : I_\Gamma(f) < \infty, \|f\|_q > 1\}.$$

Hence, we can obtain

$$\inf_{\|f\|_q > 1} I_\Gamma(f) \leqslant \inf_{v\in(0,1),\varepsilon>0} I_\Gamma(h_{v,\varepsilon}) = \inf_{0<v<1} \frac{2q^2(v)}{v(1-v)}.$$

Therefore, the proof is complete. □

Proof of Corollary 4.4. Let

$$q_1(t) = \begin{cases} \hat{q}(t), & 0 \leqslant t \leqslant \frac{1}{2}, \\ \hat{q}(\frac{1}{2}), & \frac{1}{2} \leqslant t \leqslant 1, \end{cases}$$

and

$$q_2(t) = \begin{cases} \hat{q}(\frac{1}{2}), & 0 \leqslant t \leqslant \frac{1}{2}, \\ \hat{q}(t), & \frac{1}{2} \leqslant t \leqslant 1, \end{cases}$$

Then, q_1, q_2 satisfy the same conditions as our usual q that we used since the beginning of Section 2. Hence, by Corollaries 4.2, 4.3, we obtain

$$\begin{aligned} &\limsup_{u\to\infty} \frac{1}{u^2} \log P\left(\sup_{0\leqslant x\leqslant 1} \sup_{0<t<1} \frac{|\Gamma(x,t)|}{\hat{q}(t)} \geqslant u \right) \\ &\leqslant -(\inf_{0<t\leqslant 1} \frac{2q_1^2(t)}{t(1-t)}) \wedge (\inf_{0\leqslant t<1} \frac{2q_2^2(t)}{t(1-t)}) \\ &= -\inf_{0<t<1} \frac{2\hat{q}^2(t)}{t(1-t)}, \end{aligned}$$

and

$$\liminf_{u\to\infty} \tfrac{1}{u^2} \log P(\sup_{0\leqslant x\leqslant 1} \sup_{0<t<1} \tfrac{|\Gamma(x,t)|}{\hat{q}(t)} \geqslant u)$$

$$\geqslant \lim_{u\to\infty} \tfrac{1}{u^2} \log P(\sup_{0\leqslant x\leqslant 1} \sup_{0<t\leqslant 1} \tfrac{|\Gamma(x,t)|}{q_1(t)} \geqslant u)$$

$$= -\inf_{0<t<1} \tfrac{2\hat{q}^2(t)}{t(1-t)}.$$

This also completes the proof. □

Remark 4.7. *Interestingly enough, Corollary 4.5 and Remark 4.6 tell us that even "constants-wise", the weighted large deviation behaviour of a tied down Kiefer process is not very much different from that of a Brownian bridge. In view of (1.8) and (1.10) it is not surprising of course that, after taking* $\sup_{x\in\mathbb{R}}$ *as in (1.6), "shape-wise" it should end up having the same type of weighted tail behaviour that is possessed by a Brownian bridge (cf. Corollary 4.5 versus Remark 4.6).*

Remark 4.8. *It is also of interest to note that, while in (1.3) we have a characterization of the limiting behaviour of the weighted empirical change process (weighted "tied down" two-parameter empirical process)* $\{\hat{\beta}_n(x,t)/\hat{q}(t),\ x\in\mathbb{R},\ 0<t<1\}$ *in terms of the similarly weighted "tied down" Kiefer process* $\{\Gamma_F(x,t)/\hat{q}(t),\ x\in\mathbb{R},\ 0<t<1\}$*, a Gaussian process, the extreme value problem of finding norming constants* a_n *and* b_n *such that, as* $n\to\infty$,

$$a_n \sup_{0<t<1}\sup_{x\in\mathbb{R}} |\hat{\beta}_n(x,t)|/(t(1-t))^{1/2} - b_n \xrightarrow{d} \text{a nondegenerate random variable}$$

remains open (cf. M. Csörgő and R. Norvaiša 1998 for some related results). On the other hand, via D. A. Darling and P. Erdős (1956), for any fixed $x = x_0$ *with* $\tilde{a}_n = (2\log\log n)^{1/2}$, $\tilde{b}_n = 2\log\log n + \frac{1}{2}\log\log\log n - \frac{1}{2}\log\pi$ *we have (cf. M. Csörgő, L. Horváth, 1997: Theorems 2.1.2 and A.4.2)*

$$\lim_{n\to\infty} P\left\{\tilde{a}_n \sup_{0<t<1} |\hat{\beta}_n(x_0,t)|/(t(1-t))^{1/2} - \tilde{b}_n \leqslant y\right\} = \exp(-2e^{-y}),\quad y\in\mathbb{R}.$$

References

[1] R. F. Bass, Probability estimates for multiparameter Brownian processes, *Ann. Probab.* **16** (1988) 251–264.

[2] M. Beibel and H. R. Lerche, A refined large deviation principle for Brownian motion and its application to boundary crossing, *Stochastic Process. Appl.* **51** (1994) 269–276.

[3] M. Csörgő, S. Csörgő, L. Horváth and D. M. Mason, Weighted empirical and quantile processes, *Ann. Probab.* **14** (1986) 31–85.

[4] M. Csörgő and L. Horváth, *Weighted Approximations in Probability and Statistics*, Wiley, Chichester, 1993.

[5] M. Csörgő and L. Horváth, *Limit Theorems in Change-Point Analysis*, Wiley, Chichester, 1997.

[6] M. Csörgő, L. Horváth and B. Szyszkowicz, Integral tests for suprema of Kiefer processes with application, *Statists & Decisions* **15** (1997) 365–377.

[7] M. Csörgő and R. Norvaiša, Standardized sequential empirical processes, *Studia Sci. Math. Hung.* **34** (1998) 51–69.

[8] M. Csörgő, R. Norvaiša and B. Szyszkowicz, Convergence of weighted partial sums when the limiting distribution is not necessarily Radon, *Stochastic Process. Appl.* **81** (1999) 81–101.

[9] M. Csörgő and P. Révész, *Strong Approximations in Probability and Statistics*, Academic Press, New York, 1981.

[10] M. Csörgő and B. Szyszkowicz, Applications of multi-time parameter processes to change-point analysis, In: *Probability Theory and Mathematical Statistics* 1994 - Proceedings of the Sixth Vilnius Conference, pp. 159–222, B. Grigelionis et al. (Eds.), VSP/TEV 1994.

[11] B. S. Darkhovsky, On two problems of estimating the moments of changes in probabilistic characteristics of a random sequence (in Russian), *Teor. Veroyatn. Primen.* **29** (1984) 464–473.

[12] A. D. Darling and P. Erdős, A limit theorem for the normalized sums of independent random variables, *Duke Math. J.* **23** (1956) 143–155.

[13] J. Deshayes and D. Picard, Off-line statistical analysis of change-point models using nonparametric and likelihood methods, in: M. Thoma and A. Wyner, eds., M. Basseville and A. Beneveniste, guest eds., *Detection of Abrupt Changes in Signals and Dynamical Systems*, pp. 103–168, Springer, Berlin, 1986.

[14] J. D. Deuschel and D. W. Stroock, *Large Deviations*, Academic Press, Boston, 1989.

[15] D. Picard, Testing and estimating change-points in time series, *Adv. Appl. Probab.* **17** (1985) 126–135.

[16] W. E. Pruitt and S. Orey, Sample functions of the N-parameter Wiener process, *Ann. Probab.* **1** (1973) 138–163.

[17] B. Szyszkowicz, Empirical type processes and contiguity, *C.R. Math. Rep. Acad. Sci. Canada* **13** (1991) 161–166.

[18] B. Szyszkowicz, Weak convergence of weighted empirical type processes under contiguous and changepoint alternatives, *Stochastic Process. Appl.* **50** (1994) 281–313.

[19] B. Szyszkowicz, Weighted sequential empirical type processes with applications to change-point problems. In *Handbook of Statistics*, Vol. **16** (N. Balakrishnan and C.R. Rao, eds.), pp. 573–630, Elsevier Science B.V. 1998.

[20] J. Yeh, Cameron-Martin translation theorems in the Wiener space of functions of two variables, *Trans. Am. Math. Soc.* **107** (1963) 409–420.

Intersection Local Times: Large Deviations and Laws of the Iterated Logarithm *

Xia Chen

Department of Mathematics, The University of Tennessee
Knoxville, TN 37996, USA
E-mail: xchen@math.utk.edu

Abstract

Sample path intersection has been of interest to physicists for many years, due to its connections to renormalization group methods for quantum field theory, self-avoiding random walks and random polymers. In particular, the large deviations arising from sample path intersections have been applied to solve some hard problems raised by physicists such as the identification of certain critical exponents in polymer models and the investigation of intermittency in parabolic Anderson models. On the mathematical side, the study of the large deviations provides tail estimates needed for the strong laws in probability such as the law of the iterated logarithm. In this article we present a survey of some developments on the large deviations for intersection local times and related models.

1 Introduction

Let $W(t)$ be a d-dimensional Brownian motion. Assume for a moment that for any integer $p \geqslant 2$ we can define the quantity

$$\begin{aligned}&\beta\big([0,t]^p_<\big)\\ &= \int\cdots\int_{\{0\leqslant s_1<\cdots<s_p\leqslant t\}} \prod_{j=2}^p \delta_0\big(W(s_j)-W(s_{j-1})\big)ds_1\cdots ds_p,\end{aligned} \tag{1.1}$$

where the notation $[0,t]^p_<$ represents the set

$$[0,t]^p_< = \big\{(s_1,\cdots,s_p)\in[0,t]^p;\quad s_1<\cdots<s_p\big\}.$$

Then $\beta\big([0,t]^p_<\big)$ measures the amount of p-multiple self-intersection of the random path $\{W(s);\ 0\leqslant s\leqslant t\}$. When $d=1$, we have

$$\beta\big([0,t]^p_<\big) = \frac{1}{p!}\int_{\mathbb{R}} L^p(t,x)dx \tag{1.2}$$

*The author is partly supported by NSF grant DMS-0405188.

where $L(t,x)$ is the local time of the 1-dimensional Brownian motion $W(t)$. When $d \geqslant 2$, we will soon find out that $\beta\big([0,t]^p_<\big)$ can not be defined properly. For our convenience, however, we continue our discussion based on the false information that $\beta\big([0,t]^p_<\big)$ is properly defined whenever needed and we will take care of the difficulty later in section 4.

Let $W_1(t), \cdots, W_p(t)$ be independent d-dimensional Brownian motions. For any $t_1, \cdots, t_p > 0$, the quantity formally written as

$$\begin{aligned}&\alpha\big([0,t_1]\times\cdots\times[0,t_p]\big)\\&=\int_0^{t_1}\cdots\int_0^{t_p}\prod_{j=2}^p \delta_0\big(W_{j-1}(s_{j-1})-W_j(s_j)\big)ds_p\cdots ds_1\end{aligned} \tag{1.3}$$

measures the amount of mutual intersection of the independent paths

$$\{W_1(s);\ 0\leqslant s\leqslant t_1\},\cdots,\{W_p(s);\ 0\leqslant s\leqslant t_p\}$$

In literature, the quantities $\beta\big([0,t]^p_<\big)$ and $\alpha\big([0,t_1]\times\cdots\times[0,t_p]\big)$ are called, respectively, p-multiple self-intersection local time and p-multiple mutual intersection local time. One can also consider the intersection local times in the setting of random walks. Given a $\mathbb{Z}^d$-valued random walk $\{S(n)\}$, the p-multiple intersection local time is defined as

$$Q_n=\sum_{1\leqslant k_1<\cdots<k_p\leqslant n} 1_{\{S(k_1)=\cdots=S(k_p)\}}. \tag{1.4}$$

Let $\{S_1(n)\},\cdots,\{S_p(n)\}$ be independent copies of $\{S(n)\}$. The mutual intersection local time run by $\{S_1(n)\},\cdots,\{S_p(n)\}$ is defined as

$$I_n=\sum_{k_1,\cdots,k_p=1}^n 1_{\{S_1(k_1)=\cdots=S_p(k_p)\}}. \tag{1.5}$$

Some closely related random quantities are the range

$$\#\big\{S(0,n]\big\}=\#\{S(1),\cdots,S(n)\} \tag{1.6}$$

and the intersection of the ranges

$$\begin{aligned}J_n&=\#\big\{S_1(0,n]\cap\cdots\cap S_p(0,n]\big\}\\&=\big\{x\in\mathbb{Z}^d;\ \ x=S_1(k_1)=\cdots=S_p(k_p)\\&\quad\text{for some } k_1,\cdots,k_p\in(0,n]\big\}.\end{aligned} \tag{1.7}$$

The study of intersection local times is partially motivated by the needs from physics. Some problems in this area were suggested by physicists interested in applications to Euclidean quantum field theory (see Fernández, Fröhlich and Sokal (1992)), the growth of polymers (see ven der Hofstad and König (2001), den Hollander (1996), Vanderzande (1998), Westwater (1980, 1981, 1982)) and the polaron problem (see Donsker and Varadhan (1981, 1983), Mansmann (1991)). In the model of random walk in random environment, the dependence of a moving particle and a random environment frequently comes from the particle's ability to revisit sites with an attractive environment, and therefore measures of self-intersection quantify the degree of dependence between the movement and environment. We give the following partial list of literature on this topic: Carmona and Mochanov (1994), Cranston, Mountford and Shiga (2005), Ganternt, König and Shi (2007), van der Hofstad, König and Mörters (2006).

In this survey paper, we focus our attention on the large deviations for intersection local times and related models. More specifically, we are interested in the probabilities that the intersection local times take large values. Such probabilities are called tail probabilities in literature and often have exponential decay rates. Most of tail probabilities discussed in this article lead to the law of the iterated logarithm through some standard (or nearly standard) procedure of Borel-Cantelli lemma. Indeed, the relation between the tail probability and the law of the iterated logarithm is so close that we feel there is no need to distinguish one from another in our following discussion.

In addition to its application to the law of the iterated logarithm, the large deviations for intersection local times connect to the problems in some other disciplines. We list three connections below.

Connection to the study of polymer: Physicists concern about geometric shape of the polymer which is often described by a suitable random path (such as a Brownian curve). The geometry of a polymer is decided by the intensity that the random path intersects itself. The case when $\beta([0,t]^2_<)$ is large corresponds to a "contracting" polymer; while the case when $\beta([0,t]^2_<)$ is small corresponds to a "spread-out" polymer. The geometric shape of a polymer is often influenced by the environment (media). If the environment encourages attraction among the molecules, then the polymer is contracting. In this case, the polymer is called self-attracting polymer. In the opposite case, the polymer is spread-out and is called self-repelling polymer.

In physics, the probability measure $\widehat{P}_\lambda$ on $C\{[0,t],\mathbb{R}^2\}$ defined as

$$\widehat{P}_\lambda(A)=\widehat{C}_\lambda^{-1}\mathbb{E}\left(\exp\left\{\lambda\beta([0,t]^2_<)\right\}1_{\{W(\cdot)\in A\}}\right) \tag{1.8}$$

is regarded as the distribution of the curve representing a self-attracting polymer, where $\beta\big([0,t]^2_<\big)$ is given in (1.1) with $p=2$, $A\subset C\{[0,t],\mathbb{R}^2\}$, $\lambda>0$ represents the temperature and $\widehat{C}_\lambda$ is the normalizing constant making $\widehat{P}_\lambda$ a probability measure. Indeed, the curves with high self-intersection are given more distributional weight and the degree of this favoritism is decided by the value of λ.

Similarly, a self-repelling polymer is modeled by the distribution

$$\widetilde{P}_\lambda(A)=\widetilde{C}_\lambda^{-1}\mathbb{E}\Big(\exp\Big\{-\lambda\beta\big([0,t]^2_<\big)\Big\}1_{\{W(\cdot)\in A\}}\Big). \tag{1.9}$$

One of the problems that physicists try to understand is the behaviors of the partition function

$$\mathbb{E}\exp\Big\{\pm\lambda\iint_{\{0\leqslant r<s\leqslant t\}}\delta_0\big(W(r)-W(s)\big)drds\Big\}.$$

As we will see in section 4, the large deviations is the right tool to the investigations of this problem.

Connection to Parabolic Anderson model: Parabolic Anderson model is described by the system

$$\begin{cases}\partial_t u(t,x)=\kappa\triangle u(t,x)+\xi(x)u(t,x),\\ u(0,x)=1,\end{cases}$$

where $\xi(x)$ is a random field. For example, one can take $\xi(x)$ $(x\in\mathbb{Z}^d)$ as i.i.d. random variables; or one can take $\xi(x)=\dot{W}(x)$ with $W(x)$ $(x\in\mathbb{R}^d)$ being a Brownian sheet. We refer the reader to the references Carmona and Mochanov (1994), Cranston, Mountford and Shiga (2005), van der Hofstad, König and Mörters (2006).

In a parabolic Anderson model, the quantity $u(t,x)$ is interpreted as expected total mass at time t carried by a particle initially placed at the site x with a unit mass on it. The particle diffuses like a simple random walk (discrete setting) or like a Brownian motion (continuous setting) with generator $\kappa\triangle$; when present at site x, its mass is increased/decreased by an infinitesimal amount at rate $\pm\xi(x)\vee 0$. Of particular interest is the phenomenon of *intermittency*, which is often reflected by the asymptotic order of the quantity

$$\log\langle u^p(t,0)\rangle$$

as $t\to\infty$, where the notation "$\langle\cdot\rangle$" is used for the expectation with respect to the random potential $\xi(x)$. In the case $\xi(x)=\dot{W}(x)$ with

$W(x)$ ($x \in \mathbb{R}$) being a 1-dimensional Brownian motion, for example, by Feynman-Kac formula we have

$$\langle u^p(t,0)\rangle = \mathbb{E}\exp\left\{\frac{\kappa}{2}\int_0^t \dot{W}(B(s))ds\right\} = \mathbb{E}\exp\left\{\frac{\kappa^2}{8}\int_{\mathbb{R}} L^2(t,x)dx\right\}.$$

So the problem is reduced to the tail behaviors of intersection local times.

Random walk in random scenery: The asymptotic behavior of self-intersection local times contributes in a fundamental way to the limit theorems for the random walks in random sceneries. We refer the reader to Kesten and Spitzer (1979), Khoshnevisan and Lewis (1998), Csáki, König and Shi (1999), Révész and Shi (2000), Ganternt, König and Shi (2007), Asselah and Castell (2006) for the links to this problem. To established this connection, let $S(n)$ be a symmetric random walk taking values in $\mathbb{Z}^d$ and consider the random sequence

$$R_n = \sum_{k=1}^n \xi(S(k)), \qquad n = 1, \cdots$$

where $\xi(x)$ ($x \in \mathbb{Z}^d$) is often assumed to be an independent (of the random walks) family of i.i.d. random variable with mean zero and strong enough integrability. In the case $p = 1$, this model is known as random walk in random scenery. Let $l(n,x)$ be the local time of the random walk:

$$l(n,x) = \sum_{k=1}^n 1_{\{S(k)=x\}}, \qquad x \in \mathbb{Z}^d, \quad n = 1, \cdots .$$

Notice that

$$R_n = \sum_{x\in\mathbb{Z}^d} \xi(x) l(n,x), \qquad n = 1, 2, \cdots .$$

By the fact that conditionally on $l(n,x)$, R_n has the variance given by the quadratic form

$$\sum_{x\in\mathbb{Z}^d} l^2(n,x)$$

which is essentially twice of self-intersection local time of $\{S(n)\}$.

The systematic study of sample path intersection goes back at least to the work by Dvoretzky, Erdös and Kakutani (1950, 1954) on the multiple points of Brownian paths. The list of the references is too long to be included here. We mention some pioneering works by Varadhan (1969), Geman-Horowitz-Rosen (1984), Dynkin (1988), Yor (1985), Calais-Yor (1987), Le Gall (1986a, b), Rosen-Le Gall (1991) on the renormalization, construction of intersection local times, and (or) on the laws of weak convergence for the intersection local times and related models. The remarkable papers by Donsker-Varadhan (1975a, 1975b, 1976, 1979) on the large deviations created profound influence in the study of large deviations for intersection local times and related models. A key ingredient in their approach is Feynman-Kac Formula. A partial list of the references on this line are Mansmann (1991), Bolthausen (1999), van den Berg-Bolthausen-den Hollander (2001, 2004), Kesten-Hamana (2001, 2002), Bass-Chen (2004) Bass-Chen-Rosen (2006a, b), Chen (2004), Chen-Li (2004) and Chen-Li-Rosen (2005). In their paper on the upper tails for intersection local times, König and Mörters (2002) proposed a method computing high moment of intersection local times. This approach started to show its effectiveness in dealing with some difficult problems in the area of large deviations for intersection local times and for some related models. We refer the reader to section 7 for discussion on this method and for some applications.

We mention some remarkable contributions from Xianyin Zhou (1963-1996), whose life was cut short by sudden disease. In his short academic career, Zhou produced more than 50 high quality papers. Significant part of his work deals with sample path intersection and its connection to polymers. We refer the reader to a special memorial volume (Zhou (2002)) for a collection of Zhou's work.

The literature is vast and one has to make a selection of topics. What is presented here does reflect the author's interest and preference. Apart from the topics addressed in this article, we mention the works in some recent important development in the closely related areas. (a).The solution of the intersection exponent problem and its associations with self-avoiding random walks and with the stochastic Loewner evolution: The reader is referred to the papers by Lawler, Scheraman and Werner (2001a, 2001b, 2002), the book by Lawler (2005). (b). The large deviations for the random walks and Brownian motions under the polymer structure in the spirit of (1.8) and (1.9): The reader is referred to van der Hofstad, den Hollander and König (2003). (c). The large deviations for Brownian intersection local times confined in a bounded domain and running up to the exit times of the domain and, their applications to the problem of finding the Hausdorff dimension spectrum for the thick points of the intersection: The reader is referred to König and Mörters (2002).

Little has been known beyond the models with independent increments. To the best of the author's knowledge, the large deviations for the intersection local times of general Markov processes remain essentially open. In this category, the models of particular interest are diffusion processes, super processes and random walks and Brownian motions on graphs and fractals. Perhaps an even harder problem is to ask the same question to the non-Markovian processes such as fractional Brownian motions and Gaussian processes in general. Here we refer the reader to the paper by Hu and Nualart (2005) for renormalization of the self-intersection local times of fractional Brownian motions, and to the paper by Xiao (2006) for the recent development of the method known as *local non-determinism* and its applications on the tail probabilities for local times and intersection local times of Gaussian processes.

We end this introduction with an outline of the rest of the paper. In section 2 we investigate exponential asymptotics for the smoothed intersection local times by Feynmen-Kac formula. In section 3, we discuss the large deviations for the mutual intersection local time $\alpha([0,t]^p)$ run by independent Brownian motions and for the local times of additive Brownian motions. In section 4, we summarize the large deviations for self-intersection local time of a single Brownian. In section 5 and 6, our attention is switched to the intersections of random walks in lattice spaces. In section 7, we discuss a new method in computing high moment of intersection local times.

2 Exponential asymptotics for smoothed intersection local times

A substantial amount of our work is fighting against singularity, which increases with the dimension. When $d = 1$, the intersection local times can be written in terms of the local times, as we have seen in (1.2). When $d \geqslant 2$, the local time no longer exist. We introduce

$$L(t,x,\epsilon) = \int_0^t h_\epsilon\big(W(s) - x\big)ds,$$

instead, where

$$h_\epsilon(x) = \epsilon^{-d} h(\epsilon^{-1}x), \qquad x \in \mathbb{R}^d, \quad \epsilon > 0,$$

and $h(x)$ is a suitable probability density on $\mathbb{R}^d$. Our strategy is to establish exponential asymptotics for the smoothed quantities

$$\int_{\mathbb{R}^d} L^p(t,x,\epsilon)dx, \qquad \int_{\mathbb{R}^d} \prod_{j=1}^p L_j(t,x,\epsilon)dx,$$

and to approximate intersection local times using these quantities by letting $\epsilon \to 0^+$. To this end we write, for $\theta > 0$,

$$M_\epsilon(\theta) = \sup_{f\in\mathcal{F}_d} \left\{ \theta\left(\int_{\mathbb{R}^d} \big[(f^2 * h_\epsilon)(x)\big]^p dx \right)^{1/p} - \frac{1}{2}\int_{\mathbb{R}^d} |\nabla f(x)|^2 dx \right\}, \tag{2.1}$$

$$N_\epsilon(\theta) = \sup_{f\in\mathcal{F}_d}\left\{\theta\left(\int_{\mathbb{R}^d}\left[(f^2 * h_\epsilon)(x)\right]^p dx\right)^{1/p} - \frac{p}{2}\int_{\mathbb{R}^d}|\nabla f(x)|^2 dx\right\}, \tag{2.2}$$

where

$$\mathcal{F}_d = \left\{f\in\mathcal{L}^2(\mathbb{R}^d);\ \ ||f||_2 = 1 \text{ and } \nabla f\in\mathcal{L}^2(\mathbb{R}^d)\right\}. \tag{2.3}$$

The following result is given in Chen (2004).

Theorem 2.1. *For any $\theta > 0$ and $p \geqslant 2$,*

$$\lim_{t\to\infty}\frac{1}{t}\log\mathbb{E}\exp\left\{\theta\left(\int_{\mathbb{R}^d}L^p(t,x,\epsilon)dx\right)^{1/p}\right\} = M_\epsilon(\theta), \tag{2.4}$$

$$\lim_{t\to\infty}\frac{1}{t}\log\mathbb{E}\exp\left\{\theta\left(\int_{\mathbb{R}^d}\prod_{j=1}^p L_j(t,x,\epsilon)dx\right)^{1/p}\right\} = N_\epsilon(\theta). \tag{2.5}$$

One of the main ingredients in our approach for Theorem 2.1 is the well known Feynman-Kac formula, which claim that for a nice function $f\in\mathbb{R}^d$, the semi-group $\{T_t\}_{t\geqslant 0}$ of self-adjoint linear operators on $\mathcal{L}^2(\mathbb{R}^d)$

$$T_t g(x) = \mathbb{E}_x\exp\left\{\int_0^t f\big(W(s)\big)ds\right\}g\big(W(t)\big), \qquad g\in\mathcal{L}^2(\mathbb{R}^d)$$

has infinitesimal generator

$$\mathcal{A}g(x) = \frac{1}{2}\triangle g(x) + f(x)g(x).$$

The relation is formally given as $T_t = e^{t\mathcal{A}}$. Consequently,

$$\begin{aligned}\mathbb{E}\exp\left\{\int_0^t f\big(W(s)\big)ds\right\} &\approx \sup_{g\in\mathcal{F}_d}\langle g, T_t g\rangle = \exp\left\{t\sup_{g\in\mathcal{F}_d}\langle g,\mathcal{A}g\rangle\right\}\\ &= \exp\left\{t\sup_{g\in\mathcal{F}_d}\left(\int_{\mathbb{R}^d}f(x)g^2(x)dx - \frac{1}{2}\int_{\mathbb{R}^d}|\nabla g(x)|^2dx\right)\right\}.\end{aligned}$$

Thus

$$\begin{aligned}&\lim_{t\to\infty}\frac{1}{t}\log\mathbb{E}\exp\left\{\int_0^t f\big(W(s)\big)ds\right\}\\ &= \sup_{g\in\mathcal{F}_d}\left\{\int_{\mathbb{R}^d}f(x)g^2(x)dx - \frac{1}{2}\int_{\mathbb{R}^d}|\nabla g(x)|^2dx\right\}.\end{aligned} \tag{2.6}$$

The reader is referred to Remillard (2000) for the details of the proof of (2.6).

Let $q > 1$ be the conjugate number of p defined by $p^{-1} + q^{-1} = 1$. By Hölder inequality, for any f with $|f||_q = 1$

$$\left(\int_{\mathbb{R}^d}L^p(t,x,\epsilon)dx\right)^{1/p} \geqslant \int_{\mathbb{R}^d}L(t,x,\epsilon)f(x)dx = \int_0^t (f * h_\epsilon)\big(W(s)\big)ds.$$

Replacing f by $\theta f * h_\epsilon$ in (2.6),

$$\begin{aligned}
&\liminf_{t\to\infty}\frac{1}{t}\log\mathbb{E}\exp\left\{\theta\Big(\int_{\mathbb{R}^d}L^p(t,x,\epsilon)dx\Big)^{1/p}\right\}\\
&\geqslant \sup_{g\in\mathcal{F}_d}\left\{\theta\int_{\mathbb{R}^d}(f*h_\epsilon)(x)g^2(x)dx-\frac{1}{2}\int_{\mathbb{R}^d}|\nabla g(x)|^2dx\right\}\\
&=\sup_{g\in\mathcal{F}_d}\left\{\theta\int_{\mathbb{R}^d}f(x)(g^2*h_\epsilon)(x)dx-\frac{1}{2}\int_{\mathbb{R}^d}|\nabla g(x)|^2dx\right\}.
\end{aligned}$$

Taking supremum over $||f||_q = 1$ on the right hand side gives the lower bound for (2.4). The upper bound for (2.4) follows, in spirit, also from the Feynman-Kac large deviation given in (2.6). The argument is much harder than the one for the lower bound and the technical difficulty here is the absence of exponential tightness of the model.

The upper bound of (2.5) follows from the deterministic relation

$$\Big(\int_{\mathbb{R}^d}\prod_{j=1}^p L_j(t,x,\epsilon)dx\Big)^{1/p}\leqslant\frac{1}{p}\sum_{j=1}^p\Big(\int_{\mathbb{R}^d}L_j^p(t,x,\epsilon)dx\Big)^{1/p} \tag{2.7}$$

and (2.4). The idea behind the lower bound of (2.5) is that the inequality (2.7) is essentially reversible on the event

$$\Big\{L_1(t,\cdot,\epsilon)\approx\cdots\approx L_p(t,\cdot,\epsilon)\Big\},$$

and that decay of the probability of this set is slower than exponential rate.

3 Intersection of independent Brownian motions

Let $W_1(t),\cdots,W_p(t)$ be independent d-dimensional Brownian motions. If we allow $W_j(\cdot)$ run up to time t_j $(j=1,\cdots,p)$, a natural question is to ask how much time is spent for the p independent trajectories $W_1(t),\cdots,W_p(t)$ to intersect. In other words, we are interested in the time set

$$\Big\{(s_1,\cdots,s_p)\in[0,t_1]\times\cdots\times[0,t_p];\quad W_1(s_1)\approx\cdots\approx W_p(s_p)\Big\}.$$

If properly defined, the Lebesgue measure of this set is called the intersection local time of $W_1(t),\cdots,W_p(t)$ and is denoted by $\alpha\big([0,t_1]\times\cdots\times[0,t_p]\big)$.

One of the fascinating phenomena in the area of sample path intersection is dimension dependence. It is well known (Dvoretzky-Erdös-Kakutani (1950, 1954)) that

$$W_1(0,\infty)\cap\cdots\cap W_p(0,\infty)\neq\phi$$

if and only if $p(d-2)<d$. That is, it is hard for trajectories to intersect in high dimensional space.

Under $p(d-2)<d$, there are two equivalent ways to construct Brownian intersection local time in the multi-dimensional case. The first approach (Geman, Horowitz and Rosen (1984)) corresponds to the notation given in (1.3). Geman, Horowitz and Rosen (1984) proved that under $p(d-2)<d$, the occupation measure on $\mathbb{R}^{d(p-1)}$ given by

$$\mu_A(B)=\int_A 1_B\big(W_1(s_1)-W_2(s_2),\cdots,W_{p-1}(s_{p-1})-W_p(s_p)\big)ds_1\cdots ds_p$$

($B\subset\mathbb{R}^{d(p-1)}$) is absolutely continuous, with probability 1, with respect to Lebesgue measure on $\mathbb{R}^{d(p-1)}$ for any Borel set $A\subset(\mathbb{R}^p)^+$ (in particular, for $A=[0,t_1]\times\cdots\times[0,t_p]$) and, the density $\alpha(x,A)$ of such measure can be chosen in such a way that the function

$$(x,t_1,\cdots,t_p)\longmapsto\alpha\big(x,[0,t_1]\times\cdots\times[0,t_p]\big)$$

($x\in\mathbb{R}^{d(p-1)}$, $(t_1,\cdots,t_p)\in(\mathbb{R}^+)^p$) is jointly continuous. The random measure $\alpha(\cdot)$ on $(\mathbb{R}^p)^+$ is defined as

$$\alpha(A)=\alpha(0,A),\qquad\forall\text{ Borel set }A\subset(\mathbb{R}^p)^+.$$

Another approach (Le Gall (1992)) constitutes the notation

$$\alpha\big([0,t_1]\times\cdots[0,t_p]\big)=\int_{\mathbb{R}^d}\bigg[\prod_{j=1}^p\int_0^{t_j}\delta_0\big(W(s)-x\big)ds\bigg]dx.\tag{3.1}$$

Let $f(x)$ be a nice probability density function on $\mathbb{R}^d$. Given $\epsilon>0$, write $f_\epsilon(x)=\epsilon^{-d}f(\epsilon^{-1}x)$ and define

$$\alpha_\epsilon\big([0,t_1]\times\cdots\times[0,t_p]\big)=\int_{\mathbb{R}^d}\bigg[\prod_{j=1}^p\int_0^{t_j}f_\epsilon\big(W(s)-x\big)ds\bigg]dx.$$

Under $p(d-2)<d$, Le Gall (1992) proved that there is a random variable $\alpha\big([0,t_1]\times\cdots\times[0,t_p]\big)$ such that

$$\lim_{\epsilon\to0^+}\alpha_\epsilon\big([0,t_1]\times\cdots\times[0,t_p]\big)=\alpha\big([0,t_1]\times\cdots\times[0,t_p]\big)$$

holds in L^m-norm for any $m \geqslant 1$ and for any $t_1, \cdots, t_p > 0$. By the scaling property of Brownian motions

$$\alpha\big([0,t]^p\big) \stackrel{d}{=} t^{\frac{2p-d(p-1)}{2}} \alpha\big([0,1]^p\big). \tag{3.2}$$

In the special case $d = 1$, let $L_1(t,x), \cdots, L_p(t,x)$ be the local times of $W_1, \cdots, W_p$, respectively. By Le Gall's construction, one can see that

$$\alpha\big([0,t_1] \times \cdots \times [0,t_p]\big) = \int_{-\infty}^{\infty} \prod_{j=1}^{p} L_j(t,x)dx.$$

The following theorem is given in Chen (2004).

Theorem 3.1. *Under* $p(d-2) < d$,

$$\lim_{t\to\infty} t^{-\frac{2}{d(p-1)}} \log \mathbb{P}\Big\{\alpha\big([0,1]^p\big) \geqslant t\Big\} = -\frac{p}{2}\kappa(d,p)^{-\frac{4p}{d(p-1)}}, \tag{3.3}$$

where $\kappa(d,p)$ *is the best constant of the Gagliardo-Nirenberg inequality*

$$||f||_{2p} \leqslant C||\nabla f||_2^{\frac{d(p-1)}{2p}} ||f||_2^{1-\frac{d(p-1)}{2p}}, \qquad f \in W^{1,2}(\mathbb{R}^d) \tag{3.4}$$

and

$$W^{1,2}(\mathbb{R}^d) = \big\{f \in \mathcal{L}^2(\mathbb{R}^d); \ \ \nabla f \in \mathcal{L}^2(\mathbb{R}^d)\big\}.$$

Finding the best constant of Gagliardo-Nirenberg inequalities remains open in general. It has been attracting considerable attention partially due to its connection to some problems in physics. The best constant for Nash inequality, which is a special case of Gagliardo-Nirenberg inequalities, was found by Carlen and Loss (1993). See also papers by Cordero-Erausquim, Nararet and Villani (2004) and by Del Pino and Dolbeault (2002) for recent progress on the best constants for a class of Gagliardo-Nirenberg inequalities. See the paper Del Pino and Dolbeault (2003) for a connection between the best constants for a class of Gagliardo-Nirenberg inequalities and logarithmic Sobolev inequalities. Two papers are directly related to $\kappa(2,3)$: Levine (1980) obtained the sharp estimate

$$\sqrt[3]{\frac{1}{4.6016}} < \kappa(2,3) < \sqrt[3]{\frac{1}{4.5981}}. \tag{3.5}$$

He conjectured that $\kappa(2,3) = \pi^{-4/9}$. Another numerical solution was obtained by Weinstein (1983). He obtained

$$\kappa(2,2) \approx \sqrt[4]{\frac{1}{\pi \times 1.86225\cdots}}. \tag{3.6}$$

Here is the basic idea behind Theorem 3.1: As $\epsilon \to 0^+$, $\alpha([0,t]^p)$ is so closed to

$$\int_{\mathbb{R}^d} \prod_{j=1}^{p} L_j(t,x,\epsilon)dx$$

given in Theorem 2.1 that (2.5) implies replaced by

$$\begin{aligned}
&\lim_{t\to\infty} \frac{1}{t} \log \mathbb{E} \exp\Big\{\theta\Big(\alpha([0,t]^p)\Big)^{1/p}\Big\}\\
&= \sup_{f\in\mathcal{F}_d} \Big\{\theta\Big(\int_{\mathbb{R}^d} |f(x)|^{2p}dx\Big)^{1/p} - \frac{p}{2}\int_{\mathbb{R}^d} |\nabla f(x)|^2 dx\Big\}\\
&= \theta^{\frac{2p}{2p-d(p-1)}} p^{-\frac{d(p-1)}{2p-d(p-1)}} \sup_{f\in\mathcal{F}_d} \Big\{\Big(\int_{\mathbb{R}^d} |f(x)|^{2p}dx\Big)^{1/p} - \frac{1}{2}\int_{\mathbb{R}^d} |\nabla f(x)|^2 dx\Big\}
\end{aligned}$$

where the last step follows from a suitable rescaling. Finally, Theorem 3.1 follows from scaling identity (2.10), a standard application of Gärtner-Ellis theorem on large deviation, and the relation (Lemma A.2, Chen (2004))

$$\begin{aligned}
&\sup_{f\in\mathcal{F}_d} \Big\{\Big(\int_{\mathbb{R}^d} |f(x)|^{2p}dx\Big)^{1/p} - \frac{1}{2}\int_{\mathbb{R}^d} |\nabla f(x)|^2 dx\Big\}\\
&= \frac{2p-d(p-1)}{2p}\Big(\frac{d(p-1)}{p}\Big)^{\frac{d(p-1)}{2p-d(p-1)}} \kappa(d,p)^{\frac{4p}{2p-d(p-1)}}.
\end{aligned}$$

□

From the view point of Geman, Horowitz and Rosen (1984), the intersection local time $\alpha([0,t]^p)$ is the local time of the multi-parameter process

$$\widetilde{W}(t_1,\cdots,t_p) = \Big(W_1(t_1) - W_2(t_2), \cdots, W_{p-1}(t_{p-1}) - W_p(t_p)\Big)$$

as indicated by (1.3). In this regard, Theorem 3.1 contributes solution to a problem on the tail probability for the local times of multi-parameter processes.

Perhaps the most important multi-parameter process is Brownian sheet. We refer the reader to the book by Khoshnevisan (2002) for the overall information about the Brownian sheets and multi-parameter processes, among which it has been known that a Brownian sheet $B(t_1,\cdots,t_p)$ with space dimension d and time dimension $p \geqslant 2$ has local time if and only if $d < 2p$. The exact tail of the local time of Brownian sheet remains unknown — an indication how little we know about the tails of the local times of multi-parameter processes.

In the rest of the section, we discuss some recent progress on a model which resembles Brownian sheet in many ways but is more amenable to analysis than Brownian sheet. We hope it will give us more insight and sharpen our tools for the case of Brownian sheets. The model is known as *additive Brownian motion* and is defined by

$$\widetilde{B}(t_1,\cdots,t_p)=W_1(t_1)+\cdots+W_p(t_p),\qquad t_1,\cdots,t_p\geqslant 0.$$

We concern about the large deviation problem for the local time

$$\eta^x(I)=\int_I\delta_x\big(W_1(s_1)+\cdots+W_p(s_p)\big)ds_1\cdots ds_p,\qquad x\in\mathbb{R}^d,\quad I\subset(\mathbb{R}^+)^p$$

of $\widetilde{B}(t_1,\cdots,t_p)$ and rely on two recent papers by Khoshnevisan, Xiao and Zhong (2003a, b) for the constructions of the local time $\eta^x(I)$. In their papers, Khoshnevisan, Xiao and Zhong (2003a, b) consider a more general multi-parameter random field named additive Lévy process, which is generated by independent Lévy processes. In their construction, $\eta^x(I)$ is defined as the density function of the occupation measure μ_I:

$$\mu_I(A)=\int_I\delta_{W_1(s_1)+\cdots+W_p(s_p)}(A)ds_1\cdots ds_p,\qquad A\subset\mathbb{R}^d$$

in the case when μ_I is absolutely continuous with respect to the Lebesgue measure on $\mathbb{R}^d$. Not surprising at all, the necessary and sufficient condition for existence of $\eta^x(I)$ is $d<2p$, same as the one for existence of the local time of Brownian sheet. Under this condition, the local time

$$\eta^x\big([0,t]^p\big),\qquad (x,t)\in\mathbb{R}^d\times\mathbb{R}^+$$

is jointly continuous in (x,t) (Corollary 3.3, Khoshnevisan, Xiao and Zhong (2003b)).

By Brownian scaling property,

$$\eta\big([0,t]^p\big)\overset{d}{=}t^{\frac{2p-d}{2}}\eta\big([0,1]^p\big).\tag{3.7}$$

The local time $\eta^x\big([0,t]^p\big)$ connects to the intersection local time $\alpha\big([0,1]^p\big)$ when $p=2$, in which case

$$\eta^0\big([0,t]^2\big)\overset{d}{=}\alpha\big([0,t]^2\big).$$

To state the large deviations for $\eta^x\big([0,t]^p\big)$, we define

$$\rho_1=\sup_{||f||_2=1}\int_{\mathbb{R}^d}\left[\int_{\mathbb{R}^d}\frac{f(\lambda+\gamma)f(\gamma)}{\sqrt{1+2^{-1}|\lambda+\gamma|^2}\sqrt{1+2^{-1}|\gamma|^2}}d\gamma\right]^p d\lambda$$

$$\rho_2 = \sup_{||f||_2=1} \int_{\mathbb{R}^d} \left[\int_{\mathbb{R}^d} \frac{f(\lambda+\gamma)f(\gamma)}{\sqrt{1+2^{-1}|\lambda+\gamma|^2}\sqrt{1+2^{-1}|\gamma|^2}} d\gamma \right]^{2p} d\lambda.$$

By essentially Hölder inequality, one can prove that under $d < 2p$

$$0 < \rho_1 \leqslant \int_{\mathbb{R}^d} \frac{1}{\left(1+2^{-1}|\gamma|^2\right)^p} d\lambda < \infty,$$

$$0 < \rho_2 \leqslant \int_{\mathbb{R}^d} \frac{1}{\left(1+2^{-1}|\gamma|^2\right)^{2p}} d\lambda < \infty.$$

The following result is given in Chen (2006a, 2007a).

Theorem 3.2. *Under $d < 2p$,*

$$\lim_{t\to\infty} t^{-2/d} \log \mathbb{P}\Big\{ \eta^0\big([0,1]^p\big) \geqslant t \Big\} = -(2\pi)^2 \frac{d}{2}\Big(1-\frac{d}{2p}\Big)^{\frac{2p-d}{d}} \rho_1^{-2/d} \tag{3.8}$$

$$\lim_{t\to\infty} t^{-2/d} \log \mathbb{P}\Big\{ \sup_{x\in\mathbb{R}^d} \eta^x\big([0,1]^p\big) \geqslant t \Big\} = -(2\pi)^2 \frac{d}{2}\Big(1-\frac{d}{2p}\Big)^{\frac{2p-d}{d}} \rho_1^{-2/d} \tag{3.9}$$

$$\lim_{t\to\infty} t^{-2/d} \log \mathbb{P}\Big\{ \int_{\mathbb{R}^d} \big[\eta^x\big([0,1]^p\big)\big]^2 dx \geqslant t \Big\} = -\frac{d}{4}(2\pi)^2 \Big(1-\frac{d}{4p}\Big)^{\frac{4p-d}{d}} \rho_2^{-2/d}. \tag{3.10}$$

Theorem 3.2 can be extended to the setting of additive Lévy processes and additive random walks. We refer the reader to Chen (2007b) for details.

It is interesting to see that $\sup_{x\in\mathbb{R}^d} \eta^x\big([0,1]^p\big)$ has exactly same up tail as $\eta^0\big([0,1]^p\big)$. The proof of (3.10) is based on (3.8) and on a chaining argument. To explain the idea behind (3.8), we recall the following lemma given in König and Mörters (2002).

Lemma 3.3. *Let Y be any non-negative random variable and let $\theta > 0$ be fixed. Assume that*

$$\lim_{m\to\infty} \frac{1}{m} \log \frac{1}{(m!)^\theta} \mathbb{E}\, Y^m = -\kappa \tag{3.11}$$

for some $\kappa \in \mathbb{R}$. Then we have

$$\lim_{t\to\infty} t^{-1/\theta} \log \mathbb{P}\{Y \geqslant t\} = -\theta e^{\kappa/\theta}. \tag{3.12}$$

To prove (3.8), therefore, one needs to establish

$$\begin{aligned}&\lim_{m\to\infty}\frac{1}{m}\log(m!)^{-d/2}\mathbb{E}\Big[\eta([0,1]^p)^m\Big]\\&=\log\Big(\frac{2p}{2p-d}\Big)^{\frac{2p-d}{2}}+\log\frac{\rho_1}{(2\pi)^d}.\end{aligned}\tag{3.13}$$

By Fourier transform, for any $t_1,\cdots,t_p\geqslant 0$,

$$\eta([0,t_1]\times\cdots\times[0,t_p])=\frac{1}{(2\pi)^d}\int_{\mathbb{R}^d}d\lambda\prod_{j=1}^p\int_0^{t_j}\exp\Big\{i\lambda\cdot W_j(s)\Big\}ds.$$

For any integer $m\geqslant 1$, by time rearrangement and by increment independence,

$$\begin{aligned}&\mathbb{E}\Big[\eta^0([0,t_1]\times\cdots\times[0,t_p])^m\Big]\\&=\frac{1}{(2\pi)^{dm}}\int_{(\mathbb{R}^d)^m}d\lambda_1\cdots d\lambda_m\prod_{j=1}^p\sum_{\sigma\in\Sigma_m}\int_{\{0\leqslant s_1\leqslant\cdots\leqslant s_m\leqslant t_j\}}\\&\times\prod_{k=1}^m\exp\Big\{-\frac{s_k-s_{k-1}}{2}\Big|\sum_{j=1}^k\lambda_{\sigma(j)}\Big|^2\Big\}ds_1\cdots ds_m.\end{aligned}\tag{3.14}$$

where Σ_m is the permutation group on $\{1,\cdots,m\}$ and $s_0=0$.

To simplify the above representation, we replace $t_1\cdots,t_p$ by the i.i.d. exponential times $\tau_1,\cdots,\tau_p$. We assume that the exponential distribution has parameter 1 and that $\tau_1,\cdots,\tau_p$ are independent of $W_1(t),\cdots,W_p(t)$. From (3.14)

$$\begin{aligned}&\mathbb{E}\Big[\eta([0,\tau_1]\times\cdots\times[0,\tau_p])^m\Big]\\&=\frac{1}{(2\pi)^{dm}}\int_{(\mathbb{R}^d)^m}d\lambda_1\cdots d\lambda_m\Big[\sum_{\sigma\in\Sigma_m}\int_0^\infty e^{-t}dt\int_{\{0\leqslant s_1\leqslant\cdots s_m\leqslant t\}}\\&\times\prod_{k=1}^m\exp\Big\{-\frac{s_k-s_{k-1}}{2}\Big|\sum_{j=1}^k\lambda_{\sigma(j)}\Big|^2\Big\}ds_1\cdots ds_m\Big]^p\\&=\frac{1}{(2\pi)^{dm}}\int_{(\mathbb{R}^d)^m}d\lambda_1\cdots d\lambda_m\\&\times\Big[\sum_{\sigma\in\Sigma_m}\prod_{k=1}^m\int_0^\infty e^{-t}\exp\Big\{-\frac{t}{2}\Big|\sum_{j=1}^k\lambda_{\sigma(j)}\Big|^2\Big\}dt\Big]^p\\&=\frac{1}{(2\pi)^{dm}}\int_{(\mathbb{R}^d)^m}d\lambda_1\cdots d\lambda_m\Big[\sum_{\sigma\in\Sigma_m}\prod_{k=1}^m Q\Big(\sum_{j=1}^k\lambda_{\sigma(j)}\Big)\Big]^p,\end{aligned}\tag{3.15}$$

where $Q(\lambda) = (1+2^{-1}|\lambda|^2)^{-1}$.

A crucial step is to establish

$$\lim_{m\to\infty}\frac{1}{m}\log\frac{1}{(m!)^p}\int_{(\mathbb{R}^d)^m} d\lambda_1\cdots d\lambda_m \times\Big[\sum_{\sigma\in\Sigma_m}\prod_{k=1}^m Q\Big(\sum_{j=1}^k \lambda_{\sigma(j)}\Big)\Big]^p = \log\rho_1, \tag{3.16}$$

which is left to the discussion in section 7.

Let $t_1,\cdots,t_p \geqslant 0$. By (3.14), by Hölder inequality and by the scaling (3.7),

$$\mathbb{E}\Big[\eta([0,t_1]\times\cdots\times[0,t_p])^m\Big] \leqslant \prod_{j=1}^p\Big\{\mathbb{E}\Big[\eta\big([0,t_j]^p\big)^m\Big]\Big\}^{1/p}$$
$$= (t_1\cdots t_p)^{\frac{2p-d}{2p}m}\mathbb{E}\Big[\eta^0\big([0,1]^p\big)^m\Big].$$

Thus,

$$\mathbb{E}\Big[\eta([0,\tau_1]\times\cdots\times[0,\tau_p])^m\Big]$$
$$= \int_0^\infty\cdots\int_0^\infty e^{-(t_1+\cdots+t_p)}\mathbb{E}\Big[\eta([0,t_1]\times\cdots\times[0,t_p])^m\Big]dt_1\cdots dt_p$$
$$\leqslant \mathbb{E}\Big[\eta([0,1]^p)^m\Big]\int_0^\infty\cdots\int_0^\infty (t_1\cdots t_p)^{\frac{2p-d}{2p}m}e^{-(t_1+\cdots+t_p)}dt_1\cdots dt_p$$
$$= \mathbb{E}\Big[\eta([0,1]^p)^m\Big]\Big[\Gamma\Big(\frac{2p-d}{2p}m+1\Big)\Big]^p.$$

By (3.16) and Stirling formula,

$$\liminf_{m\to\infty}\frac{1}{m}\log(m!)^{-d/2}\mathbb{E}\Big[\eta([0,1]^p)^m\Big] \geqslant \log\Big(\frac{2p}{\alpha p-d}\Big)^{\frac{2p-d}{2}} + \log\frac{\rho_1}{(2\pi)^d}. \tag{3.17}$$

On the other hand, notice that $\bar\tau \equiv \min\{\tau_1,\cdots,\tau_p\}$ has the exponential distribution with the parameter p. Hence,

$$\mathbb{E}\Big[\eta\big([0,\tau_1]\times\cdots\times[0,\tau_p]\big)^m\Big] \geqslant \mathbb{E}\Big[\eta\big([0,\bar\tau]^p\big)^m\Big]$$
$$= \mathbb{E}\bar\tau^{\frac{2p-d}{2}m}\mathbb{E}\Big[\eta\big([0,1]^p\big)^m\Big]$$
$$= p^{-\frac{2p-d}{2}m-1}\Gamma\Big(1+\frac{2p-d}{2}m\Big)\mathbb{E}\Big[\eta\big([0,1]^p\big)^m\Big]$$

where the second step follows from (3.7). By Stirling formula and (3.16) again we have

$$\begin{aligned}&\limsup_{m\to\infty}\frac{1}{m}\log(m!)^{-d/2}\mathbb{E}\left[\eta([0,1]^p)^m\right]\\&\leqslant\log\left(\frac{2p}{2p-d}\right)^{\frac{2p-d}{2}}+\log\frac{\rho_1}{(2\pi)^d}.\end{aligned}\tag{3.18}$$

Combining (3.17)) and (3.18)) gives (3.13). □

4 Self-intersection of single Brownian path

First we consider the case $d=1$. By (1.2) the self-intersection local time is essentially defined by the quantity

$$\int_{\mathbb{R}}L^p(t,x)dx.\tag{4.1}$$

The tail probability of the above quantity was computed in Mansmann (1991) in the special case $p=2$. He essentially proved that

$$\lim_{t\to\infty}t^{-2}\log\mathbb{P}\Big\{\int_{\mathbb{R}}L^2(1,x)dx\geqslant t\Big\}=-\frac{3}{2}.$$

We consider the general case in the following discussion. The following nice property of this model have been exploit extensively in Chen and Li (2004) and in Chen, Li and Rosen (2005): For any $s,t>0$, the quantity

$$\int_{\mathbb{R}}\big[L(s+t,x)-L(s,x)\big]^pdx$$

is independent of the family $\{W(u);\ u\leqslant s\}$ and equal in law to the quantity given in (4.1). By this property and by triangular inequality, one can prove that

$$\mathbb{E}\exp\left\{\theta\Big(\int_{\mathbb{R}}L^p(t,x)dx\Big)^{1/p}\right\}<\infty\qquad\forall\theta,t>0.$$

Further, for any $s,t>0$,

$$\begin{aligned}&\mathbb{E}\exp\left\{\theta\Big(\int_{\mathbb{R}}L^p(s+t,x)dx\Big)^{1/p}\right\}\\&\leqslant\mathbb{E}\exp\left\{\theta\Big(\int_{\mathbb{R}}L^p(s,x)dx\Big)^{1/p}\right\}\mathbb{E}\exp\left\{\theta\Big(\Big(\int_{\mathbb{R}}L^p(t,x)dx\Big)^{1/p}\right\}.\end{aligned}$$

By standard sub-additivity argument, the limit

$$\lim_{t\to\infty}\frac{1}{t}\log\mathbb{E}\exp\left\{\theta\Big(\int_{\mathbb{R}}L^p(t,x)dx\Big)^{1/p}\right\}$$

exists for any $\theta > 0$.

This observation gives, in comparison with (2.4), a strong indication as what to expect. Indeed, by a approximation via (2.4), or by applying the argument for (2.4) to the local time $L(t,x)$ instead of $L(t,x,\epsilon)$, we can show that the above limit is equal to

$$\begin{aligned}&\sup_{f\in\mathcal{F}_1}\left\{\theta\Big(\int_{\mathbb{R}}|f(x)|^{2p}dx\Big)^{1/p}-\frac{1}{2}\int_{\mathbb{R}}|f'(x)|^2dx\right\}\\&=\theta^{\frac{2p}{p+1}}\sup_{f\in\mathcal{F}_1}\left\{\Big(\int_{\mathbb{R}}|f(x)|^{2p}dx\Big)^{1/p}-\frac{1}{2}\int_{\mathbb{R}}|f'(x)|^2dx\right\}.\end{aligned}$$

According to the identity given in Lemma 7.2 of Chen and Li (2004),

$$\begin{aligned}&\sup_{f\in\mathcal{F}_1}\left\{\Big(\int_{\mathbb{R}}|f(x)|^{2p}dx\Big)^{1/p}-\frac{1}{2}\int_{\mathbb{R}}|f'(x)|^2dx\right\}\\&=p^{-\frac{2p}{p+1}}\left(\frac{\sqrt{2}}{(p-1)(p+1)}B\Big(\frac{1}{p-1},\frac{1}{2}\Big)\right)^{\frac{2(p-1)}{p+1}}\end{aligned}$$

where $B(\cdot,\cdot)$ is beta function. By the scaling property

$$\int_{\mathbb{R}}L^p(t,x)dx \overset{d}{=} t^{\frac{p+1}{2}}\int_{\mathbb{R}}L^p(1,x)dx$$

and by Gärtner-Ellis theorem, we obtain the following result (Chen and Li (2004)).

Theorem 4.1. *Let $L(t,x)$ be the local time of an 1-dimensional Brownian motion. Then for any $p \geqslant 2$*

$$\begin{aligned}&\lim_{t\to\infty}t^{\frac{2}{p-1}}\log\mathbb{P}\Big\{\int_{\mathbb{R}}L^p(1,x)dx\geqslant t\Big\}\\&=-\frac{1}{4(p-1)}\Big(\frac{p+1}{2}\Big)^{\frac{3-p}{p-1}}B\Big(\frac{1}{p-1},\frac{1}{2}\Big)^2.\end{aligned}\tag{4.2}$$

We mention the fact that before Theorem 4.1 was obtained, Csörgö, Shi and Yor (1999) had established the exactly same form of large deviation for the local time of 1-dimensional Brownian bridge. Their argument is essentially based on excursion analysis and some distributional identities.

The self-intersection local time given in (1.1) can not be properly defines as $d \geqslant 2$. Take $p = 2$ for example. A simple way to see this is to do the following symbolic calculation:

$$\begin{aligned}\mathbb{E}\,\beta([0,t]^2_<) &= \iint_{\{0\leqslant r<s\leqslant t\}} \mathbb{E}\,\delta_0\big(W(r)-W(s)\big)drds \\ &= \iint_{\{0\leqslant r<s\leqslant t\}} \frac{1}{2\pi(s-r)}drds = \infty.\end{aligned}$$

It suggests that the quantity $\beta([0,t]^2_<)$ blows up due to the contribution from the region close to the diagonal $\{(r,s)|\ r=s\}$. In other words, the problem is caused by the unbalance between short-range intersection and long-range intersection in the case $d \geqslant 2$ in which the short range self-intersection is too strong compared with long-range intersection.

This problem can be fixed for $d = 2$ by renormalization. In this following discussion, we let $d = p = 2$. Based on above observation, one can construct (by the usual way of approximation) the quantity

$$\beta(B) = \iint_B \delta_0\big(W(r)-W(s)\big)drds$$

if $B \subset \{(s,t)|\ r \leqslant s\}$ is substantially away from the diagonal. In particular, if $B = [a,b]\times[b,c]$ for $0 \leqslant a < b < c$

$$\begin{aligned}&\beta\big([a,b]\times[b,c]\big) \\ &= \int_a^b\int_b^c \delta_0\Big(\big(W(r)-W(b)\big)-\big(W(s)-W(b)\big)\Big)dsdr \\ &= \int_0^{b-a}\int_0^{c-b} \delta_0\Big(\big(W(b-r)-W(b)\big)-\big(W(b+s)-W(b)\big)\Big)dsdr \\ &= \int_0^{b-a}\int_0^{c-b} \delta_0\big(W_1(r)-W_2(s)\big)dsdr\end{aligned}$$

where

$$W_1(r) = -\Big(W(b)-W(b-r)\Big) \quad (0 \leqslant r \leqslant b-a)$$

and

$$W_2(s) = W(b+s)-W(b) \qquad (s \geqslant 0)$$

are independent Brownian motions. Therefore, we have proved that

$$\beta\big([a,b]\times[b,c]\big) \stackrel{d}{=} \alpha\big([0,b-a]\times[0,c-b]\big) \tag{4.3}$$

where $\alpha([0,s]\times[0,t])$ is the intersection local time of two independent planar Brownian motions discussed in section 3.

We now construct the *renormalized self-intersection local time* formally given as

$$\gamma([0,t]^2_<) = \iint_{\{0\leqslant r<s\leqslant t\}} \delta_0\big(W(r)-W(s)\big)drds - \mathbb{E}\iint_{\{0\leqslant r<s\leqslant t\}} \delta_0\big(W(r)-W(s)\big)drds.$$

The renormalization procedure we give in the following essentially belongs to Varadhan (1969). Let $t>0$ be fixed and set

$$A_l^k = \Big[\frac{2l}{2^{k+1}}t, \frac{2l+1}{2^{k+1}}t\Big) \times \Big(\frac{2l+1}{2^{k+1}}t, \frac{2l+2}{2^{k+1}}t\Big],$$

where $l=0,1,\cdots,2^k-1$ and $k=0,1,\cdots$. An important fact is that the family

$$\Big\{A_l^k;\ \ l=0,1,\cdots,2^k-1,\ \ k=0,1,\cdots\Big\}$$

forms a partition of the triangular region $\{(r,s)|\ 0\leqslant r<s\leqslant t\}$. It will be reasonable to define

$$\gamma([0,t]^2_<) = \sum_{k=0}^\infty \bigg\{\sum_{l=0}^{2^k-1}\Big(\beta(A_l^k)-\mathbb{E}\,\beta(A_l^k)\Big)\bigg\} \tag{4.4}$$

if the right hand side converges in proper sense. Indeed, we have

Theorem 4.2. *The series*

$$\sum_{k=0}^\infty \bigg\{\sum_{l=0}^{2^k-1}\Big(\beta(A_l^k)-\mathbb{E}\,\beta(A_l^k)\Big)\bigg\}$$

convergences a.s. and in L^2.

Proof. The proof we give here comes from the monograph Le Gall (1992). By (4.3) we have that

$$\beta(A_l^k) \stackrel{d}{=} \alpha\big([0,2^{-(k+1)}]^2\big) \qquad l=0,1,\cdots,2^k-1,\quad k=0,1,\cdots.$$

In addition, for each $k\geqslant 0$, the finite sequence

$$\beta(A_l^k), \qquad l=0,1,\cdots,2^k-1$$

is independent. Consequently, by the scaling property (3.2) (with $d=p=2$)

$$\mathrm{Var}\bigg(\sum_{l=0}^{2^k-1}\beta(A_l^k)\bigg) = 2^k\mathrm{Var}\Big(\alpha\big([0,2^{-(k+1)}]^2\big)\Big)$$
$$= 2^{-k-2}\mathrm{Var}\Big(\alpha\big([0,1]^2\big)\Big) = C2^{-k}.$$

Finally, the conclusion follows from the estimate

$$\left\{\mathbb{E}\left[\sum_{k=0}^{\infty}\left|\sum_{l=0}^{2^k-1}\left(\beta(A_l^k)-\mathbb{E}\,\beta(A_l^k)\right)\right|\right]^2\right\}^{1/2}$$
$$\leqslant \sum_{k=0}^{\infty}\left\{\mathrm{Var}\left(\sum_{l=0}^{2^k-1}\beta(A_l^k)\right)\right\}^{1/2}<\infty.$$

□

Now $\gamma\big([0,t]_<^2\big)$ is defined by (4.4) and is called renormalized self-intersection local time. It can be proved that for any $t>0$,

$$\gamma\big([0,t]_<^2\big)\overset{d}{=}t\gamma\big([0,1]_<^2\big). \tag{4.5}$$

Returning the polymer models discussed in (1.8) and (1.9). To remedy the problem caused by the ill-definition of $\beta\big([0,1]_<^2\big)$ we replace $\beta\big([0,1]_<^2\big)$ by $\gamma\big([0,1]_<\big)$. That is, we intend to redefine

$$\widehat{P}_\lambda(A)=\widehat{C}_\lambda^{-1}\mathbb{E}\,\exp\left\{\lambda\gamma\big([0,1]_<^2\big)\right\}1_{\{W(\cdot)\in A\}} \tag{4.6}$$

and

$$\widetilde{P}_\lambda(A)=\widetilde{C}_\lambda^{-1}\mathbb{E}\,\exp\left\{-\lambda\gamma\big([0,1]_<^2\big)\right\}1_{\{W(\cdot)\in A\}} \tag{4.7}$$

as the distributions of self-attraction polymer and self-repelling polymer, respectively. Beyond mathematical technicality, nothing significant has been changed— all we do is to renormalize model by the "constant" multiplier

$$\exp\left\{\pm\lambda\mathbb{E}\,\beta\big([0,1]_<^2\big)\right\}.$$

Nevertheless, this idea works only if

$$\mathbb{E}\,\exp\left\{\lambda\gamma\big([0,1]_<^2\big)\right\}<\infty \tag{4.8}$$

at least for λ in a neighborhood of 0. This fact was established by Le Gall (1994).

Theorem 4.3. *There is a $\lambda_0>0$ such that*

$$\mathbb{E}\,\exp\left\{\lambda\gamma\big([0,1]_<^2\big)\right\}\begin{cases}<\infty, & \lambda<\lambda_0,\\ \\ =\infty, & \lambda>\lambda_0.\end{cases} \tag{4.9}$$

In particular,

$$\mathbb{E}\,\exp\left\{\lambda\gamma\big([0,1]_<^2\big)\right\}<\infty,\qquad\forall\lambda<0. \tag{4.10}$$

Proof. Take $t = 1$ in the definition of A_l^k and write $\gamma(A) = \beta(A) - \mathbb{E}\,\beta(A)$. By Theorem 3.1 (with $d = p = 2$) there is a $\lambda_1 > 0$ such that

$$\mathbb{E}\exp\Big\{\lambda_1\alpha\big([0,1]^2\big)\Big\} < \infty.$$

By Taylor expansion, therefore, there is a $C > 0$ such that

$$\mathbb{E}\exp\Big\{\lambda\Big(\alpha\big([0,1]^2\big) - \mathbb{E}\,\alpha\big([0,1]^2\big)\Big)\Big\} \leqslant e^{C\lambda^2}, \qquad \lambda \leqslant \lambda_1. \tag{4.11}$$

Fix $a \in (0,1)$. For each $N \geqslant 1$ set

$$\theta_N = 2\lambda_1\prod_{j=2}^{N}\Big(1 - 2^{-a(j-1)}\Big)$$

$(\theta_1 = 2\lambda_1)$. By Hölder inequality

$$\begin{aligned}
&\mathbb{E}\exp\Big\{\theta_N\sum_{k=0}^{N}\sum_{l=0}^{2^k-1}\Big(\beta(A_l^k) - \mathbb{E}\,\beta(A_l^k)\Big)\Big\}\\
&\leqslant \Bigg[\mathbb{E}\exp\Big\{\theta_{N-1}\sum_{k=0}^{N-1}\sum_{l=0}^{2^k-1}\Big(\beta(A_l^k) - \mathbb{E}\,\beta(A_l^k)\Big)\Big\}\Bigg]^{1-2^{-a(N-1)}}\\
&\quad\times\Bigg[\mathbb{E}\exp\Big\{2^{a(N-1)}\theta_N\sum_{l=0}^{2^N-1}\Big(\beta(A_l^N) - \mathbb{E}\,\beta(A_l^N)\Big)\Big\}\Bigg]^{2^{-a(N-1)}}\\
&\leqslant \mathbb{E}\exp\Big\{\theta_{N-1}\sum_{k=0}^{N-1}\sum_{l=0}^{2^k-1}\Big(\beta(A_l^k) - \mathbb{E}\,\beta(A_l^k)\Big)\Big\}\\
&\quad\times\Bigg[\mathbb{E}\exp\Big\{2^{a(N-1)}\theta_N\Big(\beta(A_0^N) - \mathbb{E}\,\beta(A_0^N)\Big)\Big\}\Bigg]^{2^{(1-a)(N-1)}}.
\end{aligned}$$

By the fact that $\beta(A_0^N) \stackrel{d}{=} 2^{-(N+1)}\alpha\big([0,1]^2\big)$ and by (4.11)

$$\begin{aligned}
&\mathbb{E}\exp\Big\{2^{a(N-1)}\theta_N\Big(\beta(A_0^N) - \mathbb{E}\,\beta(A_0^N)\Big)\Big\}\\
&= \mathbb{E}\exp\Big\{2^{-N-1+a(N-1)}\theta_N\Big(\alpha\big([0,1]^2\big) - \mathbb{E}\,\alpha\big([0,1]^2\big)\Big)\Big\}\\
&\leqslant \exp\Big\{C\theta_N^2 2^{-2N+2a(N-1)}\Big\}.
\end{aligned}$$

Notice that $\theta_N \leqslant 2\lambda_1$. Summarizing what we have

$$\begin{aligned}&\mathbb{E}\exp\left\{\theta_N\sum_{k=0}^{N}\sum_{l=0}^{2^k-1}\Big(\beta(A_l^k)-\mathbb{E}\,\beta(A_l^k)\Big)\right\}\\&\leqslant \mathbb{E}\exp\left\{\theta_{N-1}\sum_{k=0}^{N-1}\sum_{l=0}^{2^k-1}\Big(\beta(A_l^k)-\mathbb{E}\,\beta(A_l^k)\Big)\right\}\exp\Big\{C'2^{(a-1)N}\Big\}.\end{aligned}$$

Repeating above procedure gives

$$\begin{aligned}&\mathbb{E}\exp\left\{\theta_N\sum_{k=0}^{N}\sum_{l=0}^{2^k-1}\Big(\beta(A_l^k)-\mathbb{E}\,\beta(A_l^k)\Big)\right\}\\&\leqslant \exp\Big\{C'\sum_{k=0}^{N}2^{(a-1)k}\Big\}\leqslant\exp\Big\{C'\Big(1-2^{a-1}\Big)^{-1}\Big\}<\infty.\end{aligned}$$

Notice that

$$\theta_\infty = 2\lambda_1\prod_{j=2}^{\infty}\Big(1-2^{-a(j-1)}\Big)>0.$$

By Fatou lemma we have

$$\mathbb{E}\exp\Big\{\theta_\infty\gamma\big([0,1]_<^2\big)\Big\}\leqslant\exp\Big\{C'\Big(1-2^{a-1}\Big)^{-1}\Big\}<\infty.$$

On the other hand, notice that Theorem 3.2 also implies that there is a $\lambda_2>0$ such that

$$\mathbb{E}\exp\Big\{\lambda_2\alpha\big([0,1]^2\big)\Big\}=\infty.$$

By a comparison argument one can prove that

$$\mathbb{E}\exp\Big\{\lambda\gamma\big([0,1]_<^2\big)\Big\}=\infty$$

if $\lambda>0$ is large enough. □

In physics, the critical exponent $\lambda_0>0$ given in Theorem 4.3 represents the critical temperature between two different states of polymers. There are some concerns from physicists about the value of λ_0. From mathematical side, this problem is related to the tail of $\gamma\big([0,1]_<^2\big)$. The following result was given in Chen and Bass (2004).

Theorem 4.4.

$$\lim_{t\to\infty}\frac{1}{t}\log\mathbb{P}\Big\{\gamma\big([0,1]_<^2\big)\geqslant t\Big\}=-\kappa(2,2)^{-4}, \tag{4.12}$$

where $\kappa(2,2) > 0$ is the best constant of Gagliardo-Nirenberg inequality given in (3.4). In addition, there is a constant $0 < L < \infty$ such that for any $\theta > 0$

$$\lim_{t\to\infty} t^{-2\pi\theta} \log \mathbb{P}\Big\{ -\gamma\big([0,1]^2_<\big) \geqslant \theta \log t\Big\} = -L. \tag{4.13}$$

The upper tail behavior stated in (4.12) implies that $\lambda_0 = \kappa(2,2)^{-4}$. By the numerical approximation given in (3.6)

$$\lambda_0 \approx \pi \times 1.86225\cdots \approx 5.85043\cdots .$$

This is very close to a conjecture made by Duplantier (private communication).

Another interesting point about Theorem 4.4 is that despite of the fact that $\mathbb{E}\,\gamma\big([0,1]^2_<\big) = 0$, $\gamma\big([0,1]^2_<\big)$ has non-symmetric tails. Indeed, the lower tail given in (4.13) is much thiner than the upper tail given in (4.12).

The main ingredient in the proof for the upper tail (4.12) is to compare $\gamma\big([0,1]^2_<\big)$ with $\alpha\big([0,1]^2\big)$, the mutual intersection local time of two independent planar Brownian motions. To illustrate the idea, we only prove the upper bound. Consider the decomposition

$$\begin{aligned}\gamma\big([0,1]^2_<\big) &= \gamma\big([0,1/2]^2_<\big) + \gamma\big([1/2,1]^2_<\big) + \beta\big([0,1/2]\times[1/2,1]\big)\\ &\quad - \mathbb{E}\,\beta\big([0,1/2]\times[1/2,1]\big).\end{aligned}$$

We have that $\gamma\big([0,1/2]^2_<\big)$ and $\gamma\big([1/2,1]^2_<\big)$ are independent and have the common distribution same as $(1/2)\gamma\big([0,1]^2_<\big)$. By (4.3),

$$\beta\big([0,1/2]\times[1/2,1]\big) \overset{d}{=} \alpha\big([0,1/2]^2\big) \overset{d}{=} \frac{1}{2}\alpha\big([0,1]^2\big). \tag{4.14}$$

Given $\epsilon > 0$, by triangular inequality

$$\begin{aligned}&\mathbb{P}\Big\{\gamma\big([0,1]^2_<\big) \geqslant t\Big\}\\ &\leqslant \mathbb{P}\Big\{\gamma\big([0,1/2]^2_<\big) + \gamma\big([1/2,1]^2_<\big) \geqslant \frac{1+\epsilon}{2}t\Big\}\\ &+ \mathbb{P}\Big\{\alpha\big([0,1]^2\big) - \mathbb{E}\,\alpha\big([0,1]^2\big) \geqslant (1-\epsilon)t\Big\}.\end{aligned}$$

Consequently,

$$\begin{aligned}&\limsup_{t\to\infty}\frac{1}{t}\log P\Big\{\gamma\big([0,1]^2_<\big) \geqslant t\Big\}\\ &\leqslant \max\bigg\{\limsup_{t\to\infty}\frac{1}{t}\log\mathbb{P}\Big\{\gamma\big([0,1/2]^2_<\big) + \gamma\big([1/2,1]^2_<\big) \geqslant \frac{1+\epsilon}{2}t\Big\},\\ &\qquad \limsup_{t\to\infty}\frac{1}{t}\log\mathbb{P}\Big\{\alpha\big([0,1]^2\big) - \mathbb{E}\,\alpha\big([0,1]^2\big) \geqslant (1-\epsilon)t\Big\}\bigg\}.\end{aligned}$$

By (4.14) and Theorem 3.1 (with $d = p = 2$) we have that the second limsup on the right hand side is equal to $-(1-\epsilon)\kappa(2,2)^{-4}$.

Let $\lambda_0 > 0$ be the critical exponent given in Theorem 4.3. We have

$$\limsup_{t\to\infty} \frac{1}{t} \log P\Big\{\gamma\big([0,1]^2_<\big) \geqslant t\Big\} = -\lambda_0. \tag{4.15}$$

Notice that

$$\mathbb{E} \exp\Big\{\lambda\Big(\gamma\big([0,1/2]^2_<\big) + \gamma\big([1/2,1]^2_<\big)\Big)\Big\} = \Big[\mathbb{E} \exp\Big\{\frac{\lambda}{2}\gamma\big([0,1]^2\big)\Big\}\Big]^2 < \infty$$

for every $\lambda < 2\lambda_0$. By Chebyshev inequality, this gives

$$\limsup_{t\to\infty} \frac{1}{t} \log \mathbb{P}\Big\{\gamma\big([0,1/2]^2_<\big) + \gamma\big([1/2,1]^2_<\big) \geqslant \frac{1+\epsilon}{2}t\Big\} \leqslant -(1+\epsilon)\lambda_0.$$

Summarizing our argument, we have obtained

$$-\lambda_0 \leqslant \max\Big\{-(1+\epsilon)\lambda_0, \quad -(1-\epsilon)\kappa(2,2)^{-4}\Big\}.$$

So we must have $\lambda_0 \geqslant (1-\epsilon)\kappa(2,2)^{-4}$. Letting $\epsilon \to 0$ we have $\lambda_0 \geqslant \kappa(2,2)^{-4}$. Thus the upper bound of (4.12) follows from (4.15).

The idea for the lower tail (4.13) is sub-additivity. One needs only to prove (4.13) in the case $\theta = (2\pi)^{-1}$. That is,

$$\lim_{t\to\infty} \frac{1}{t} \log \mathbb{P}\Big\{-\gamma\big([0,1]^2_<\big) \geqslant (2\pi)^{-1}\log t\Big\} = -L. \tag{4.16}$$

Indeed, the general statement will follow if we substitute t by $t^{2\pi\theta}$.

Given $s, t > 0$,

$$\begin{aligned}
&\gamma\big([0,s+t]^2_<\big) = \gamma\big([0,s]^2_<\big) + \gamma\big([s,s+t]^2_<\big) \\
&+ \beta\big([0,s]\times[s,s+t]\big) - \mathbb{E}\,\beta\big([0,s]\times[s,s+t]\big) \\
&\geqslant \gamma\big([0,s]^2_<\big) + \gamma\big([s,s+t]^2_<\big) - \mathbb{E}\,\beta\big([0,s]\times[s,s+t]\big).
\end{aligned}$$

By a simple computation,

$$\begin{aligned}
&\mathbb{E}\,\beta\big([0,s]\times[s,s+t]\big) = \int_0^s\int_s^{s+t} \frac{1}{2\pi}\frac{1}{v-u}dvdu \\
&= \frac{1}{2\pi}\big[(s+t)\log(s+t) - s\log s - t\log t\big].
\end{aligned}$$

By independence between $\gamma([0,s]^2_<)$ and $\gamma([s,s+t]^2_<)$, therefore,

$$(s+t)^{-(s+t)}\mathbb{E}\exp\Big\{-2\pi\gamma\big([0,s+t]^2_<\big)\Big\}$$
$$\leqslant\Big(s^{-s}\mathbb{E}\exp\Big\{-2\pi\gamma\big([0,s]^2_<\big)\Big\}\Big)\Big(t^{-t}\mathbb{E}\exp\Big\{-2\pi\gamma\big([0,t]^2_<\big)\Big\}\Big)$$

or, $a(s+t)\leqslant a(s)+a(t)$, where

$$a(t)=\log\Big(t^{-t}\mathbb{E}\exp\Big\{-2\pi\gamma\big([0,t]^2_<\big)\Big\}\Big).$$

This sub-additivity property implies that the limit

$$\lim_{t\to\infty}\frac{a(t)}{t}=\inf_{t>0}\Big\{\frac{a(t)}{t}\Big\}$$

exists and is finite. Let $t=n$ be integer. By the scaling $\gamma([0,n]^2_<)\overset{d}{=}n\gamma([0,1]^2_<)$ and by Stirling formula,

$$\lim_{n\to\infty}\frac{1}{n}\log\Big(\frac{1}{n!}\mathbb{E}\exp\Big\{-2\pi n\gamma\big([0,1]^2_<\big)\Big\}\Big)=1+\inf_{t>0}\Big\{\frac{a(t)}{t}\Big\}.$$

Applying Lemma 3.3 to the non-negative random variable

$$Y=\exp\Big\{-2\pi\gamma\big([0,1]^2_<\big)\Big\}$$

gives (4.16) with

$$L=\exp\Big\{-1-\inf_{t>0}\Big\{\frac{a(t)}{t}\Big\}\Big\}>0.$$

A minorization estimate of the probability

$$\mathbb{P}\Big\{-\gamma\big([0,1]^2_<\big)\geqslant(2\pi)^{-1}\log t\Big\}$$

gives that $L<\infty$. □

Most of the results that we have stated by far have been extended from the setting of Brownian motions to the setting of symmetric stable processes. We refer reader to Bass, Chen and Rosen (2005), Chen, Li and Rosen (2005), Chen and Rosen (2005) for these extensions. Sometimes, the stable case presents some unusual patterns which are sharply contrary to those found in the Brownian case. Take the lower tail behavior of the renormalized 2-multiple self-intersection local time as an example. Given a d-dimensional stable process with index $\alpha\in(0,2]$, its

2-multiple self-intersection local time has to be renormalized and can be renormalized if and only if $2d/3 < \alpha \leqslant d$. As $d = \alpha$, which corresponds to 2-dimensional Brownian case and 1-dimensional Cauchy case, it has been found that the lower tail of the renormalized self-intersection local time $\gamma([0,t]^2_{\leqslant})$ falls into the pattern described by (4.13). In the non-critical cases defined by $2d/3 < \alpha < d$, on the other hand, it was proved in Bass, Chen and Rosen (2005) that there is a $0 < b < \infty$, such that

$$\lim_{t\to\infty} \frac{1}{t} \log \mathbb{P}\{ -\gamma([0,1]^2_{\leqslant}) \geqslant t^{d/\alpha - 1}\} = -b.$$

The problem on the tail probabilities for renormalized self-intersection local times with intersection multiple $p \geqslant 3$ remains open. Given a 2-dimensional Brownian motion $W(t)$, the renormalized p-multiple self intersection local time is formally defined (Rosen (1986)) as

$$\gamma([0,t]^p_<) = \int\cdots\int_{\{0\leqslant s_1<\cdots<s_p\leqslant t\}} \overline{\prod_{j=2}^p \delta_0\big(W(s_j) - W(s_{j-1})\big)} ds_1\cdots ds_p$$

where for any random variable X, we use the notation $\overline{X}$ for its centralization $X - \mathbb{E}\, X$. It can be shown that

$$\gamma([0,t]^p_<) \stackrel{d}{=} t\gamma([0,1]^p_<).$$

To the best of the author's knowledge, nothing has been known about the tail behaviors of $\gamma([0,t]^p_<)$ in the case $p \geqslant 3$, to which the argument used in Theorem 4.3 and Theorem 4.4 is no longer applicable. Here we conjecture that there is a $\lambda_0 > 0$ such that

$$\mathbb{E}\exp\Big\{\lambda|\gamma([0,1]^p_<)|^{\frac{1}{p-1}}\Big\}\begin{cases} < \infty, & \lambda < \lambda_0, \\ = \infty, & \lambda > \lambda_0. \end{cases}$$

We also ask for the indentification of the critical value λ_0.

In the case $d \geqslant 3$, the self-intersection local time of a d-dimensional Brownian motion can not be renormalized. The problem is to study the limiting behavior of the quantity

$$\beta_\epsilon([0,t]^2_<) = \iint_{0\leqslant r<s\leqslant t} f_\epsilon\big(W(s) - W(r)\big) dr ds$$

as $\epsilon \to 0^+$, where $f(x)$ is a nice density function on $\mathbb{R}^d$ and $f_\epsilon(x) = \epsilon^{-d} f(\epsilon^{-1}x)$. Yor (1985) and Calais and Yor (1987) have proved that random variables

$$\begin{cases} \big(\log(1/\epsilon)\big)^{-1/2}\Big\{\beta_\epsilon([0,1]^2_<) - \mathbb{E}\,\beta_\epsilon([0,1]^2_<)\Big\}, & \text{if } d = 3, \\ \epsilon^{\frac{d-3}{2}}\Big\{\beta_\epsilon([0,1]^2_<) - \mathbb{E}\,\beta_\epsilon([0,1]^2_<)\Big\}, & \text{if } d \geqslant 4. \end{cases}$$

weakly converge to symmetric normal distributions. Consequently, a Gaussian tail is expected for $\beta_\epsilon([0,1]^2_<) - \mathbb{E}\,\beta_\epsilon([0,1]^2_<)$. The reader is referred to section 6 for its implementation on the tail behaviors for self-intersection local times and ranges of random walks with dimension $d \geqslant 3$.

5 Intersection of independent random walks

Let $\{S_1(n)\}, \cdots, \{S_p(n)\}$ be independent symmetric random walks on $\mathbb{Z}^d$. Through this section, we assume that they have same distribution and that their smallest supporting group is $\mathbb{Z}^d$. Let Γ be their covariance matrix. It is well known that

$$\#\big\{S_1(0,\infty)\cap\cdots\cap S_p(0,\infty)\big\} = \infty, \qquad \text{a.s.}$$

if any only if $p(d-2) \leqslant d$. where for given $A \subset \mathbb{Z}^+$, $S_j(A) = \{S_j(k);\ k \in A\}$. Under $p(d-2) \leqslant d$, we are interested in the asymptotic behaviors of

the intersection local time I_n and the intersection J_n of the independent ranges, where I_n and J_n are defined by (1.5) and (1.7), respectively. I_n counts the times that the random walks intersect in n steps while J_n counts the sites where the intersection takes place. An easy observation is that $J_n \leqslant I_n$, where the inequality is caused by the possibility that intersection takes place at same site more than once. Since the random walks in high dimension ($d \geqslant 3$) can only intersect finitely many times at the same site, we expect that I_n and J_n differ only by a constant multiplier in this case.

In the case $p(d-2) > d$,

$$I_\infty = \#\Big\{(k_1,\cdots,k_p) \in (0,\infty)^d;\ \ S_1(k_1) = \cdots = S_p(k_p)\Big\} < \infty, \tag{5.1}$$

$$J_\infty = \#\big\{S_1(0,n]\cap\cdots\cap S_p(0,n]\big\} < \infty, \tag{5.2}$$

with probability 1. The question is on the tail probabilities of I_∞ and J_∞.

In the following discussion, we consider three very different situations: $p(d-2) < d$, $p(d-2) = d$ and $p(d-2) > d$, which are referred as sub-critical dimensions, critical dimensions and super-critical dimensions, respectively.

The critical dimensions consist of two cases: the case $d = 4$, $p = 2$ and the case $d = p = 3$. Le Gall (1986b) proved that as $d = 4$ and $p = 2$

$$(\log n)^{-1} I_n \xrightarrow{\ d\ } (2\pi)^{-2}\det(\Gamma)^{-1/2} U^2, \tag{5.3}$$

$$(\log n)^{-1}J_n \xrightarrow{d} \gamma_S^2(2\pi)^{-2}\det(\Gamma)^{-1/2}U^2, \tag{5.4}$$

where U is a $N(0,1)$ random variable and

$$\gamma_S = \mathbb{P}\{S(n) \neq 0 \ \forall n \geqslant 1\}. \tag{5.5}$$

In the same paper, Le Gall also proved that when $d = p = 3$,

$$(\log n)^{-1}I_n \xrightarrow{d} (2\pi)^{-2}\frac{1}{\det(\Gamma)}V, \tag{5.6}$$

$$(\log n)^{-1}J_n \xrightarrow{d} (2\pi)^{-2}\frac{\gamma_S^3}{\det(\Gamma)}V, \tag{5.7}$$

where the random variable V has the gamma distribution with parameter $1/4$.

Therefore, it is reasonable to expect that I_n and J_n have gamma tails in the critical dimensions. This was confirmed in Marcus-Rosen (1997) and in Rosen (1997) in the form of law of the iterated logarithm.

Theorem 5.1. *Under $d = 4$ and $p = 2$,*

$$\begin{aligned}&\limsup_{n\to\infty}(\log n)^{-1}(\log\log\log n)^{-1}I_n\\&= (2\pi^2)^{-1}\det(\Gamma)^{-1/2}, \qquad \text{a.s.}\end{aligned} \tag{5.8}$$

$$\begin{aligned}&\limsup_{n\to\infty}(\log n)^{-1}(\log\log\log n)^{-1}J_n\\&= \gamma_S^2(2\pi^2)^{-1}\det(\Gamma)^{-1/2}, \qquad \text{a.s.}\end{aligned} \tag{5.9}$$

Under $d = p = 3$,

$$\limsup_{n\to\infty}(\log n)^{-1}(\log\log\log n)^{-1}I_n = \frac{1}{\pi\det(\Gamma)}, \qquad \text{a.s.} \tag{5.10}$$

$$\limsup_{n\to\infty}(\log n)^{-1}(\log\log\log n)^{-1}J_n = \frac{\gamma_S^3}{\pi\det(\Gamma)}, \qquad \text{a.s.} \tag{5.11}$$

In the sub-critical dimensions $p(d-2) < d$, Le Gall (1986a) proved that

$$n^{-\frac{2p-d(p-1)}{2}}I_n \xrightarrow{d} \det(\Gamma)^{-\frac{p-1}{2}}\alpha\big([0,1]^p\big). \tag{5.12}$$

where $\alpha\big([0,1]^p\big)$ is the intersection local time of p independent d-dimensional Brownian motions.

The weak laws for J_n are more complicated and they further break the sub-critical dimensions into three cases: the case $d = 1$, $p \geqslant 2$; the case $d = 2$, $p \geqslant 2$; and the case $d = 3$, $p = 2$. The following results were established in Le Gall (1986a) and Rosen (1990): As $d = 1$ and $p \geqslant 2$,

$$\frac{1}{\sqrt{n}}J_n \xrightarrow{d} \sigma\Big(\min_{1\leqslant j\leqslant p}\max_{1\leqslant t\leqslant 1} W_j(t) - \max_{1\leqslant j\leqslant p}\min_{1\leqslant t\leqslant 1} W_j(t)\Big) \tag{5.13}$$

where $W_1(t),\cdots,W_p(t)$ are independent 1-dimensional Brownian motions.

As $d=2$ and $p\geqslant 2$,

$$\frac{(\log n)^p}{n}J_n \xrightarrow{d} (2\pi)^p\sqrt{\det(\Gamma)}\alpha\big([0,1]^p\big) \tag{5.14}$$

where $\alpha\big([0,1]^p\big)$ is intersection local time run by p independent planar Brownian motions.

As $d=3$ and $p=2$,

$$\frac{1}{\sqrt{n}}J_n \xrightarrow{d} \gamma_S^2\det(\Gamma)^{-\frac{1}{2}}\alpha\big([0,1]^2\big) \tag{5.15}$$

where $\alpha\big([0,1]^p\big)$ is intersection local time between two 3-dimensional independent Brownian motions.

To investigate the tail probabilities of I_n and J_n in the sub-critical dimensions, we begin with the critical integrability of I_n and J_n. By Theorem 3.1, the critical integrability of $\alpha\big([0,1]^2\big)$ is described as following:

$$\mathbb{E}\exp\left\{\theta\Big(\alpha\big([0,1]^p\big)\Big)^{\frac{2}{d(p-1)}}\right\}\begin{cases} <\infty, & \theta < \frac{p}{2}\kappa(d,p)^{-\frac{4p}{d(p-1)}}, \\ =\infty, & \theta > \frac{p}{2}\kappa(d,p)^{-\frac{4p}{d(p-1)}}. \end{cases} \tag{5.16}$$

The following theorem was given in Bass, Chen and Rosen (2006, 2007).

Theorem 5.2. *Under $p(d-2)<d$, there is a $\theta>0$ such that*

$$\sup_{n\geqslant 1}\mathbb{E}\exp\left\{\theta n^{-\frac{2p-d(p-1)}{d(p-1)}}I_n^{\frac{2}{d(p-1)}}\right\}<\infty. \tag{5.17}$$

In the case $d=1$ and $p\geqslant 2$,

$$\sup_{n\geqslant 1}\mathbb{E}\exp\left\{\theta n^{-1}J_n^2\right\}<\infty. \tag{5.18}$$

In the case $d=2$ and $p\geqslant 2$,

$$\sup_{n\geqslant 1}\mathbb{E}\exp\left\{\theta\frac{(\log n)^p}{n}J_n\right\}<\infty. \tag{5.19}$$

In the case $d=3$ and $p=2$,

$$\sup_{n\geqslant 1}\mathbb{E}\exp\left\{\theta n^{-1/3}J_n^{2/3}\right\}<\infty. \tag{5.20}$$

The integrability established for I_n and J_n are best possible in the sense that there is a $\theta_0 > 0$ such that left hand sides of (5.17)—(5.20) become infinity if $\theta > \theta_0$. This can be seen from Le Gall's weak laws and from the critical integrability of $\alpha([0,1]^p)$ given in (5.16).

The approach for Theorem 5.2 relies on two steps of moment estimate. The moment estimate in the first step goes back at least to Le Gall and Rosen (1991). Write

$$I_n = \sum_{x\in\mathbb{Z}^d}\prod_{j=1}^p l_j(n,x) \quad\text{and}\quad J_n = \sum_{x\in\mathbb{Z}^d}\prod_{j=1}^p 1_{\{T_x^j\leqslant n\}} \tag{5.21}$$

where $T_x^j = \inf\{k\geqslant 1;\ S_j(k)=x\}$. For each integer $m\geqslant 1$,

$$\begin{aligned}
\mathbb{E}\, I_n^m &= \sum_{x_1,\cdots,x_m\in\mathbb{Z}^d}\Big[\mathbb{E}\prod_{k=1}^m l(n,x_k)\Big]^p \\
&\leqslant \sum_{x_1,\cdots,x_m\in\mathbb{Z}^d}\Big[\sum_{\sigma\in\Sigma_m}\ \sum_{1\leqslant i_1\leqslant\cdots\leqslant i_m\leqslant n}\mathbb{E}\prod_{k=1}^m 1_{\{S(i_k)=x_{\sigma(k)}\}}\Big]^p \\
&= \sum_{x_1,\cdots,x_m\in\mathbb{Z}^d}\Big[\sum_{\sigma\in\Sigma_m}\ \sum_{1\leqslant i_1\leqslant\cdots\leqslant i_m\leqslant n} \\
&\qquad\times\prod_{k=1}^m\mathbb{P}\{S(i_k-i_{k-1})=x_{\sigma(k)}-x_{\sigma(k-1)}\}\Big]^p
\end{aligned}$$

where Σ_m is the permutation group on $\{1,\cdots,m\}$. Notice that

$$\begin{aligned}
&\sum_{\sigma\in\Sigma_m}\ \sum_{1\leqslant i_1\leqslant\cdots\leqslant i_m\leqslant n}\ \prod_{k=1}^m\mathbb{P}\{S(i_k-i_{k-1})=x_{\sigma(k)}-x_{\sigma(k-1)}\} \\
&\leqslant\prod_{k=1}^m\mathbb{E}\,\widetilde{l}(n,x_{\sigma(k)}-x_{\sigma(k-1)})
\end{aligned}$$

where $\widetilde{l}(n,x)$ is the augmented local time

$$\widetilde{l}(n,x)=\sum_{k=0}^n 1_{\{S(k)=x\}}.$$

Hence,

$$
\begin{aligned}
\mathbb{E}\, I_n^m &\leqslant \sum_{x_1,\cdots,x_m\in\mathbb{Z}^d}\Big[\sum_{\sigma\in\Sigma_m}\prod_{k=1}^m \mathbb{E}\,\widetilde{l}(n, x_{\sigma(k)}-x_{\sigma(k-1)})\Big]^p \\
&\leqslant (m!)^{p-1}\sum_{x_1,\cdots,x_m\in\mathbb{Z}^d}\sum_{\sigma\in\Sigma_m}\prod_{k=1}^m\Big[\mathbb{E}\,\widetilde{l}(n, x_{\sigma(k)}-x_{\sigma(k-1)})\Big]^p \qquad (5.22)\\
&= (m!)^p\bigg(\sum_{x\in\mathbb{Z}^d}\Big[\mathbb{E}\,\widetilde{l}(n,x)\Big]^p\bigg)^m = (m!)^p\Big(\mathbb{E}\,\widetilde{I}_n\Big)^m
\end{aligned}
$$

where $\widetilde{I}_n$ is the augmented intersection local time

$$\widetilde{I}_n = \#\Big\{(k_1,\cdots,k_p)\in[0,n]^p;\ \ S_1(k_1)=\cdots=S_p(k_p)\Big\}.$$

Similarly,

$$\mathbb{E}\, J_n^m \leqslant (m!)^p\Big(\mathbb{E}\,\widetilde{J}_n\Big)^m. \qquad (5.23)$$

Unfortunately, (5.22) and (5.23) are not sharp enough for Theorem 5.3. We need to reduce the power on $m!$ by the second step of moment estimate. To this end we introduce the following moment inequalities originally given in Chen (2004, 2005).

Theorem 5.3. *For any integers $m\geqslant 1$, $a\geqslant 2$ and $n_1,\cdots,n_a\geqslant 1$*

$$
\begin{aligned}
&\Big(\mathbb{E}\, I^m_{n_1+\cdots+n_a}\Big)^{1/p}\\
&\leqslant \sum_{\substack{k_1+\cdots+k_a=m\\ k_1,\cdots,k_a\geqslant 0}}\frac{m!}{k_1!\cdots k_a!}\Big(\mathbb{E}\, I^{k_1}_{n_1}\Big)^{1/p}\cdots\Big(\mathbb{E}\, I^{k_a}_{n_a}\Big)^{1/p} \qquad (5.24)
\end{aligned}
$$

$$
\begin{aligned}
&\Big(\mathbb{E}\, J^m_{n_1+\cdots+n_a}\Big)^{1/p}\\
&\leqslant \sum_{\substack{k_1+\cdots+k_a=m\\ k_1,\cdots,k_a\geqslant 0}}\frac{m!}{k_1!\cdots k_a!}\Big(\mathbb{E}\, J^{k_1}_{n_1}\Big)^{1/p}\cdots\Big(\mathbb{E}\, J^{k_a}_{n_a}\Big)^{1/p}. \qquad (5.25)
\end{aligned}
$$

We improve the estimate (5.22) by showing that there is a constant $C>0$ such that

$$\mathbb{E}\, I_n^m \leqslant C^m (m!)^{\frac{d(p-1)}{2}} n^{\frac{2p-d(p-1)}{2}m}, \qquad m,n=1,2,\cdots. \qquad (5.26)$$

We first deal with the case $m\leqslant n$. Take $a=m$ and $n_1=\cdots=n_m=[n/m]+1$ in (5.24). By (5.22) and by the fact $m\leqslant n$,

$$\mathbb{E}\, I^{k_i}_{[n/m]+1} \leqslant c_1^{k_i}(k_i!)^p\Big(\frac{n}{m}\Big)^{\frac{2p-d(p-1)}{2}k_i}, \qquad i=1,\cdots,m.$$

By (5.24),

$$\Big(\mathbb{E}\,I_n^m\Big)^{1/p}\leqslant \sum_{\substack{k_1+\cdots+k_m=m\\ k_1,\cdots,k_m\geqslant 0}}\frac{m!}{k_1!\cdots k_m!}c_1^{m/p}k_1!\cdots k_m!\Big(\frac{n}{m}\Big)^{\frac{2p-d(p-1)}{2p}m}$$
$$=m!c_1^{m/p}(m^m)^{-\frac{2p-d(p-1)}{2p}}n^{\frac{2p-d(p-1)}{2p}m}\binom{2m-1}{m}.$$

Therefore, (5.26) follows from the easy bounds $m^m\geqslant m!$ and

$$\binom{2m-1}{m}\leqslant\sum_{k=0}^{2m-1}\binom{2m-1}{k}=2^{2m-1}.$$

As for the case $m>n$, by the fact that $I_n\leqslant n^p$ we have the trivial bound

$$\mathbb{E}\,I_n^m\leqslant n^{pm}\leqslant m^{\frac{d(p-1)}{2}m}n^{\frac{2p-d(p-1)}{2}m}\leqslant C^m(m!)^{\frac{d(p-1)}{2}}n^{\frac{2p-d(p-1)}{2}m}$$

where the last step follows from Stirling formula. Clearly, (5.26) implies (5.17) via Taylor expansion. The rest of Theorem 5.3 follows from the similar treatment applied to the moment of J_n. □

By the fact that Theorem 5.2 gives the optional integrability to I_n and J_n. It is natural to believe that it provides sharp (up to constant) bounds for the tail probabilities of I_n and J_n, by a simple use of Chebyshev inequality. In the case of I_n, that means that the estimate

$$\mathbb{P}\Big\{I_n\geqslant n^{\frac{2p-d(p-1)}{2}}b_n^{\frac{d(p-1)}{2}}\Big\}\leqslant e^{-\theta b_n}\mathbb{E}\exp\Big\{\theta n^{-\frac{2p-d(p-1)}{d(p-1)}}I_n^{\frac{2}{d(p-1)}}\Big\}$$

suggests that for suitable b_n, the tail probability

$$\mathbb{P}\Big\{I_n\geqslant n^{\frac{2p-d(p-1)}{2}}b_n^{\frac{d(p-1)}{2}}\Big\}$$

has a decay rate $e^{-\theta b_n}$ for some $\theta>0$. More precisely, we have (Chen (2004)) the following theorem.

Theorem 5.4. *Under $p(d-2)<d$,*

$$\lim_{n\to\infty}\frac{1}{b_n}\log\mathbb{P}\{I_n\geqslant n^{\frac{2p-d(p-1)}{2}}b_n^{\frac{d(p-1)}{2}}\}$$
$$=-\frac{p}{2}\det(\Gamma)^{1/d}\kappa(d,p)^{-\frac{4p}{d(p-1)}}\tag{5.27}$$

for any positive sequence $\{b_n\}$ satisfying

$$b_n\longrightarrow\infty\quad\text{and}\quad b_n=o(n)\qquad(n\to\infty)$$

where $\kappa(d,p)$ is the best constant of Gagliardo-Nirenberg inequality given in (3.4).

For the intersection of J_n of the independent ranges in sub-critical dimensions, we have (Chen (2005, 2006b)) the following theorem.

Theorem 5.5. *Let b_n be a positive sequence satisfying $b_n \to \infty$ and the restrictions in each of the following cases.*
(1). *As $d = 1$ and $p \geqslant 2$,*

$$\lim_{n\to\infty} \frac{1}{b_n} \log \mathbb{P}\{J_n \geqslant \sqrt{nb_n}\} = -\frac{p}{2\sigma^2} \tag{5.28}$$

for $b_n = o(n)$, where σ^2 is the variance of the random walks.
(2). *As $d = 2$ and $p \geqslant 2$,*

$$\begin{aligned} &\lim_{n\to\infty} \frac{1}{b_n} \log \mathbb{P}\Big\{J_n \geqslant \frac{n}{(\log n)^p} b_n^{p-1}\Big\} \\ &= -\frac{p}{2}(2\pi)^{-\frac{p}{p-1}} \det(\Gamma)^{-\frac{1}{2(p-1)}} \kappa(2,p)^{-\frac{2p}{p-1}} \end{aligned} \tag{5.29}$$

for $b_n = o\big((\log n)^{2/3}\big)$.
(3). *As $d = 3$ and $p = 2$,*

$$\lim_{n\to\infty} \frac{1}{b_n} \log \mathbb{P}\Big\{J_n \geqslant \sqrt{nb_n^3}\Big\} = -\det(\Gamma)^{1/3} \gamma^{-4/3} \kappa(3,2)^{-8/3} \tag{5.30}$$

for $b_n = o\big(n^{2/9}\big)$.

By comparing Theorem 5.4 and 5.5 to Theorem 3.1, one can see how the weak laws pass the large deviation for $\alpha([0,1]^p)$ to I_n and J_n, except the case $d = 1$ from J_n (which is the discrete version of the tail for the intersection of the ranges of independent 1-dimensional Brownian motions). This fact is also reflected by the proof of Theorem 5.4 and 5.5. In the following we try to sketch some of the key ideas behind Theorem 5.4.

We first recall the following lemma, which was given in Chen (2005).
Lemma 5.6. *Let $\{Y_n\}$ be a sequence of non-negative random variables and let $\{b_n\}$ be a positive sequence with $b_n \to \infty$ such that for any $\theta > 0$,*

$$\lim_{n\to\infty} \frac{1}{b_n} \log \sum_{m=0}^{\infty} \frac{\theta^m}{m!} b_n^m \Big(\mathbb{E}\, Y_n^m\Big)^{1/p} = \Psi(\theta) \tag{5.31}$$

where $\Psi(\theta)$ is essentially smooth in its domain. Then

$$\lim_{n\to\infty} \frac{1}{b_n} \log \mathbb{P}\{Y_n \geqslant \lambda\} = -I(\lambda) \quad (\lambda > 0) \tag{5.32}$$

where

$$I(\lambda) = p \sup_{\theta>0} \Big\{ \lambda^{1/p}\theta - \Psi(\theta) \Big\}.$$

To prove Theorem 5.4, we need only to verify that for any $\theta > 0$

$$\begin{aligned} &\lim_{n\to\infty} \frac{1}{b_n} \log \sum_{m=0}^{\infty} \frac{\theta^m}{m!} \Big(\frac{b_n}{n}\Big)^{\frac{2p-d(p-1)}{2p}m} \Big(\mathbb{E}\, I_n^m\Big)^{1/p} \\ &= 2^{\frac{d(p-1)}{2p-d(p-1)}} \frac{2p-d(p-1)}{2p} \Big(\frac{d(p-1)}{2p}\Big)^{\frac{d(p-1)}{2p-d(p-1)}} \\ &\times \det(\Gamma)^{-\frac{p-1}{2p-d(p-1)}} \kappa(d,p)^{\frac{4p}{2p-d(p-1)}} \theta^{\frac{2p}{2p-d(p-1)}}. \end{aligned} \tag{5.33}$$

Notice that (5.24) in Theorem 5.3 implies that for any $\theta > 0$,

$$\sum_{m=0}^{\infty} \frac{\theta^m}{m!} \Big(\mathbb{E}\, I_{n_1+\cdots+n_a}^m\Big)^{1/p} \leqslant \prod_{i=1}^{a} \sum_{m=0}^{\infty} \frac{\theta^m}{m!} \Big(\mathbb{E}\, I_{n_i}^m\Big)^{1/p}.$$

Let $t > 0$ be fixed and let $t_n = [tn/b_n]$ and $\gamma_n = [n/t_n]$. Then $n \leqslant t_n(\gamma_n + 1)$. By taking $a = \gamma_n + 1$,

$$\begin{aligned} &\sum_{m=0}^{\infty} \frac{1}{m!} \theta^m \Big(\frac{b_n}{n}\Big)^{\frac{2p-d(p-1)}{2p}m} \big(\mathbb{E}\, I_n^m\big)^{1/p} \\ &\leqslant \bigg(\sum_{m=0}^{\infty} \frac{1}{m!} \theta^m \Big(\frac{b_n}{n}\Big)^{\frac{2p-d(p-1)}{2p}m} \big(\mathbb{E}\, I_{t_n}^m\big)^{1/p}\bigg)^{\gamma_n+1}. \end{aligned}$$

By the weak convergence given in (5.12) we have

$$\begin{aligned} &\sum_{m=0}^{\infty} \frac{1}{m!} \theta^m \Big(\frac{b_n}{n}\Big)^{\frac{2p-d(p-1)}{2p}m} \big(\mathbb{E}\, I_{t_n}^m\big)^{1/p} \\ &\longrightarrow \sum_{m=0}^{\infty} \frac{1}{m!} \theta^m t^{\frac{2p-d(p-1)}{2p}m} \det(\Gamma)^{-\frac{p-1}{2p}m} \big(\mathbb{E}\, \alpha([0,1]^p)^m\big)^{1/p} \end{aligned}$$

as $n \to \infty$. Hence,

$$\begin{aligned} &\limsup_{n\to\infty} \frac{1}{b_n} \log \bigg(\sum_{m=0}^{\infty} \frac{1}{m!} \theta^m \Big(\frac{b_n}{n}\Big)^{\frac{2p-d(p-1)}{2p}m} \big(\mathbb{E}\, I_n^m\big)^{1/p}\bigg) \\ &\leqslant \frac{1}{t} \log \bigg(\sum_{m=0}^{\infty} \frac{1}{m!} \theta^m t^{\frac{2p-d(p-1)}{2p}m} \det(\Gamma)^{-\frac{p-1}{2p}m} \Big(\mathbb{E}\, \alpha\big([0,1]^p\big)^m\Big)^{1/p}\bigg). \end{aligned}$$

By an inverse of the general result given in Lemma 5.6, Theorem 3.1 implies that

$$\begin{aligned}
&\lim_{t\to\infty}\frac{1}{t}\log\Big(\sum_{m=0}^{\infty}\frac{1}{m!}\theta^m t^{\frac{2p-d(p-1)}{2p}m}\det(\Gamma)^{-\frac{p-1}{2p}m}\Big(\mathbb{E}\,\alpha\big([0,1]^p\big)^m\Big)^{1/p}\Big)\\
&=2^{\frac{d(p-1)}{2p-d(p-1)}}\frac{2p-d(p-1)}{2p}\Big(\frac{d(p-1)}{2p}\Big)^{\frac{d(p-1)}{2p-d(p-1)}}\\
&\times\det(\Gamma)^{-\frac{p-1}{2p-d(p-1)}}\kappa(d,p)^{\frac{4p}{2p-d(p-1)}}\theta^{\frac{2p}{2p-d(p-1)}}.
\end{aligned}$$

Thus, we have established the upper bound for (5.33).

We now come to the lower bound of (5.33). Use $[x]$ for the lattice part of $x\in\mathbb{R}^d$. Notice that

$$\begin{aligned}
I_n&=\sum_{x\in\mathbb{Z}^d}\prod_{j=1}^p l_j(n,x)=\int_{\mathbb{R}^d}\prod_{j=1}^p l_j(n,[x])dx\\
&=\Big(\frac{n}{b_n}\Big)^{d/2}\int_{\mathbb{R}^d}\prod_{j=1}^p l_j\big(n,[\sqrt{n/b_n}x]\big)dx.
\end{aligned}$$

For any integer $m\geqslant 1$,

$$\begin{aligned}
\mathbb{E}\,I_n^m&=\Big(\frac{n}{b_n}\Big)^{\frac{md}{2}}\mathbb{E}\int_{(\mathbb{R}^d)^m}dx_1\cdots dx_m\prod_{j=1}^p\prod_{k=1}^m l_j\big(n,[\sqrt{n/b_n}x_k]\big)\\
&=\Big(\frac{n}{b_n}\Big)^{\frac{md}{2}}\int_{(\mathbb{R}^d)^m}dx_1\cdots dx_m\Big[\prod_{k=1}^m l\big(n,[\sqrt{n/b_n}x_k]\big)\Big]^p.
\end{aligned}$$

Let $q>1$ be given by $p^{-1}+q^{-1}=1$ and let f be a nice function on $\mathbb{R}^d$ such that $||f||_q=1$.

$$\begin{aligned}
&\big(\mathbb{E}\,I_n^m\big)^{1/p}\\
&\geqslant\Big(\frac{n}{b_n}\Big)^{\frac{md}{2p}}\int_{(\mathbb{R}^d)^m}dx_1\cdots dx_m\Big[\prod_{k=1}^m f(x_k)\Big]\Big[\prod_{k=1}^m l\big(n,[\sqrt{n/b_n}x_k]\big)\Big]\\
&=\Big(\frac{n}{b_n}\Big)^{\frac{md}{2p}}\mathbb{E}\Big[\int_{\mathbb{R}^d}f(x)l\big(n,[\sqrt{n/b_n}x]\big)dx\Big]^m\\
&=\Big(\frac{b_n}{n}\Big)^{\frac{md(p-1)}{2p}}\mathbb{E}\Big[\int_{\mathbb{R}^d}f\Big(\sqrt{\frac{b_n}{n}}x\Big)l(n,[x])dx\Big]^m.
\end{aligned}$$

Therefore,

$$\begin{aligned}&\sum_{m=0}^{\infty}\frac{1}{m!}\theta^m\Big(\frac{b_n}{n}\Big)^{\frac{2p-d(p-1)}{2p}m}\big(\mathbb{E}\, I_n^m\big)^{1/p}\\ &\geqslant \mathbb{E}\exp\Big\{\theta\frac{b_n}{n}\int_{\mathbb{R}^d}f\Big(\sqrt{\frac{b_n}{n}}x\Big)l(n,[x])dx\Big\}\\ &\sim \mathbb{E}\exp\Big\{\theta\frac{b_n}{n}\sum_{k=0}^{n}f\Big(\sqrt{\frac{b_n}{n}}S(k)\Big)\Big\}.\end{aligned}$$

To deal with the right hand side, we need a discrete version (Theorem 4.1, Chen and Li (2004)) of the Feynman-Kac large deviation. Similar to (2.6),

$$\begin{aligned}&\liminf_{n\to\infty}\frac{1}{b_n}\log\mathbb{E}\exp\Big\{\theta\frac{b_n}{n}\sum_{k=0}^{n}f\Big(\sqrt{\frac{b_n}{n}}S(k)\Big)\Big\}\\ &\geqslant \sup_{g\in\mathcal{F}_d}\Big\{\theta\int_{\mathbb{R}^d}f(x)g^2(x)dx-\frac{1}{2}\int_{\mathbb{R}^d}\langle\nabla g(x),\Gamma\nabla g(x)\rangle dx\Big\}.\end{aligned}$$

Consequently,

$$\begin{aligned}&\liminf_{n\to\infty}\frac{1}{b_n}\log\sum_{m=0}^{\infty}\frac{\theta^m}{m!}\Big(\frac{b_n}{n}\Big)^{\frac{2p-d(p-1)}{2p}m}\Big(\mathbb{E}\, I_n^m\Big)^{1/p}\\ &\geqslant \sup_{g\in\mathcal{F}_d}\Big\{\theta\int_{\mathbb{R}^d}f(x)g^2(x)dx-\frac{1}{2}\int_{\mathbb{R}^d}\langle\nabla g(x),\Gamma\nabla g(x)\rangle dx\Big\}.\end{aligned}$$

Taking supremum over $||f||_q = 1$ on the right hand side gives

$$\begin{aligned}&\liminf_{n\to\infty}\frac{1}{b_n}\log\sum_{m=0}^{\infty}\frac{\theta^m}{m!}\Big(\frac{b_n}{n}\Big)^{\frac{2p-d(p-1)}{2p}m}\Big(\mathbb{E}\, I_n^m\Big)^{1/p}\\ &\geqslant \sup_{g\in\mathcal{F}_d}\Big\{\theta\Big(\int_{\mathbb{R}^d}|g(x)|^{2p}dx\Big)^{1/p}-\frac{1}{2}\int_{\mathbb{R}^d}\langle\nabla g(x),\Gamma\nabla g(x)\rangle dx\Big\}.\end{aligned}$$

Finally, the lower bound of (5.33) follows from the analytic fact (Lemma A.2, Chen (2004)) that the right hand side of the above inequality is equal to the right hand side of (5.33). □

To compare with the law of the iterated logarithm (LIL) given in the critical dimensions, we apply Theorem 5.4 and 5.5 to the LIL by taking $b_n = \lambda \log\log n$. The technical involvement is a essentially standard practice of Borel-Cantelli lemma.

Theorem 5.7. *Assume $p(d-2) < d$. For the intersection local time I_n,*

$$\limsup_{n\to\infty} n^{-\frac{2p-d(p-1)}{2}} (\log\log n)^{-\frac{d(p-1)}{2}} I_n = \left(\frac{2}{p}\right)^{-\frac{d(p-1)}{2}} \det(\Gamma)^{-\frac{p-1}{2}} \kappa(d,p)^{2p}, \quad \text{a.s.} \tag{5.34}$$

For the range intersection J_n, we have

(1). *As $d = 1$ and $p \geqslant 2$*

$$\limsup_{n\to\infty} \frac{1}{\sqrt{n\log\log n}} J_n = \sqrt{\frac{2}{p}}\sigma, \quad \text{a.s.} \tag{5.35}$$

(2). *As $d = 2$ and $p \geqslant 2$,*

$$\limsup_{n\to\infty} \frac{(\log n)^p}{n(\log\log n)^{p-1}} J_n = (2\pi)^p \left(\frac{2}{p}\right)^{p-1} \sqrt{\det(\Gamma)}\kappa(2,p)^{2p}, \quad \text{a.s.} \tag{5.36}$$

(3). *As $d = 3$ and $p = 2$,*

$$\limsup_{n\to\infty} \frac{1}{\sqrt{n(\log\log n)^3}} J_n = \gamma_S^2 \det(\Gamma)^{-1/2}\kappa(3,2)^4, \quad \text{a.s.} \tag{5.37}$$

We now come to the super-critical dimensions defined by $p(d-2) > d$ and discuss the tail behaviors of I_∞ and J_∞ defined by (5.1) and (5.2), respectively. The investigation of I_∞ and J_∞ in the super-critical dimensions goes back to an influential paper by Khanin, Mazel, Shlosman and Sinai (1994) where it is shown that in the special case $p = 2$, $d \geqslant 5$,

$$\exp\{-c_1\lambda^{1/2}\} \leqslant \mathbb{P}\{I_\infty \geqslant \lambda\} \leqslant \exp\{-c_2\lambda^{1/2}\}, \tag{5.38}$$

$$\exp\Big\{-\lambda^{\frac{d-2}{d}+\delta}\Big\} \leqslant \mathbb{P}\{J_\infty \geqslant \lambda\} \leqslant \exp\Big\{-\lambda^{\frac{d-2}{d}-\delta}\Big\}, \tag{5.39}$$

for large $\lambda > 0$. The discovery of Khanin *et al* sharply contrasts the results stated in Theorem 5.1, and the "$d = 3$" part in Theorem 5.4 and 5.5, which suggest that the intersection of independent ranges behaves like a constant multiple of the intersection local time in the high dimension ($d \geqslant 3$). The challenge lies in understanding the difference of these behaviors, providing sharp estimates for the tails, and understanding the underlying "optimal strategies".

The exact tail of I_∞ was recently obtain in Chen and Mörters (2007). For comparison we also consider the tail for the random walks with lattice value but continuous time. Let $X_1(t), \cdots, X_p(t)$ be independent symmetric random walks on $\mathbb{Z}^d$ with same distribution. Under $p(d-2) > d$ we write

$$\mathcal{I}_\infty = \int_0^\infty \cdots \int_0^\infty 1_{\{X_1(t_1)=\cdots=X_p(t_p)\}} dt_1 \cdots dt_p.$$

Theorem 5.8. *Under $p(d-2) > d$,*

$$\lim_{\lambda\to\infty} \frac{1}{\lambda^{1/p}} \log \mathbb{P}\{\mathcal{I}_\infty \geqslant \lambda\} = -\frac{p}{\rho} \tag{5.40}$$

while

$$\lim_{\lambda\to\infty} \frac{1}{\lambda^{1/p}} \log \mathbb{P}\{I_\infty \geqslant \lambda\} = -p\gamma^{-1}(1) \tag{5.41}$$

where

$$\rho = \sup\left\{ \sum_{x,y\in\mathbb{Z}^d} G(x-y)f(x)f(y);\quad \sum_{x\in\mathbb{Z}^d} |f(x)|^{\frac{2p}{2p-1}} = 1\right\} \tag{5.42}$$

$$G(x) = \int_0^\infty \mathbb{P}\{X_1(t) = x\}dt, \qquad x\in\mathbb{Z}^d; \tag{5.43}$$

and where $\gamma^{-1}(b)$ is the inverse of the function

$$\begin{aligned}\gamma(\theta) &= \sup\left\{ \sum_{x,y\in\mathbb{Z}^d} G_1(x-y)\sqrt{\left(e^{\theta f(x)}-1\right)\left(e^{\theta f(y)}-1\right)}g(x)g(y);\right. \\ &\qquad \left. f,g\geqslant 0 \text{ and } \sum_{x\in\mathbb{Z}^d} f^{\frac{p}{p-1}}(x) = \sum_{x\in\mathbb{Z}^d} g^2(x) = 1\right\}\end{aligned} \tag{5.44}$$

for $\theta > 0$ and

$$G_1(x) = \sum_{k=1}^\infty \mathbb{P}\{S_1(k) = x\}, \qquad x\in\mathbb{Z}^d. \tag{5.45}$$

It is interesting to see some difference in variation between the case of continuous time and the case of continuous time. The approach for Theorem 5.8 is the method of high moment asymptotics, which will be discussed in section 7.

The problem on the exact tail for J_∞ is harder and still remains open. Perhaps the most important progress since the paper Khanin *et al* was made by van den Berg, Bolthausen and den Hollander (2004): To apply the famous Donsker-Varadhan large deviations, they consider the intersection of two independent Brownian sausages $W_1^\epsilon(t)$ and $W_2^\epsilon(t)$

$$W_i^\epsilon(t) = \bigcup_{0\leqslant s\leqslant t} \{x \in \mathbb{R}^d;\ |x - W_i(s)| \leqslant \epsilon\}, \qquad i = 1, 2$$

instead of $S_1(0,\infty)$ and $S_2(0,\infty)$ and show that as $d \geqslant 5$, the volume $|W_1^\epsilon(\theta t) \cap W_2^\epsilon(\theta t)|$ of the intersection of $W_1^\epsilon(t)$ and $W_2^\epsilon(t)$ has the the following tail behavior: as

$$\lim_{t\to\infty} t^{-\frac{d-2}{d}} \log \mathbb{P}\Big\{|W_1^\epsilon(\theta t) \cap W_2^\epsilon(\theta t)| \geqslant t\Big\} = -I_d^\epsilon(\theta) \tag{5.46}$$

where the rate function $I_d^\epsilon(\cdot)$ is given in terms of the variation

$$I_d^\epsilon(\theta) = \theta \inf_{\phi\in\Phi_d(\theta)} \int_{\mathbb{R}^d} |\nabla\phi(x)|^2 dx \tag{5.47}$$

and

$$\begin{aligned}\Phi_d(\theta) = \Big\{\phi \in W^{1,2}(\mathbb{R}^d);\ \int_{\mathbb{R}^2} \phi^2(x)dx = 1, \text{ and}\\ \int_{\mathbb{R}^d} \Big(1 - \exp\Big\{-2\epsilon^{d-2}\pi^{d/2}\theta\Gamma\Big(\frac{d-2}{2}\Big)^{-1}\phi^2(x)\Big\}\Big)^2 dx \geqslant 1\Big\}.\end{aligned} \tag{5.48}$$

They also show that there exists a critical θ^* such that

$$\begin{aligned}I_d^\epsilon(\theta) =& \frac{\Gamma\Big(\dfrac{d-2}{d}\Big)}{2\epsilon^{d-2}\pi^{d/2}} \inf\Big\{||\nabla\psi||_2^2;\ \ ||\psi||_2 = 1\\ &\text{and} \int_{\mathbb{R}^d} \big(1 - e^{-\psi^2(x)}\big)^2 dx = 1\Big\}\end{aligned}$$

for all $\theta \geqslant \theta^*$. This strongly suggests (conjectured in van den Berg *et al* (2004)) that

$$\lim_{t\to\infty} t^{-\frac{d-2}{d}} \log \mathbb{P}\Big\{|W_1^\epsilon(\infty) \cap W_2^\epsilon(\infty)| \geqslant t\Big\} = -I_d^\epsilon(\theta^*). \tag{5.49}$$

Turning back to the setting of random walks, it becomes natural to conjecture that for the intersection J_∞ of p independent ranges given in (5.2), the limit

$$\lim_{t\to\infty} t^{-\frac{d-2}{d}} \log \mathbb{P}\{J_\infty \geqslant t\} = -C(d,p) \tag{5.50}$$

is non-trivial, and to ask for the identification of the constant $C(d,p)$.

6 Self-intersection local time and range of single random walk

Through this section, $\{S(n)\}$ is a symmetric, square integrable random walk on $\mathbb{Z}^d$. We assume that the smallest subgroup of $\mathbb{Z}^d$ that supports $\{S(n)\}$ is $\mathbb{Z}^d$ itself. We use the notation Γ for the covariance matrix of $\{S(n)\}$, and σ^2 for the variance of $\{S(n)\}$ in the case $d = 1$. We set $\gamma_S = \mathbb{P}\{S(n) \neq 0 \ \forall n \geqslant 1\}$ when $d \geqslant 3$.

The self-intersection is closely related to the range. Roughly speaking, the more intense the self-intersection is, the less spread out the sample path becomes. As $d = 1$, the p-multiple self-intersection local time is essentially same as

$$\frac{1}{p!}\sum_{x\in\mathbb{Z}} l^p(n,x),$$

where $l(n,x)$ is the local time of the random walk $\{S(n)\}$. The following result appears as a discrete version Theorem 4.1 and was obtained by Chen and Li (2004).

Theorem 6.1. *Let $d = 1$. For any $p \geqslant 2$ and for any positive sequence satisfying*

$$b_n \longrightarrow \infty \text{ and } b_n = o(n) \quad (n \to \infty). \tag{6.1}$$

$$\begin{aligned}\lim_{n\to\infty}\frac{1}{b_n}\log\mathbb{P}\Big\{\sum_{x\in\mathbb{Z}} l^p(n,x) \geqslant n^{\frac{p+1}{2}} b_n^{\frac{p-1}{2}}\Big\}\\ = -\sigma^2\frac{1}{4(p-1)}\Big(\frac{p+1}{2}\Big)^{\frac{3-p}{p-1}} B\Big(\frac{1}{p-1},\frac{1}{2}\Big).\end{aligned} \tag{6.2}$$

As for the range, we have (Chen (2006))

Theorem 6.2. *Let $d = 1$. For any positive sequence satisfying (6.1),*

$$\lim_{n\to\infty}\frac{1}{b_n}\log\mathbb{P}\Big\{\#\{S[0,n]\} \geqslant \sqrt{nb_n}\Big\} = -\frac{1}{2\sigma^2}, \tag{6.3}$$

$$\lim_{n\to\infty}\frac{1}{b_n}\log\mathbb{P}\Big\{\#\{S[0,n]\} \leqslant \sqrt{\frac{n}{b_n}}\Big\} = -\frac{\pi^2\sigma^2}{2}. \tag{6.4}$$

To see how the self-intersection connect to the range in the case $d = 1$, we perform the following rough but simple comparison:

$$n = \sum_{x\in\mathbb{Z}} l(n,x) = \sum_{x\in S[0,n]} l(n,x) \leqslant \Big(\#\{S[0,n]\}\Big)^{1/q}\Big(\sum_{x\in\mathbb{Z}} l^p(n,x)\Big)^{1/p}.$$

In view of (6.2) and (6.4), the above comparison captures the relation between self-intersection and range with accuracy up to constant.

In the case $d \geqslant 2$, concentration becomes a dominating force. Dvoretzky and Erdös proved a strong law of large numbers

$$\#\{S[0,n]\}/\mathbb{E}\,\#\{S[0,n]\} \longrightarrow 1, \qquad \text{a.s.} \tag{6.5}$$

Due to the difficulty brought by the dimension multiplicity, the current study of self-intersection local time in the case $d \geqslant 2$ focuses on the case $p = 2$. From now on

$$Q_n = \sum_{1\leqslant j<k\leqslant n} 1_{\{S(j)=S(k)\}}. \tag{6.6}$$

The law of the large numbers also holds for Q_n. The reason behind is that when the dimension gets higher, the short-range intersection gets bigger share in total intersection. When $d \geqslant 2$, one can pick a suitable $l = l(n)$ such that $l = o(n)$, $l(n) \to \infty$ as $n \to \infty$ and that in both decompositions (by the classic inclusion-exclusion formula)

$$\begin{aligned}\#\{S(0,n]\} = &\sum_{i=1}^{l} \#\Big\{S\Big(\frac{i-1}{l}n, \frac{i}{l}n\Big]\Big\} \\ &- \sum_{i=2}^{l} \#\Big\{S\Big(0, \frac{i-1}{l}n\Big] \cap S\Big(\frac{i-1}{l}n, \frac{i}{l}n\Big]\Big\}\end{aligned} \tag{6.7}$$

and

$$\begin{aligned}Q_n = &\sum_{i=1}^{l} \sum_{\frac{i-1}{l}n<j,k\leqslant\frac{i}{l}n} 1_{\{S(j)=S(k)\}} \\ &+ \sum_{i=2}^{l} \sum_{j=1}^{\frac{i-1}{l}n} \sum_{k=\frac{i-1}{l}n+1}^{\frac{i}{l}n} 1_{\{S(j)=S(k)\}}\end{aligned} \tag{6.8}$$

the first summation (short-range intersection) dominates. Notice that the terms in the first summation are independent. So the classic law of large numbers essentially applies here.

Due to concentration phenomenon, the renormalized quantities

$$\#\{S(0,n]\} - \mathbb{E}\,\#\{S(0,n]\} \quad \text{and} \quad Q_n - \mathbb{E}\,Q_n$$

naturally become the objects of the investigation. We now list the expected values of $\#\{S(0,n]\}$ and Q_n below for comparison

$$\mathbb{E}\,Q_n \sim \begin{cases} \dfrac{1}{2\pi\sqrt{\det(\Gamma)}} n\log n, & d = 2, \\[2ex] (\gamma_S^{-1} - 1)n, & d \geqslant 3; \end{cases} \tag{6.9}$$

$$\mathbb{E}\,\#\{S(0,n]\} \sim \begin{cases} 2\pi\sqrt{\det(\Gamma)}\dfrac{n}{\log n}, & d=2, \\ \\ \gamma_S n, & d \geqslant 3. \end{cases} \tag{6.10}$$

Renormalization reduces the contribution from short-range intersection. In the case $d \geqslant 3$, however, the long range intersection is so weak that the short-range intersection still dominates even after being renormalized. Consequently, the classic central limit theorem implies that the sequences

$$\frac{\#\{S(0,n]\} - \mathbb{E}\,\#\{S(0,n]\}}{\sqrt{\operatorname{Var}\left(\#\{S(0,n]\}\right)}} \quad \text{and} \quad \frac{Q_n - \mathbb{E}\,Q_n}{\sqrt{\operatorname{Var}\left(Q_n\right)}}$$

are asymptotically standard normal. The reader is referred to the papers Orey and Jain for the case $d \geqslant 5$, and Jain and Pruitt (1971) for the case $d = 3, 4$.

Historically, the study in the case $d \geqslant 3$ has been focused on the renormalized range $\#\{S(0,n]\} - \mathbb{E}\,\#\{S(0,n]\}$ due to the fact that the quantities

$$\operatorname{Var}\left(\#\{S(0,n]\}\right) \quad \text{and} \quad \operatorname{Var}\left(Q_n\right)$$

differ asymptotically only by a constant multiplier (it seems that this fact has been known by the people in the area for a long time without being explicitly stated, see Lemma 5.6, Chen (2007c) for a proof). Consequently, the renormalized range and renormalized self-intersection local time differ asymptotically by a constant multiplier when $d \geqslant 3$. The computation of the variance of the range can be found in Jain and Orey (1968), Jain and Pruitt (1971). $\operatorname{Var}\left(\#\{S(0,n]\}\right)$ is of the order $n\log n$ when $d = 3$ and of the order n when $n \geqslant 4$. Not surprisingly, $\#\{S(0,n]\} - \mathbb{E}\,\#\{S(0,n]\}$ yields a Gaussian tail. This fact was proved in Jain-Pruitt (1972) and Bass-Kumagai (2002) in the form of law of the iterated logarithm or almost sure invariance principle.

Theorem 6.3. [Jain and Pruitt] *When $d \geqslant 4$,*

$$\limsup_{n\to\infty} \frac{\#\{S(0,n]\} - \mathbb{E}\,\#\{S(0,n]\}}{\sqrt{2n\log\log n}} = \gamma_S\lambda_0, \qquad \text{a.s.} \tag{6.11}$$

$$\liminf_{n\to\infty} \frac{\#\{S(0,n]\} - \mathbb{E}\,\#\{S(0,n]\}}{\sqrt{2n\log\log n}} = -\gamma_S\lambda_0, \qquad \text{a.s.} \tag{6.12}$$

where

$$\lambda_0 = \sqrt{G^2(0) + G(0) + 2\sum_{x\in\mathbb{Z}^d} G^3(x)}$$

and

$$G(x) = \sum_{k=1}^{p} \mathbb{P}\{S(k) = x\}, \qquad x \in \mathbb{Z}^d.$$

Theorem 6.4. [Bass and Kumagai (2002)] *When $d = 3$,*

$$\limsup_{n\to\infty} \frac{\#\{S(0,n]\} - \mathbb{E}\,\#\{S(0,n]\}}{\sqrt{n\log n\log\log n}} = \frac{\gamma_S}{\pi}\sqrt{\det\Gamma}, \qquad \text{a.s.} \tag{6.13}$$

$$\liminf_{n\to\infty} \frac{\#\{S(0,n]\} - \mathbb{E}\,\#\{S(0,n]\}}{\sqrt{n\log n\log\log n}} = -\frac{\gamma_S}{\pi}\sqrt{\det\Gamma}, \qquad \text{a.s.} \tag{6.14}$$

In the same paper, Bass and Kumagai also obtained a partial result in the case $d = 2$. They prove under additional condition that

$$\limsup_{n\to\infty} \frac{(\log n)^2}{n\log\log\log n}\Big(\#\{S(0,n]\} - \mathbb{E}\,\#\{S(0,n]\}\Big) = C, \qquad \text{a.s.} \tag{6.15}$$

with the unidentified constant C.

The case $d = 2$ is the most challenging and interesting case. First, renormalization changes the whole dynamics of the system. The long-range intersection represented by the second summations in (6.7) and (6.8) becomes the dominating force after renormalization. Consequently, the renormalized self-intersection local time and renormalized range have quite different limiting behaviors. In addition to the difference in magnitude caused by recurrence, there is a sign switch due to the difference in sign of the second terms in the decompositions (6.7) and (6.8). Notice also that the terms in the second summations of both (6.7) and (6.8) are strongly correlated so the long-range intersection parts yield a non-Gaussian limit. Remarkably, Le Gall (1986a) observed that the long-range intersection part in both (6.7) and (6.8) are essentially attracted by the renormalized self-intersection local time $\gamma\big([0,1]^2_<\big)$ of a planar Brownian motion (see section 4). Consequently, Le Gall proved that

$$\frac{Q_n - \mathbb{E}\,Q_n}{n} \xrightarrow{d} \det(\Gamma)^{-1/2}\gamma\big([0,1]^2_<\big) \tag{6.16}$$

$$\begin{aligned} &\frac{(\log n)^2}{n}\Big(\#\{S(0,n]\} - \mathbb{E}\,\#\{S(0,n]\}\Big) \\ &\xrightarrow{d} -4\pi^2\sqrt{\det(\Gamma)}\,\gamma\big([0,1]^2_<\big). \end{aligned} \tag{6.17}$$

The tail probabilities for the renormalized self-intersection local time and for the renormalized range were computed in Bass, Chen and Rosen (2006, 2007). To compare them with Theorem 6.3 and Theorem 6.4, we state them in terms of the law of the iterated logarithm.

Theorem 6.5. *When $d = 2$,*

$$\limsup_{n\to\infty} \frac{Q_n - \mathbb{E}\,Q_n}{n \log\log n} = \frac{\kappa(2,2)^4}{\sqrt{\det(\Gamma)}}, \qquad \text{a.s.} \tag{6.18}$$

$$\liminf_{n\to\infty} \frac{Q_n - \mathbb{E}\,Q_n}{n \log\log\log n} = \frac{1}{2\pi\sqrt{\det(\Gamma)}}, \qquad \text{a.s.} \tag{6.19}$$

$$\begin{aligned}&\limsup_{n\to\infty} \frac{(\log n)^2}{n \log\log\log n}\Big(\#\{S(0,n]\} - \mathbb{E}\,\#\{S(0,n]\}\Big)\\ &= 2\pi\sqrt{\det(\Gamma)}, \qquad \text{a.s.}\end{aligned} \tag{6.20}$$

$$\begin{aligned}&\liminf_{n\to\infty} \frac{(\log n)^2}{n \log\log n}\Big(\#\{S(0,n]\} - \mathbb{E}\,\#\{S(0,n]\}\Big)\\ &= -(2\pi)^2\sqrt{\det(\Gamma)}\kappa(2,2)^4, \qquad \text{a.s.}\end{aligned} \tag{6.21}$$

where $\kappa(2,2)$ is the best constant of Gagliardo-Nirenberg inequality given in (3.4) (with $d = p = 2$).

The connection between Theorem 4.4 and Theorem 6.5 is clearly visible. In particular, it is not hard to understand where the non-symmetric behavior appearing in Theorem 6.5 comes from. It is also interesting to see that the limsup for $Q_n - \mathbb{E}\,Q_n$ is relevant to the liminf for $\#\{S[0,n]\} - \mathbb{E}\,\#\{S[0,n]\}$, while the limsup for $\#\{S[0,n]\} - \mathbb{E}\,\#\{S[0,n]\}$ is relevant to the liminf for $Q_n - \mathbb{E}\,Q_n$.

Based on the integrability of I_n and J_n established in Theorem 5.3 (with $d = p = 2$), a discrete version of Theorem 4.3 was obtained in Bass, Chen and Rosen (2006). The approach appears to be a modification of the one used for Theorem 4.3.

Theorem 6.6. *Let $d = 2$. There is $\theta > 0$ such that*

$$\sup_{n\geqslant 1} \mathbb{E}\exp\Big\{\theta\frac{1}{n}\big(Q_n - \mathbb{E}\,Q_n\big)\Big\} < \infty \tag{6.22}$$

$$\sup_{n\geqslant 1} \mathbb{E}\exp\Big\{-\theta\frac{(\log n)^2}{n}\Big(\#\{S(0,n]\} - \mathbb{E}\,\#\{S(0,n]\}\Big)\Big\} < \infty. \tag{6.23}$$

On the other hand, by the fact (Theorem 4.4) that

$$\mathbb{E}\exp\Big\{\theta\gamma\big([0,1]^2_<\big)\Big\} = \infty, \qquad \theta > \kappa(2,2)^{-4}$$

and by the weak laws given in (6.16) and (6.17), we have

$$\sup_{n\geqslant 1} \mathbb{E}\exp\Big\{\theta\frac{1}{n}\big(Q_n - \mathbb{E}\,Q_n\big)\Big\} = \infty$$

for $\theta > \sqrt{\det(\Gamma)}\kappa(2,2)^{-4}$; and

$$\sup_{n\geqslant 1}\mathbb{E}\exp\Big\{-\theta\frac{(\log n)^2}{n}\Big(\#\{S(0,n]\}-\mathbb{E}\,\#\{S(0,n]\}\Big)\Big\}=\infty$$

for every $\theta > (4\pi^2)^{-1}\det(\Gamma)^{-1/2}\kappa(2,2)^{-4}$.

A natural question is to find the critical exponents. We may ask, for example, if $\theta = \sqrt{\det(\Gamma)}\kappa(2,2)^{-4}$ is the critical exponent for $Q_n - \mathbb{E}\,Q_n$. The following example (originally given in Bass, Chen and Rosen (2006)) gives a negative answer.

Example 6.1. Let N be an arbitrarily large integer and write $\epsilon = 2/N^2$. Let the random walk $\{S(n)\}$ is the partial sum of the i.i.d sequence of random vectors in $\mathbb{Z}^d$ that take the values $(N,0),(-N,0),(0,N),(0,-N)$ with probability $\epsilon/4$, and take the value $(0,0)$ with probability $1-\epsilon$. Then Γ is identity matrix. The event that $S(k) = 0$ for all $k \leqslant n$ has probability at least $(1-\epsilon)^n$. On this event $Q_n = n(n-1)/2$. Hence,

$$\mathbb{E}\exp\Big\{\theta\frac{1}{n}(Q_n-\mathbb{E}\,Q_n)\Big\}\geqslant\exp\Big\{-\theta\frac{\mathbb{E}\,Q_n}{n}\Big\}(1-\epsilon)^n\exp\Big\{\theta\frac{n-1}{2}\Big\}.$$

So the critical exponent for $Q_n - \mathbb{E}\,Q_n$ is no more than $2\log(1-\epsilon)^{-1}$. Since ϵ can be arbitrarily small, so is the critical exponent.

This example shows the critical exponent in the continuous setting does not pass to the discrete setting. The identification of these critical exponents in the case of random walks remains open.

Some recent papers have shown interest in the tail behaviors of the self-intersection local times which are not governed by the weak law of convergence. As is well known that $\mathbb{E}\,Q_n$ is of order n as $d \geqslant 3$. Therefore, the investigation of the tail probability

$$\mathbb{P}\{Q_n \geqslant yn\}$$

is related to the law of large numbers given in (6.5). Some recent investigation suggests an interesting dimension dependence pattern. Asselah (2006) recently proved that as $d = 3$

$$\exp\{-c_1y^{2/3}n^{1/3}\}\leqslant\mathbb{P}\Big\{\sum_{x\in\mathbb{Z}^d}l^2(n,x)\geqslant yn\Big\}\leqslant\exp\{-c_2y^{-2/3}n^{1/3}\}$$

for y large enough. As $d \geqslant 5$, it is shown in Asselah and Castell (2006) that for y large enough there are $c_1, c_2 > 0$ (depending on y) such that

$$\exp\{-c_1\sqrt{n}\}\leqslant\mathbb{P}\Big\{\sum_{x\in\mathbb{Z}^d}l^2(n,x)\geqslant yn\Big\}\leqslant\exp\{-c_2\sqrt{n}\}.$$

The case $d = 4$ remains open. Further, one may ask for a sharper estimate of the above tails.

We now mention a recent work (Chen (2007c)) on a model called charged polymer. A lattice polymer is often described by physicists as interpolation line segment with the vertexes given as the n-step lattice (simple) random walk

$$\{S(1), \cdots, S(n)\}.$$

By placing independent, identically distributed electric charges $\omega_k = \pm 1$ to each vertex of the polymer, Kantor and Kardar (1991) consider a model of polymers with random electrical charges associated with the Hamiltonian

$$H_n = \sum_{1 \leqslant j < k \leqslant n} \omega_j \omega_k 1_{\{S(j)=S(k)\}}. \tag{6.24}$$

If we assume that when two charges meet, the pair with opposite sign gives negative contribution while the pair with same sign gives positive contribution, then H_n represents the total electrical interaction charge of the polymer $\{S(1), \cdots, S(n)\}$.

We point out some other works by physicists in this direction. In Derrida, Griffiths and Higgs (1992), the charges are i.i.d. Gaussian variables. In Derrida and Higgs (1994), the charges take $0-1$ values. We also refer the reader to Martinez and Petritis (1996); Buffet and Pulé (1997) for the continuous versions of the polymer with random charges. In particular, we cite the comment by Martinez and Petritis (1996): "It is argued that a protein molecule is very much like a random walk with random charges attached at the vertexes of the walk; these charges are interacting through local interactions mimicking Lennard-Jones or hydrogen-bond potentials". The central limit theorems and moderate deviations for H_n have been established in the recent work Chen (2007c). The central limit theorems state: When $d = 1$,

$$\frac{1}{n^{3/4}} H_n \xrightarrow{d} (2\sigma)^{-1/2} \left(\int_{-\infty}^{\infty} L^2(1, x) dx \right)^{1/2} U, \tag{6.25}$$

where U is a random variable with standard normal distribution, $L(t, x)$ is the local time of the 1-dimensional Brownian motion $W(t)$ such that U and $W(t)$ are independent. When $d = 2$,

$$\frac{1}{\sqrt{n \log n}} H_n \xrightarrow{d} \frac{1}{\sqrt{2\pi} \sqrt[4]{\det \Gamma}} U. \tag{6.26}$$

When $d \geqslant 3$,

$$\frac{1}{\sqrt{n}} H_n \xrightarrow{d} \sqrt{\gamma s} U. \tag{6.28}$$

Not surprisingly, it is the limiting random variables in the laws of of weak convergence who decide the tail behaviors of H_n. Therefore we have the following theorem on the moderate deviations for H_n (Chen (2007c)).

Theorem 6.7. *As $d = 1$,*

$$\begin{aligned} &\lim_{n\to\infty} \frac{1}{b_n} \log \mathbb{P}\Big\{ \pm H_n \geqslant \lambda (nb_n)^{3/4} \Big\} \\ &= -\frac{1}{2}\sigma^{2/3}(3\lambda)^{4/3}, \qquad (\lambda > 0) \end{aligned} \tag{6.29}$$

for any positive sequence $\{b_n\}$ satisfying

$$b_n \longrightarrow \infty \quad and \quad b_n = o(\sqrt[7]{n}), \qquad (n \to \infty).$$

As $d = 2$,

$$\begin{aligned} &\lim_{n\to\infty} \frac{1}{b_n} \log \mathbb{P}\Big\{ \pm H_n \geqslant \lambda \sqrt{n(\log n) b_n} \Big\} \\ &= -\pi\sqrt{\det(\Gamma)}\lambda^2, \qquad (\lambda > 0) \end{aligned} \tag{6.30}$$

for any positive sequence $\{b_n\}$ satisfying

$$b_n \longrightarrow \infty \quad and \quad b_n = o(\log n), \qquad (n \to \infty).$$

As $d \geqslant 3$,

$$\lim_{n\to\infty} \frac{1}{b_n} \log \mathbb{P}\Big\{ \pm H_n \geqslant \lambda \sqrt{nb_n} \Big\} = -\frac{\lambda^2}{2\gamma}, \qquad (\lambda > 0) \tag{6.31}$$

for any positive sequence $\{b_n\}$ satisfying

$$b_n \longrightarrow \infty \quad and \quad b_n = o\Big(\frac{n^{1/3}}{(\log n)^{4/3}}\Big), \qquad (n \to \infty).$$

7 Method of high moment asymptotics

The conventional approach dealing with the tail probabilities of intersection local times is Feynman-Kac formula. This method is also closely related to the results and tools developed along the line of Donsker-Varadhan large deviation theory. In this section we provide a new alternative known as the *method of high moment asymptotics.* Moment method has been extensively used in the study of limit laws of intersection local times and related models, due to the fact that the integer moments of intersection local times often have reasonably nice representations — see (3.15) for example. In Le Gall (1986a, 1986b), Rosen (1990), Le Gall and Rosen (1991), the law of weak convergence is obtained by establishing the convergence between the moments. In this case the power of the moment is arbitrary but fixed. According to Lemma 3.3, on the other hand, the power of the moment goes to infinity when it comes to large deviations. We call this type of the problems as high moment asymptotics.

Here we take the proof of (5.40) in Theorem 5.8 as example. The conventional technique does not allow the treatment of infinite time horizon. So we compute the high moment of I_∞. Write

$$\mathcal{I}_\infty = \sum_{x\in\mathbb{Z}^d}\prod_{j=1}^p \int_0^\infty 1_{\{X_j(t)=x\}}dt.$$

We have that for any $m \geqslant 1$,

$$\begin{aligned}
&\mathbb{E}\mathcal{I}_\infty^m\\
&= \sum_{x_1,\cdots,x_m\in\mathbb{Z}^d}\left[\int_0^\infty\cdots\int_0^\infty 1_{\{X(t_1)=x_1,\cdots,X(t_m)=x_m\}}dt_1\cdots dt_m\right]^p\\
&= \sum_{x_1,\cdots,x_m\in\mathbb{Z}^d}\left[\sum_{\sigma\in\Sigma_m}\int_{\{t_1\leqslant\cdots\leqslant t_m\}} 1_{\{x(t_1)=x_{\sigma(1)},\cdots,X(t_m)=x_{\sigma(m)}\}}\right.\\
&\qquad\qquad \left.\times\, dt_1\cdots dt_m\right]^p \qquad\qquad (7.1)\\
&= \sum_{x_1,\cdots,x_m\in\mathbb{Z}^d}\left[\sum_{\sigma\in\Sigma_m}\prod_{k=1}^m G(x_{\sigma(k)}-x_{\sigma(k-1)})\right]^p
\end{aligned}$$

where Σ_m is the permutation group on $\{1,\cdots,m\}$. By Lemma 3.3 we need to establish

$$\lim_{m\to\infty}\frac{1}{m}\log\sum_{x_1,\cdots,x_m\in\mathbb{Z}^d}\left[\frac{1}{m!}\sum_{\sigma\in\Sigma_m}\prod_{k=1}^m G(x_{\sigma(k)}-x_{\sigma(k-1)})\right]^p \qquad (7.2)$$
$$= p\log\rho.$$

The method of high moment asymptotics first appeared in König-Mörters (2002). The following Theorem 7.1 and 7.2 are modified versions given in in Chen and Mörters (2007).

Theorem 7.1. *Let $p \geqslant 1$ be a constant, let $(\Omega, \mathcal{A}, \pi)$ be a measure space and let $K\colon \Omega \times \Omega \longrightarrow \mathbb{R}^+$ be a measurable, non-negative function satisfying the following assumptions:* (1). *Symmetry: $K(x,y) = K(y,x)$ for any $x, y \in \Omega$.* (2). *Irreducibility: For any $x \in \Omega$,*

$$\pi\Big(\big\{y \in \Omega;\ \ G(x,y) = 0\big\}\Big) = 0.$$

(3). *Double integrability: For any $g \in \mathcal{L}^{\frac{2p}{2p-1}}(\Omega, \mathcal{A}, \pi)$,*

$$\iint_{\Omega\times\Omega} K(x,y)g(x)g(y)\pi(dx)\pi(dy) < \infty.$$

Then

$$\begin{aligned}\liminf_{m\to\infty} \frac{1}{m} \log \int_{\Omega^m} &\pi(dx_1)\cdots\pi(dx_m)\\ &\times \left[\frac{1}{m!}\sum_{\sigma\in\Sigma_m}\prod_{k=1}^m K(x_{\sigma(k-1)}, x_{\sigma(k)})\right]^p \\ \geqslant p \log \sup_g \iint_{\Omega\times\Omega} &K(x,y)g(x)g(y)\pi(dx)\pi(dy),\end{aligned} \tag{7.3}$$

where the supremum is taken over all functions g on Ω satisfying

$$\int_\Omega |g(x)|^{\frac{2p}{2p-1}}\pi(dx) = 1.$$

The key in the proof of Theorem 7.1 is to remove the permutations. By Hölder inequality:

$$\begin{aligned}&\left\{\int_{\Omega^m} \pi(dx_1)\cdots\pi(dx_m)\left[\frac{1}{m!}\sum_{\sigma\in\Sigma_m}\prod_{k=1}^m K(x_{\sigma(k-1)}, x_{\sigma(k)})\right]^p\right\}^{1/p}\\ &\geqslant \int_{\Omega^m}\pi(dx_1)\cdots\pi(dx_m)\Big(\prod_{k=1}^m f(x_k)\Big)\frac{1}{m!}\sum_{\sigma\in\Sigma_m}\prod_{k=1}^m K(x_{\sigma(k-1)}, x_{\sigma(k)})\\ &= \int_{\Omega^m}\pi(dx_1)\cdots\pi(dx_m)\prod_{k=1}^m K(x_{k-1}, x_k)f(x_k),\end{aligned}$$

where the test function $f \geqslant 0$ satisfies $||f||_q = 1$ with $q > 1$ conjugate to p. In this way, the proof is reduced to a typical principal eigenvalue problem followed by optimization of f.

The upper bound is much harder. Indeed, we are able to establish it only when the state space Ω is finite.

Theorem 7.2. *Let Ω be a finite set and $\pi\colon \Omega \longrightarrow \mathbb{R}^+$ and $K\colon \Omega\times\Omega \longrightarrow \mathbb{R}^+$. Assume that $K(x,y) = K(y,x)$. Then*

$$\begin{aligned}\limsup_{m\to\infty}\frac{1}{m}\log &\sum_{x_1,\cdots,x_n\in\Omega}\Big(\prod_{k=1}^m \pi(x_k)\Big)\\ &\times\Big[\frac{1}{m!}\sum_{\sigma\in\Sigma_m}\prod_{k=1}^m K(x_{\sigma(k-1)},x_{\sigma(k)})\Big]^p \leqslant p\log\hat{\rho},\end{aligned}\tag{7.4}$$

where Σ_m is the permutation group on $\{1,\cdots,m\}$ and

$$\hat{\rho} = \sup\Big\{\sum_{x,y\in\Omega}K(x,y)g(x)g(y)\pi(x)\pi(y);\ \sum_{x\in\Omega}|g(x)|^{\frac{2p}{2p-1}}\pi(x) = 1\Big\}.$$

The lower bound given in Theorem 7.1 has enough generality for practical use. The proof of Theorem 7.2 is combinatorial and the assumption on the finite state space is essential to the argument currently used for Theorem 7.2. The challenge we face is to extend Theorem 7.2 to the setting of reasonable generality.

Going back to the proof of Theorem 5.8, we need only to establish the upper bound of (7.2). By performing a projection on a torus, we have that for any integer $N \geqslant 1$,

$$\begin{aligned}&\sum_{x_1,\cdots,x_m\in\mathbb{Z}^d}\Big[\sum_{\sigma\in\Sigma_m}\prod_{k=1}^m G(x_{\sigma(k)}-x_{\sigma(k-1)})\Big]^p\\ &\leqslant \sum_{y_1,\cdots,y_m\in[-N,N)^d}\Big[\sum_{\sigma\in\Sigma_m}\prod_{k=1}^m \widetilde{G}_N(y_{\sigma(k)}-y_{\sigma(k-1)})\Big]^p,\end{aligned}\tag{7.5}$$

where

$$\widetilde{G}_N(y) = \Big(\sum_{z\in\mathbb{Z}^d}G^p(2Nz+y)\Big)^{1/p}.$$

Applying Theorem 7.2 to the right hand side of (7.5) gives

$$\limsup_{m\to\infty}\frac{1}{m}\log\sum_{x_1,\cdots,x_m\in\mathbb{Z}^d}\Big[\frac{1}{m!}\sum_{\sigma\in\Sigma_m}\prod_{k=1}^m G(x_{\sigma(k)}-x_{\sigma(k-1)})\Big]^p \leqslant p\log\widetilde{\rho}_N$$

where

$$\widetilde{\rho}_N = \sup\Big\{\sum_{x,y\in[-N,N)^d}\widetilde{G}_N(x-y)g(x)g(y);\ \sum_{x\in[-N,N)^d}|g(x)|^{\frac{2p}{2p-1}} = 1\Big\}.$$

Finally, the upper bound for (7.2) follows from the fact that

$$\limsup_{N\to\infty} \widetilde{\rho}_N \leqslant \rho.$$

The proof of (5.41) in Theorem 5.8 is more delicate. Indeed, the moment representation of I_∞ takes a much more complicated form than (7.1). To state it, we introduce the following notations. For any $1 \leqslant l \leqslant m$, let $\pi = (\pi_1, \cdots, \pi_l)$ represent a partition of $\{1, \cdots, m\}$, where $\pi_1, \cdots, \pi_l$ are dijoint, non-empty subsets of $\{1, \cdots, m\}$ satisfying $\{1, \cdots, m\} = \pi_1 \cup \cdots \cup \pi_l$. Set

$$\mathcal{A}(\pi) = \Big\{(x_1, \cdots, x_m) \in \mathbb{Z}^d; \quad x_i = x_j \text{ for any } i, j \in \pi_k,\ k = 1, \cdots l\Big\}.$$

For any $(x_1, \cdots, x_m) \in \mathcal{A}(\pi)$ and for any $1 \leqslant k \leqslant l$, we use x_{π_k} for the common value of the points $\{x_j;\ j \in \pi_k\}$. For $\leqslant l \leqslant m$, write $\mathcal{E}_l$ for the set of the partitions π which divide $\{1, \cdots, m\}$ into l non-empty groups. By combinatorics

$$\#\{\mathcal{E}_l\} = \frac{1}{l!} \sum_{\substack{j_1+\cdots+j_l=m \\ j_1,\cdots,j_l\geqslant 1}} \frac{m!}{j_1! \cdots j_l!}.$$

A computation comparable to (7.1) gives the following moment representation:

$$\mathbb{E}\, I_\infty^m = \sum_{x_1,\cdots,x_m\in\mathbb{Z}^d} \Bigg[\sum_{l=1}^m \sum_{\pi\in\mathcal{E}_l} 1_{\{(x_1,\cdots,x_m)\in\mathcal{A}(\pi)\}} \times \sum_{\sigma\in\Sigma_l} \prod_{k=1}^l G_1(x_{\pi_{\sigma(k)}} - x_{\pi_{\sigma(k-1)}})\Bigg]^p \tag{7.6}$$

where G_1 is defined in (5.45). The increase of complexity in (7.6) is caused by the difficulty in counting the time vectors $(i_1, \cdots, i_m)$ with non-distinct components.

By an argument harder than the one use for (7.2), the representation (7.6) leads to

$$\lim_{m\to\infty} \frac{1}{m} \log \frac{1}{(m!)^p} \mathbb{E}\, I_\infty^m = -p \log \gamma^{-1}(1). \tag{7.7}$$

Finally, the desired (5.41) follows from Lemma 3.3.

Another successful application of the method of high moment asymptotics is the proof of (3.8) in Theorem 3.2, which has been reduced to the proof of (3.16). Like the previous case, we can directly prove the lower bound

$$\liminf_{m\to\infty}\frac{1}{m}\log\int_{(\mathbb{R}^d)^m} d\lambda_1\cdots d\lambda_m \times\Big[\frac{1}{m!}\sum_{\sigma\in\Sigma_m}\prod_{k=1}^m Q\Big(\sum_{j=1}^k\lambda_{\sigma(j)}\Big)\Big]^p \geqslant \log\rho_1. \tag{7.8}$$

As for the upper bound, we have the following discrete version.

Theorem 7.3. *Let $\pi(x)$ and $Q(x)$ be two non-negative functions on $\mathbb{Z}^d$ such that π is locally supported, $\pi(-x)=\pi(x)$ for all $x\in\mathbb{Z}^d$, and that $Q(x)\longrightarrow 0$ as $|x|\to\infty$. Then*

$$\limsup_{m\to\infty}\frac{1}{m}\log\sum_{x_1,\cdots,x_n\in\mathbb{Z}^d}\Big(\prod_{k=1}^m\pi(x_k)\Big) \times\Big[\frac{1}{m!}\sum_{\sigma\in\Sigma_m}\prod_{k=1}^m Q\Big(\sum_{j=1}^k x_{\sigma(j)}\Big)\Big]^p \leqslant \log\widehat{\rho} \tag{7.9}$$

where

$$\widehat{\rho}=\sup_{|f|_2=1}\sum_{x\in\mathbb{Z}^d}\pi(x)\Big[\sum_{y\in\mathbb{Z}^d}\sqrt{Q(x+y)}\sqrt{Q(y)}f(x+y)f(y)\Big]^p \tag{7.10}$$

and

$$|f|_2=\Big(\sum_{x\in\mathbb{Z}^d}f^2(x)\Big)^{1/2}.$$

In order to apply Theorem 7.3 to the proof of the upper bound of (3.16), the major challenge is descretization. This has been done by the tool of Fourier analysis. □

In summary, the method of high moment asymtotics is at its beginning stage and there are considerable challenges are to be expected. On the other hand, it opens up new avenues for the treatment of a wide range of problems arisen from area of sample path intersections, especially those out of the reach of the conventional approaches.

References

[1] Asselah, A. and Castell, F. (2006). Self-intersection local times for random walk, and random walk in random scenery in dimension $d\geqslant 5$. *Probab. Theor. Rel. Fields* (to appear).

[2] Asselah, A. (2006). Large deviation estimates for self-intersection local times for simple random walk in $\mathbb{Z}^3$. (preprint).

[3] Bass, R. F. and Chen, X. (2004). Self-intersection local time: critical exponent and laws of the iterated logarithm. *Ann. Probab.* **32** 3221–3247.

[4] Bass, R. F., Chen, X. and Rosen, J. (2005). Large deviations for renormalized self-intersection local times of stable processes. *Ann. Probab.* **33** 984–1013.

[5] Bass, R. F., Chen, X. and Rosen, J. (2006). Moderate deviations and laws of the iterated logarithm for the renormalized self-intersection local times of planar random walks. *Electron. J. Probab.* **11** 933–1030.

[6] Bass, R. F., Chen, X. and Rosen, J. (2007). Moderate deviations for the range of planar random walks. *Memoirs of AMS.* (to appear).

[7] Bass, R. F. and Kumagai, T. (2002). Laws of the iterated logarithm for the range of random walks in two and three dimensions. *Ann. Probab.* **30** 1369–1396.

[8] van den Berg, M., Bolthausen, E. and den Hollander, F. (2004). On the volume of the intersection of two Wiener sausages. *Ann. Math.* **159** 741–785.

[9] Biscup, M. and König, W. (2001). Long-time tails in the parabolic Anderson model with bounded potential. *Ann. Probab.* **29** 636–682.

[10] Bolthausen, E. (1999). Large deviations and interacting random walks. *Lecture Notes in Math.* **1781** 1–124.

[11] Buffet, E. and Pulé, J. V. (1997). A model of continuous polymers with random charges. *J. Math. Phys.* **38** 5143–5152.

[12] Calais, J. Y. and Yor, M. (1987). Renormalization et convergence en loi pour certaines intégrales multiples associées au mouvement brownien dans $\mathbb{R}^d$. *Séminaire de Probabilités XXI. Lecture Notes in Math.* **1247** 375–403. Springer, Berlin.

[13] Carlen, E. A. and Loss, M. (1993). Sharp constant in Nash's inequality. Internat. *Math. Res. Notices* **1993** 213–215.

[14] armona, R. and Molchanov, S. A. (1994). Parabolic Anderson Problem and Intermittency. *Am. Math. Soc.,* Providence, RI.

[15] Chen, X. (2004). Exponential asymptotics and law of the iterated logarithm for intersection local times of random walks. *Ann. Probab.* **32** 3248–3300.

[16] Chen, X. (2005). Moderate deviations and law of the iterated logarithm for intersections of the range of random walks. *Ann. Probab.* **33** 1014–1059.

[17] Chen, X. (2006a). Self-intersection local times of additive processes: large deviation and law of the iterated logarithm. *Stoc. Proc. Appl.* **116** 1236–1253.

[18] Chen, X. (2006b). Moderate and small deviations for the ranges of one-dimensional random walks. *J. Theor. Probab.* (to appear).

[19] Chen, X. (2007a). Large deviations and laws of the iterated logarithm for the local times of additive stable processes. *Ann. Probab.* (to appear).

[20] Chen, X. (2007b). Moderate deviations and laws of the iterated logarithm for the local times of additive Levy processes and additive random walks. *Ann. Probab.* (to appear).

[21] Chen, X. (2007c). Limit laws for the energy of a charged polymer. (preprint).

[22] Chen, X. and Li, W. (2004).Large and moderate deviations for intersection local times. *Probab. Theor. Rel. Fields.* **128** 213–254.

[23] Chen, X., Li, W. and Rosen, J. (2005). Large deviations for local times of stable processes and stable random walks in 1 dimension. *Electron. J. Probab.* **10** 577–608.

[24] Chen, X. and Mörters, P. (2007). Upper tails for intersections of random walks in supercritical dimensions. (preprint).

[25] Chen, X., and Rosen, J. (2005). Exponential asymptotics for intersection local times of stable processes and random walks. *Ann. Inst. H. Poincaré* **41** 901–928.

[26] Cordero-Erausquin, D., Nazaret, B. and Villani, C. (2004). A mass-transportation approach to sharp Sobolev and Gagliardo-Nirenberg inequalities. *Adv. Math.* **182** 307–332.

[27] Cranston, M., Mountford, T. S. and Shiga, T. (2005). Lyapunov exponent for the parabolic Anderson model with Lévy noise. *Probab. Theory Related Fields* **132** 321–355.

[28] Csáki, E., König, W. and Shi Z. (1999). An embedding for Kesten-Spitzer random walk in random scenery. *Stochastic Process. Appl.* **82** 283–292.

[29] Csörgö, M., Shi, Z. and Yor, M. (1999). Some asymptotic properties of the local time of the uniform empirical process. *Bernoulli* **5** 1035–1058.

[30] Del Pino, M. and Dolbeault, J. (2002). Best constants for Gagliardo-Nirenberg inequalities and applications to nonlinear diffusions. *J. Math. Pures Appl.* **81** 847–875.

[31] Del Pino, M. and Dolbeault, J. (2003). The optimal Euclidean L^p-Sobolev logarithmic inequality. *J. Funct. Anal.* **197** 151–161.

[32] Dembo, A. and Zeitouni, O. (1998). *Large Deviations Techniques and Applications. (2nd ed.)* Springer, New York.

[33] Derrida, B., Griffiths, R. B. and Higgs, R. G. (1992). A model of directed walks with random self interactions. *Europhys. Lett.* **18** 361–366.

[34] Derrida, B.and Higgs, P. G. (1994) Low-temperature properties of directed walks with random self-interactions. *J. Phys. A* **27** 5485–5493.

[35] Donsker, M. D. and Varadhan, S. R. S. (1975a). Asymptotic evaluation of certain Markov process expectations for large time. I. *Comm. Pure Appl. Math.* **28** 1–47, 279–301.

[36] Donsker, M. D. and Varadhan, S. R. S. (1975b). Asymptotics for the Wiener sausage. *Comm. Pure Appl. Math.* **28** 525–565

[37] Donsker, M. D. and Varadhan, S. R. S. (1976). Asymptotic evaluation of certain Markov process expectations for large time. III. *Comm. Pure Appl. Math.* **29** 389–461.

[38] Donsker, M. D. and Varadhan, S. R. S. (1979). On the number of distinct sites visited by a random walk. *Comm. Pure Appl. Math.* **32** 721–747.

[39] Donsker, M. D. and Varadhan, S. R. S. (1981). The polaron problem and large deviations. *New stochastic methods in physics. Phys. Rep.* **77** 235–237.

[40] Donsker, M. D. and Varadhan, S. R. S. (1983). Asymptotics for the polaron. *Comm. Pure Appl. Math.* **36** 505–528.

[41] Dvoretzky, A., Erdös, P. and Kakutani, S. (1950). Double points of paths of Brownian motions in n-space. *Acta Sci. Math. Szeged* **12**, Leopoldo, Fejér et Frederico Riesz LXX annos natis dedicatus, Pars B, 75–81.

[42] Dvoretzky, A., Erdös, P. and Kakutani, S. (1954). Multiple points of paths of Brownian motion in the plane. *Bull. Res. Council Israel.* **3**, 364–371

[43] Dynkin, E. B. (1988). Regularized self-intersection local times of planar Brownian motion. *Ann. Probab.* **16** 58–74.

[44] Fernández, R., Fröhlich, J. and Sokal, A. D. (1992).*Random Walks, Critical Phenomena, and Triviality in quantum field theory.* Springer, New York.

[45] Ganternt, R., König, W. and Shi, Z. (2007). Annealed deviations of random walk in random scenery. *Ann. Inst. H. Poincaré* **43** 47–76.

[46] Geman, D., Horowitz, J. and Rosen, J. (1984). A local time analysis of intersections of Brownian paths in the plane. *Ann. Probab.* **12** 86–107.

[47] Hamana, Y. and Kesten, H. (2001). A large-deviation result for the range of random walk and for the Wiener sausage. *Probab. Theory Related Fields* **120** 183–208.

[48] Hamana, Y. and Kesten, H. (2002). Large deviations for the range of an integer valued random walk. *Ann. Inst. H. Poincaré Probab. Statist.* **38** 17–58.

[49] van der Hofstad, R., den Hollander, F. and König, W. (2003). Large deviations for the one-dimensional Edwards model. *Ann. Probab.* **31** 2003–2039.

[50] van der Hofstad, R. and König, W. (2001). A survey of one-dimensional random polymers. *J. Stat. Phys.* **103** 915–944.

[51] van der Hoftad, R., König, W. and Mörters, P. (2006). The universality classes in the parabolic Anderson model. Comm. *Math. Phys.* **267** 307–353.

[52] den Hollander, F. (1996). Random polymers. *Stat. Nederl* **50** 136–145.

[53] Hu, Y. and Nualart, D. (2005). Renormalized self-intersection local time for fractonal Brownian motion. *Ann. Probab.* **33** 948–983.

[54] Jain, N. C. and Orey, S. (1968). On the range of random walk. *Israel J. Math.* **6** 373–380.

[55] Jain, N.C. and Pruitt, W.E. (1971). The range of transient random walk. *J. Anal. Math.* **24** 369–393.

[56] Jain, N.C. and Pruitt, W.E. (1972). The law of the iterated logarithm for the range of random walk. *Ann. Math. Stat.* **43** 1692–1697.

[57] Kantor, Y. and Kardar, M. (1991). Polymers with self-interactions. *Europhys. Lett.* **14** 421–426.

[58] Kesten, H. and Spitzer, F. (1979). A limit theorem related to a new class of self-similar processes. *Z. Wahrsch. Verw. Gebiete* **50** 2–25.

[59] Khanin, K. M., Mazel, A. E., Sholosman, S. B. and Sinai, Ya. G. (1994). Loop condensation effect in the behavior of random walks. *Markov process and their applications (Freidlin, edu), Progr. Probab.* **34**, Birkhäuser, Boston, p.167–184.

[60] Khoshnevisan, D. (2002). *Multiparameter Processes — An Introduction to Random Fields.* Springer, New York.

[61] Khoshnevisan, D. and Lewis, T. M. (1998). A law of the iterated logarithm for stable processes in random scenery. *Stochastic Proc. Appl.* **74** 89–121.

[62] Khoshnevisan, D., Xiao, Y. and Zhong, Y. (2003a). Local times of additive Lévy processes. *Stoch. Proc. Their Appl.* **104** 193–216.

[63] Khoshnevisan, D., Xiao, Y. and Zhong, Y. (2003b). Measuring the range of an additive Lévy process. *Ann. Probab.* **31** 1097–1141..

[64] König, W. and Mörters, P. (2002). Brownian intersection local times: Upper tail asymptotics and thick points. *Ann Probab.* **30** 1605–1656.

[65] Lawler, G. F. (2005). *Conformally Invariant Processes in the Plane. Mathematical Surveys and Monographs,* **114**. American Mathematical Society, Providence, RI.

[66] Lawler, G. F., Schramm, O. and Werner, W. (2001). Values of Brownian intersection exponents. I. Half-plane exponents. *Acta Math.* **187** 237–273.

[67] Lawler, G. F., Schramm, O. and Werner, W. (2001). Values of Brownian intersection exponents. II. Plane exponents. *Acta Math.* **187** 275–308.

[68] Lawler, G. F., Schramm, O. and Werner, W. (2002). Values of Brownian intersection exponents. III. Two-sided exponents. *Ann. Inst. H. Poincaré Probab. Stat.* **38** 109–123.

[69] Le Gall, J-F. (1986a). Propriétés d'intersection des marches aléatoires. I. Convergence vers le temps local d'intersection. *Comm. Math. Phys.* **104**, 471–507,

[70] Le Gall, J-F. (1986b). Propriétés d'intersection des marches aléatoires. II. Étude des cas critiques. *Comm. Math. Phys.* **104**, 509–528.

[71] Le Gall, J-F. (1992). Some properties of planar Brownian motion. *École d'Été de Probabilités de Saint-Flour XX—1990*, 111–235. Springer, Berlin.

[72] Le Gall, J-F. (1994). Exponential moments for the renormalized self-intersection local time of planar Brownian motion. *Séminaire de Probabilités, XXVIII*, 172–180. Springer, Berlin.

[73] Le Gall, J-F. and Rosen, J. (1991). The range of stable random walks. *Ann. Probab.* **19** 650–705.

[74] Levine, H. A. (1980). An estimate for the best constant in a Sobolev inequality involving three integral norms. *Ann. Mat. Pura Appl.* **124** 181–125.

[75] Madras, N. and Slade, G. (1993). *The Self-avoiding Walk.* Birkhäuser, Boston.

[76] Mansmann, U. (1991). The free energy of the Dirac polaron, an explicit solution. *Stochastics & Stochastics Report.* **34** 93–125.

[77] Martanez, S. and Petritis, D. (1996). Thermodynamics of a Brownian bridge polymer model in a random environment. *J. Phys. A* **29** (1996), 1267–1279.

[78] Remillard, B. (2000). Large deviations estimates for occupation time integrals of Brownian motion. *Stochastic models (Ottwa, ON)* 375-398, *CMS Cof. Proc.* **26** Amer. Soc. Providence. RI.

[79] Révész, P. and Shi, Z. (2000). Strong approximation of spatial random walk in random scenery. *Stochastic Proc. Appl.* **88** 329–345.

[80] Rosen, J. (1986). A renormalized local time for multiple intersections of planar Brownian motion. *Séminaire de Probabilités, XX, 1984/85, Lecture Notes in Math.,* **1204** 515–531.

[81] Rosen, J. (1990). Random walks and intersection local time. *Ann. Probab.* **18** 959–977.

[82] Vanderzande, C. (1998) *Lattice models of Polymers.* Cambridge University Press, Cambridge.

[83] Weinstein, M. I. (1983). Nonlinear Shrödinger equations and sharp interpolation estimates. *Comm. Math. Phys.* **87** 567–576.

[84] Westwater, J. (1980). On Edwards' model for long polymer chains. *Comm. Math. Phys.* **72** 131–174

[85] Westwater, J. (1981). On Edwards' model for polymer chains. II. The self-consistent potential. *Comm. Math. Phys.* **79** 53–73

[86] Westwater, J. (1982). On Edwards' model for polymer chains. III. Borel summability. *Comm. Math. Phys.* **84** 459–470.

[87] Xiao, Y. (2006). Properties of local nondeterminism of Gaussian and stable random fields and their applications. (preprint).

[88] Yor, M. (1985). Renormalization et convergence en loi pour les temps locaux d'intersection du mouvement brownien dans $\mathbb{R}^3$. *Séminaire de Probabilités XIX. Lecture Notes in Math.* **1123** 350–365. Springer, Berlin.

[89] Zhou, X. Y. (2002). *Collection Papers of Zhou Xianyin (I & II).* Beijing Normal University Modern Mathematics Series, Bnumms **10**, Beijing Normal Univ. Press Beijing.

URL: http://www.math.utk.edu/~xchen

Limit Theorems for U-Statistics

Bing-Yi JING

Department of Mathematics

Hong Kong University of Science and Technology

Clear Water Bay, Kowloon, Hong Kong, China

E-mail: majing@ust.hk

Abstract

The U-statistics are one of the universal objects of the modern probability theory. There have been tremendous advances in their developments since the fundamental work of Hoeffding (1948). Most of the studies, however, are concerned with standardized U-statistics. It is of both theoretical and practical interest to investigate studentized U-statistics since they are more relevant in statistical inference. The purpose of this paper is to present some of the results on asymptotic normality, Berry-Esséen bounds, and Edgeworth expansions for both standardized and studentized U-statistics in a unified approach under very weak or minimal moment conditions. Some nonparametric methods to U-statistics are also discussed.

1 Introduction

Suppose that we are interested in the distribution of some statistic (or its standardized version with mean zero and variance one), $T_n = T(X_1, \cdots, X_n)$, where $X_1, \cdots, X_n$ is a sequence of independent and identically distributed (i.i.d.) real random variables. Typically, one can use the delta method to show that T_n converges in distribution to a normal distribution. The rates of convergence to normality is usually of the order $n^{-1/2}$ and can be described by Berry-Esséen bounds. To get a better approximation than asymptotic normality, one can develop first-order or even higher-order Edgeworth expansions under appropriate conditions.

On the other hand, from a statistical point of view, it is of more interest to study the studentized version of T_n. That is, the statistics under study can be written in the form of T_n/S_n, where T_n as the statistics of interest and S_n^2 as the normalizing variable. For simplicity, we shall refer to T_n/S_n loosely as the *studentized* statistic when S_n is a random variable, and as the *standardized* statistics when $S_n = 1$. There are several reasons to consider such a class of statistics, which can be represented by T_n/S_n. First of all, a great many of commonly seen statistics all belong to this class. These include (standardized or studentized) U-statistics, function of sample means and some L-statistics as considered later in the paper. By applying the Hoeffding's decomposition, we can easily see that this class of statistics is sometimes related to the symmetric statistics considered, for instance, by von Zwet (1984), Lai and Wang (1993), Bentkus, Götze, van Zwet (1997), and Putter and von Zwet (1998). Another reason for considering statistics of type T_n/S_n is to unify the treatment for both standardized and studentized statistics under the same framework.

A quick glance at the literature reveals that the moment conditions in the limit theorems, such as Berry-Esséen bounds and Edgeworth expansions, for the studentized statistics are typically stronger than the corresponding standardized statistics. This may not be surprising since it is usually more difficult to handle studentized statistics than standardized ones. One notable exception is the case of the sample mean, where the same third moment assumption is enough in establishing Berry-Esséen bounds and Edgeworth expansions for both the standardized mean and Student's t-statistic, see Hall (1987) and Bentkus and Götze (1996), for instance. This begs the question whether the same phenomenon is also true for U-statistics or even some more general classes of statistics.

Throughout the paper, we assume that both T_n and S_n are symmetric statistics, i.e., they are invariant under the permutation of $X_1, \cdots, X_n$. As a consequence, T_n/S_n is a symmetric statistic as well. Therefore, one could apply the Hoeffding decomposition to it. There are two ways to do this. The first one is to apply Hoeffding's decomposition directly to the whole statistic T_n/S_n, but this proves to be difficult to carry out in most cases. The second way is to deal with T_n and S_n separately. Assuming that $ET_n = 0$, one can apply Hoeffding's decomposition to T_n, resuling in

$$T_n = \sum_{i=1}^{n} T^{(1)}(X_i) + \sum_{1 \leqslant i < j \leqslant n} T^{(2)}(X_i, X_j) + V_{1n},$$

where $T^{(1)}(X_i) = E(T_n|X_i)$, $T^{(2)}(X_i, X_j) = E(T_n|X_i, X_j) - E(T_n|X_i) - E(T_n|X_j)$, V_{1n} is the remainder term, and the terms on the right hand side of (1.1) are all pairwise uncorrelated. On the other hand, the normalizing variable S_n^2 is typically a symmetric statistic itself, and has the following stochastic expansion

$$S_n^2 = \sigma_n^2 + \sum_{1 \leqslant i < j \leqslant n} s(X_i, X_j) + V_{2n},$$

where $s(x, y)$ is symmetric in x and y, and V_{2n} is the remainder term.

Note that in considering limit theorems for the normalized statistics T_n/σ_n, such as Berry-Esséen bounds and Edgeworth expansions, if one could ignore the remainder term V_{1n}, T_n/σ_n can be approximated by a U-statistics (of order 2). Similarly, when dealing with the studentized statistics T_n/S_n, one might be able to ignore the remainder terms V_{1n} and V_{2n}. Then, applying Taylor expansion to $1/S_n$, one could approximate T_n/S_n by a U-statistics of sufficiently high order. Therefore, the problem reduces to finding limit theorems for U-statistics. Thus, it suffices to consider U-statistics from now on.

The theory of U-statistics goes back to Halmos (1946) and Hoeffding (1948), who introduced the class of U-statistics and proved their central limit theorem. Since then, there have been tremendous advances in their developments. There are several factors that are unique to the class of U-statistics. First of all, U-statistics are generalizations of sums of independent random variables, thus these developments on the theory of U-statistics are strongly influenced by the classical theory on sums of independent random variables. Secondly, the special dependence structure in the U-statistics allow one to employ the martingale properties extensively. Thirdly, many commonly used statistics are or can be approximated by U-statistics, so they are much more applicable than first appears. Finally, many of the techniques developed for U-statistics are also essential in studying even more general classes of statistics, such as the symmetrical statistics.

U-statistics generalize the concept of the sample mean of independent identically distributed (i.i.d.) random variables. The statistical interest in U-statistics stems from the fact that they form a class of unbiased estimators of a certain parameter with minimal variances. Many studies have been devoted to U-statistics, and by now, the properties for U-statistics are pretty much well understood. On the other hand, the properties for studentized U-statistics are, by comparison, only recently attaining a satisfactory degree of completeness. One of the purposes in this paper is to present some of the results for U-statistics, studentized or otherwise, under the same unified framework.

The paper is arranged as follows. Section 2 gives a general framework for our discussion in this paper. Sections 3–5 discuss asymptotic normality, Berry-Esséen bounds, Edgeworth expansions, respectively for both the standardized and studentized U-statistics. In Section 6, we discuss some of nonparametric methods to U-statistics.

2 Basic Framework

Let $X_1, X_2, \cdots$ be i.i.d. random variables with common distribution function $F(x)$. Let $m \geqslant 1$ and let $h : \mathbb{R}^m \to \mathbb{R}$ be a real-valued Borel measurable function, symmetric in its arguments with $Eh(X_1, \cdots, X_m) = \theta$. The U-statistic with kernel h is defined by

$$U_n := U_n(h) = \binom{n}{m}^{-1} \sum_{1 \leqslant i_1 < \cdots < i_m} h(X_{i_1}, \cdots, X_{i_m}).$$

The kernel h is called degenerate with respect to $F(x)$ if $Eh(x, X_2, \cdots, X_n) = 0$.

For simplicity, we shall only discuss U-statistics of degree 2 for estimation of θ with kernel $h(x, y)$, i.e.,

$$U_n = \binom{n}{2}^{-1} \sum_{i<j} h(X_i, X_j).$$

Define

$$g(x) = Eh(X_1, x) - \theta, \qquad \psi(x, y) = h(x, y) - g(x) - g(y) - \theta.$$

The Hoeffding decomposition of U_n in this case is given by

$$\begin{aligned} U_n - \theta &= \frac{2}{n} \sum_{i=1}^{n} g(X_i) + \binom{n}{2}^{-1} \sum_{1 \leqslant i < j \leqslant n} [h(X_i, X_j) - g(X_i) - g(X_j)], \\ &= \frac{2}{n} \sum_{i=1}^{n} g(X_i) + \binom{n}{2}^{-1} \sum_{1 \leqslant i < j \leqslant n} \psi(X_i, X_j), \end{aligned}$$

or equivalently,

$$\begin{aligned} T_n &:= \sqrt{n}(U_n - \theta) \\ &= \frac{2}{\sqrt{n}} \sum_{i=1}^{n} g(X_i) + \frac{2}{\sqrt{n}(n-1)} \sum_{1 \leqslant i < j \leqslant n} \psi(X_i, X_j) \\ &= \frac{2}{\sqrt{n}} \sum_{i=1}^{n} g(X_i) + \frac{2}{\sqrt{n}(n-1)} \sum_{1 \leqslant i < j \leqslant n} \psi(X_i, X_j) \\ &:= T^{(1)} + T^{(2)}. \end{aligned} \tag{2.1}$$

Since all parts of the Hoeffding decomposition are uncorrected, we have

$$\begin{aligned} \sigma^2 &:= Var(T_n) \\ &= Var(T^{(1)}) + Var(T^{(2)}) \\ &= Var(2g(X_1)) + \frac{2}{n-1} Var(\psi(X_1, X_2)) \\ &:= 4\sigma_g^2 + \frac{2}{n-1} \sigma_\psi^2. \end{aligned}$$

We see that $T^{(1)}$ is the linear part of T_n while $T^{(2)}$ is the quadratic part of T_n, and that the first liner term stochastically dominates the second quadratic term. Therefore, we can now define the distribution functions of the standardized and studentized U-statistic respectively by

$$\begin{aligned} G_0(x) &= P\left(\sqrt{n}(U_n - \theta)/\sigma \leqslant x\right), \\ G_1(x) &= P\left(\sqrt{n}(U_n - \theta)/(2\sigma_g) \leqslant x\right), \\ G_2(x) &= P\left(\sqrt{n}(U_n - \theta)/S_n \leqslant x\right), \end{aligned}$$

where S_n^2 is a consistent estimator of $\sigma^2(U_n)$ or σ_g^2, e.g., the following jackknife variance estimator (see Arvesen, 1969)

$$S_n^2 = \frac{4(n-1)}{(n-2)^2} \sum_{i=1}^{n} \left(\frac{1}{n-1} \sum_{\substack{j=1 \\ j \neq i}}^{n} h(X_i, X_j) - U_n \right)^2. \tag{2.2}$$

We shall discuss the asymptotic properties of G_i, $i = 0, 1, 2$ below, including asymptotic normality, Berry-Esséen bounds, Edgeworth expansion.

3 Asymptotic Normality

From the Hoeffding's decomposition (2.1), we see that the normalized U-statistic $T_n = \sqrt{n}(U_n - \theta)$ was decomposed into the linear term $T^{(1)}$ plus some smaller order noise term $T^{(2)}$. So under appropriate conditions, we quickly establish the asymptotic normalities for G_0 and G_1. The result for G_1 follows easily from Slutsky's theorem. These results are summarized below.

Theorem 3.1. *Suppose that $Eh^2(X_1, X_2) < \infty$ and $\sigma_g^2 > 0$, then*

$$\sup_{x \in \mathbb{R}} |G_i(x) - \Phi(x)| \to 0, \qquad i = 0, 1, 2.$$

Asymptotic normalities of $G_1(x)$ and $G_2(x)$ were established by Hoeffding (1948) and Arvesen (1969), respectively. Note that $G_0(x)$ involves the variance of U_n, which requires that $Eh^2(X_1, X_2) < \infty$. If we focus only on $G_1(x)$ and $G_2(x)$, the asymptotic normality will still hold under somewhat weaker moment condition; see, e.g., Koroljuk and Borovskich (1989, Theorem 4.2.4, page 129).

Theorem 3.2. *Suppose that $\sigma_g^2 \in (0, \infty)$ and $E|h(X_1, X_2)|^{4/3} < \infty$, then*

$$\sup_{x \in \mathbb{R}} |G_i(x) - \Phi(x)| \longrightarrow 0, \qquad i = 1, 2.$$

The above weaker moment condition on h can further be weakened, yielding the following result.

Theorem 3.3. *Suppose that* $P(|h(X_1, X_2)| > t) = o(t^{-4/3})$ *as* $t \to \infty$ *and* $\sigma_g^2 > 0$, *then*

$$\sup_{x \in \mathbb{R}} |G_i(x) - \Phi(x)| \longrightarrow 0, \qquad i = 1, 2.$$

To see why the condition on h is indeed weaker than the previous theorem, we note that the condition $P(|h(X_1, X_2)| > t) = o(t^{-4/3})$ as $t \to \infty$ implies that $E|h(X_1, X_2)|^{4/3-\epsilon} < \infty$ for any $\epsilon > 0$, but may not lead to $E|h(X_1, X_2)|^{4/3} < \infty$. The proof of this statement is similar to that of a more general result on Berry-Esséen bounds as given in Jing and Zhou (2006).

4 Berry-Esséen Bounds

In order to describe the convergence rates to normality for G_i's, one needs to study the Berry-Esséen bounds. In recent years, there has been considerable interest in obtaining rates of convergence to normality for standardized and studentized U-statistics.

For standardized U-statistics, for instance, this has been investigated by Grams and Serfling (1973), Bickel (1974) and Chan and Wierman (1977). A sharper Berry-Esséen bound was given by Callaert and Janssen (1978), which is given in Theorem 4.1 below for the case $i = 1$.

For studentized U-statistics, on the other hand, Berry-Esséen bounds were given by Callaert and Veraverbeke (1981) and Helmers (1985) among others. Zhao (1983) sharpened the work of Callaert and Veraverbeke (1981) and obtained the classical rate $O(n^{-1/2})$ under the condition $E|h(X_1, X_2)|^4 < \infty$ and $\sigma_g^2 > 0$. A sharper Berry-Esséen bound was given by Wang, Jing and Zhao (2000), which is given in Theorem 4.1 below for the case $i = 2$.

Theorem 4.1. *Assume that* $E|h(X_1, X_2)|^3 < \infty$ *and* $\sigma_g^2 > 0$, *then for* $n \geqslant 3$,

$$\sup_{x} |G_i(x) - \Phi(x)| \leqslant A\sigma_g^{-3} E|h(X_1, X_2)|^3 n^{-1/2}, \qquad i = 0, 1, 2.$$

Similar versions of Theorem 4.1 have also been given in the literature. We list some of the results below.

Theorem 4.2. *Assume that* $E|h(X_1, X_2)|^{2+\delta} < \infty$ *for* $0 < \delta \leqslant 1$, *and* $\sigma_g^2 > 0$, *then for* $n \geqslant 3$,

$$\sup_x |G_i(x) - \Phi(x)| \leqslant A\sigma_g^{-(2+\delta)} E|h(X_1, X_2)|^{2+\delta} n^{-\delta/2}, \quad i = 0, 1, 2.$$

The result in Theorem 4.2 for $i = 2$ was obtained by Lin (2000) while the case for $i = 1$ can be shown similarly. Clearly, when $\delta = 1$, Theorem 4.2 reduces to Theorem 4.1. Conversely, using truncation techniques, one can use Theorem 4.1 to prove Theorem 4.2. Furthermore, we point out that, when $0 < \delta < 1$, Theorem 4.2 may not be optimal. To see why this is the case, let us take a special case $h(x, y) = (x + y)/2$, in which case Bentkus and Gotze (1996) showed that

$$\sup_x |G_2(x) - \Phi(x)| = o(n^{-\delta/2}).$$

In other words, the Berry-Esséen bound is of size $o(n^{-\delta/2})$ here, which is smaller than that of $O(n^{-\delta/2})$ when $0 < \delta < 1$.

Theorem 4.1 is given under the moment condition imposed on the kernel function h. As in the asymptotic normality, these moment conditions are not the optimal. For standardized U-statistics, however, that the sharpest Berry-Esséen bound of order $O(n^{-1/2})$ comes from Friedrich (1989), and Koroljuk and Borovskich (1984). On the other hand, the best available Berry-Esséen bound of order $O(n^{-1/2})$ for studentized U-statistics so far was obtained by Bentkus, Jing and Zhou (2006). See the next theorem.

Theorem 4.3. *Assume that* $E|g(X_1)|^3 < \infty$, *and* $\sigma_g^2 > 0$, *and* $n \geqslant 3$.

(a) *If* $E|h(X_1, X_2)|^{5/3} < \infty$, *then*

$$\sup_x |G_1(x) - \Phi(x)| \leqslant A \left(\sigma_g^{-5/3} E|h(X_1, X_2)|^{5/3} + \sigma_g^{-3} E|g(X_1)|^3\right) n^{-1/2},$$

(b) *If* $E|h(X_1, X_2)|^{5/3+\varepsilon} < \infty$ *for any arbitrary* $\varepsilon > 0$, *then*

$$\sup_x |G_2(x) - \Phi(x)| \leqslant A \left(\sigma_g^{-(5/3+\varepsilon)} E|h(X_1, X_2)|^{5/3+\varepsilon} + \sigma_g^{-3} E|g(X_1)|^3\right) n^{-1/2}.$$

Indeed, Bentkus, Götze and Zitikis (1994) showed that the moment conditions of Friedrich (1989) are the weakest for standardized U-statistics.

To see why, we note that, first of all, the moment condition, $E|g(X_1)|^3 < \infty$, is optimal. Secondly, Bentkus, Götze and Zitikus (1994) and Bentkus and Götze (1995) showed that the moment condition, $E|h(X, X_1)|^{5/3} < \infty$, is also "*final*" in the sense that, if $E|h(X, X_1)|^{5/3-\epsilon} < \infty$ for some $\epsilon > 0$, an example can be constructed such that

$$\sup_x |G_n(x) - \Phi(x)| \geqslant c(\epsilon)n^{-1/2+\epsilon}$$

holds with a constant $c(\epsilon) > 0$. Therefore, the *moment* conditions in Theorem 4.3(a) are optimal.

We conjecture that ε in (b) of Theorem 4.3 can be removed, which would give us a unified Berry-Esséen bounds for both standardized and studentized U-statistics. Efforts are still being made to prove the conjecture.

Before we finish the discussion of Berry-Esséen bounds, we mention the work of Alberink and Bentkus (2001), who gave one even more general forms of Berry-Esséen bound for the standardized U-statistics, which also has a very nice interpretation. First we state the result.

Theorem 4.4. *Assume that* $E|g(X_1)|^3 < \infty$, $E|h(X_1, X_2)|^{5/3} < \infty$ *and* $\sigma_g^2 > 0$, *and* $n \geqslant 3$, *then*

$$\sup_x |G_1(x) - \Phi(x)| \leqslant A\left(\sigma_g^{-3}E|g(X_1)|^3 n^{-1/2} + \sigma_g^2 E|h(X_1, X_2)|^2 n^{-1}\right).$$

It can be easily shown that this bound does generalizes the previous bounds for the standardized U-statistics. Now rewrite the Hoeffding decomposition (2.1) as

$$\begin{aligned} U_n &= \frac{2}{n}\sum_{i=1}^n g(X_i) + \binom{n}{2}^{-1} \sum_{1\leqslant i<j\leqslant n} [h(X_i, X_j) - g(X_i) - g(X_j)] \\ &:= L + Q \end{aligned}$$

where L and Q stand for the linear and quadratic terms, respectively. As mentioned before, the linear term L is the dominant term and the quadratic term Q may be regarded as the "noise" term. Then, it can be easily shown, as given in Alberink and Bentkus (2001) that the inequality in Theorem 4.4 may be rewritten as

$$\sup_x |G_1(x) - \Phi(x)| \leqslant A\left(\sigma_g^{-3}E|g(X_1)|^3 n^{-1/2} + Var(Q)\right).$$

In other words, the Berry-Esséen bounds for the standardized U-statistics is simply the usual Berry-Esséen bounds for the linear term in U-statistics plus the variance of the noise term.

Efforts have been spent to prove similar results to Theorem 4.4 for the studentized U-statistics. However, it remains an open question.

5 Edgeworth Expansions

Berry-Esséen bounds obtained in the last section shows that the errors in normal approximations of the U-statistics are of size $O(n^{-1/2})$. This indicates that it might be possible to improve the normal approximations by including an extra term of order $n^{-1/2}$. This is achieved by the Edgeworth expansions.

5.1 First-order Edgeworth Expansions

We focus on one-term Edgeworth expansions for both standardized and studentized U-statistics in this section. First we define

$$E_1(x) = \Phi(x) - \frac{\phi(x)}{6\sqrt{n}\sigma^3}(x^2-1)\left\{Eg^3(X) + 3Eg(X)g(X_1)h(X,X_1)\right\},$$

$$E_2(x) = \Phi(x) + \frac{\phi(x)}{6\sqrt{n}\sigma^3}\left\{(2x^2+1)Eg^3(X_1) \right.$$
$$\left. +3(x^2+1)Eg(X_1)g(X_2)h(X_1,X_2)\right\},$$

where Φ is a standard normal distribution function.

The development of Edgeworth expansions closely parallels that of Berry-Esséen bounds. The corresponding result to Theorem 4.1 is as follows.

Theorem 5.1. *Assume that*

(i) $E|h(X_1,X_2)|^3 < \infty$ *and* $\sigma_g^2 > 0$;
(ii) *the d.f. of* $g(X_1)$ *is nonlattice.*

Then, we have

$$\sup_x |G_i(x) - E_i(x)| = o(n^{-1/2}), \qquad i = 1, 2.$$

The result for $i = 1$ was established by Bickel, Gotze and van Zwet (1986) while the result for $i = 2$ was obtained by Helmers (1991) and later improved by Jing and Wang (2003). As in the Berry-Esséen bounds, the moment condition in Theorem 5.1 can be further relaxed.

Theorem 5.2. *Assume that* $E|g(X_1)|^3 < \infty$, $\sigma_g^2 > 0$, *and the d.f. of* $g(X_1)$ *is nonlattice.*

(a) *If* $E|h(X_1,X_2)|^{5/3} < \infty$, *then*

$$\sup_x |G_1(x) - E_1(x)| = o(n^{-1/2}).$$

(b) *If* $E|h(X_1, X_2)|^{5/3+\varepsilon} < \infty$ *for any arbitrary* $\varepsilon > 0$, *then*

$$\sup_x |G_2(x) - E_2(x)| = o(n^{-1/2}).$$

The result (a) in Theorem 5.2 provides the best result on one-term Edgeworth expansions for the standardized U-statistics under the optimal moment condition. Its proof was given by Jing and Wang (2003). The result (b) cantains ε because we can not get rid of ε in part (b) of Theorem 4.3. Once we can prove part (b) of Theorem 4.3, we could eliminate ε in (b) of Theorem 5.2, which would give us a unified one-term Edgeworth expansions for both standardized and studentized U-statistics.

The optimal *moment* condition in part (a) of Theorem 5.2 can be further weakened to the more natural and more precise tail *probability* condition: $P(|h(X_1, X_2)| > t) = o(t^{-5/3})$ as $t \to \infty$. The precise statement is given below. See Jing and Zhou (2005) for further details.

Theorem 5.3. *Assume that*

(i) $\sigma^2 > 0$, $E|g(X_1)|^3 < \infty$, *and* $P(|h(X_1, X_2)| > t) = o(t^{-5/3})$ *as* $t \to \infty$;

(ii) *the d.f. of* $g(X_1)$ *is nonlattice.*

Then, we have

$$\sup_x |G_1(x) - E_1(x)| = o(n^{-1/2}), \qquad \text{as } n \to \infty.$$

5.2 Higher-order Edgeworth Expansions

Under further conditions, one can establish high-order Edgeworth expansions. For simplicity, we are only concerned with second-order expansions with remainder term of size $o(n^{-1})$. To be specific, let

$$\kappa_3 = \sigma_g^{-3}\{Eg^3(X_1) + 3Eg(X_1)g(X_2)h(X_1, X_2)\},$$
$$\kappa_4 = \sigma_g^{-3}\{Eg^4(X_1) - 3\sigma_g^4 + 12Eg^2(X_1)g(X_2)\psi(X_1, X_2) + 12Eg(X_1)g(X_2)\psi(X_1, X_3)\psi(X_2, X_3)\}.$$

Define

$$E_2^{(2)}(x) = \Phi(x) - \phi(x)\left\{\frac{\kappa_3}{6\sqrt{n}}(x^2 - 1) + \frac{\kappa_4}{24n}(x^3 - 3x) + \frac{\kappa_3^2}{72n}(x^5 - 10x^3 + 15x)\right\}.$$

The following result was obtained in Bickel *et al.* (1986):

Theorem 5.4. *Suppose that there exist a number $r > 2$ and integer k such that $(r-2)(k-4) > 8$ and that the following assumptions are satisfied:*

(a) $E|\psi(X_1, X_2)|^r < \infty$.

(b) $E|g(X_1)|^4 < \infty$.

(c) $\limsup_{|t|\to\infty} |Ee^{itg(X_1)}| < 1$.

(d) *ψ possesses k nonzero eigenvalues with respect to the underlying probability measure.*

Then we have

$$\sup_x |G_2(x) - E_2^{(2)}(x)| = o(n^{-1/2}).$$

6 Resampling methods for U-statistics

In this section we shall briefly illustrate how to use nonparametric method in the statistical inference for the parameter $\theta = Eh(X_1, X_2)$. If one knows the distribution function

$$G_2(x) = P\left(\sqrt{n}(U_n - \theta)/S_n \leqslant x\right),$$

then an exact $(1-\alpha)$ level confidence interval for θ would be

$$\left(U_n - \frac{S_n}{\sqrt{n}} G_2^{-1}(1-\alpha),\ U_n - \frac{S_n}{\sqrt{n}} G_2^{-1}(\alpha)\right),$$

However, in practice, one typically does not know G_2, therefore, approximation becomes necessary. Below we shall introduce several commonly used approaches in this context: asymptotic, bootstrap, and empirical likelihood methods.

6.1 Asymptotic methods

The first approach to be introduced is the asymptotic normality method. From Theorem 3.2, one can approximate G_2 by the standard normal distribution Φ, resulting in an approximate $(1-\alpha)$ level confidence interval for θ as follows:

$$\left(U_n - \frac{S_n}{\sqrt{n}} \Phi^{-1}(1-\alpha),\ U_n - \frac{S_n}{\sqrt{n}} \Phi^{-1}(\alpha)\right),$$

where S_n is the jackknife variance estimate given in (2.2).

A more accurate approximation than the asymptotic normality can be given by the empirical Edgeworth expansion (EEE), obtained by the

Edgeworth expansion in Theorem 5.1 by replacing the unknown population coefficients by their empirical counterparts. This results in another approximate $(1-\alpha)$ confidence interval for θ:

$$\left(U_n - \frac{S_n}{\sqrt{n}}\hat{E}_2^{-1}(1-\alpha),\ U_n - \frac{S_n}{\sqrt{n}}\hat{E}_2^{-1}(\alpha)\right),$$

where

$$\hat{E}_2(x) = \Phi(x) + \frac{\phi(x)}{6\sqrt{n}}\left\{(2x^2+1)a_n + 3(x^2+1)b_n\right\},$$

with a_n and b_n are empirical estimates of $Eg^3(X_1)/\sigma_g^3$ and $Eg(X_1)g(X_2)$ $h(X_1,X_2)/\sigma_g^3$. To be more precise, denote $h_{ij} = h(X_i, X_j)$, we have

$$a_n = \frac{n^{-1}\sum_{i=1}^n \left(n^{-1}\sum_{j=1}^n h_{ij} - U_n\right)^3}{\left(n^{-1}\sum_{i=1}^n \left(n^{-1}\sum_{j=1}^n h_{ij} - U_n\right)^2\right)^{3/2}}$$

and

$$b_n = \frac{n^{-2}\sum_{i=1}^n \sum_{j=1}^n \left(n^{-1}\sum_{k=1}^n h_{ik} - U_n\right)\left(n^{-1}\sum_{l=1}^n h_{jl} - U_n\right) h_{ij}}{\left(n^{-1}\sum_{i=1}^n \left(n^{-1}\sum_{j=1}^n h_{ij} - U_n\right)^2\right)^{3/2}}.$$

Although, in theory, under certain conditions, we have

$$\sup_x |G_2(x) - \hat{E}_2(x)| = o(n^{-1/2}).$$

which gives the same level of accuracy as $E_2(x)$. However, the performance of the EEE may not be as accurate as one hopes, particularly when the sample size is not large. This might be due to high estimation error of a_n and b_n when the sample size is not large. In order to avoid estimating a_n and b_n directly, we shall introduce below two nonparametric methods for the inference of θ: the bootstrap method and empirical likelihood method.

6.2 Bootstrap method to U-statistics

The bootstrap was first introduced by Efron in 1989. Since then it has become one of the useful techniques in statisticians' toolbox. For a better review, we refer the reader to some of the monographs on bootstrap, including Efron and Tibshirani (1993), Hall (1992), Shao and Tu (1995), and Davison and Hinkley (1997).

Here we consider employing the bootstrap Studentized U-statistic. Let $X_1^*, \cdots, X_n^*$ be a bootstrap sample, taken randomly from the original data, $X_1, \cdots, X_n$, with replacement. Therefore, the corresponding bootstrap statistic to $n^{1/2}(U_n - \theta)/S_n$ is given by

$$n^{1/2}(U_n^* - U_n)/S_n^*,$$

where U_n^* and S_n^* are obtained from U_n and S_n simply by replacing the X_i's by the X_i's. Then the bootstrap approximation to $G_2(x) = P\left(\sqrt{n}(U_n - \theta)/S_n \leqslant x\right)$ is given by

$$G_2^*(x) = P^*\left(\sqrt{n}(U_n^* - U_n)/S_n^* \leqslant x\right),$$

where P^* is the conditional probability measure given the original data $X_1, \cdots, X_n$. Under the usual condition to guarantee the asymptotic normality of G_2, plus the further condition $Eh(X_1, X_1)^2 < \infty$, Athreya *el al.* (1984) proved the strong consistency of $G_2^*(x)$ to $G_2(x)$, i.e., $\sup_x |G_2^*(x) - G_2(x)| = o(1)$ as $n \to \infty$. Helmers (1991) proves its $\sqrt{n}$-consistency under some further conditions; see also Putter and van Zwet (1998).

Theorem 6.1. *Suppose that following conditions hold,*

(1) $E|g(X_1)|^{3+\epsilon} < \infty$ *for some* $\epsilon > 0$,
(2) $E|h(X_1, X_2)|^{2+\epsilon} < \infty$ *for some* $\epsilon > 0$,
(3) $E|h(X_1, X_1)|^{1+\epsilon/2} < \infty$,
(4) *the d.f. of* $g(X_1)$ *is nonlattice.*

Then, as $n \to \infty$, *we have*

$$\sup_x |G_2^*(x) - G_2(x)| = o_p(n^{-1/2}).$$

The above theorem shows that the bootstrap approximation G_2^* provides a better approximation to G_2 than the normal approximation Φ does. As a consequence, a bootstrap-based approximate $(1-\alpha)$ confidence interval for θ given by

$$\left(U_n - \frac{S_n}{\sqrt{n}} {G_2^*}^{-1}(1-\alpha),\ U_n - \frac{S_n}{\sqrt{n}} \hat{G_2^*}^{-1}(\alpha)\right),$$

should provide better performances than the normal approximation method in terms of coverage probabilities.

6.3 Empirical likelihood method to U-statistics

The empirical likelihood method was first introduced by Owen (1988, 1990) for constructing confidence regions. Hall and LaScala (1990) has summarized its advantages over the bootstrap: the empirical likelihood regions are shaped automatically by the sample, Bartlett correctable, range preserving and transformation respecting. For these reasons, the empirical likelihood has found many applications such as in smooth functions of means (DiCiccio, Hall and Romano (1991)), in nonparametric density and regression function estimation (Owen (1991), Chen (1996), Chen and Qin (2000)), in quantile related estimation (Chen and Hall (1993)) and so on. For a more thorough review of the empirical likelihood method and its applications, the reader is referred to the monograph by Owen (2001).

Here we are interested in applying the empirical likelihood method to U-statistics. Let $X, X_1, \cdots, X_n, n \geqslant 2$, be independent and identically distributed (i.i.d.) random variables with common distribution function $F(x)$. Recall that a U-statistic of degree $m \geqslant 2$ with a symmetric kernel h is defined to be

$$U_n = \binom{n}{m}^{-1} \sum_{1 \leqslant i_1 < \cdots < i_m \leqslant n} h(X_{i_1}, \cdots, X_{i_m}),$$

where $\theta = Eh(X_1, \cdots, X_m)$ is a parameter of interest.

The straightforward application of Owen's empirical likelihood in this context can be described as follows. Denote F_q to be the empirical distribution function which assigns probability q_i to the observation X_i. Then, the empirical likelihood, evaluated at the true parameter value θ, can be defined by

$$\widetilde{L}(\theta) = \max\left\{\prod_{i=1}^{n} q_i : \sum_{i=1}^{n} q_i = 1,\ \widetilde{\theta}(F_q) = \theta\right\},$$

where

$$\widetilde{\theta}(F_q) = \binom{n}{m}^{-1} \sum_{1 \leqslant i_1 < \cdots < i_m \leqslant n} n^m q_{i_1} \cdots q_{i_m} h(X_{i_1}, \cdots, X_{i_m}).$$

Note that, subject to the sole constraint $\sum_{i=1}^n q_i = 1$, $\prod_{i=1}^n q_i$ attains its maximum n^{-n} at $q_i = n^{-1}$. Then, the empirical likelihood ratio at θ is given by

$$\widetilde{R}(\theta) = \frac{\widetilde{L}(\theta)}{n^{-n}} = \max\left\{\prod_{i=1}^{n} (nq_i) : \sum_{i=1}^{n} q_i = 1,\ \widetilde{\theta}(F_q) = \theta\right\}.$$

As mentioned in Wood *et al.* (1996), Wilks's theorem holds under mild conditions in this case, i.e.,

$$-2\log \widetilde{R}(\theta) \longrightarrow_d \chi_1^2,$$

where $\longrightarrow_d$ means convergence in distribution as $n \to \infty$, and χ_1^2 denotes the chi square distribution with one degree of freedom. This can be used to construct confidence intervals for the parameter θ. We shall refer to this procedure as Owen's *direct or "exact" empirical likelihood method* to U-statistics.

The major drawback of Owen's direct empirical likelihood method is its computational difficulty due to the presence of nonlinear constraints in the underlying optimization problem. Wood *et al.* (1996) proposed a so-called *sequential linearization method* for empirical likelihood methods with nonlinear constraints, and applied it to U-statistics. They found in their simulation studies that a single iteration of the linearization procedure may not be enough to achieve reliable coverage probabilities, and suggested to employ multiple (three to ten) iterations of the linearization procedure or bootstrap calibration in practice in order to improve coverage probabilities.

A different empirical likelihood method was proposed by Jing, Yuan and Zhou (2006), which is applicable to a very general class of statistics, including U-statistics. To illustrate the idea, it is best to describe the approach in the context of U-statistics. The key idea of our method is to turn the U-statistic into a sample mean based on Tukey's pseudo-values (Quenouille 1956, Tukey 1958), and then simply apply Owen's empirical likelihood for the mean functional. To be specific, define Tukey's pseudo-values defined by

$$\widehat{V}_i = nU_n - (n-1)U_{n-1}^{(-i)},$$

where $U_{n-1}^{(-i)}$ is the statistic U_{n-1} computed on the sample of $n-1$ variables formed from the original data set by deleting the ith observation X_i. The jackknifed version $\widehat{U}_n$, which coincides exactly with U_n, is simply the average of the pseudo-values:

$$U_n = \widehat{U}_n = \frac{1}{n}\sum_{i=1}^{n} \widehat{V}_i. \tag{6.1}$$

The relation (6.1) can be either checked directly or found in Theorem 3 of Lee (1990, page 167). For the special case of the mean where $h(x,y) = (x+y)/2$, we have $\widehat{V}_i = X_i$ $(1 \leqslant i \leqslant n)$, which are independent random variables. Generally speaking, $\widehat{V}_i$'s are typically dependent random variables, although they can be shown to be asymptotically independent in many cases, see Shi (1984). This motivates us to employ

Owen's empirical likelihood method directly to these pseudo-values $\widehat{V}_i$. The resulting method is refereed to as the *pseudo-empirical likelihood method*. To do this, let $p = (p_1, \cdots, p_n)$ be a probability vector, i.e., $\sum_{i=1}^{n} p_i = 1$ and $p_i \geqslant 0$ for $1 \leqslant i \leqslant n$. Let G_p be the distribution function which assigns probability p_i at the ith pseudo-value $\widehat{V}_i$, and consider the mean functional $\theta(G_p) = \sum_{i=1}^{n} p_i \widehat{V}_i$. Then, the empirical likelihood, evaluated at θ, is given by

$$L(\theta) = \max \left\{ \prod_{i=1}^{n} p_i : \sum_{i=1}^{n} p_i = 1, \ \theta(G_p) = \theta \right\}.$$

Note that $\prod_{i=1}^{n} p_i$, subject to $\sum_{i=1}^{n} p_i = 1$, attains its maximum n^{-n} at $p_i = n^{-1}$. So we define the pseudo-empirical likelihood ratio at θ by

$$R(\theta) = \frac{L(\theta)}{n^{-n}} = \max \left\{ \prod_{i=1}^{n} (np_i) : \sum_{i=1}^{n} p_i = 1, \ \theta(G_p) = \theta \right\}. \tag{6.2}$$

Using Lagrange multipliers, when $\min\limits_{1\leqslant i\leqslant n} \widehat{V}_i < \theta < \max\limits_{1\leqslant i\leqslant n} \widehat{V}_i$, we have

$$p_i = \frac{1}{n} \frac{1}{1 + \lambda(\widehat{V}_i - \theta)},$$

where λ satisfies

$$g(\lambda) \equiv \frac{1}{n} \sum_{i=1}^{n} \frac{\widehat{V}_i - \theta}{1 + \lambda(\widehat{V}_i - \theta)} = 0.$$

After plugging the p_i's back into (6.2) and taking the logarithm of $R(\theta)$, we get the nonparametric pseudo-log-likelihood ratio

$$\log R(\theta) = -\sum_{i=1}^{n} \log\{1 + \lambda(\widehat{V}_i - \theta)\}.$$

The next theorem shows that Wilks's theorem holds here under a mild condition.

Theorem 6.2. *Assume that* $Eh^2(X_1, \cdots, X_m) < \infty$ *and* $\sigma_g^2 = Var(g(X)) > 0$, *where* $g(x) = Eh(x, X_1, \cdots, X_{m-1}) - \theta$. *Then, we have*

$$-2 \log R(\theta) \longrightarrow_d \chi_1^2.$$

Based on this theorem, one can construct an approximate $(1 - \alpha)$ level confidence interval for θ by

$$\Re_c = \{\theta : -2 \log R(\theta) \leqslant c\},$$

where c is chosen to satisfy $P(\chi_1^2 \leqslant c) = 1 - \alpha$. From Theorem 6.2, we have

$$\lim_{n\to\infty} P\{\theta \in \Re_c\} = P(\chi_1^2 \leqslant c) = 1 - \alpha.$$

In other words, the interval $\Re_c$ gives asymptotically correct coverage probability.

The key idea of combining Owen's empirical likelihood for a mean functional with jackknife pseudo-values could work more generally than just for U-statistics. To describe this, let $Z_1, \cdots, Z_n$ be independent (but not necessarily identically distributed r.v.'s. Let

$$T_n = T(Z_1, \cdots, Z_n)$$

be a consistent estimator of the parameter θ, where T may not be symmetric in its arguments. Define the pseudo-values by

$$\widehat{V}_i = nT_n - (n-1)T_{n-1}^{(-i)},$$

where $T_{n-1}^{(-i)}$ is the statistic T_{n-1} computed on the sample of $n-1$ variables formed from the original data set by deleting the ith data value. Observe that, in general, we may not necessarily have $T_n = n^{-1}\sum_{i=1}^n \widehat{V}_i$ unless T_n is a U-statistic. However, we can then apply the pseudo-empirical likelihood method of Section 2.1 to $Z_1, \cdots, Z_n$. Of course, one needs to check that Wilks's theorem holds or not in each individual case. Such an example will be given later in Section 3.2.

Another most attractive feature of the pseudo-empirical likelihood method is its **simplicity**. The simplicity of our approach is made particularly obvious when it is compared with the direct application of Owen's empirical likelihood to U-statistics, and the sequential linearization method of Wood *et al.* (1996).

7 Conclusions

We list some of the known results for both standardized and studentized U-statistics, in particular, the asymptotic normality, Berry-Esséen bounds, one-term Edgeworth expansions. The central theme is unification. Other topics, such as non-uniform Berry-Esséen bounds, large-deviations, etc. are excluded since there is not much unified treatment for both types of statistics. Lack of the coverage in these topics indicates that there are still many interesting open problems in the area. Typically, one starts with an easier case of the standardized U-statistics, and then investigates whether the results obtained will continue to hold for the studentized U-statistics. In light of the results for the sample mean, it is interesting to ask whether one could obtain large deviation results for studentized U-statistics under much weaker conditions than their counterparts, the standardized U-statistics. There are many other interesting problems to explore. There will be new and exciting developments to be made in the years to come.

References

[1] Alberink, I. B. and Bentkus, V. (2001). Berry-Esséen bounds for von-Mises and U-statistics. *Lith. Math. J.*, **41**, 1–16.

[2] Arvesen, J. N. (1969). Jackknifing U-statistics. *Ann. Math. Stat.*, **6**, 417–421.

[3] Athreya, K. B., Ghosh, M., Low, L. Y. and Sen, P. K. (1984). Laws of large numbers for bootstrapped U-statistics. *J. Stat. Plann. Infer.*, **9**, 185–194.

[4] Bentkus, V. and Götze, F. (1995). On minimal moment assumptions in Berry-Esséen theorems for U-statistics. *Theor. Probab. Appl.*, **40**, 430–445.

[5] Bentkus, V. and Götze, F. (1996). The Berry-Esséen bound for Student's statistic. *Ann. Probab.* **24** 491–501.

[6] Bentkus, V., Götze, F. and van Zwet, W. R. (1997). An Edgeworth expansion for symmetric statistics. *Ann. Stat.* **25**, 851–896.

[7] Bentkus, V., Götze, F. and Zitikis, M. (1994). Lower estimates of the convergence rate for U-statistics. *Ann. Probab.*, **22**, 1707–1714.

[8] Bentkus, V., Jing, B.-Y. and Zhou, W. (2006). Berry-Esséen bounds for studentized U-statistics. Manuscript.

[9] Bickel, P. J. (1974). Edgeworth expansions in nonparametric statistics. *Ann. Stat.*, **2**, 1–20.

[10] Bickel, P. J., Gotze, F. and van Zwet, W. R. (1986). The Edgeworth expansion for U-statistics of degree 2. *Ann. Stat.*, **14**, 1463–1484.

[11] Chan, Y. K. and Wierman, J. (1977). On the Berry-Esséen theorem for U-statistics. *Ann. Probab.*, **5**, 136–139.

[12] Callaert, H. and Janssen, P. (1978). The Berry-Esséen theorem for U-statistics. *Ann. Stat.*, **6**, 417–421.

[13] Callaert, H. and Veraverbeke, N. (1981). The order of the normal approximation for a studentized statistic. *Ann. Stat.*, **9**, 194–200.

[14] Chen, S. X. (1996). Empirical likelihood confidence intervals for nonparametric density estimation. *Biometrika*, **83**, 329–341.

[15] Chen, S. and Hall, P. (1993). Smoothed empirical likelihood confidence intervals for quantiles. *Ann. Stat.*, **21**, 1166–1181.

[16] Chen, S. X. and Qin, Y. S. (2000). Empirical likelihood confidence intervals for local linear smoothers. *Biometrika*, **87**, 946–953.

[17] Davison, A. C. and Hinkley, D. V. (1997). *Bootstrap Methods and their Applications.* Cambridge University Press.

[18] DiCiccio, T. S., Hall, P. and Romano, J. (1991). Empirical likelihood is Bartlett correctable. *Ann. Stat.*, **19**, 1053–1061.

[19] Efron, B. and Tibshirani, R. J. (1993). *An Introduction to the Bootstrap.* Chapman and Hall, New York.

[20] Friedrich, K. O. (1989). A Berry-Esséen bound for functions of independent random variables. *Ann. Stat.* **17**, 170–183.

[21] Grams, W. F. and Serfling, R.J. (1973). Convergence rates for U-statistics and related statistics. *Ann. Stat.*, **1**, 153–160.

[22] Hall, P. (1987). Edgeworth expansion for student's t-statistic under minimal conditions. *Ann. Probab.* **15**, 920–931.

[23] Hall, P. (1988). On the effect of random norming on the rate of convergence in the central limit theorem. *Ann. Probab.* **16**, 1265–1280.

[24] Hall, P. (1992). *The Bootstrap and Edgeworth Expansion.* New York, Springer.

[25] Hall, P. and LaScala, B. (1990). Methodology and Algorithms of empirical likelihood. *International Statistical Review*, **58**, 109–127.

[26] Halmos, P. (1946). The theory of unbiased estimation. *Ann. Math. Stat.*, **1**, 34–43.

[27] Helmers, R. (1985). The Berry-Esséen bound for studentized U-statistics. *Canad. J. Stat.*, **13**, 79–82.

[28] Hoeffding, W. (1948). A class of statistics with asymptotically normal distribution. *Ann. Math. Stat.*, **19**, 293–325.

[29] Helmers, R. (1991). On the Edgeworth expansion and the bootstrap approximation for a studentized U-statistic. *Ann. Stat.*, **19**, 470–484.

[30] Jing, B. Y. and Wang, Q. Y. (2003). Edgeworth expansion for U-statistics under minimal conditions. *Ann. Stat.*, **31**, 1376–1391.

[31] Jing, B.-Y. and Zhou, W. (2005). A note on Edgeworth expansions for U-statistics under minimal conditions. *Liet. matem. rink.*, **45**, 435–440.

[32] Koroljuk, V. S. and Borovskich, Yu. V. (1984). *Asymptotic Analysis of Distributions of Statistics.* Naukova Dumka, Kiev.

[33] Koroljuk, V. S. and Borovskich, Yu. V. (1989). *Theory of U-Statistics.* Kluwer Academic Publishers.

[34] Lai, T. L. and Wang, J. Q. (1993). Edgeworth Expansions for symmetric statistics with applications to bootstrap methods. *Statistica Sinica*, **3**, 517–542.

[35] Lee, A. J. (1990). *U-statistics, Theory and Practice.* New York: Marcel Dekker, Inc.

[36] Lin, Z. Y. (2000). The Berry-Esséen bound for Studentized U-statistics. *Sci. China Ser. A* **43**, 1154–1163.

[37] Owen, A. B. (1988). Empirical likelihood ratio confidence intervals for a single functional. *Biometrika*, **75**, 237–249.

[38] Owen, A. B. (1990). Empirical likelihood ratio confidence regions. *Ann. Stat.*, **18**, 90–120.

[39] Owen, A. B. (2001). *Empirical Likelihood.* Chapman and Hall, London.

[40] Putter, H. and van Zwet, W. R. (1998). Empirical Edgeworth expansions for symmetric statistics. *Ann. Stat.* **26**, 1540–1569.

[41] Shao, J. and Tu, D. (1995). *The Jackknife and Bootstrap.* New York, Springer.

[42] Shi, X. (1984). The approximate independence of jackknife pseudo-values and the bootstrap methods. *J. Wuhan Inst. Hydra. Elect. Engin.*, **2**, 83–90.

[43] Wang, Q. Y., Jing, B.-Y. and Zhao, L. C. (2000). The Berry-Esséen bound for studentized statistics. *Annals of Probability* **28**, 511–535.

[44] Wood, A. T. A., Do, K. A., and Broom, N. M. (1996). Sequential linearization of empirical likelihood constraints with application to U-statistics. *J. Comput. Graph. Stat.*, **5**, 365–385.

[45] van Zwet W. R. (1984). A Berry-Esséen Bound for symmetric statistics. *Z-wahrsch. verw. Gebiete*, **66**, 425–440.

[46] Zhao, L. (1983). The rate of the normal approximation for a studentized U-statistic. *Science Exploration*, **3**, 45–52.

Part II: Statistics and Applications

On the Inverse Problem for the t-Statistic

Zhenhai Yang*
College of Applied Sciences, Beijing University of Technology
Beijing, 10002, China
E-mail: zhyang@bjpu.edu.cn

Kai-Tai Fang
BNU-HKBU United International College
Zhuhai Campus of Beijing Normal University
Jinfeng Road, Zhuhai 519085, China
E-mail: ktfang@uic.edu.hk

Samuel Kotz
Department of Engineering Management
George Washington University, USA
E-mail: kotz@seas.gwu.edu

Abstract

If a random sample of size n from a normal population with mean zero, its t-statistic follows the t-distribution with $n-1$ degrees of freedom. The inverse problem of the above fact can characterize normality, i.e., a characterization problem of finding all the distribution for which the t-statistic follows the $t_{(n-1)}$-distribution. This paper provides rather weak conditions on a inverse problem for the t-statistic.The results sharpen the earlier conclusions of Mauldon (1956) and the more recent advances due to Bondesson (1983).

1 Introduction

Since W. S. Gosset ("Student") in his famous paper in 1908 proposed the t-statistic and the t-distribution, there were a lot of studies on the t-distribution and t-statistic. Let $X_1, \cdots, X_n$ be a random sample from a cumulative distribution function (cdf) $F(\cdot)$ with mean zero and let

$$T_n = T(\mathbf{X}_n) = \frac{\sqrt{n}\bar{X}_n}{s_n} \tag{1.1}$$

*The author is supported by Chinese Naturel Science Foundation with contract 10371005 and Beijing Naturel Science Foundation.

be its t-statistic, where $\mathbf{X}_n = (X_1, \cdots, X_n)^\top, \bar{X}_n = \sum_{i=1}^n X_i/n$, and $s_n^2 = \sum_{i=1}^{n}(X_i - \bar{X}_n)^2/(n-1)$. It is well-known that if a random sample comes from a normal distribution, $N(0, \sigma^2)$, then T_n has the Student t-distribution with $n-1$ degrees of freedom. In our recent paper, Yang, Fang and Kotz (2006), we show many interesting examples where $T(\mathbf{X}_n)$ follows the t-distribution with $n-1$ degrees of freedom, but $X_1, \cdots, X_n$ are either dependent or have different marginal distributions.

Mauldon (1956) raised the question for which cdf's the t-statistic, defined by (1.1), is t-distributed with $n-1$ degrees of freedom. This is a particular case of the characterization problem of finding all the cdf's for which a certain statistic possesses the required property. Many authors, among them Kagan, Linnik and Rao (1973) and Bondesson (1974, 1983), have tackled Mauldon's problem.

Under the assumption that all the moments exist, Bondesson (1983) pointed out that if the t-statistic $T(\mathbf{X}_n)$ of the sample follows the t-distribution with $n-1$ degrees of freedom, then the sample comes from a normal distribution. The purpose of this paper is to solve the inverse problem under a rather weaker condition which does not requires existence of any moment. Our main result in this paper is presented in the following theorem.

Theorem 1.1. *Let $\mathbf{X}_n^\top = (X_1, \cdots, X_n)$ be a random sample from the population with probability density $f(\cdot)$, where $f(\cdot)$ has a continuous derivative, is decreasing for $x > 0$ and increasing for $x < 0$ and $EX_1 = 0$. If the t-statistic $T(\mathbf{X}_n)$ follows the t-distribution with $n-1$ degrees of freedom, then the sample comes from a normal distribution $N(0, \sigma^2)$ for some $\sigma > 0$.*

In order to prove the theorem, we need several lemmas in our previous paper [8] that will be presented in Section 2. Two new concepts, *solid set* and *bound function*, will be defined in Section 3. In Section 4 we shall present the main theorem in a parametric form. We then provide a lemma on independent variables in Section 5.

Throughout the paper the norm $\|\cdot\|$ is the L_2-norm, $L_k(A)$ is the Lebesgue measure of A in $\mathbb{R}^k$ and for a $(k-1)$-dimension manifold in $\mathbb{R}^k$, $\bar{L}_k(B)$ denotes the Lebesgue measure of B on this manifold. The uniform distribution on A is denoted by $U(A)$.

2 Preparation Knowledge

For the need of our approach to the inverse problem for t-statistic, we adopt some notation and results in the paper [8]. Consider a class of random n-vectors, $\mathbf{X}_n^\top = (X_1, \cdots, X_n)$, designated by $\mathcal{F}$, where $\mathbf{X}_n$

has a continuous density function $f(\mathbf{x}_n)$, $\mathbf{x}_n \in \mathbb{R}^n$, and satisfying the following conditions:

1. $f_0 = f(\mathbf{0}_n) = \sup\{f(\mathbf{x}_n) : \mathbf{x}_n \in \mathbb{R}^n\} < \infty$, $\mathbf{0}_n^\top = (0, \cdots, 0) \in \mathbb{R}^d$;
2. $E(\sum\limits_{k=1}^{n} X_k) = 0$.

Let $f(\mathbf{x}_n)$, $\mathbf{x}_n \in \mathbb{R}^n$ be a density function defined on $\mathbb{R}^n$. Set

$$D_{n,[f]} = \{\mathbf{x}_{n+1} = (\mathbf{x}_n, x_{n+1}) : 0 < x_{n+1} \leqslant f(\mathbf{x}_n)\},$$

$$D_{n,[f]}(v) = \{\mathbf{x}_n : f(\mathbf{x}_n) \geqslant v\}, 0 < v \leqslant f_0 = \sup\{f(\mathbf{x}_n), \mathbf{x}_n \in \mathbb{R}^n\}. \tag{2.1}$$

If $f \in \mathcal{F}$, then we have $\mathbf{0}_n \in D_{n,[f]}(v)$, $\forall\ v \in (0, f_0]$. The following lemmas in [8] are based on the theory in Fang, Kotz and Ng (1990) and Troutt, Pang and Hou (2004) as well as Fang, Yang and Kotz (2001) and Yang and Kotz (2003).

Lemma 2.1. *Let* $(X_1, \cdots, X_n, X_{n+1}) = (\mathbf{X}_n^\top, X_{n+1})$ *be a random* $(n+1)$*-vector that is uniformly distributed on* $D_{n,[f]}$. *Then we have*

1). *The density function of* X_{n+1} *is*

$$L_n(D_{n,[f]}(v)),\ \ 0 < v \leqslant f_0 = \sup\{f(\mathbf{x}) : \mathbf{x} \in \mathbb{R}^n\} \textit{ and zero otherwise;}$$

2). *The density function of* $\mathbf{X}_n$ *is* $f(\cdot)$;

3). *Given* $X_{n+1} = v$, *the conditional distribution of* $\mathbf{X}_n$ *is the uniform distribution on* $D_{n,[f]}(v)$, *i.e.,* $U(D_{n,[f]}(v))$.

Furthermore, 1) *and* 3) *imply* 2).

For any random variable $R > 0$ with a density function $f_R(r)$, we have

$$P(\mathbf{X}_n \leqslant \mathbf{x}) = \int_0^\infty P(\mathbf{X}_n \leqslant \mathbf{x} | R = r) f_R(r) dr,\ \mathbf{x} \in \mathbb{R}^n. \tag{2.2}$$

In particular, we choose

$$R = R^* = \sqrt{-2\ln(X_{n+1}/f_0)},\ \ 0 < X_{n+1} \leqslant f_0, \tag{2.3}$$

and it follows that $X_{n+1} = f_0 \exp\{-\frac{1}{2}(R^*)^2\}$. The density of R^* is given by

$$f_{R^*}(r) = f_0 L_n(D_{n,[f]}(f_0 \exp(-\frac{1}{2}r^2))) \cdot re^{-\frac{1}{2}r^2},\quad 0 \leqslant r < \infty. \tag{2.4}$$

Utilizing (2.2) the density function of $\mathbf{X}_n$ can be represented as

$$\begin{aligned} f(\mathbf{x}) &= \int_{\{r:\mathbf{x}\in D_{n,[f]}(f_0 \exp(-\frac{1}{2}r^2))\}} \frac{f_{R^*}(r)}{L_n(D_{n,[f]}(f_0 \exp(-\frac{1}{2}r^2)))} dr \\ &= f_0 \int_{\{r:\ \mathbf{x}\in D_{n,[f]}(f_0 \exp(-\frac{1}{2}r^2))\}} re^{-\frac{1}{2}r^2} dr. \end{aligned} \tag{2.5}$$

Lemma 2.2. *Let $\mathbf{X}_n$ be a random vector in $\mathcal{F}$. Its cdf can be represented as (2.2), where*

1) R is a non-negative random variable with the density function

$$g(r) = L_n(D_{n,[f]}(f_0 \exp(-\frac{1}{2}r^2)))re^{-\frac{1}{2}r^2},\ 0 \leqslant r < \infty;$$

2) Given $R = r$, the conditional distribution of $\mathbf{X}_n$ is the uniform distribution $U(D^*_{n,[f]}(r))$, where

$$D^*_{n,[f]}(r) = D_{n,[f]}(f_0 \exp(-\frac{1}{2}r^2)). \tag{2.6}$$

Lemma 2.3. *Let $\mathbf{X}_n$ be a random vector in $\mathcal{F}$ and $\mathbf{U}_n = \frac{\mathbf{X}_n}{\|\mathbf{X}_n\|}$. Then the density function of $\mathbf{U}_n$ on the sphere surface $\|\mathbf{u}_n\| = 1$ is given by*

$$p(\mathbf{u}_n) = \int_0^\infty q(\mathbf{u}_n; r)re^{-\frac{1}{2}r^2}\,dr, \tag{2.7}$$

where

$$\begin{aligned} q(\mathbf{u}_n; r) &= \int_{A(\mathbf{u}_n;r)} s^{n-1}ds, \\ A(\mathbf{u}_n, r) &= \{s : s\mathbf{u}_n \in D^*_{n,[f]}(r)\},\ \ \forall\ \|\mathbf{u}_n\| = 1. \end{aligned} \tag{2.8}$$

Let $\mathbf{Q}$ be a n by n orthogonal matrix with $(\frac{1}{\sqrt{n}}, \cdots, \frac{1}{\sqrt{n}})$ as its last row. Set

$$\mathbf{V}_n = \mathbf{Q}\mathbf{U}_n. \tag{2.9}$$

Obviously $\|\mathbf{V}_n\| = 1$. The density function of $\mathbf{V}_n$ is given by

$$p^*(\mathbf{v}_n) = p(\mathbf{Q}^\top \mathbf{v}_n),\ \|\mathbf{v}_n\| = 1.$$

Denote

$$h_n(v) = \int_{\{\|\mathbf{v}\|=1\}} p^*(\sqrt{1-v^2}\mathbf{v}, v)\bar{L}_{n-1}(d\mathbf{v}), \tag{2.10}$$

where $\mathbf{v} \in \mathbb{R}^{n-1}$.

Lemma 2.4. *Let $\mathbf{X}_n, n \geqslant 3$ be a random vector in $\mathcal{F}$. The t-statistic $T(\mathbf{X}_n)$ has the t-distribution with $n-1$ degrees of freedom if and only if*

$$h_n(v) = \frac{\Gamma(\frac{n}{2})}{\sqrt{\pi}\Gamma(\frac{n-1}{2})}, -1 < v < 1, \tag{2.11}$$

where $h_n(v)$ is defined by (2.10)

3 A Solid set and its bound function

We shall consider a unit sphere surface $\bar{S}_n^2 = \{\mathbf{x} : \|\mathbf{x}\| = 1\}$ in $\mathbb{R}^n$ as a parameter to construct a certain kind of closed sets.

Definition 3.1. *A finite closed set $A \in \mathbb{R}^n$ is called solid if $\mathbf{x} \in A$ implies $\{r\mathbf{x}, 0 \leqslant r \leqslant 1\} \subseteq A$.*

Obviously, a solid set A is a closed compact set. Denote the boundary set of A by B_A. Mapping each point $\mathbf{x}$ on B_A into the unit sphere surface $\bar{S}_n^2$, we obtain a set on $\bar{S}_n^2$, denoted by $S_A = \{\mathbf{z} : \mathbf{z} = \mathbf{x}/\|\mathbf{x}\|, \mathbf{x} \in B_A\}$. The set S_A is called the projection set of A on $\bar{S}_n^2$ as S_A can be expressed as $S_A = \{\mathbf{z} : \mathbf{z} = \mathbf{x}/\|\mathbf{x}\|, \mathbf{x} \in A\}$. However, for any point $\mathbf{z}$ on S_A we may find many points $\mathbf{x}$'s on A such that $\mathbf{z} = \mathbf{x}/\|\mathbf{x}\|$. If we want to map a point form S_A to B_A, we need the following concept:

Definition 3.2. *Lat A be a finite solid set in $\mathbb{R}^n$ and let B_A and S_A be its boundary set and projection set on $\bar{S}_n^2$, respectively. The function $b(\mathbf{z}) = \mathbf{x}$, where $\mathbf{x} \in B_A$, $\mathbf{z} \in S_A$ and $\mathbf{z} = \mathbf{x}/\|\mathbf{x}\|$, is called boundary function of A.*

Clearly, $b(\mathbf{z})$ gives the Euclidean distance from the origin to $\mathbf{x} \in B_A$ such that $\mathbf{z} = \mathbf{x}/\|\mathbf{x}\|$. Conversely, given a set Z on $\bar{S}_n^2$ and a boundary function $b(\cdot)$ on Z we can define a solid set A such that its projection set is just Z. $A = \{\mathbf{x} : \|\mathbf{x}\| \leqslant b(\mathbf{z}), \mathbf{z} \in Z\}$. The set A is completely determined by its boundary function.

If the set $D^*_{n,[f]}(r)$ defined in (2.6) is solid, its boundary function will be denoted by $b(\mathbf{z}_n; r)$ to emphasize that the boundary function depends on r. In this case the density function of $\mathbf{U}_n = \frac{\mathbf{X}_n}{\|\mathbf{X}_n\|}$ can be expressed as a different way from (2.7).

Lemma 3.3. *Let $\mathbf{X}_n$ be a random vector with the density $f(\cdot)$ such that*

$$D^*_{n,[f]}(r) \text{ is closed solid set}, \forall\, r > 0,$$

then the density of $\mathbf{U}_n = \frac{\mathbf{X}_n}{\|\mathbf{X}_n\|}$ is

$$p(\mathbf{u}_n) = \frac{1}{n} \int_0^\infty r e^{-\frac{1}{2}r^2} [b(\mathbf{u}_n; r)]^n dr, \tag{3.1}$$

where $b(\cdot; r)$ is the boundary function of $D^*_{n,[f]}(r)$ and latter is defined by (2.6).

If we replace r by a positive random variable R into $D^*_{n,[f]}(r)$ and let $\mathbf{X}_n = U_{D^*_{n,[f]}}(R)$. Suppose the density of R is $g^*(\cdot)$, then the density of

$\mathbf{U}_n = \frac{\mathbf{X}_n}{\|\mathbf{X}_n\|}$ *is*

$$p^*(\mathbf{u}_n) = \frac{1}{n} \int_0^\infty \frac{g^*(r) b^n(\mathbf{u}_n; r)}{L_n(D^*_{n,[f]}(r))} dr, \tag{3.2}$$

where $\bar{L}_n(\cdot)$ *is the Lebesgue measure on the surface of a unit sphere in* $\mathbb{R}^n$.

Proof. Observe that $A(\mathbf{u}_n, r) = [0, b(\mathbf{u}_n, r)]$, and the lemma follows from Lemma 2.3 □

4 Lemma 2.4 in a parametric form

We now restate Lemma 2.4 in terms of a boundary function. Specifically, what is a necessary and sufficient condition on the boundary function $b(\mathbf{x}_n, r)$, $\|\mathbf{x}_n\| = 1, r > 0$, for $h_n(v)$ to be $\frac{\Gamma(\frac{n}{2})}{\sqrt{\pi}\Gamma(\frac{n-1}{2})}$? Let $\mathbf{V}_n$ and $\mathbf{Q}$ be defined in (2.9) and $b^*(\mathbf{v}_n, r) = b(\mathbf{Q}^\top \mathbf{v}_n, r)$.

Theorem 4.1. *Let* $f(\mathbf{x}_n)$, $\mathbf{x}_n \in \mathbb{R}^n$, *be the density of the random vector* $\mathbf{X}_n$ *and*

$$D^*_{n,[f]}(r) \text{ be a solid closed set with the boundary function } b(\mathbf{x}_n, r),\ \|\mathbf{x}_n\| = 1, \tag{4.1}$$

then the corresponding t-statistic is t-distributed with $n-1$ *degrees of freedom if and only if*

$$\int_0^\infty r e^{-\frac{1}{2}r^2} \int_{\|\mathbf{v}_{n-1}\|=1} [b^*((\sqrt{1-v^2}\ \mathbf{v}_{n-1}, v), r)]^n \bar{L}_{n-1}(d\mathbf{v}_{n-1}) dr = \frac{\Gamma(\frac{n}{2})}{\sqrt{\pi}\Gamma(\frac{n-1}{2})}. \tag{4.2}$$

Corollary 4.2. *Suppose* $\mathbf{X}_n = R\mathbf{V}_n$ *and* R *and* $\mathbf{X}_n$ *are independent,* R *is a nonnegative random variable and* $\mathbf{V}_n$ *is uniformly distributed on*

$$D_b = \{\mathbf{x}_n : \|\mathbf{x}_n\| \leqslant b\left(\frac{\mathbf{x}_n}{\|\mathbf{x}_n\|}\right)\}.$$

Then the corresponding t-statistic $T(\mathbf{X}_n)$ *has the t-distribution with* $n-1$ *degrees of freedom if and only if for all* $-1 \leqslant v \leqslant 1$,

$$\int_{\{\mathbf{v}_{n-1}=1\}} [b^*((\sqrt{1-v^2}\ \mathbf{v}_{n-1}, v), r)]^n \bar{L}_{n-1}(d\mathbf{v}_{n-1}) = \frac{\Gamma(\frac{n}{2})}{\sqrt{\pi}\Gamma(\frac{n-1}{2})}. \tag{4.3}$$

Proof. It follows from (2.10)(the definition of $h_n(v_n)$) and (3.1)(the density of $\mathbf{U}_n$ in terms of a boundary function) that

$$h_n(v) = \int\limits_{\|\mathbf{v}_{n-1}\|=1} [\int\limits_0^{\infty} re^{-\frac{1}{2}r^2}[b^*((\sqrt{1-v^2}\ \mathbf{v}_{n-1}, v), r)]^n dr]\bar{L}_{n-1}(d\mathbf{v}_{n-1}). \tag{4.4}$$

Applying Lemma 2.4 to (4.4), we complete the proof. □

5 Independent random variables

In the above study the distribution class $\mathcal{F}$ does not require independence of $X_1, \cdots, X_n$. When $X_1, \cdots, X_n$ are mutually independent, we may find some specific properties. The following theorem shows a sufficient and necessary condition for independence of the components of $\mathbf{X}_n$ in $\mathcal{F}$.

Theorem 5.1. *Let $\mathbf{X}_n \in \mathcal{F}$ with a density $f(\mathbf{x}_n)$, $\mathbf{x}_n \in \mathbb{R}^n$. Then its components are mutually independent if and only if*

(a). *The set of $D^*_{d,[f]}(r)$ can be expressed as*

$$D^*_{d,[f]}(r) = \{\mathbf{x}_n^\top = (x_1, \cdots, x_n) : \sum_{i=1}^n h_i^2(x_i) \leqslant r^2\} \overset{def}{=} S(r),$$

where h_i is a Borel function;

(b). *There exits a random variable $R > 0$ with the density $g(\cdot) = L_n(S(r))re^{-\frac{1}{2}r^2}$ such that the conditional distribution of $\mathbf{X}_n$ for given $R = r$ is $U(S(r))$.*

Proof. If the conditions (a) and (b) hold, then by representation (2.5), the density of $\mathbf{X}_n$ is given by

$$f(\mathbf{x}_n) = f_0 \int\limits_{\{\sum_{i=1}^n h_i^2(x_i)\}}^{\infty} r^{-\frac{1}{2}r^2} dr = f_0 \prod_{i=1}^n e^{-h_i^2(x_i)},$$

and evidently $e^{-h_i^2(x_i)}$ is proportional to the marginal density of X_i, $i = 1, \cdots, n$. This completes proof of the sufficiency.

Suppose now that the components of a random vector $\mathbf{X}_n \in \mathcal{F}$ are independent of each other and the density $f(\cdot)$ can be written as

$$f(\mathbf{x}_n) = \prod_{i=1}^n f_i(x_i),$$

where $f_i(\cdot)$ is the marginal density of the ith component. In this case

$$\begin{aligned} D^*_{d,[f]}(r) &= \{\mathbf{x}_n^\top = (x_1, \cdots, x_n) : \prod_{i=1}^n f_i(x_i) \geqslant f_0 e^{-\frac{1}{2}r^2}\} \\ &= \{\mathbf{x}_n^\top = (x_1, \cdots, x_n) : \sum_{i=1}^n [\sqrt{-2\ln(\tfrac{f_i(x_i)}{f_{0,i}})}]^2 \leqslant r^2\} \\ &= \{\mathbf{x}_n^\top = (x_1, \cdots, x_n) : \sum_{i=1}^n h_i^2(x_i) \leqslant r^2\}, \end{aligned}$$

where $f_0 = \prod_{i=1}^n f_{0,i}$ and $f_{0,i} = \sup\{f_i(x), x \in \mathbb{R}, i = 1, \cdots, n\}$. This completes proof of the necessity by Theorem 5.1. □

Proof of Theorem 1.1

Recall that

$$\begin{aligned} D^*_{n,[f]}(r) &= \{\mathbf{u}_n = (u_1, \cdots, u_n)^\top : \sum_{k=1}^n h^2(u_k) \leqslant r^2\} \\ &= \{\mathbf{u}_n : \|\mathbf{u}_n\| \leqslant b(\tfrac{\mathbf{u}_n}{\|\mathbf{u}_n\|}; r)\}, \end{aligned}$$

where $b(\mathbf{u}_n; r)$ is the bound function of $D^*_{n,[f]}(r)$. Again, let $\mathbf{Q} = (\mathbf{q}_1, \cdots, \mathbf{q}_n) = (q_{i,j})$ be an orthogonal matrix with the last row $(\frac{1}{\sqrt{n}}, \cdots, \frac{1}{\sqrt{n}})$ and $\mathbf{v}_n = \mathbf{Q}\mathbf{u}_n$. We have

$$b^*(\mathbf{v}_n; r) = b(\mathbf{Q}^\top \mathbf{v}_n; r),$$

$$\begin{aligned} D^*_{n,[f]}(r) &= \{\mathbf{x}_n : \sum_{k=1}^n h^2(\mathbf{q}_k^\top \mathbf{x}_n) \leqslant r^2\} \\ &= \{\mathbf{x}_n : \|\mathbf{x}_n\| \leqslant b^*(\tfrac{\mathbf{x}_n}{\|\mathbf{x}_n\|}; r)\}. \end{aligned}$$

The bound function satisfies the relation

$$\sum_{k=1}^n h^2(b^*(\mathbf{v}_n; r)\mathbf{q}_k' \mathbf{v}_n)) = r^2, \ \forall \ r > 0, \ \|\mathbf{v}_n\| = 1. \tag{5.1}$$

Next

$$h_n(v_n) = \int_{\{|\mathbf{v}_n|=1\}} \left\{ \int_0^\infty r e^{-r^2} [b^*((\sqrt{1 - v_n^2}\ \mathbf{v}_{n-1}, v_n); r)]^n dr \right\} \bar{L}_{n-1}(d\mathbf{v}_{n-1}).$$

Differentiating with respect to v_n we obtain

$$h_n'(v_n) \tag{5.2}$$
$$= \int_{\{|\mathbf{v}_n|=1\}} \left\{ \int_0^\infty re^{-r^2} \frac{d[b^*((\sqrt{1-v_n^2}\,\mathbf{v}_{n-1}, v_n); r)]^n}{dv_n} dr \right\} \bar{L}_{n-1}(d\mathbf{v}_{n-1}),$$

and

$$\frac{d[b^*((\sqrt{1-v_n^2}\,\mathbf{v}_{n-1}, v_n); r)]^n}{n\,dv_n} \tag{5.3}$$
$$= [b^*((\sqrt{1-v_n^2}\,\mathbf{v}_{n-1}, v_n); r)]^{n-1} \left(\frac{-v_n}{\sqrt{1-v_n^2}} \sum_{k=1}^{n-1} v_k \frac{\partial b}{\partial v_k} + \frac{\partial b}{\partial v_n} \right).$$

If $h'(x) = constant = \frac{1}{\sqrt{2}\sigma}$, then $b(\mathbf{v}_n; r) = \sigma r$ and $h(x) = \frac{x}{\sqrt{2}\sigma}$. In this case, by Theorem 5.1, the sample comes from a normal distribution $N(0, \sigma^2)$. However if $h'(x) = constant = \frac{1}{\sqrt{2}\sigma}$ is not true, then, differentiating (5.1) with respect to v_i, we obtain

$$\sum_{k=1}^n h(b^*\mathbf{q}_k^\top \mathbf{v}_n) h'(b^*\mathbf{q}_k^\top \mathbf{v}_n) \mathbf{q}_k^\top \mathbf{v}_n \frac{\partial b^*}{\partial v_i} + b^* \sum_{k=1}^n h(b^*\mathbf{q}_k^\top \mathbf{v}_n) h'(b^*\mathbf{q}_k^\top \mathbf{v}_n) q_{i,k} = 0$$

$$\frac{\partial b^*}{\partial v_i} = -\frac{b^* \sum_{k=1}^n h(b^*\mathbf{q}_k^\top \mathbf{v}_n) h'(b^*\mathbf{q}_k^\top \mathbf{v}_n) q_{i,k}}{\sum_{k=1}^n h(b^*\mathbf{q}_k^\top \mathbf{v}_n) h'(b^*\mathbf{q}_k^\top \mathbf{v}_n) \mathbf{q}_k^\top \mathbf{v}_n}, \quad i = 1, \cdots, n, \tag{5.4}$$

and from (5.2),(5.3)and (5.4) we have that

$$\frac{dh_n(v_n)}{dv_n} = n \int_{\{\|\mathbf{v}_{n-1}\|=1\}} \left\{ \int_0^\infty re^{-r^2} [b^*((\sqrt{1-v_n^2}\,\mathbf{v}_{n-1}, v_n); r)]^{n-1} \times \right.$$
$$\left. \left(\frac{-v_n}{\sqrt{1-v_n^2}} \sum_{i=1}^{n-1} v_i \frac{\partial b}{\partial v_i} + \frac{\partial b}{\partial v_n} \right) dr \right\} \bar{L}_{n-1}(d\mathbf{v}_{n-1})$$
$$= n \int_{\{\|\mathbf{v}_{n-1}\|=1\}} \left\{ \int_0^\infty re^{-r^2} [b^*]^{n-1} \times \right.$$
$$\left. \left[\frac{v_n b^*}{\sqrt{1-v_n^2}} \frac{\sum_{k=1}^n h(b^*\mathbf{q}_k^\top \mathbf{v}_n) h'(b^*\mathbf{q}_k^\top \mathbf{v}_n) \sum_{i=1}^{n-1} q_{k,i} v_i}{\sum_{k=1}^n h(b^*\mathbf{q}_k^\top \mathbf{v}_n) h'(b^*\mathbf{q}_k^\top \mathbf{v}_n) \mathbf{q}_k^\top \mathbf{v}_n} + \frac{\partial b^*}{\partial v_n} \right] dr \right\} \bar{L}_{n-1}(d\mathbf{v}_{n-1})$$
$$= n \int_{\{\|\mathbf{v}_{n-1}\|=1\}} \left\{ \int_0^\infty re^{-r^2} \times \right.$$
$$\left. [b^*]^{n-1} \left[\frac{v_n b^*}{\sqrt{1-v_n^2}} + \left(\frac{v_n^2}{\sqrt{1-v_n^2}} + 1 \right) \frac{\partial b^*}{\partial v_n} \right] dr \right\} \bar{L}_{n-1}(d\mathbf{v}_{n-1})$$
$$= \frac{nv_n}{\sqrt{1-v_n^2}} h_n(v_n) + n(1 + \frac{v_n^2}{\sqrt{1-v_n^2}}) m(v_n),$$

where

$$m(v_n) \tag{5.5}$$

$$= \int\limits_{\{\|\mathbf{v}_{n-1}\|=1\}} \left\{ \int\limits_0^\infty re^{-r^2} [b^*((\sqrt{1-v_n^2}\ \mathbf{v}_{n-1}, v_n); r)]^{n-1} \frac{\partial b^*}{\partial v_n} dr \right\} \bar{L}_{n-1}(d\mathbf{v}_{n-1})$$

$$= - \int\limits_{\{\|\mathbf{v}_{n-1}\|=1\}} \left\{ \int\limits_0^\infty re^{-r^2} [b^*]^n [\frac{\sum\limits_{k=1}^n h(b^* \mathbf{q}_k^\top \mathbf{v}_n) h'(b^* \mathbf{q}_k^\top \mathbf{v}_n) q_{n,k}}{\sum\limits_{k=1}^n h(b^* \mathbf{q}_k^\top \mathbf{v}_n) h'(b^* \mathbf{q}_k^\top \mathbf{v}_n) \mathbf{q}_k^\top \mathbf{v}_n}] dr \right\} \times$$

$$\bar{L}_{n-1}(d\mathbf{v}_{n-1}) \tag{5.6}$$

and $\mathbf{v}_n = (\sqrt{1-v_n^2}\ \mathbf{v}_{n-1}, v_n)$. From the above equation for the derivative of $h_n(v_n)$ we have that $h_n(\cdot)$ is a solution of the following differential equation

$$\frac{dh_n(v_n)}{dv_n} = \frac{nv_n}{\sqrt{1-v_n^2}} h_n(v_n) + n(1 + \frac{v_n^2}{\sqrt{1-v_n^2}}) m(v_n). \tag{5.7}$$

We assert that solutions of differential equation (5.7) are not constant. If it is true, we have

$$h_n(v_n) = a \text{ (a constant)}, \quad -1 \leqslant v_n \leqslant 1,$$

then

$$m(v_n) = -\frac{av_n}{v_n^2 + \sqrt{1-v_n^2}}.$$

Substituting it into (5.7), we obtain that $h_n(\cdot)$ satisfies

$$\frac{dh_n(v_n)}{dv_n} = \frac{nv_n}{\sqrt{1-v_n^2}} h_n(v_n) - a\frac{nv_n}{\sqrt{1-v_n^2}}.$$

This is equivalent to the equation:

$$\frac{d(h_n(v_n) - a)}{dv_n} = \frac{nv_n}{\sqrt{1-v_n^2}} (h_n(v_n) - a)$$

from which it follows that

$$h_n(v_n) = a + Ae^{-n\sqrt{1-v_n^2}}, -1 < v_n < 1. \tag{5.8}$$

Consequently $h_n(\cdot)$ can not be a constant and this contradicts the assertion that $h_n(\cdot)$ is a constant when t-statistic has a t-distribution with $n-1$ degrees of freedom. This is derived from assumption that h' is not constant. Thus we must have

$$h'(u) = constant \neq 0.$$

for $h_n(v_n)$ to be a constant which is a necessary condition for a t-statistic to be t-distributed with $n-1$ degrees of freedom. This completes the proof. □

References

[1] Bondesson, L., When is the t-statistic t-distributed, *Sankhyā, Series A*, 45 (1983), 338–345.

[2] Fang, K. T., Ng, K. W. and Kotz, S., *Symmetric Multivariate and Related Distributions*, Chapman and Hall, London. 1990.

[3] Fang K. T., Yang, Z. H. and Kotz, S., Generation of multivariate distributions by vertical density representation, *Statistics*, 35 (2001), 281–293.

[4] Mauldon, J. G., Characterizing properties of statistical distributions, *Quart. J. Math.* Oxford, Series 2, 7 (1956), 155–160.

[5] Kagan, A. M., Linnik, Yu. V. and Rao, C. R., *Characterization Problems in Mathematical Statistics*, Wiley, New York. 1973.

[6] "Student" On the probable error of the mean, *Biometrika*, 6 (1908), 1–25.

[7] Troutt, M. D., Pang, W. K. and Hou, S. H., *Vertical Density Representation and Its Applications*, World Scientific Publishing Co. Pte. Ltd., Singarpore. 2004.

[8] Yang Z. H., Fang, K.T. and Kotz, S., On the student's t-distribution and the t-statistic, *J. Multivariate Anal.*, 2006, to appear.

[9] Yang, Z. H. and Kotz, S., Center-similar distributions with applications in multivariate analysis. *Statistics and Probability Letters*, 64 (2003), 335–345.

Statistical Analysis for Rounded Data*

Zhidong Bai
Key Laboratory for Applied Statistics of MOE
and Department of Statistics, Northeast Normal University
Changchun 130024, China
E-mail: baizd@nenu.edu.cn

Shurong Zheng
Key Laboratory for Applied Statistics of MOE
and Department of Statistics, Northeast Normal University
Changchun 130024, China
E-mail: zhengsr@nenu.edu.cn

Baoxue Zhang
Key Laboratory for Applied Statistics of MOE
and Department of Statistics, Northeast Normal University
Changchun 130024, China
E-mail: bxzhang@nenu.edu.cn

Guorong Hu
Key Laboratory for Applied Statistics of MOE
and Department of Statistics, Northeast Normal University
Changchun 130024, China
E-mail: hugr@nenu.edu.cn

Abstract

Unless the model is discrete, rounding of data is unavoidable in practical measurement. However, the errors caused by rounding of data are almost ignored by all classical statistical theories. Although some pioneers have noticed this problem, few suitable approaches were proposed to deal with this error. In this paper, by some concrete examples when measurements are rounded to some extent, we propose to use MLE or approximated MLE (AMLE) to estimate the parameters and discuss the properties of them and tests based on the new estimators. In particular, as an

*The authors are partly supported by NSF of China (No.10571020) and Science Foundation for Young Teachers of Northeast Normal University (No.20050107).

example, we shall discuss the limiting properties of the new estimator of parameters in an AR(p) model and MA(q) model when the observations are rounded.

1 Introduction

It is so popular that nowhere observations are not rounded, unless the model is discrete. The rounding of data may be due to the precision of the measuring instruments in experiments or that of the recording/storage mechanism, say the computer precision, or due to the unnecessary need of high accuracy, say, the blood pressure on a patient is not needed to be measured to decimals. In some cases, the rounding errors may be very large, especially in social survey, *e.g.* the GDP of a large country may be rounded at tens of million dollars, the annual income may be rounded at thousands of dollars. In some cases, the rounding error may be small, *e.g.* in high technology, the rounding errors may be as small as millimeters, or even 10^{-8} meters. No matter the precision is high or low, no observations are not rounded. In general, the rounding scheme is to record the measurements to the nearest multiples of the precision unit. With suitable normalization, we may assume that the rounding is up to integers, say, if the $X = 0.6$ or $10/3$, then the rounded values will be $\widetilde{X} = 1$ or 3 respectively.

It seemed not a big problem in old days when the precision of instruments is high, because the sample size is small and the rounding errors may be negligible. But nowadays, it is not the case because, in many cases, statisticians are facing huge data sets due to the rapid development of computer technologies which makes it possible to collect, store and deal with huge data sets, *e.g.* in biological sciences, economics and finance, the wireless communications *etc.* It is natural to ask how serious the rounding errors will affect the accuracy of statistical inference.

As a simple example, we consider the case where $\{x_1, x_2, \cdots, x_n\}$ is a simple random sample drawn from $N(\mu, \sigma^2)$ and our data set is $\{\tilde{x}_1, \tilde{x}_2, \cdots, \tilde{x}_n\}$ which are the rounded values of $\{x_1, x_2, \cdots, x_n\}$. The rounding scheme is as follows

$$\begin{cases} x \in [\tilde{x} - 0.5, \tilde{x} + 0.5), & x > 0, \\ x \in (\tilde{x} - 0.5, \tilde{x} + 0.5], & x < 0, \end{cases}$$

where $\tilde{x}$ is the rounded datum of x. If the rounded data are regarded as the true sample from the normal population, the parameters μ and σ^2 should be estimated by the sample mean $\hat{\mu} = \frac{1}{n}\sum_{i=1}^{n} \tilde{x}_i$ and the sample variance $\hat{\sigma}^2 = \frac{1}{n-1}\sum_{i=1}^{n}(\tilde{x}_i - \hat{\mu})^2$. Unfortunately, unlike what claimed by statistical theory, the estimators $\hat{\mu}$ and $\hat{\sigma}^2$ are not consistent at all. The theoretical proof of this phenomenon is not hard but we omitted it for it is not the main aim of this paper. We just present the following simple simulation results as an illustration. As a simple example, we consider the rounded data $\{\tilde{x}_1, \tilde{x}_2, \cdots, \tilde{x}_n\}$ which are from $\{x_1, x_2, \cdots, x_n\} \overset{i.i.d.}{\sim} N(\mu, \sigma^2)$. Now, we consider the following two tests:

$$H_0 : \mu = \mu_0 \quad v.s. \quad H_1 : \mu \neq \mu_0$$

and

$$H_0 : \sigma = \sigma_0 \quad v.s. \quad H_1 : \sigma \neq \sigma_0.$$

Let Rejec-Prob denote the rejection probability

$$P_{\mu_0,\sigma_0}\left(\frac{\sqrt{n}|\hat{\mu} - \mu_0|}{\hat{\sigma}} \geqslant t_{n-1,0.975}\right) \tag{1.1}$$

or

$$P_{\mu_0,\sigma_0}\left(\frac{(n-1)\hat{\sigma}^2}{\sigma_0^2} \geqslant \chi^2_{n-1,0.975} \;\; or \;\; \frac{(n-1)\hat{\sigma}^2}{\sigma_0^2} \leqslant \chi^2_{n-1,0.025}\right), \tag{1.2}$$

where $t_{n-1,0.975}$ is the 97.5% quantile of t-distribution with the degree of $n-1$, $\chi^2_{n-1,0.0.025}$ is the 2.5% quantile of χ^2-distribution with the degree of $n-1$ and $\chi^2_{n-1,0.975}$ is the 97.5% quantile of χ^2-distribution with the degree of $n-1$.

	(μ, σ^2)	$(2.25, 0.25^2)$	$(2.25, 1)$	$(2.25, 4)$
n=10000	$(\hat{\mu}, \sqrt{MSE})$	(2.157, 0.093)	(2.25, 0.010)	(2.25, 0.020)
	$(\hat{\sigma}^2, \sqrt{MSE})$	(0.135, 0.073)	(1.083, 0.085)	(4.084, 0.102)
	Rejec-Prob of [(1.1), (1.2)]	(1.0, 1.0)	(0.051, 1.0)	(0.051, 0.31)
n=200	$(\hat{\mu}, \sqrt{MSE})$	(2.157, 0.096)	(2.25, 0.074)	(2.25, 0.143)
	$(\hat{\sigma}^2, \sqrt{MSE})$	(0.135, 0.075)	(1.083, 0.134)	(4.081, 0.417)
	Rejec-Prob of [(1.1), (1.2)]	(0.91, 1.0)	(0.051, 0.129)	(0.05, 0.058)
n=50	$(\hat{\mu}, \sqrt{MSE})$	(2.157, 0.106)	(2.25, 0.147)	(2.251, 0.285)
	$(\hat{\sigma}^2, \sqrt{MSE})$	(0.135, 0.081)	(1.083, 0.232)	(4.081, 0.823)
	Rejec-Prob of [(1.1), (1.2)]	(0.452, 0.930)	(0.05, 0.070)	(0.05, 0.051)
n=20	$(\hat{\mu}, \sqrt{MSE})$	(2.158, 0.123)	(2.25, 0.234)	(2.25, 0.453)
	$(\hat{\sigma}^2, \sqrt{MSE})$	(0.135, 0.093)	(1.080, 0.357)	(4.071, 1.315)
	Rejec-Prob of [(1.1), (1.2)]	(0.36, 0.67)	(0.05, 0.060)	(0.05, 0.05)
n=10	$(\hat{\mu}, \sqrt{MSE})$	(2.158, 0.149)	(2.248, 0.331)	(2.25, 0.641)
	$(\hat{\sigma}^2, \sqrt{MSE})$	(0.135, 0.111)	(1.083, 0.513)	(4.079, 1.914)
	Rejec-Prob of [(1.1), (1.2)]	(0.18, 0.67)	(0.044, 0.059)	(0.05, 0.05)

The above simulation results show that the estimates $\hat{\mu}$ and $\hat{\sigma}^2$ are inconsistent. As $n \to +\infty$, it is easily to reject the tests:

$$H_0 : \mu = \mu_0 \ \ v.s. \ \ H_1 : \ \mu \neq \mu_0$$

or

$$H_0 : \sigma = \sigma_0 \ \ v.s. \ \ H_1 : \sigma \neq \sigma_0.$$

This shows that some problems between statistical theory and its applications are coming forth which require the statisticians specify the original model according to true observed values.

From the above simulation results, one also sees that the estimation for $n = 10000$, is the same as $n = 200$ and only has a small improvement than $n = 20$. It is not hard to imagine that, when rounding errors are involved in the data, there is a given value of the sample sizes (depends on the accuracy of the rounding scheme) such that additional samples will not improve the accuracy of the traditional estimation when the sample size is beyond this value. So, it is natural to ask what we should do when we have already had a large sample? If a small portion of the sample gives the estimation of the same accuracy as the whole sample, why we need to use the whole sample? Also, do we have alternative ways to improve the estimation and statistical inference? In recent one or two decades, the topic of data mining becomes very hot. However, few discussions on the rounding errors were found in the literature, though rounding errors are strongly influential to the inference. The aim of this paper is to explore new methodologies of higher performance accuracy of statistical inference when rounding errors are involved.

Some pioneer workers have noticed that the rounding errors may cause serious effect to statistical inference. The earliest discussion on rounded data dates back to at least Sheppard (1898). Recent work can be found in Peter Hall (1982) who examined the influence of rounding errors on nonparametric density. Furthermore, Dempster and Rubin (1983) compared three simple approaches in dealing with rounding errors in a linear regression model, the usual least squares estimate and those with BRB and Sheppard corrections. The latter two methods were developed by assuming that the rounding errors are uniformly distributed over a symmetric interval with the unit length of the rounding precision and are independent of the rounded data or the original sample respectively. However, it is not difficult to examine the invalidity of these assumptions. Heitjan (1989) gave a detailed review of the most popular approaches in dealing with rounded data. Heitjan and Rubin (1991) presented a general model for coarsened data, including rounded, heaped, censored and missing data. All earlier works had not provided consistent estimators until the recent works Lee and Vardeman (2001, 2002, 2003) who investigated confidence interval estimation of the parameters μ and σ^2 when a rounded sample comes from the $N(\mu, \sigma^2)$ and the extension to the balanced one-way ANOVA with random effects. However, they didn't obtain simultaneous interval estimation of the parameter μ and σ^2 since their method is based on an unjustified inversion of rounded data likelihood ratio test. Wright and Bray (2003) showed the danger of ignoring heaping before presenting a case-study of the ultrasound measurements. They analyzed a mixture model for rounded data by Bayes approach implemented by using the Gibbs sampler. The general issue of rounding continuous data has been indispensable for practical experts. See Tricker, Coates and Okell (1998) and Vardeman (2004, 2005) for recent references. Early discussions and bibliographies are referred to Fisher (1936) and Haitovsky (1982).

The rounding errors also exist in time series data. However, the statistical inference would become much more complicated than *iid* cases. Jones (1980) theoretically discussed the problem of maximum likelihood estimation of ARMA models when data are rounded, Machak and Rose (1984, 1985) reported on studies of the problem of rounding errors in ARMA models, Stam and Cogger (1993) consider the effects of rounding errors in parameter estimation of autoregressive (AR) models, deriving appropriate adjustments for the estimates of the parameters when using rounded data, and Rose (1993) addressed the impact of rounding errors on model identification and parameter estimation for ARMA$(1, 1)$.

In this paper, we proposed to use MLE or approximated MLE (AMLE) to estimate the unknown parameters based on the rounded data and study the statistical inference for the cases where the rounded data sets are from an independent and identically distributed sample of size n, having a continuous distribution function $F(x, \boldsymbol{\theta})$ with a density $p(x, \boldsymbol{\theta})$ and especially those from time series models such as $AR(p)$ and $MA(q)$. The organization of the paper is as follows. In Section 2, we consider the case where the rounded data are from an i.i.d. sample and discuss the properties of the MLE of the unknown parameters. In Section 3, we assume that the rounded data are from an AR(p) or MA(q) model. We propose to use AMLE to estimate the unknown parameters and prove the consistency and asymptotic normality of new estimators. Some simulation results will be presented in Section 4 and some technical proofs are contained in the Appendix.

2 Rounding an I.I.D. Sample

In this section, as mentioned in the introduction section, without loss of generality, we assume that the observations are rounded to integers. Suppose that the original sample is drawn from a population $F(x, \boldsymbol{\theta})$

with a density function $p(x, \boldsymbol{\theta})$. Under this assumption, the probability that the n rounded observations $X_1, X_2, \cdots, X_n$ take the integer values $x_1, x_2, \cdots, x_n$, or the likelihood, is

$$\begin{aligned} L(\mathbf{x}; \boldsymbol{\theta}) &= P(X_1 = x_1, X_2 = x_2, \cdots, X_n = x_n) \\ &= \prod_{i=1}^{n} [F(x_i + 0.5, \boldsymbol{\theta}) - F(x_i - 0.5, \boldsymbol{\theta})] \\ &= \prod_{j=-\infty}^{\infty} [F(j + 0.5, \boldsymbol{\theta}) - F(j - 0.5, \boldsymbol{\theta})]^{n_j}, \end{aligned} \tag{2.1}$$

where n_j is the frequency of $\{x_1, \cdots, x_n\}$ at the value j.

It may be more convenient to work with the natural logarithm of the likelihood function $L(\mathbf{x}; \boldsymbol{\theta})$ in (2.1), and we thus define the log-likelihood function by

$$l(\mathbf{x}; \boldsymbol{\theta}) = n \sum_{i=-\infty}^{\infty} \widehat{p}_i \log(p_i),$$

where $\widehat{p}_i = \frac{n_i}{n}, p_i = p_i(\boldsymbol{\theta}) = F(i + 0.5, \boldsymbol{\theta}) - F(i - 0.5, \boldsymbol{\theta})$.

It is easily seen that the problem reduces to the ordinary parameter estimation of a discrete model. Therefore, under usual regularity conditions, it is easy to verify that the MLE of $\boldsymbol{\theta}$ exists and is unique, consistent and asymptotically normal. From the result in Burridge (1981), and Silvapulle and Burridge (1986), the MLE satisfies the following equations

$$\sum_{i=-\infty}^{\infty} \frac{\widehat{p}_i}{p_i} \frac{\partial p_i}{\partial \boldsymbol{\theta}} = 0.$$

For the location-scale model, it reduces to

$$\begin{cases} \displaystyle\sum_{i=-\infty}^{\infty} \frac{\widehat{p}_i}{p_i} \frac{\partial p_i}{\partial \mu} = 0, \\ \displaystyle\sum_{i=-\infty}^{\infty} \frac{\widehat{p}_i}{p_i} \frac{\partial p_i}{\partial \sigma} = 0, \end{cases}$$

where

$$\begin{cases} \dfrac{\partial p_i}{\partial \mu} = -\dfrac{1}{\sigma} p\left(\dfrac{i + 0.5 - \mu}{\sigma}\right) + \dfrac{1}{\sigma} p\left(\dfrac{i - 0.5 - \mu}{\sigma}\right), \\ \dfrac{\partial p_i}{\partial \sigma} = -(\dfrac{i + 0.5 - \mu}{\sigma^2}) p\left(\dfrac{i + 0.5 - \mu}{\sigma}\right) + (\dfrac{i - 0.5 - \mu}{\sigma^2}) p\left(\dfrac{i - 0.5 - \mu}{\sigma}\right), \end{cases}$$

We simply list the regularity conditions as follows:

1. The parameter space $\boldsymbol{\theta}$ is compact.
2. At the true value $\boldsymbol{\theta}_0$, $\sum_{i=-\infty}^{\infty} p_i(\boldsymbol{\theta}_0)\log p_i(\boldsymbol{\theta})$ exists for all $\boldsymbol{\theta} \in \boldsymbol{\theta}$.
3. The true parameter $\boldsymbol{\theta}_0$ is the only maximizer of the expected log-likelihood, *i.e.*,

$$\boldsymbol{\theta}_0 = arg\max_{\boldsymbol{\theta}} \sum_{i=-\infty}^{\infty} p_i(\boldsymbol{\theta}_0)\log p_i(\boldsymbol{\theta}).$$

4. The likelihood function is such that $\frac{1}{n}l(\mathbf{x};\boldsymbol{\theta})$ uniformly and almost surely (or in probability) converges to $\sum_{i=-\infty}^{\infty} p_i(\boldsymbol{\theta}_0)\log p_i(\boldsymbol{\theta})$ in $\boldsymbol{\theta}$.
5. The value $\boldsymbol{\theta}_0$ is an inner point of $\boldsymbol{\theta}$.
6. In a neighborhood of $\boldsymbol{\theta}_0$, $p(x, \boldsymbol{\theta})$ is continuously differentiable in $\boldsymbol{\theta}$.
7. The summation in i and the differentiation in $\boldsymbol{\theta}$ are interchangeable.
8. The Fisher information matrix is non-singular.

By the general theory of MLE, we have the following theorem.

Theorem 2.1. *Under regularity conditions 1 – 4, we have*

$$\widehat{\boldsymbol{\theta}} \to \boldsymbol{\theta}_0, \qquad \textit{almost surely (a.s.)}.$$

Remark 2.2. *The regularity condition 1 can be satisfied by choosing a compact subset of the original parameter space which contains the true value of the parameter. The regularity conditions 2 and 3 are satisfied by almost all usually used continuous distribution families, say, normal, Cauchy, Gamma, student t, etc. To guarantee the validity of Theorem 2.1, the only condition needed to verify is the regularity condition 4. By the strong law of large numbers, for each fixed* $\boldsymbol{\theta}$,

$$\frac{1}{n}l(\boldsymbol{x};\boldsymbol{\theta}) \to \sum_{i=-\infty}^{\infty} p_i(\boldsymbol{\theta}_0)\log p_i(\boldsymbol{\theta}), \ a.s.$$

Then the condition 4 holds if $p_i(\boldsymbol{\theta})$ *is concave or analytic. This can by proved by Rockefellar theorem (see Rockefellar (1970)) or Vitalli lemma (see Titchmarsh (1939)) respectively for the two cases.*

So, Theorem 2.1 holds for almost all usually used continuous distribution families.

Remark 2.3. *By the strong law of large numbers, for any M, N, we have*

$$\lim_{n\to+\infty}(\widehat{p}_i,\ -M<i<N)=(p_i,\ -M<i<N),\quad a.s.$$

From this, it is easy to verify that for any $r \geqslant 1$,

$$\sum_{i=-\infty}^{\infty}|\hat{p}_i-p_i|^r \leqslant \sum_{i=-M}^{N}|\hat{p}_i-p_i|^r+\Big(2-\sum_{-M}^{N}(\hat{p}_i+p_i)\Big)$$

$$\to 2(1-\sum_{-M}^{N}p_i),\ a.s..$$

Noting that the right hand side can be made arbitrarily small, thus, we have

$$\lim_{n\to+\infty}(\widehat{p}_i,\ -\infty<i<\infty)=(p_i,\ -\infty<i<\infty),\quad a.s.\ in\ L_r.$$

If we set k (the dimension of $\boldsymbol{\theta}$) equations $\hat{p}_i=p_i(\boldsymbol{\theta})$, $i=j_1,\cdots,j_k$ and take the solution as an estimator of $\boldsymbol{\theta}$, by the implicit function theorem, we conclude that this estimator is consistent. Intuitively, MLE is the "best" estimator and thus it is believed to be consistent.

Again, using the general theory of MLE, we have the following theorem.

Remark 2.4. *Under regularity conditions 1 – 8,*

$$\sqrt{n}(\hat{\boldsymbol{\theta}}-\boldsymbol{\theta})\xrightarrow{d}N(0,I^{-1}(\boldsymbol{\theta}))$$

where the Fisher information matrix

$$I(\boldsymbol{\theta})=\left(\sum_{i=-\infty}^{\infty}\frac{\partial_l p_i(\boldsymbol{\theta})\partial_j p_i(\boldsymbol{\theta})}{p_i(\boldsymbol{\theta})}\right)_{l,j=1}^{k},$$

and $\partial_j p_i(\boldsymbol{\theta})=\partial p_i(\boldsymbol{\theta})/\partial\boldsymbol{\theta}_j$.

Remark 2.5. *The regularity conditions 1 – 8 are satisfied if there is a constant sequence a_i such that for all $\ell=0,1,2$,*

$$\sup_{x\in[i-0.5,i+0.5]} p^{(\ell)}(x,\boldsymbol{\theta})\leqslant a_i$$

and

$$\sum_{i=-\infty}^{\infty}a_i p_i(\boldsymbol{\theta}_0)<\infty.$$

3 Rounded Data from Time Series Models

It has been seen that for the rounded data of i.i.d. observations, the analysis is very similar to the grouped data, except that there are infinitely many groups for which some more complex analysis is needed, and the MLE method and theory can be employed. However, it is not the case for rounded time series data. For simplicity, we first consider the MA(p) models because the time series is a p-dependent sequence, and hence simpler analysis will be done.

3.1. MA(p) model

Now, we consider MA(p) model which is

$$X_t = c + \epsilon_t + \phi_1\epsilon_{t-1} + \cdots + \phi_p\epsilon_{t-p}$$

where $\epsilon_t \overset{i.i.d.}{\sim} N(0,\sigma^2)$ for $t = 1,\cdots,n$. Let $\boldsymbol{\phi}= (\phi_1,\cdots,\phi_p)$. Here, $(X_1,\cdots,X_n)'$ is normally distributed as $N(c\mathbf{1},\mathbf{\Sigma})$ where $\mathbf{\Sigma} = (\sigma_{ij})_{n\times n}$, $\sigma_{ij} = \gamma_{|i-j|}$,

$$\mathbf{\Sigma} = \begin{bmatrix} \gamma_0 & \gamma_1 & \cdots & \gamma_p & \cdots & \gamma_{n-1} \\ \gamma_1 & \gamma_0 & \cdots & \gamma_{p-1} & \cdots & \gamma_{n-2} \\ \vdots & \vdots & \vdots & \vdots & \vdots & \vdots \\ \gamma_{n-1} & \gamma_{n-2} & \cdots & \gamma_{n-p-1} & \cdots & \gamma_0 \end{bmatrix}$$

and

$$\gamma_i = \begin{cases} \sigma^2(\phi_i + \phi_{i+1}\phi_1 + \phi_{i+2}\phi_2 + \cdots + \phi_p\phi_{p-i}), \; i = 0,1,\cdots,p, \\ 0, \qquad\qquad i > p. \end{cases}$$

where $\phi_0 = 1$. Here, only rounded data $\widetilde{X}_1,\cdots,\widetilde{X}_n$ are observed. For simplicity, let

$$p_{\mathbf{i}} \triangleq P(i_j - 0.5 \leqslant X_j < i_j + 0.5, j = 1,\cdots,p+1),$$

where $\mathbf{i} = (i_1,\cdots,i_{p+1})$ and i_j takes integer values. Let $A_{\mathbf{i}}$ denote the rectangle $A_{\mathbf{i}} = \prod_{j=1}^{p+1}[i_j - 0.5, i_j + 0.5)$ and define $p+1$ subsets of the the rounded data as follows:

Sub-samples:

(1)	$(\widetilde{X}_1\cdots\widetilde{X}_{p+1})$	$\cdots$	$(\widetilde{X}_{(m-1)(p+1)+1}\cdots\widetilde{X}_{m(p+1)})$;
(2)	$(\widetilde{X}_2\cdots\widetilde{X}_{p+2})$	$\cdots$	$(\widetilde{X}_{(m-1)(p+1)+2}\cdots\widetilde{X}_{m(p+1)+1})$;
	$\cdots$	$\cdots$	$\cdots$
$(p+1)$	$(\widetilde{X}_{p+1}\cdots\widetilde{X}_{2p+1})$	$\cdots$	$(\widetilde{X}_{m(p+1)}\cdots\widetilde{X}_{m(p+1)+p})$;

where $m = [(n-p)/(p+1)]$.

Estimation Procedure based on $\widetilde{X}_1, \cdots, \widetilde{X}_n$:

Step (1). Note that $(\widetilde{X}_1 \cdots \widetilde{X}_{p+1}) \cdots (\widetilde{X}_{(m-1)(p+1)+1} \cdots \widetilde{X}_{m(p+1)})$ form a sample of m iid. $p+1$-dimensional random vectors. Denote by $n_{\mathbf{i}}$ the frequency of $\mathbf{i}$ in this sub-sample. Then based on this sub-sample, the MLE of parameters $(c, \sigma^2, \boldsymbol{\phi})$ can be obtained by maximizing $\sum n_{\mathbf{i}} \log p_{\mathbf{i}}$. Denote the MLE by $(\widehat{c}_1, \widehat{\sigma}_1^2, \widehat{\boldsymbol{\phi}}_1)$;

Step (2). Similarly, by the j-th subset of the sample, we can construct an MLE of the parameters $(\widehat{c}_j, \widehat{\sigma}_j^2, \widehat{\boldsymbol{\phi}}_j)$, $j = 2, \cdots, p+1$.

Step (3). Take the averages of the MLE's as our estimators of parameters, *i.e.*

$$\widehat{c} = \sum_{i=1}^{p+1} \widehat{c}_i/(p+1), \quad \widehat{\boldsymbol{\phi}} = \sum_{i=1}^{p+1} \widehat{\boldsymbol{\phi}}_i/(p+1), \quad \widehat{\sigma}^2 = \sum_{i=1}^{p+1} \widehat{\sigma}_i^2/(p+1).$$

Theorem 3.1. *The estimates $(\widehat{c}, \widehat{\boldsymbol{\phi}}, \widehat{\sigma}^2)$ obtained by proposed estimation procedure based on $\widetilde{X}_1, \cdots, \widetilde{X}_n$ are consistent.*

Theorem 3.2. *The estimates $(\widehat{c}, \widehat{\boldsymbol{\phi}}, \widehat{\sigma}^2)$ are asymptotically multivariate normally distributed, that is,*

$$\sqrt{n} \begin{pmatrix} \widehat{c} - c \\ \widehat{\boldsymbol{\phi}} - \boldsymbol{\phi} \\ \widehat{\sigma}^2 - \sigma^2 \end{pmatrix} \sim N(0, \mathbf{G})$$

where $\mathbf{G} = \mathbf{I}^{-1}(\boldsymbol{\theta})\mathbf{V}_p\mathbf{I}^{-1}(\boldsymbol{\theta})$ and $I(\boldsymbol{\theta})$ and for some matrices $V_p(\boldsymbol{\theta})$ which will be specied in the proof of the theorem.

Remark 3.3. *It seems that the MLE method and theory are also applicable to the MA(p) model. However, it is practically impossible because the the whole sequence are not independent. If the construct the cells by the whole sequence, the probability for each cell will be too small to calculate. Thus, the AMLE is applicable and performs excellently.*

3.2 AR(1) model

Suppose that $\{X_t\}$ is the causal the AR(p) process,

$$X_{t+1} = c + \phi_1 \cdot X_t + \cdots + \phi_p X_{t-p+1} + \epsilon_{t+1},$$

where $\epsilon_t \overset{i.i.d.}{\sim} N(0,\sigma^2)$ for $t = 1,\cdots,n$. The causality assumption means that $1 - \phi_1 z - \cdots - \phi_p z^p \neq 0$ for $\mid z \mid \leqslant 1$. Then $(X_1,\cdots,X_n)$ are normally distributed as

$$(X_1,\cdots,X_n)' \sim N\left(\mu\mathbf{1}_n, \mathbf{\Sigma}_{n\times n}\right)$$

where $\mu = c/(1 - \phi_1 - \cdots - \phi_p)$.

Now, we assume that only rounded data are available, that is, $\widetilde{X}_1,\cdots,\widetilde{X}_n$ are observed where $\widetilde{X}_i$ are the rounded data of X_i for $i = 1,\cdots,n$. As expected, if the rounding errors were ignored, the usual estimates such as the estimators using Yule-Walker equation are not consistent. Some simulation results will be presented in Section 4. In this section, we will look for better estimates which are consistent and asymptotically normal.

Our problem is to find new estimates of the parameter vector $\phi = (c,\phi_1,\cdots,\phi_p)'$, and the innovation variance σ^2. We propose a new procedure, called the *approximated MLE*, to estimate the parameters $(c,\phi_1,\cdots,\phi_p,\sigma^2)$ based on the observed rounded data. For simplicity, we only give the case where $p = 1$, that is, the AR(1) model. Estimation procedures for the general case where $p \geqslant 2$ may be similarly derived.

In a causal AR(1) model

$$X_{t+1} = c + \phi \cdot X_t + \epsilon_t,$$

where $|\phi| < 1$ and $\epsilon_t \overset{i.i.d.}{\sim} N(0,\sigma^2)$ for $t = 1,\cdots,n-1$. Under this set up, we have

$$(X_1,\cdots,X_n)' \sim N\left(\mu\cdot\mathbf{1}_n, \mathbf{\Sigma}_{n\times n}\right)$$

where $\mu = c/(1-\phi)$ and

$$\mathbf{\Sigma}_{n\times n} = \frac{\sigma^2}{1-\phi^2}\cdot\begin{bmatrix} 1 & \phi & \phi^2 & \cdots & \phi^{n-1} \\ \phi & 1 & \phi & \cdots & \phi^{n-2} \\ \phi^2 & \phi & 1 & \cdots & \phi^{n-3} \\ \vdots & \vdots & \vdots & \vdots & \vdots \\ \phi^{n-1} & \phi^{n-2} & \phi^{n-3} & \cdots & 1 \end{bmatrix}. \tag{3.1}$$

Especially, we have

$$\begin{pmatrix} X_t \\ X_{t+1} \end{pmatrix} \sim N\left(\begin{pmatrix} \mu \\ \mu \end{pmatrix}, \frac{\sigma^2}{1-\phi^2}\cdot\begin{pmatrix} 1 & \phi \\ \phi & 1 \end{pmatrix}\right)$$

where ϕ is the correlation coefficient of X_t and X_{t+1}. Then for any integer i and j, the probability of the event $\tilde{x}_1 = i, \tilde{x}_2 = j$ is given by

$$\begin{aligned} p_{ij} &\triangleq P(\widetilde{x}_1 = i, \widetilde{x}_2 = j) \\ &= P(i - 0.5 \leqslant X_1 < i + 0.5, j - 0.5 \leqslant X_2 < j + 0.5) \\ &= \int_{i-\frac{1}{2}}^{i+\frac{1}{2}} \int_{j-\frac{1}{2}}^{j+\frac{1}{2}} \frac{1}{2\pi|\boldsymbol{\Sigma}_{2\times 2}|^{\frac{1}{2}}} \exp\left[-\frac{1}{2}\begin{pmatrix} x_1-\mu \\ x_2-\mu \end{pmatrix}' \boldsymbol{\Sigma}_{2\times 2}^{-1} \begin{pmatrix} x_1-\mu \\ x_2-\mu \end{pmatrix}\right] dx_2 dx_1 \\ &= \int_{i-\frac{1}{2}-\mu}^{i+\frac{1}{2}-\mu} \int_{j-\frac{1}{2}-\mu}^{j+\frac{1}{2}-\mu} \frac{\sqrt{1-\phi^2}}{2\pi\sigma^2} \exp\left(-\frac{x_1^2+x_2^2-2\phi x_1 x_2}{2\sigma^2}\right) dx_2 dx_1. \end{aligned} \tag{3.2}$$

Because $|\phi| < 1$, $|\phi|^k$ may be as small as desired by choosing k large. That is, when k is large enough, (X_i, X_{i+1}) and (X_{i+k}, X_{i+k+1}) are approximately independent. So the rounded samples $(\widetilde{X}_i, \widetilde{X}_{i+1})$ and $(\widetilde{X}_{i+k}, \widetilde{X}_{i+k+1})$ are also approximately independent. For example, when $\phi = 0.3$, then $\phi^{10} = 5.90 \times 10^{-6}$. Then when $\phi = \pm 0.3$, $(\widetilde{X}_i, \widetilde{X}_{i+1})$ and $(\widetilde{X}_{i+11}, \widetilde{X}_{i+12})$ may be considered to be mutually independent. Similarly, $0.5^{20} = 9.54 \times 10^{-7}$; $0.75^{40} = 1.0 \times 10^{-5}$; $0.9^{100} = 2.66 \times 10^{-5}$.

Based on the above argument, we propose the following estimation procedure to obtain the AMLE of parameters c, ϕ and σ^2. Assume that k is large integer. Split the rounded data as the following k subgroups of size size $m = [(n-1)/k]$ (overlapping between groups is allowed):

$$\begin{array}{lll} (\widetilde{X}_1, \widetilde{X}_2) & (\widetilde{X}_{k+1}, \widetilde{X}_{k+2}) \cdots, & (\widetilde{X}_{(m-1)k+1}, \widetilde{X}_{(m-1)k+2}); \\ (\widetilde{X}_2, \widetilde{X}_3) & (\widetilde{X}_{k+2}, \widetilde{X}_{k+3}) \cdots, & (\widetilde{X}_{(m-1)k+2}, \widetilde{X}_{(m-1)k+3}); \\ \cdots, & \cdots, & \cdots, \cdots \\ (\widetilde{X}_k, \widetilde{X}_{k+1}) & (\widetilde{X}_{2k}, \widetilde{X}_{2k+1}) \cdots, & (\widetilde{X}_{mk}, \widetilde{X}_{mk+1}); \end{array}$$

Estimation Procedure based on $\widetilde{X}_1, \cdots, \widetilde{X}_n$:

Step (1). Considering $(\widetilde{X}_1, \widetilde{X}_2), \cdots, , (\widetilde{X}_{(m-1)k+1}, \widetilde{X}_{(m-1)k+2})$ as if a sample of iid. 2-dimensional random vectors and maximizing the approximated log-likelihood $\sum_{ij} n_{ij}^{(1)} \log p_{ij}$, we then obtain an AMLE of parameters (c, σ^2, ϕ), denoted by $(\widehat{c}_1, \widehat{\sigma}_1^2, \widehat{\phi}_1)$, where $n_{ij}^{(1)}$ is the frequency of (i, j) in the sub-sample;

Step (2). Similarly, by the j-th subgroup of the data, we can obtain an AMLE of parameters (c, σ^2, ϕ), denoted by $(\widehat{c}_j, \widehat{\sigma}_j^2, \widehat{\phi}_j)$, $j = 2, \cdots, k$.

Step (3). Set our estimators of parameters (c, σ^2, ϕ) as the averages of these AMLE's, *i.e.*,

$$\widehat{c} = \sum_{i=1}^{k} \widehat{c}_i/k, \quad \widehat{\phi} = \sum_{i=1}^{k} \widehat{\phi}_i/k, \quad \widehat{\sigma}^2 = \sum_{i=1}^{k} \widehat{\sigma}_i^2/k$$

To compute the AMLE, as an example, consider Step 1. The estimating equations are

$$\begin{cases} \sum_{ij} \dfrac{n_{ij}^{(1)}}{p_{ij}} \dfrac{\partial p_{ij}}{\partial c} = 0 \\ \sum_{ij} \dfrac{n_{ij}^{(1)}}{p_{ij}} \dfrac{\partial p_{ij}}{\partial \phi} = 0 \\ \sum_{ij} \dfrac{n_{ij}^{(1)}}{p_{ij}} \dfrac{\partial p_{ij}}{\partial \sigma^2} = 0 \end{cases}$$

where

$$\frac{\partial p_{ij}}{\partial c} = \int_{i-\frac{1}{2}-\mu}^{i+\frac{1}{2}-\mu} \int_{j-\frac{1}{2}-\mu}^{j+\frac{1}{2}-\mu} \frac{\sqrt{1-\phi^2}(x_1+x_2)}{2\pi\sigma^4} \exp\left(-\frac{x_1^2+x_2^2-2\phi x_1 x_2}{2\sigma^2}\right) dx_2 dx_1 \tag{3.3}$$

$$\frac{\partial p_{ij}}{\partial \phi} = \int_{i-\frac{1}{2}-\mu}^{i+\frac{1}{2}-\mu} \int_{j-\frac{1}{2}-\mu}^{j+\frac{1}{2}-\mu} \left(\frac{\sqrt{1-\phi^2}x_1x_2}{\sigma^4} - \frac{\phi}{\sigma^2\sqrt{1-\phi^2}}\right) \times \frac{1}{2\pi} \exp\left(-\frac{x_1^2+x_2^2-2\phi x_1 x_2}{2\sigma^2}\right) dx_2 dx_1$$

and

$$\frac{\partial p_{ij}}{\partial \sigma^2} = \int_{i-\frac{1}{2}-\mu}^{i+\frac{1}{2}-\mu} \int_{j-\frac{1}{2}-\mu}^{j+\frac{1}{2}-\mu} \left(\frac{\sqrt{1-\phi^2}(x_1^2+x_2^2-2\phi x_1x_2)}{2\sigma^6} - \frac{\sqrt{1-\phi^2}}{\sigma^4}\right) \times \frac{1}{2\pi} \exp\left(-\frac{x_1^2+x_2^2-2\phi x_1 x_2}{2\sigma^2}\right) dx_2 dx_1.$$

Theorem 3.4. *Choosing* $k = k_n \to \infty$ *with order* $o(n^{2/3})$, *then the AMLE* $(\widehat{c}, \widehat{\phi}, \widehat{\sigma}^2)$ *obtained by proposed estimation procedure based on* $\widetilde{X}_1, \cdots, \widetilde{X}_n$ *are weakly consistent. The AMLE is aslo strongly consistent if* $-(1+\delta)\log n/\log|\phi| \leqslant k \leqslant n^{(1-\delta)/3}$ *for some positive constant* δ.

Theorem 3.5. *If $k = k_n$ is selected to satisfy the condition for strong consistency of Theorem 3.4, then the AMLE $(\widehat{c}, \widehat{\phi}, \widehat{\sigma}^2)$ are asymptotically multivariate normally distributed, that is,*

$$\sqrt{\frac{n}{m}} \begin{pmatrix} \widehat{c} - c \\ \widehat{\phi} - \phi \\ \widehat{\sigma}^2 - \sigma^2 \end{pmatrix} \sim N(0, \mathbf{G}).$$

3.3 The general AR(p) model

For the general AR(p) model

$$X_t = c + \phi_1 X_{t-1} + \cdots + \phi_p X_{t-p} + \epsilon_t$$

where $\epsilon_t \overset{i.i.d.}{\sim} N(0, \sigma^2)$ for $t = p+1, \cdots, n$. Let $\boldsymbol{\theta} = (c, \phi_1, \cdots, \phi_p, \sigma^2) = (c, \boldsymbol{\phi}, \sigma^2)$. Here, only rounded data $\widetilde{X}_1, \cdots, \widetilde{X}_n$ are available. Similar to the $AR(1)$ model, we may use a similar procedure to obtain the estimates $(\hat{c}, \hat{\boldsymbol{\phi}}, \hat{\sigma}^2)$ of parameters $(c, \boldsymbol{\phi}, \sigma^2)$. The sample can be split into the following k groups:

Sub-samples:

$$\begin{array}{ll} (1) & \widetilde{X}_1 \cdots \widetilde{X}_{p+1} \ \cdots \ \widetilde{X}_{(m-1)k+1} \cdots \widetilde{X}_{(m-1)k+p+1}; \\ (2) & \widetilde{X}_2 \cdots \widetilde{X}_{p+2} \ \cdots \ \widetilde{X}_{(m-1)k+2} \cdots \widetilde{X}_{(m-1)k+p+2}; \\ & \cdots \qquad\quad \cdots \ \cdots \\ (k) & \widetilde{X}_k \cdots \widetilde{X}_{k+p} \ \cdots \ \widetilde{X}_{mk} \cdots \widetilde{X}_{mk+p}; \end{array}$$

where $m = [(n-p)/k]$.

Estimation Procedure based on $\widetilde{X}_1, \cdots, \widetilde{X}_n$:

Step (1). Considering $(\widetilde{X}_1 \cdots \widetilde{X}_{p+1}), \cdots, (\widetilde{X}_{mk+1} \cdots \widetilde{X}_{mk+p+1})$ as if a sample of iid. p-dimensional random vectors and maximizing the approximated log-likelihood $\sum_{ij} n_{ij}^{(1)} \log p_{ij}$, we then obtain an AMLE of parameters $(c, \boldsymbol{\phi}, \sigma^2)$, denoted by $(\widehat{c}_1, \widehat{\boldsymbol{\phi}}_1, \widehat{\sigma}_1^2,)$, where $n_{ij}^{(1)}$ is the frequency of (i, j) in the sub-sample;

Step (2). Similarly, by the j-th subgroup of the data, we also obtain an AMLE of parameters $(c, \boldsymbol{\phi}, \sigma^2)$, denoted by $(\widehat{c}_j, \widehat{\boldsymbol{\phi}}_j, \widehat{\sigma}_j^2)$, $j = 2, \cdots, k$.

Step (3). Set our estimators of parameters $(c, \boldsymbol{\phi}, \sigma^2)$ as the averages of these approximated MLE's, *i.e.*,

$$\widehat{c} = \sum_{i=1}^{k} \widehat{c}_i / k, \quad \widehat{\boldsymbol{\phi}} = \sum_{i=1}^{k} \widehat{\boldsymbol{\phi}}_i / k, \quad \widehat{\sigma}^2 = \sum_{i=1}^{k} \widehat{\sigma}_i^2 / k$$

Remark 3.6. *The consistency and asymptotic normality of the obtained estimates can be established similar to Theorems 3.1 and 3.2.*

4 An Example

We shall reanalyze the data in Table Series C from the book *Time Series Analysis (Forecasting and Control)* (Box *et al.* 1994, Third Edition, P544) to illustrate our estimation procedures. The 226 data are records of chemical process temperature readings per minute. From Table Series C, we can see that the 226 chemical temperature data take values at most to one digit after decimal point. But we know that temperature is continuous. Thus, the 226 data are in fact rounded data. If the conventional methods are directly used to analyze the rounded data, this will cause some serious errors, as mentioned in Section 1. In order to avoid these problems, we use our estimation procedures to deal with the rounded data. As pointed out in the book (P189) that data in Table series C is suggested to be an AR(1) process about $\phi = 0.8$ after taking the first difference. That is, the data satisfy the following model

$$\triangledown x_t = \phi \triangledown x_{t-1} + \epsilon_t$$

where $\triangledown$ denotes the first difference notation, that is, $\triangledown x_t = x_t - x_{t-1}$. By using our estimation procedure for rounded data, the estimate of ϕ is 0.716. The original estimate of ϕ in the book *Time Series Analysis* is 0.8. Then we can obtain that the difference between our estimate and the estimate in this book is 0.084. The reason causing the difference is whether we use conventional methods to directly analyze the rounded data or not. We have shown that it often leads to inconsistent estimates to use directly conventional methods to rounded data. But our estimate procedure can obtain consistent estimators which are also asymptotically normally distributed.

5 Simulation Results

Without loss of generality, AR(1) model

$$X_t = \phi X_{t-1} + \epsilon_t$$

where $\epsilon_t \sim N(0, \sigma^2)$ (*i.i.d.*) is used for simulation. $X_1, \cdots, X_n$ is simulated from the AR(1) model and let the corresponding rounded data be $\tilde{X}_1, \cdots, \tilde{X}_n$. Different configuration of parameters are used to obtain the samples. The estimates of parameters based on the above estimation procedure and the corresponding mean square errors are listed in the following tables. The simulation is based on 200 repeated procedure. The Q-Q plot based on 200 estimates of ϕ is given. From the following table 1, we can know that

(1). As the sample size becomes bigger, the MSEs of parameters estimates become smaller.

(2). In the estimation procedure, when the total sample size is the same, the choose of m is important. Although when m is large, the assumption of independence is more suitable. However, $k = [n/m]$ becomes smaller. It causes the MSEs becomes bigger.

Furthermore, the following figures 5.1,5.2,5.3 and 5.4 show that the estimates of parameters are asymptotically normal, which ensures Theorem 3.2.

Table 1. Simulation Results

	$(\phi = 0.3, \sigma^2 = 1.0)$	
$n(m)$	$\hat{\phi}$(MSE)	$\hat{\sigma}^2$(MSE)
500(5)	0.300(0.0026)	0.999(0.0012)
500(10)	0.298(0.0028)	0.9892(0.0015)
	$(\phi = 0.5, \sigma^2 = 1.0)$	
500(5)	0.499(0.0018)	0.990(0.0012)
500(20)	0.464(0.0027)	1.002(0.0015)
1000(20)	0.488(0.0012)	0.995(0.0007)
2000(20)	0.493(0.0005)	0.993(0.0003)
10000(20)	0.499(0.0001)	0.998(0.00005)
1000(10)	0.492(0.0010)	0.991(0.0008)
1005(15)	0.493(0.0009)	0.991(0.0009)
	$(\phi = 0.75, \sigma^2 = 1.0)$	
1000(5)	0.738(0.0007)	0.988(0.0008)
1000(10)	0.732(0.0009)	0.987(0.0009)
1000(20)	0.717(0.0017)	0.998(0.0008)
1000(40)	0.670(0.0072)	1.018(0.0014)
2000(5)	0.744(0.0003)	0.990(0.0005)
	$(\phi = 0.9, \sigma^2 = 1.0)$	
1000(5)	0.842(0.0064)	0.972(0.0038)
1000(10)	0.804(0.013)	0.989(0.0045)

6 Comments and Conclusions

Rounded data are often encountered in many situations, especially for continuous data cases. In fact, for many irrational number case, we can not obtain the true number so that we can only obtain its rounded data. Especially now, many instruments can only record rounded data and computers can only deal with rounded data. For these, this paper deals with rounded data in normal distribution, AR(p) and MA(p) time series models, gives estimation procedures and proves the consistency and asymptotic normality of estimates obtained. This paper seems to deal with data rounded at one digit after decimal point. In fact, it is worth to mention that the proposed estimation procedure can be easily extended to data rounded at any digit after decimal point. Although this paper has dealt with some important rounded data problems, in fact, many rounded data problems have not dealt with. For example, how to deal with rounded data in all kinds of statistical models? We need to do further research.

It is important to realize the distinction between "statistical errors for inference based on rounded data" and "the usual rounding errors". This paper focuses on the former, not the latter. As is shown in this paper, some traditional inference procedures, when applied directly to rounded data with a fixed precision, are inconsistent regardless of large sample sizes; while rounding errors can be reduced by increased precisions. t may be possible to combine the two issues in one setting, that would be presented as a nonparametric inference problem.

7 Appendix

Lemma 7.1. *If $X_1, X_2, \cdots, X_n$ is a sample of an $AR(1)$ model with autoregressive coefficient ϕ ($|\phi| < 1$) and normally distributed innovations, then*

$$\begin{aligned} &|g(\mathbf{x}_1, \mathbf{x}_2) - g(\mathbf{x}_1)g(\mathbf{x}_2)| \qquad (7.1)\\ &\leqslant Kg(\mathbf{x}_1)g(\mathbf{x}_2)|\phi|^k \exp\left\{\frac{(1-\phi^2)|\phi|^k}{2\sigma^2(1+\phi^{2k})}\Big((x_{i_2}-\mu)^2 + (x_{i_3}-\mu)^2\Big)\right\} \times \\ &\qquad \Big(1 + |x_{i_2}-\mu|^2 + |x_{i_3}-\mu|^2\Big), \end{aligned}$$

where $g(\mathbf{x}_1, \mathbf{x}_2)$ is the joint density of $(\mathbf{x}_1, \mathbf{x}_2)$ and $g(\mathbf{x}_1)$, $g(\mathbf{x}_2)$ are the joint densities of $\mathbf{x}_1 = (X_{i_1}, X_{i_1+1}, \cdots, X_{i_2})$ and $\mathbf{x}_2 = (X_{i_3}, X_{i_3+1}, \cdots, X_{i_4})$, respectively. Also, $k = i_2 - i_1$. Here K is an absolute constant depending on k and ϕ only.

Furthermore, for $\ell > 2$,

$$|g(\mathbf{x}_1, \cdots, \mathbf{x}_\ell) - \prod_{t=1}^{\ell} g(\mathbf{x}_t)|$$

$$\leqslant K \sum_{t=1}^{\ell-1} |\phi|^{k_t} g_t(\mathbf{x}_1, \cdots, \mathbf{x}_\ell) \exp\left\{\frac{(1-\phi^2)|\phi|^{k_t}}{2\sigma^2(1+\phi^{2k_t})}\Big((x_{2t}-\mu)^2 + (x_{2t+1}-\mu)^2\Big)\right\} \times$$

$$\Big[1 + |x_{2t}-\mu|^2 + |x_{2t+1}-\mu|^2\Big]$$

where

$$\begin{aligned} \mathbf{x}_t &= (x_{i_{2t-1}}, \cdots, x_{i_{2t}}), \\ k_t &= i_{2t+1} - i_{2t}, \\ g_t(\mathbf{x}_1, \cdots, \mathbf{x}_\ell) &= g(\mathbf{x}_1, \cdots, \mathbf{x}_t) g(\mathbf{x}_{t+1}) \cdots g(\mathbf{x}_\ell) \end{aligned}$$

and K is the same as in (7.1).

Lemma 7.2. *Suppose the assumptions of Lemma 1 hold with Assume that f is a k-dimensional measurable function such that $Ef(X_i) = 0$ and $Ef(X_i)f(X_i)^T = \gamma_0$ exists. If $\mathbf{V} = \gamma_0 + \sum_{j=1}^{\infty}[\gamma_j + \gamma_j^T]$ exists and is positive definite, then, with $Z_i = f(X_i)$,*

$$\frac{1}{\sqrt{n}}(Z_1 + \cdots + Z_n) \xrightarrow{L} N(\mathbf{0}, \mathbf{V})$$

where $\gamma_j = Ef(X_1)f(X_{1+j})^T = \gamma_{-j}^T$.

References

[1] Box, G. E. P., Jenkins, G. M. and Reinsel, G. C., Time series analysis: Forecasting and control. Prentice-Hall, Inc. C. 1994.

[2] Burridge, J., A note on maximum likelihood estimation for regression models using grouped data. J. Roy. Stat. Soc. Ser. B., 43 (1981), 41–45.

[3] Dempster, A. P. and Rubin, D. B., Rounding error in regression: the appropriateness of Sheppard's corrections. J. Roy. Stat. Soc. Ser. B., 45 (1983), 51–59.

[4] Fisher, R. A., The use of multiple measurements in taxonomic problems. Ann. Eugenics., 8 (1936), 179–188.

[5] Grimshaw, S. D., McDonald, J., McQueen, G. R. and Thorley, S., Estimating hazard functions for discrete lifetimes. Communications in Statistics: Simulation and Computation, 34 (2005), 451–463.

[6] Haitovsky, Y., Grouped data, in Encylopedia of Statistical science, Wiley, New York. 1982.

[7] Hall, P., The influence of rounding errors on some nonparametric estimators of a density and its derivatives. Siam. J. Appl. Math., 42 (1982),390–399.

[8] Hall, P. and Heyde, C.C. (1980), Martingale Limit Theory and Its Application, New York: Academic Press.

[9] Heitjan, D. F., Inference from grouped continuous data: a review. Stat. Sci., 4 (1989), 164–183.

[10] Heitjan, D. F. and Rubin, D. B., Ignorability and coarse data. Ann. Stat., 19 (1991), 2244–2253.

[11] Jones, R. H., Maximum Likelihood Fitting of ARMA Models to Time Series With Missing Observations. Technometrics, 22(1980), 389–395.

[12] Lee, C. S. and Vardeman, S. B., Interval Estimation of a Normal Process Mean from Rounded Data. Journal of Quality Technology, 33 (2001), 335–348.

[13] Lee, C. S. and Vardeman, S. B., Interval estimation of a normal process standard deviation from rounded data. Comm. Stat. Simulat. Comput.,31 (2002), 13–34.

[14] Lee, C. S. and Vardeman, S. B., Confidence interval based on rounded data from the balanced one-way normal random effects model. Comm. Stat. Simulat. Comput., 32 (2003), 835–856.

[15] Machak, J. A. and Rose, E. L., Uniform Additive Outliers In ARMA Models. Proceedings of the American Statistical Association, Business and Economic Statistics Section,(1984), 577–580.

[16] Machak, J. A. and Rose, E. L., Robust Identification and Estimation of Time series. Proceedings of the American Statistical Association, Business and Economic Statistics Section. (1985).

[17] Rockafellar, R. T., Convex Analysis. Princeton University Press. 1970.

[18] Sheppard, W. F., On the calculation of the most probable values of frequency constants for data arranged according to equidistant divisions of a scale. Proc. London Math. Soc., 29 (1898), 353–380.

[19] Silvapulle, M. J. and Burridge, J., Existence of maximum likelihood estimates in regression models for grouped data and ungrouped data. J. Roy. Stat. Soc. Ser. B., 48 (1986), 100–106.

[20] Stam, A. and Cogger, K. O., Rounding Errors in Autoregressive Processes.International Journal of Forecasting, 9 (1993), 487–508.

[21] Titchmarsh, E. C., The Theory of Functions, 2nd ed. Oxford Univ. Press. 1939.

[22] Tricker, A., Coates, E. and Okell, E., The effect on the R chart of precision of measurement. J. Qual. Tech., 30 (1998), 232–239.

[23] Vardeman, S. B. and Lee, C. S., Likelihood-based statistical estimation from quantization data. IEEE Trans on Instru. Measure., 54 (2005), 409–414.

[24] Vardeman, S. B., Sheppard's correction for variances and the "Quantization noise model". IEEE Trans on Instru. Measure., 54 (2005), 2117–2119.

[25] Wright, D. E. and Bray, I. A mixture model for rounded data. American. Statistician., 52 (2003),3–13.

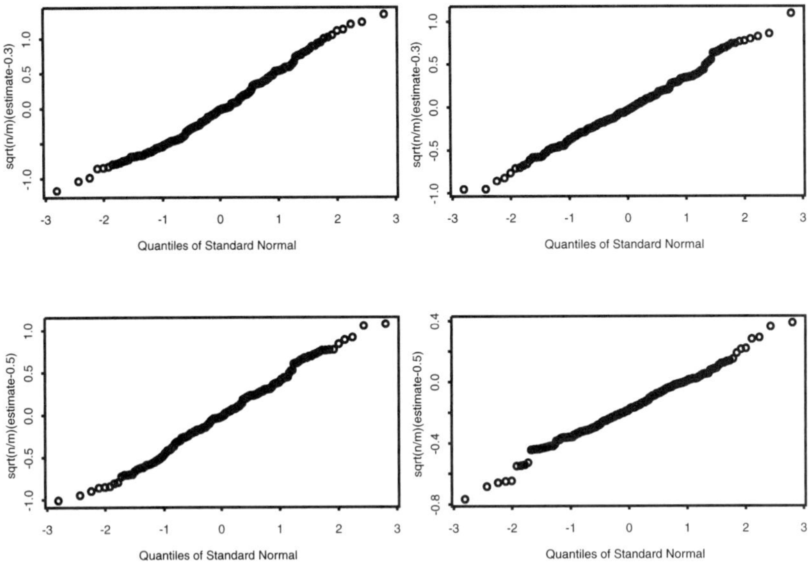

Figure 5.1:

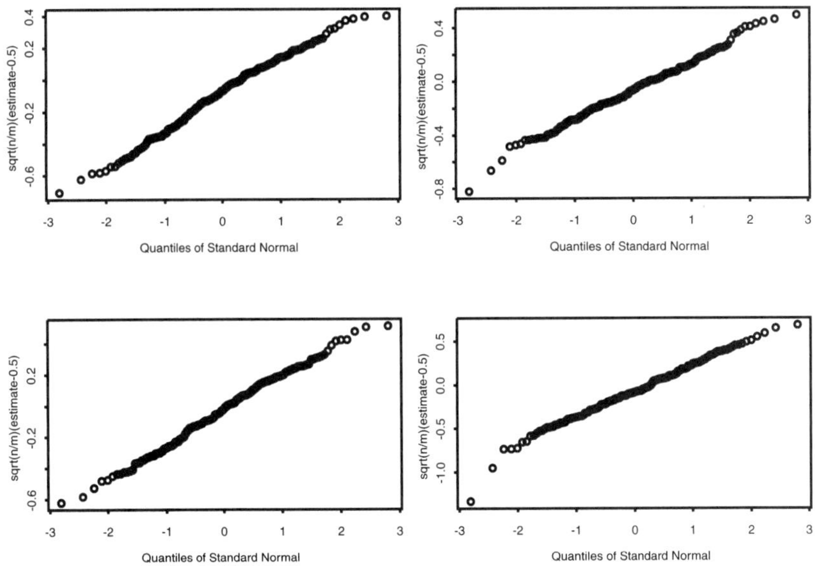

Figure 5.2:

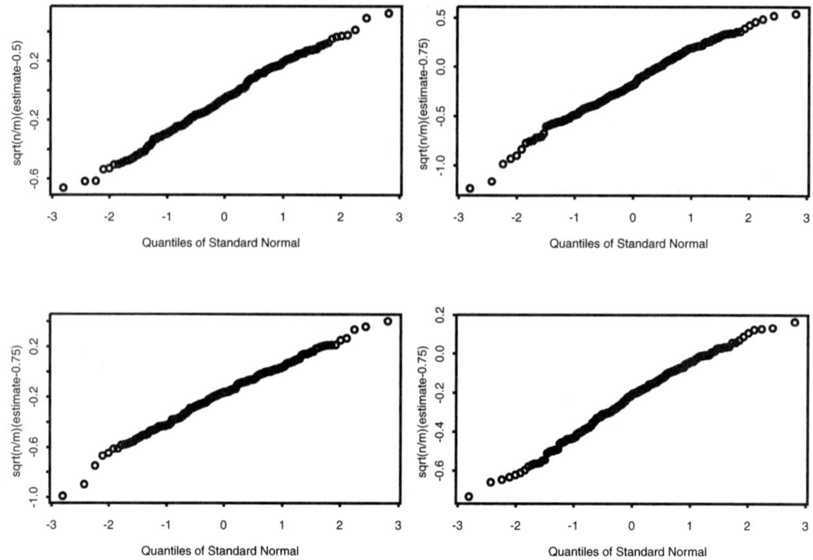

Figure 5.3:

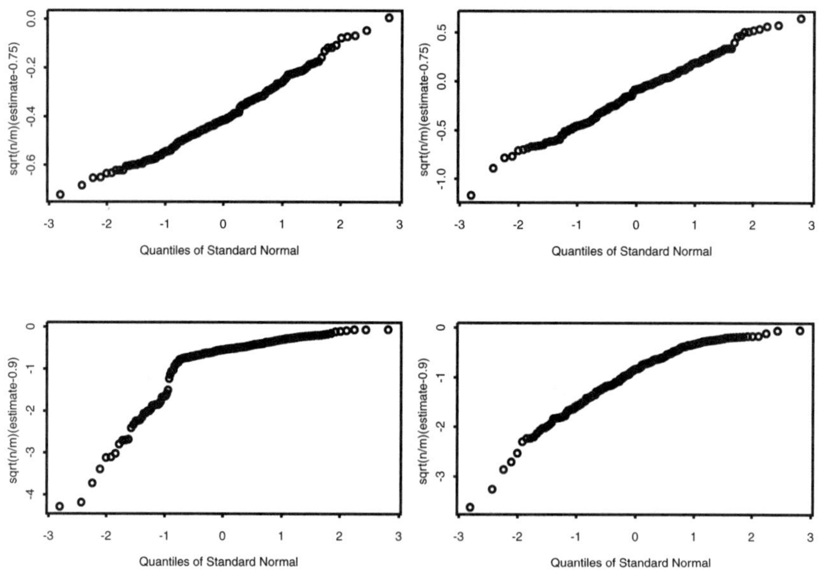

Figure 5.4:

Piecewise Regression Models: Estimation Theory and Applications

Lianfen Qian
Department of Mathematical Sciences, Florida Atlantic University
Boca Raton, FL 33431, USA
E-mail: Lqian@fau.edu

Abstract

When classical linear regression models are not adequate in modeling real data, nonlinear regression models are the natural option. Piecewise regression models are the most popular utilized nonlinear regression models in various applications. In this paper, we survey applications and theoretical research results for piecewise regression models, typically focusing on two phase linear regression models with an unknown change point. For the purpose of easy accessing and motivating reading, we first present a few concrete examples for different types of piecewise regression models. Then we continue to present hypothesis tests, estimations and limit distributions of the parameters in the piecewise regression models. Finally, we address some computational aspects and give final remarks for recent developments.

1 Introduction and Examples

In the last two decades, a considerable body of literature appears on the hypothesis tests, parameter estimation and relative computing program on change points for piecewise regression models. Various practical applications for such models have been appeared in diversified research areas such as biology (Gaylor et al. 1988; Vieth, 1989; Piegorsch and Bailer, 1997), medical science (Smith and Cook, 1980; Cox, 1987; Gaylor, et. al. 1988; Berman at. el. 1996; Piepho and Ogutu, 2003; Muggeo, 2003), epidemiology (Ulm, 1991; Pastor and Guallar, 1998), environmental science (Anderson and Nelson, 1975; Gbur, et. al. 1979; Qian and Ryu, 2003; Toms and Lesperance, 2003; Piegorsch and Bailer, 2005), software engineering (Qian and Yao, 2002), finance and econometrics

(Chow, 1960; Koul and Qian, 2002; Fiteni, 2004; Zeileis, 2006) to name a few.

One special piecewise regression model is called a two-phase linear regression model. A two-phase linear regression model is defined by two linear models over two different domains of the design variable. To be precise, let Y be the response variable and X be an univariate predictor so that (X, Y) be a bivariate random vector with $E|Y| < \infty$. The conditional expectation is denoted by $m(x) = E(Y|X = x)$.

Let $\mathbb{R}$ be the real line and $\bar{\mathbb{R}} = \mathbb{R} \cup \{-\infty, \infty\}$. Then the set $\bar{\mathbb{R}}$ is compact under the metric $d(x, y) = |\arctan x - \arctan y|$, $x, y \in \mathbb{R}$. Throughout we assume that θ is an interior point of the parameter space $\Theta = K \times \bar{\mathbb{R}}$ for a known compact set K in $\mathbb{R}^4$. A typical point in Θ will be denoted by $\vartheta' = (\beta_{10}, \beta_{11}, \beta_{20}, \beta_{21}, s) = (\vartheta_1', s)$. We define

$$m(x, \vartheta) = (\beta_{10} + \beta_{11}x)I(x \leqslant s) + (\beta_{20} + \beta_{21}x)I(x > s), \quad x \in \mathbb{R},$$

and consider a set of independent observations (X_i, Y_i), $i = 1, \cdots, n$, such that for some $\theta' = (\theta_1', \tau) = (b_{10}, b_{11}, b_{20}, b_{21}, \tau) \in \mathbb{R}^5$,

$$Y_i - m(X_i, \theta) = \varepsilon_i, \quad i = 1, \cdots, n, \tag{1.1}$$

are independent identically distributed (i.i.d.) random variables and τ is called the change point.

Bhattacharya (1990, 1994) gave a model classification. That is, there are two types of two-phase regression models: continuous (also named restricted and segmented) and discontinuous (unrestricted). By continuous, the two phases meet at the change point τ (also called joint point). By discontinuous, the two phases are separated at the change point τ with some jump. Within discontinuous models, there are further classification of fixed jump and contiguous jump. By fixed jump, the jump size at τ is a fixed number, neither equal to zero nor dependent on the sample size indicating a abrupt change, while by contiguous jump, the jump size goes to zero when the sample size goes to infinity indicating a gradually change.

Among the first studies, the hypothesis tests and parameter estimation problems under regression setting for change point problem were introduced by Quandt (1958, 1960) using maximum likelihood method under normal error. When the data pairs are observed over a discrete time points, the model changes at a change point of time. Before this change point of time, the relationship follows one linear model, while after the change point of time, the relationship follows another linear model.

Notice that when the design points $\{X_i\}$ are ordered already in model (1.1), then Quandt's model is equivalent to two-phase separate regression model, which switches at an unknown time k such that $X_k = \tau$. Hence both k and τ are called the change point.

Under a strong assumption of known change point τ, Sprent (1961) discussed hypothesis tests of two-phase continuous (segmented) regression models for several data sets with application in biometry. Under continuity constraint, Hudson (1966) obtained the estimated parameters using method of least squares. Hasselblad *el. al* (1976) applied Hudson's method to the specific case of a hockey-stick function, which assumes that the left regime of the data set is fitted by a linear regression with slope zero. The evolution of the names of hockey-stick models is very colorful. For the diversified real applications of the models, it can be very differently named. It is called *linear-plateau* model by Anderson and Nelson (1975), and *threshold response* model by Gaylor et al. (1988) and Ulm (1991), in the analysis of environmental and biological data in predicting crop yield based on the amount of nitrogen in the soil. For a *left linear plateau* model, we write

$$m(x, \theta) = b_0 I(x \leqslant \tau) + [b_0 + b_1(x - \tau)]I(x > \tau), \quad x \in \mathbb{R}, \tag{1.2}$$

where $\theta' = (b_0, b_1, \tau)$.

Example 1.1: (Determination of Osmotic Threshold) Vieth (1989) fitted the continuous two phase linear model (1.1) to plasma osmolality against arginine vasopressin (AVP) concentration in plasma of conscious dogs to determine the osmotic threshold. The detected threshold is 33.37 plasma osmolality (mOsm/kg) using least squares (LS) fit shown in Figure 1.1.

As one can see that the plasma osmolality data in Example 1.1 can also be fitted by the left plateau model (1.2). Piegorsch and Bailer (2005, p. 45) fit the data reported by Braña and Ji (2000) on embryogenic response in wall lizards (*Podarcis muralis*) by manipulating selected environmental conditions associated with their embryonic development using the *left linear plateau* model (1.2). For the data set and detail modeling, please refer to Piegorsch and Bailer (2005).

In reality, data often show patterns with more than one change point. One of the extended model of (1.2) is a multiple plateaus model. For example, a continuous model with both left and right plateaus can be expressed as

$$m(x, \theta) = \begin{cases} b_0, & \text{if } x \leqslant \tau_1, \\ b_0 + b_1(x - \tau_1), & \text{if } \tau_1 < x \leqslant \tau_2, \\ b_0 + b_1(\tau_2 - \tau_1), & \text{if } x > \tau_2, \end{cases} \tag{1.3}$$

where τ_1 and τ_2 are the change points and $\theta' = (b_0, b_1, \tau_1, \tau_2)$.

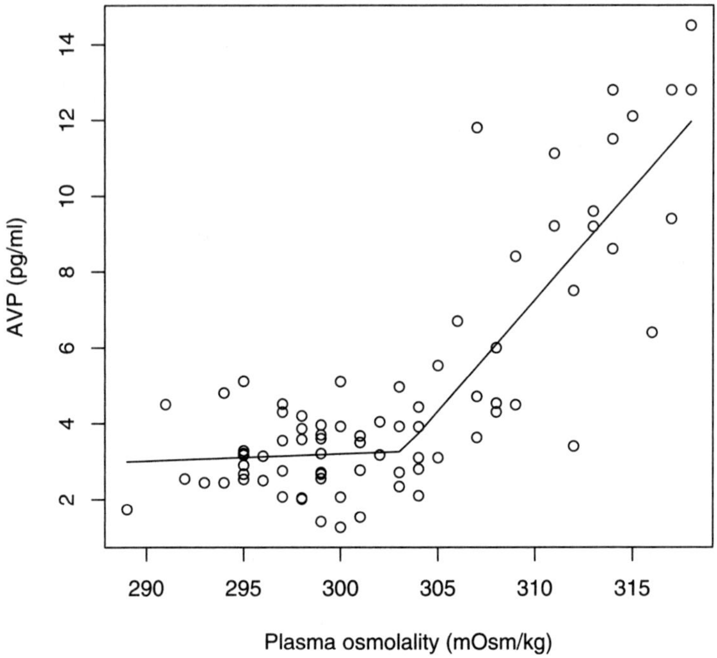

Figure 1.1: Relationship between plasma osmolality and arginine vasopressin (AVP) concentration in plasma of conscious dogs.

Example 1.2: (Pond Water Levels) Small and isolated bodies of water within larger ecosystems often exhibit substantial variation in their water levels, due ostensibly to factors associated with their supply sources (data from Piegorsch and Bailer, 2005). Blood *el. al.* (1997) studied this phenomenon through data collected from 59 small ponds in Georgia State in USA, where the water levels of those ponds are affected by the nearby deep Floridian ponds. The variables measured are Y=pond water level vs. x=aquifer level. Visualizing the data indicates that the model (1.3) with left and right plateau is adequate (Figure 1.2). The detected change points using LS fit are 37.893 and 41.079. The corresponding PROC NLIN SAS code is included in Figure 1.3 for convenience of reproducing the fits.

Abrupt (fixed jump size) change or graduate (contiguous) change may happen in real life because of policy changes, or management skill

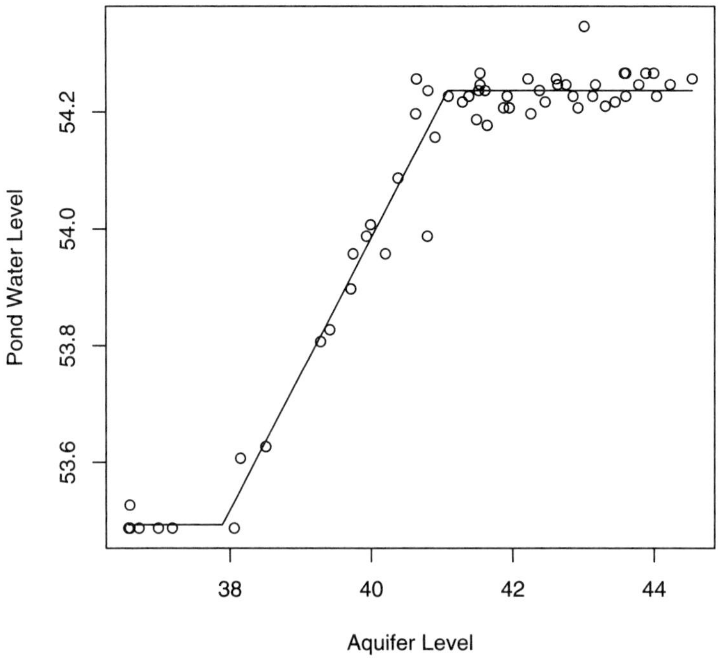

Figure 1.2: Pond water level versus aquifer level with the fitted three phase lines.

changes or the nature of the problem changes. One of the example is given below for modeling software effort using abrupt change estimation procedure.

Example 1.3: (Software Effort Cost) As Shepperd, M. and Schofield, C. (1997) pointed out, an important aspect of any software development project is to know how much it costs. In most cases the major cost factor is labor. For this reason, estimation of software project effort is one of the most important empirical modeling tasks in software engineering, as indicated by the large number of models developed over the past two decades. Shin and Goel (SG,2000) studied the modeling of the software project effort using radial basis functions. They used the well-known NASA software project data set to validate the model. The method proposed in SG is complicated and contains many parameters to be estimated.

```
/* SAS code to fit three phase pond water level vs. deep aquifer level in a wetland pond; */
/* We need to estimate the four parameters: b0, b1, tau1, tau2.                          */
data pond;
  input aquifer pond @@;
datalines;
 41.54  54.267 42.45  54.217 43.88  54.267 41.52  54.237 44.22  54.247 42.85  54.227
 40.63  54.197 42.37  54.237 41.09  54.227 41.54  54.247 43.99  54.267 43.13  54.227
 42.37  54.237 40.90  54.157 39.93  53.987 40.20  53.957 39.28  53.807 36.58  53.527
 36.56  53.487 36.98  53.487 43.60  54.227 42.61  54.257 43.31  54.210 41.29  54.217
 44.03  54.227 42.25  54.197 42.21  54.257 41.92  54.227 40.80  54.237 41.61  54.237
 43.78  54.247 43.17  54.247 41.95  54.207 40.38  54.087 39.74  53.957 40.79  53.987
 38.51  53.627 36.58  53.487 36.71  53.487 37.18  53.487 42.92  54.207 43.01  54.347
 41.87  54.207 44.53  54.257 43.45  54.217 41.64  54.177 42.63  54.247 41.38  54.227
 40.64  54.257 43.60  54.267 43.58  54.267 42.75  54.247 41.49  54.187 39.99  54.007
 39.41  53.827 39.71  53.897 38.06  53.487 36.58  53.487 38.15  53.607
;
/* We need to sort the data to make sure the fitted curve can be drawn correctly; */
proc sort data=pond threads;  /* overrides system option */
  by aquifer;
run;

proc nlin method=marquardt;
parameters b0=53.3 b1=0.1641  tau2=41.0 tau1=38;  /*initial parameters*/
if aquifer <=tau1 then do;
       model pond=b0;                          * first segment;
              der.b0=1;
              der.b1=0;
              der.tau1=0;
              der.tau2=0;
end;
if ((aquifer>tau1)& (aquifer <=tau2)) then do;        * middle segment;
       model pond=b0+b1*(aquifer-tau1);
              der.b0=1;
              der.b1=aquifer-tau1;
              der.tau1=-b1;
              der.tau2=0;
end;
if (aquifer>tau2) then do;
     model pond=b0+b1*(tau2-tau1);                   * third segment;

            der.b0=1;
            der.b1=tau2-tau1;
            der.tau1=-b1;
            der.tau2=b1;
end;
```

Figure 1.3: Sample SAS program to fit continuous three segment model to pond water level data.

The scatterplot (Figure 1.4) of the software effort cost versus the line of code (in thousands) of the projects appears to be an abrupt change in the effort. Qian and Yao (2002) re-fitted the well-known NASA software project data set. The discontinuous model (1.1) with fixed jump size is used to detected change point at 31.1 lines of codes (in thousands) shown in Figure 1.4. It is wise to do not make the continuity assumption since there is an apparent discontinuity after the value of 31.1 lines of codes (in thousands). If a continuity assumption were to imposed, either it would need to impose an unwarranted constraint on the regression function leading to a bad fit, or we would have to use more complicated functional forms that would otherwise be necessary. Furthermore, Qian

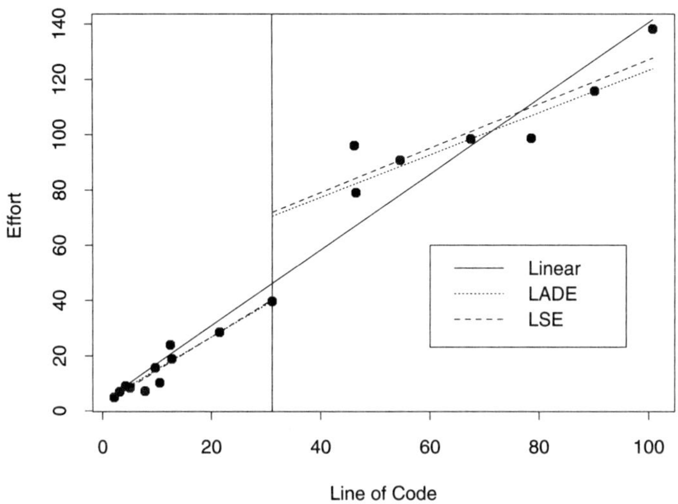

Figure 1.4: The scatterplot of Effort vs Line of Code with three fitted curves.

and Yao showed that the two-phase linear regression using least absolute deviance estimation method is the most robust model comparing with linear model, COCOMO model (Boehm 1981) and SG method.

For much earlier research applications, more examples can be found from Seber and Wild (1989, Chapter 9) and Piegorsch and Bailer (1997, section 7.1.2) which gives a good summary on threshold models and other nonlinear forms with applications in environmental sciences.

In the fixed design and restricted case, Feder (1975) gave a rigorous treatment covering the two phase regression model under i.i.d errors. Feder (1975) provided a proof that the likelihood ratio statistic is asymptotically χ^2 based on the assumption that the parameter vector remains identified under null hypothesis for i.i.d errors. It is obvious that Feder's (1975) results can not be used for testing of no change point.

Two comprehensive survey were given by Shaban(1980) and Schulze (1987). Shaban (1980) simply quoted a list of the author's summaries (abstracts) except in a few cases and classified the papers published before 1980 depending on the approach adopted by the author(s) for change-point problem and the related problem in segmented regression models. Schulze (1987) provided a collection of existing methods mainly focusing on the least squares estimation, testing of hypotheses and testing of model stability for analyzing data using multiphase segmented regression models majoring on two-phase models. Park (1978) discussed some design issues when the change points are known. Onukogu (1984) discussed question: when is a piecewise regression really necessary? In another word, it is important to test the hypothesis of having change versus no change since under no change hypothesis the change point is not identifiable. Hence the standard likelihood theory is not applicable.

Kim and Siegmund (KS, 1989) and Kim (K, 1993) studied the likelihood ratio test for a change point in simple linear regression with homogeneous and non-homogeneous errors, respectively. They consider two models: change only happens for intercept (step-type), changes on both intercept and slope.

Brodsky and Darkhovsky (1993) summarized nonparametric methods in change-point problems. Carlstein, Müller and Siegmund (1994) edited a volume of lecture notes on change-point problems for the monograph series in Institute of Mathematical Statistics. Csörgő and Horváth (1997) published a book on limit theorems in change-point analysis. Chen and Gupta (2001) published a survey on change point detection and estimation. Wu (2005) published a lecture note in statistics focusing on inference for change point and post change means after a CUSUM test.

Within regression setting, the two types of models, in terms of continuous and discontinuous, require distinctly different statistical methodologies in proving limiting property of the change point. There are three important problems to address:

(a) testing simple linear regression versus two-phase linear regression with one single unknown change point.

(b) making inference about the change point.

(c) making inference about the regression parameters and the error variances for the two phases.

The next portion of the paper is organized as follows. Section 2 addresses hypothesis testing issues based on likelihood ratio and informational based approaches. Section 3 focuses on reviewing the up-to-date research activities on asymptotic limiting theory for maximum likelihood estimators including change point estimator for continuous case. Section 4 reviews estimation methods for change points and regression parameters for discontinuous case. Section 5 addresses some computational aspects. Section 6 concludes the paper with remarks on recent developments and extensions to model other types of data such as time series and longitudinal data.

2 Hypothesis Testing

For continuous two phase regression model, Davies (1977, 1987) derived test statistics for both normal and chi-squared cases when a nuisance parameter is present only under the alternative hypothesis. In this section, we focus on the hypothesis tests for discontinuous models utilizing two approaches: Likelihood ratio based approach and informational based approach.

2.1 Likelihood Ratio Test

A likelihood ratio approach was first considered by Quandt (1958) to detect a change in a simple linear regression. Kim and Siegmund (1989) discussed analytic difficulties associated with the likelihood ratio test. Kim and Cai (1993) studied the robustness of the likelihood ratio test for a change in simple linear regression and concluded that the test statistics converge to same distribution regardless of the underlying distribution.

The hypothesis test for a single linear regression versus two-phase linear regression (1.1) with a change at a fixed observation index, say, k between 2 and $n-2$ can be formulated to test null hypothesis:

$$H_0 : E(Y_i|X_i = x_i) = b_0 + b_1 x_i, \text{ for all } i = 1, \cdots, n$$

versus the alternative hypothesis

$$H_1 : E(Y_i|X_i = x_i) = \begin{cases} b_{10} + b_{11}x_i, \ i = 1, \cdots, k \\ b_{20} + b_{21}x_i, \ i = k+1, \cdots, n. \end{cases}$$

In the fixed design case, where we use lower case letter for the design variable, $x_1 < \cdots < x_k \leqslant \tau < x_{k+1} < \cdots < x_n$ are non-stochastic and in (1.1), the coefficient vector changes after kth observations or $x_k = \tau$, so that either τ or k can be regarded as the change-point. Bhattacharya (1994) formulate the problem of testing simple linear regression versus two-phase linear regression with a single unknown change point into to test

$$H_0 : k = n$$

against the alternative

$$H_1 : 1 \leqslant k \leqslant n-1.$$

In this section, we address the hypothesis tests under general setting of the design variable X for the discontinuous models (no constraint on all parameters). That is, the design variable X in general is a random variable. For simplicity, we assume that the error term in (1.1) is normal distributed. Then for any parameter $\vartheta \in \Theta$, the conditional likelihood function of the model (1.1) is

$$L(\vartheta, \sigma^2) = \left(2\pi\sigma^2\right)^{-\frac{n}{2}} \exp\left\{-\frac{1}{2\sigma^2}\sum_{i=1}^{n}(Y_i - m(X_i;\vartheta))^2\right\}. \tag{2.1}$$

Let $X_{(i)}$ be the order statistics of $\{X_i\}$ and $Y_{(i)}$ be the concomitant of $X_{(i)}$. For a fixed $s \in \bar{\mathbb{R}}$, denote $i_s = \max\{i : X_i \leqslant s\}$, $Y_s^{(1)} = (Y_{(1)}, \cdots, Y_{(i_s)})'$ and $Y_s^{(2)} = (Y_{(i_s+1)}, \cdots, Y_{(n)})'$. The corresponding design matrices are denoted by $X_s^{(1)}$ and $X_s^{(2)}$. Then a two-step algorithm for likelihood based estimation is given below:

Step 1: For a given $s \in \overline{\mathbb{R}}$, define

$$\hat{\theta}_{1js} = \left(X_s^{(j)'} X_s^{(j)}\right)^{-1} X_s^{(j)} Y_s^{(j)}, j = 1, 2 \tag{2.2}$$

and

$$\hat{\sigma}_s^2 = \frac{S_s^{(1)} + S_s^{(2)}}{n}, \tag{2.3}$$

where $S_s^{(j)}$ is the residual sum of squares for the jth phase of the model, $j = 1, 2$. Notice that for any $1 \leqslant k < n$, and $s \in [X_{(k)}, X_{(k+1)})$, $S_s^{(j)}$ is a constant function in s for $j = 1, 2$. We also denote $S_s^{(j)}$ by $S_k^{(j)}$ for $s \in [X_{(k)}, X_{(k+1)})$. Hence the conditional likelihood function may not have unique maximizers (for discontinuous model), so we suggest to take the maximum likelihood estimator of τ to be the smallest minimizer of $\hat{\sigma}_s^2$ in (2.3), denoted as $\hat{\tau}$.

Step 2: Substitute the change point estimator $\hat{\tau}$ into the equation (2.2), we obtain the estimated regression coefficient vector

$$\hat{\theta}_1' \equiv (\hat{\theta}_{11}', \hat{\theta}_{12}') \equiv (\hat{\theta}_{11\hat{\tau}}', \hat{\theta}_{12\hat{\tau}}').$$

Then the maximum likelihood estimator (MLE) of regression coefficients $\theta_1' \equiv (\theta_{11}', \theta_{12}') = (b_{10}, b_{11}, b_{20}, b_{21})$ is denoted by

$$\hat{\theta}_1' \equiv (\hat{\theta}_{11}', \hat{\theta}_{12}') = (\hat{b}_{10}, \hat{b}_{11}, \hat{b}_{20}, \hat{b}_{21}).$$

Hence the maximum likelihood estimator of $\theta' = (\theta_1', \tau)$ is denoted by $\hat{\theta}' = (\hat{\theta}_1', \hat{\tau})$.

For a given k between 2 and $n-2$ or equivalently $s = X_{(k)}$, $S_k^{(j)} = S_s^{(j)}$ for all $s \in [X_{(k)}, X_{(k+1)})$ for $j = 1, 2$. Let S be the residual sum of squares with no change (simple linear regression model). One constructs an extra sum F-test

$$F_k = \frac{[S - (S_k^{(1)} + S_k^{(2)})]/2}{(S_k^{(1)} + S_k^{(2)})/(n-4)},$$

which has an $F_{2,n-4}$ distribution when H_0 is true. Hence for unknown change point, one intuitive test statistic is

$$F_{\max} = \max_{2 \leqslant k \leqslant n-2} F_k.$$

In the fixed design and discontinuous case, Beckman and Cook (1979) reported some simulation results on the asymptotic null distribution of $F_{\max}$. Beckman and Cook (1979) and Worsley (1983) derived simple bound of the null exceedence probability of $F_{\max}$ using Bonferroni's inequality. Instead of using 2 degrees of freedom in the numerator of F_k, Hinkley (1971) used 3 and reported strong empirical evidence that the distribution of F_k is F distribution with degrees of freedom $(3, n-4)$, hence the distribution of the likelihood ratio statistic under H_0 is asymptotically χ_3^2 under i.i.d. errors. This result has been widely adopted by applied disciplinary. Recently, Lund and Reeves (2002) experimented the degrees of freedom 3 in Hinkley's report and found that using degrees of freedom 2 would be more suitable for undocumented time under the two phase model with both step-type (intercept) and trend-type (slope) changes. They conjectured that a reasonable approximating limiting distribution of $F_{\max}$ (as $n \to \infty$) would involve the Gumbel extreme value law:

$$P[F_{\max} \leqslant x] = \exp\{-[e^{-(x-a_n)/b_n}]\},$$

for appropriately selected scaling constants a_n and b_n. This is still an open problem for readers. Wang (2003) in commenting Lund and Reeves's article pointed out that for the presence of only step-type change, it would be even more suitable to use $(1, n-3)$ degree of freedom for F_k. Both Lund and Reeves (2002), and Wang (2003) gave percentile tables for $F_{\max}$. The null hypothesis of a single regression can be rejected at the $100\alpha\%$ level if $F_{\max}$ is in the upper $100\alpha\%$ of simulated values.

2.2 Informational Approach

The other criteria for hypothesis testing is to use informational approach. The most classical criterion is Akaike Information Criterion (AIC) initially proposed by Akaike (1973). That is to minimize

$$\mathrm{AIC}(p) = -2 \log L(\hat{\theta}) + 2p$$

with respect to p, where $L(\hat{\theta})$ is the maximum likelihood function for the model with p parameters.

Schwarz (1978) proposed a modification of AIC, called Schwarz Information Criterion (SIC), taking the sample size into consideration as follow:

$$\mathrm{SIC}(n, p) = -2 \log L(\hat{\theta}) + p \log n.$$

For normal error, under H_0 and using SIC, Chen and Gupta (2001) obtain

$$\begin{aligned}\mathrm{SIC}(n) &= \mathrm{SIC}(n,3) = -\log L_0(\hat{b}_0, \hat{b}_1, \hat{\sigma}^2) + 3\log n \\ &= n\log\left[\sum_{i=1}^{n}(y_i - \hat{b}_0 - \hat{b}_1 x_i)^2\right] + n(1+\log 2\pi) + (3-n)\log n,\end{aligned}$$

where $(\hat{b}_0, \hat{b}_1, \hat{\sigma}^2)$ is the maximum likelihood estimator of the regression parameter (b_0, b_1, σ^2) under the null hypothesis H_0.

Under H_1 with a fixed change point k, for $k = 2, \cdots, n-2$, one derives

$$\begin{aligned}\mathrm{SIC}(k) &= -\log L_1(\hat{b}_{10}, \hat{b}_{11}, \hat{b}_{20}, \hat{b}_{21}, \hat{\sigma}^2) + 5\log n \\ &= n\log\left[\sum_{i=1}^{k}(y_i - \hat{b}_{10} - \hat{b}_{11} x_i)^2 + \sum_{i=k+1}^{n}(y_i - \hat{b}_{20} - \hat{b}_{21} x_i)^2\right] \\ &\quad + n(1+\log 2\pi) + (5-n)\log n,\end{aligned}$$

where $L_1(\hat{b}_{10}, \hat{b}_{11}, \hat{b}_{20}, \hat{b}_{21}, \hat{\sigma}^2)$ is the maximum likelihood function under H_1. Therefore the decision rule is to

- select the model with no change if $\mathrm{SIC}(n) \leqslant \mathrm{SIC}(k)$ for all $2 \leqslant k \leqslant n-2$;
- select the model with a change at $\hat{k}$ if

$$\mathrm{SIC}(\hat{k}) = \min\{\mathrm{SIC}(k) : 2 \leqslant k \leqslant n-2\} < \mathrm{SIC}(n).$$

In the next two sections, we present some asymptotic limiting theory for continuous and discontinuous models.

3 Asymptotic Theorems: Continuous Case

In the model (1.1), we impose the continuity condition:

$$b_{10} + b_{11}\tau = b_{20} + b_{21}\tau. \tag{3.1}$$

Denote $\Delta_0 = b_{20} - b_{10}$ and $\Delta_1 = b_{21} - b_{11}$. Then the change point $\tau = -\Delta_0/\Delta_1$ if $\Delta_1 \neq 0$. Otherwise the change point is non-identifiable. Thus within continuous model, we assume $\Delta_1 \neq 0$. Seber and Wild (1989, section 9.3) summarized the results before 1989. In this paper, we review and summarize results developed after 1989. Bhattacharya (1990, 1994) derived that the limiting distribution of the MLE for fixed design case.

More specifically, in the fixed design case, let $G_n(x) = n^{-1}\sum_1^n I(x_i \leqslant x)$ and suppose that there is a function G with $G(\tau) = p \in (0,1)$ and $g(\tau) = G'(\tau) > 0$ such that as $n \to \infty, G_n(\tau) \to G(\tau)$ and $G_n(x) - G_n(\tau) = [G(x) - G(\tau)] \cdot [1 + o(1)]$ uniformly in a neighborhood of τ. Suppose further that as $n \to \infty$,

$$(np)^{-1}\sum_{i=1}^{np}\begin{bmatrix}1 & x_i\\ x_i & x_i^2\end{bmatrix} \to \begin{bmatrix}1 & \mu_1\\ \mu_1 & \mu_1^2 + \sigma_1^2\end{bmatrix} = A_1'A_1, \tag{3.2}$$

and

$$(n-np)^{-1}\sum_{i=np+1}^{n}\begin{bmatrix}1 & x_i\\ x_i & x_i^2\end{bmatrix} \to \begin{bmatrix}1 & \mu_2\\ \mu_2 & \mu_2^2 + \sigma_2^2\end{bmatrix} = A_2'A_2, \tag{3.3}$$

where σ_1^2 and σ_2^2 are positive and that $\max_{1\leqslant i\leqslant n}|x_i| = o(n^{1/2})$. Furthermore, we assume that $x_{np} \leqslant \tau < x_{np+1}$ and $x_{np} - \tau \to 0$ as $n \to \infty$. So we can treat $x_{np} = \tau$.

Let $\delta_1' \equiv (u_0, u_1, v_0, v_1) \equiv (u', v')$. Consider any point in the parameter space

$$(\vartheta_1, k) \equiv (\theta_1 + n^{-1/2}\delta_1, np + j_n(\delta_1))$$

such that the lines $(b_{10}+n^{-1/2}u_0)+(b_{11}+n^{-1/2}u_1)x$ and $(b_{20}+n^{-1/2}v_0)+(b_{21}+n^{-1/2}v_1)x$ intersect at $x_{np+j_n(\delta_1)}$. Furthermore, the change point parameter $s(\delta_1) \equiv s_n(\delta_1)$ and the integer $j(\delta_1) \equiv j_n(\delta_1)$ satisfy

$$x_{np+j(\delta_1)} \leqslant s(\delta_1) < x_{np+j(\delta_1)+1} \text{ and } x_{np+j(\delta_1)} = s(\delta_1).$$

Under the normal error assumption and $\vartheta_1' = (\vartheta_{10}, \vartheta_{11}, \vartheta_{20}, \vartheta_{21})$, the log-likelihood ratio process is defined as:

$$R_n(\delta_1) = l_n(\vartheta_1, k) - l_n(\theta_1, np), \tag{3.4}$$

where

$$l_n(\vartheta_1, k) = \sigma^{-2}\left[\sum_{i=1}^{k}(Y_i - \vartheta_{10} - \vartheta_{11}x_i)^2 + \sum_{i=k+1}^{n}(Y_i - \vartheta_{20} - \vartheta_{21}x_i)^2\right]$$

Theorem 3.1. *For the fixed design continuous model of (1.1) with homogeneous Gaussian error terms, if conditions (3.2) and (3.3) hold, then*

$$\begin{aligned}R_n(\delta_1) \xrightarrow{\mathcal{L}} &-\left(\|W_1\|^2 + \|W_2\|^2\right)\\ &+\|p^{1/2}\sigma^{-1}A_1u - W_1\|^2 + \|(1-p)^{1/2}\sigma^{-1}A_2v - W_2\|^2,\end{aligned}$$

where W_1 and W_2 are independent bivariate normal vectors with mean vector 0 and covariance matrix I.

See Bhattacharya (1990) for details of proof. These weak convergence results indicate that $n^{1/2}(\hat{\theta}_{11}-\theta_{11})$ and $n^{1/2}(\hat{\theta}_{12}-\theta_{12})$ are asymptotically independently distributed as $\frac{\sigma}{\sqrt{p}}A_1^{-1}W_1$ and $\frac{\sigma}{\sqrt{1-p}}A_2^{-1}W_2$, respectively. Consequently,

$$n^{1/2}(\hat{\tau}-\tau) = n^{1/2}\left(\frac{\hat{b}_{10}-\hat{b}_{20}}{\hat{b}_{21}-\hat{b}_{11}} - \frac{b_{10}-b_{20}}{b_{21}-b_{11}}\right) \xrightarrow{\mathcal{L}} N(0,\alpha^2),$$

where the asymptotic variance α^2 obtained by the delta method is equal to

$$\begin{aligned}\alpha^2 = (b_{21}-b_{11})^{-2}\Big\{&\Big(Var[\hat{b}_{10}] + Var[\hat{b}_{20}]\Big)\\ &+2\tau\Big(Cov[\hat{b}_{10},\hat{b}_{11}] + Cov[\hat{b}_{20},\hat{b}_{21}]\Big)\\ &+\tau^2\Big(Var[\hat{b}_{11}] + Var[\hat{b}_{21}]\Big)\Big\}\end{aligned}$$

Denote all x_i's less than or equal to $\hat{\tau}$ as $x_{11},\cdots,x_{1n_1}$ and those bigger than $\hat{\tau}$ as $x_{21},\cdots,x_{2n_2}$ with the sample means $\bar{x}_{j\cdot} = n_j^{-1}\sum_{i=1}^{n_j} x_{ji}$.

Corollary 3.2. *For the fixed design continuous model of (1.1) with homogeneous Gaussian error terms, then for $j=1,2$,*

$$\hat{b}_{j0} \sim N\left(b_{j0}, \frac{\sigma^2\sum_i x_{ji}^2}{n_j\sum_i(x_{ji}-\bar{x}_{j\cdot})^2}\right),$$

$$\hat{b}_{j1} \sim N\left(b_{j1}, \frac{\sigma^2}{\sum_i(x_{ji}-\bar{x}_{j\cdot})^2}\right),$$

and

$$Cov[\hat{b}_{j0},\hat{b}_{j1}] = -\frac{\sigma^2\bar{x}_{j\cdot}}{\sum_i(x_{ji}-\bar{x}_{j\cdot})^2}.$$

Proof: See Seber and Wild (1989, p. 450) for details.

In the random design case, let G be the distribution function of X having a continuous Lebesgue density g at τ. For $s\in\bar{\mathbb{R}}$, denote $\bar{G}(s) = 1-G(s)$, and $\mu_j(s) = EX^jI(X\leqslant s)$, $j=1,2$. Then,

$$E\left\{(np)^{-1}\sum_{i=1}^{np}\begin{bmatrix}1 & X_i\\ X_i & X_i^2\end{bmatrix}I(X_i\leqslant\tau)\right\} \to \begin{bmatrix}G(\tau) & \mu_1(\tau)\\ \mu_1(\tau) & \mu_2(\tau)\end{bmatrix} \equiv \Sigma_{1\tau}, \quad (3.5)$$

and

$$E\left\{(n-np)^{-1}\sum_{i=np+1}^{n} E\begin{bmatrix}1 & X_i\\ X_i & X_i^2\end{bmatrix}I(X_i>\tau)\right\} \to \begin{bmatrix}\bar{G}(\tau) & \bar{\mu}_1(\tau)\\ \bar{\mu}_1(\tau) & \bar{\mu}_2(\tau)\end{bmatrix} \equiv \Sigma_{2\tau}, \quad (3.6)$$

Denote

$$\Sigma_\tau = \mathrm{diag}(\Sigma_{1\tau}, \Sigma_{2\tau}) = \begin{bmatrix} G(\tau) & \mu_1(\tau) & 0 & 0 \\ \mu_1(\tau) & \mu_2(\tau) & 0 & 0 \\ 0 & 0 & \bar{G}(\tau) & \bar{\mu}_1(\tau) \\ 0 & 0 & \bar{G}(\tau) & \bar{\mu}_2(\tau) \end{bmatrix}.$$

Theorem 3.3. *In the continuous random design case, assume the finite second moment of X and homogeneous normal error with mean zero and finite variance σ^2. Then*

$$n^{1/2}(\hat{\theta}_1 - \theta_1) \xrightarrow{\mathcal{L}} \sigma \Sigma_\tau^{-1/2} \begin{bmatrix} W_1 \\ W_2 \end{bmatrix},$$

where W_1 and W_2 are the same as in Theorem 3.1. Hence the change point estimator is the same as in the fixed design case.

Koul (2000) introduced a marked empirical process in fitting two phase linear regression models and obtained asymptotically distributed free tests in testing $M_0 : m(x) = m(x, \theta)$. For $x \in \mathbb{R}, v \in \Theta$, a marked empirical process is defined as

$$V_n(x, v) = n^{-1} \sum_{i=1}^{n} [Y_i - m(X_i, v)]\, I\,(X_i \leqslant x)\,,$$

where the residuals $Y_i - m(X_i, v)$ are the marks. For $x \in \mathbb{R}$, denote $a(x)' = (-1, -x, 1, x)$ with $a_1' = (-1, -\tau, 0, 0)$ and $a_2' = (0, 0, 1, \tau)$. Let F be the distribution function of error and $u_p(1)$ means uniformly convergent to zero in probability.

Theorem 3.4. *In the random design case of model (1.1), suppose $E|X|^2 < \infty$, g is continuous at τ and F has uniformly continuous Lebesgue density $f > 0$, a.e. Then*

$$\begin{aligned} & n^{1/2}\left[V_n(x, \hat{\theta}) - V_n(x, \theta)\right] \\ &= \theta_1' a(\tau) g(\tau) \left\{ \left[\sqrt{n}(x-\tau) I(\tau < x \leqslant \hat{\tau}) + \sqrt{n}(\hat{\tau} - \tau) I(x > \hat{\tau})\right] I(\hat{\tau} > \tau) \right. \\ &\qquad \left. - \left[\sqrt{n}(x-\tau) I(\hat{\tau} < x \leqslant \tau) - \sqrt{n}(\hat{\tau} - \tau) I(x > \hat{\tau})\right] I(\hat{\tau} \leqslant \tau) \right\} \\ & -\sqrt{n}(\hat{\theta}_1 - \theta_1)' \Gamma_\tau(x) + u_p(1). \end{aligned}$$

Consequently, in the continuous random design case, i.e. $\theta_1' a(\tau) = 0$, we have

$$n^{1/2}\left[V_n(x, \hat{\theta}) - V_n(x, \theta)\right] = -[n^{1/2}(\hat{\theta}_1 - \theta_1)]' \Gamma_\tau(x) + u_p(1),$$

where

$$\Gamma_\tau(x) - E\begin{bmatrix} I(X \leqslant x \wedge \tau) \\ XI(X \leqslant x \wedge \tau) \\ I(X > \tau, X \leqslant x) \\ XI(X > \tau, X \leqslant x) \end{bmatrix} \quad \textit{and} \quad \Gamma_\tau = \Gamma_\tau(\infty).$$

Combining Theorems 3.3 and 3.4, we obtain

$$\sqrt{n}V_n(x, \hat{\theta}) = \left\{ \frac{1}{\sqrt{n}} \sum \epsilon_i I(X_i \leqslant x) - \sigma \begin{bmatrix} W_1 \\ W_2 \end{bmatrix}' \sum_\tau^{-1/2} \Gamma_\tau(x) \right\} + u_p(1).$$

Let $Z_n(x)$ be the leading process in the right hand side of the above approximation. Then a simple calculation shows that

$$EZ_n(x)Z_n(y) = \sigma^2 \left[G(x \wedge y) + \Gamma_\tau(x)\Sigma_\tau^{-1}\Gamma_\tau(y) \right], \quad x, y \in \mathbb{R}.$$

Similarly to Koul (2000), one can construct a transformation T_n of the marked empirical process $V_n(\cdot, \hat{\theta})$ such that under null hypothesis M_0 : $m(x) = m(x, \theta), \forall\, x \in \mathbb{R}$ and for some $\theta \in \mathbb{R}^5$, the transformed process converges weakly to a $\sigma^{-1}B(G)$ before the change point, where B is a Brownian motion on $[0, \infty]$. Hence, any test of M_0 based on a continuous function of the transformed process will be asymptotically distribution free test M_0.

4 Asymptotic Theorems: Discontinuous Case

Let ν_n be a sequence of positive numbers and c_j be real numbers for $j = 1, 2$. Let

$$d_n \equiv (b_{20} - b_{10}) + (b_{21} - b_{11})\tau \equiv (c_0 + \tau c_1)/\nu_n \neq 0 \qquad (4.1)$$

be the jump size at the change point τ of the model (1.1). The model (1.1) is discontinuous since $d_n \neq 0$ for all $n \geqslant 1$. Within the discontinuous model, if $\nu_n \equiv 1$ for all $n \geqslant 1$, then the model has a fixed jump size $d \equiv d_n$ for all $n \geqslant 1$ indicating abrupt change. If $\nu_n = o(n^{1/2})$ and $\nu_n \to \infty$, then the model is contiguous indicating a gradually structural change.

Among discontinuous fixed jump models, Quandt (1958, 1960) studied the mean change problem over observation time. Chow (1960) presented two tests based on prediction intervals and F-test for the hypothesis of no change in the simple linear regression versus abrupt change with application in econometrics. Bhattacharya and Johnson (1968) investigated the non-parametric tests for step-type change (location in one sample problem) at an unknown time point. Hinkley (1970) derived the asymptotic distribution of the maximum likelihood estimator (MLE) by random-walk considerations when these regression phases are parallel (intercept change, step-type change).

Hawkins (1980) reported that the inferential theory depends strongly on whether or not continuity at the change-point is assumed. Furthermore, the likelihood ratio test for the presence of two phase regression tends to infinity. A special case is one in which a change only happens in intercept (step-type). Worsley (1983) specialized his improved Bonferroni approximation to the likelihood ratio test to this case. Null quantiles for the likelihood ratio test were calculated from simulations by Maronna and Yohai (1978). Csörgö and Horváth (1988), van de Geer (1988) and Bhattacharya (1990) reported that the asymptotic result for the contiguous case is similar to the continuous case. Bai and Perron (1998) investigated the asymptotics of the least squared estimators and the corresponding tests in multi-phase contiguous random design linear regression models when the errors satisfy some martingale type assumptions. Koul and Qian (KQ, 2002) established the consistency and the limiting distribution of the MLE in the fixed jump size two-phase random design regression models for a class of error densities that excludes the double exponential and such non-smooth densities. Koul, Qian and Surgailis (2003) extended KQ's results to M-estimation and gave out much details for proving the limiting distribution of the change point. Fiteni (2004) obtained the asymptotic distribution of the break location estimator introduced by Yohai and Zamar (1988) for a shift of fixed magnitude as well as for a shift with its magnitude converging to zero (gradual change or contiguous model) when the sample size increases under possibly contaminated distribution for both the regressors and the error term.

For data collected sequentially over time, sequential control-chart approaches to detect general changes in multiple regression parameters are discussed by Brown *et al.* (1975). CUSUM methods for detecting changes in intercept only are discussed by Schweder (1976).

Seber and Wild (1989, p.438) claimed that "we cannot hope to estimate the changepoint." Bhattacharya (1994) reported that in the fixed design case, the estimators of the regression parameters behave in the same way in discontinuous case as in continuous case, but $\hat{\tau}$ is asymptotically independent to the estimated regression parameters for contiguous case only. Koul and Qian (2002) filled up the gap for fixed jump case. The following theorem states the consistent results for both discontinuous cases.

Theorem 4.1. *Assume the error term in model* (1.1) *is Gaussian with mean zero and finite homogeneous variance and* $EX^2 < \infty$. *If the model* *is discontinuous with* (i) $\nu_n^2/n = o(1)$ *and* $\nu_n \to \infty$ *for contiguous case* (ii) $\nu_n = 1$ *for fixed jump case, then the maximum likelihood estimator of* θ *satisfies*

$$\|\hat{\theta}_1 - \theta_1\| = O_p(1), \quad \|n\nu_n^{-2}(\hat{\tau} - \tau)| = O_p(1).$$

Proof. See Bhattacharya (1990) for contiguous case, Koul and Qian (2002) for fixed jump case.

Note that the assumption of $\nu_n \to \infty$ obviously excludes the fix jump size discontinuous case. In fact, within the discontinuous model, the two types of discontinuity behave distinctly different in terms of limiting distribution of the change point estimator.

First we include results for contiguous case. Let the jump size $\nu_n \to \infty$ and $\nu_n = o(n^{1/2})$. Denote $k = np + \nu_n^2 t$, $u' = (u_0, u_1)$ and $v' = (v_0, v_1)$. Then the likelihood ratio process is

$$R_n(u, v, t) = l_n(\vartheta_1, k) - l_n(\theta_1, np)$$

with five free parameters. Bhattacharya (1990, 1994) concluded that

Theorem 4.2. (Contiguous case) *In the contiguous case of model* (1.1), *let all assumptions on the design variable in the continuous case hold. Then,*

$$\begin{aligned} R_n(u, v, t) \xrightarrow{\mathcal{L}} & -(\|W_1\|^2 + \|W_2\|^2) \\ & + \|p^{1/2}\sigma^{-1}A_1 u - W_1\|^2 + \|(1-p)^{1/2}\sigma^{-1}A_2 v - W_2\|^2 \\ & + \sigma^{-1}|c_0 + c_1\tau| \cdot \left[B(t) + \frac{1}{2}\sigma^{-1}|c_0 + c_1\tau| \cdot |t|\right], \end{aligned}$$

where W_1, W_2, A_1, A_2 *are as in Theorem 3.1 and* $B(\cdot)$ *is a standard two-sided Brownian motion which is independent of* W_1 *and* W_2.

Furthermore, the estimated regression parameters are asymptotically normal as in the continuous case and

$$n\nu_n^{-2}(\hat{\tau} - \tau) \xrightarrow{\mathcal{L}} T^*/g(\tau)$$

where $T^* = \arg\min[B(t) + \frac{1}{2}\sigma^{-1}|c_0 + c_1\tau| \cdot |t|]$, *a Brownian motion with a linear drift over* $(-\infty, \infty)$ *and* $g(\tau)$ *is the density (limiting density in fixed design case) of the design points at* τ.

In the fixed jump size,

$$d \equiv d_n = c_0 + \tau c_1 \neq 0. \tag{4.2}$$

Koul and Qian (2002) proposed likelihood ratio test under more general error density function f. Under i.i.d. errors, the part of the likelihood function of the data from (1.1) under ϑ that is relevant for making inference about θ is given by

$$L_n(\vartheta) := \Pi_1^n f(Y_i - m(X_i, \vartheta)).$$

The maximization algorithm still take two steps. Recall, $\vartheta' = (\vartheta_1', s)$ and $\theta' = (\theta_1', \tau)$. First, for each fixed s, obtain the maximizers $\vartheta_1(s)$ of $L_n(\vartheta_1, s)$ with respect to ϑ_1 over K. Notice that $\vartheta_1(s)$ is constant in s over any interval of two consecutive ordered X_i's and that the profile likelihood function $L_n(\vartheta_1(s), s)$ has only finite number of possible values with the jump points located at the ordered X_i's.

At the second stage, we compute maximizer $\hat{\tau}$ of $L_n(\vartheta_1(s), s)$ with respect to s over the sample percentiles of $\{X_i, 1 \leqslant i \leqslant n\}$. Because of (4.2), for each ϑ_1 fixed, the likelihood as a function of s changes only when the change-point parameter s passes over the sample points of the independent variable X. This maximizer may be taken to be the left end point of the interval over which it is obtained. Then the estimator $\hat{\theta} = (\vartheta_1(\hat{\tau}), \hat{\tau}) \equiv (\hat{\theta}_1, \hat{\tau})$ is the MLE of the underlying parameter θ.

The general density function f of error is required to satisfy the following conditions:

(a) f is absolutely continuous and positive everywhere on $\mathbb{R}$ with $\dot{f}$ denoting its a.e. derivative and $\varphi = \dot{f}/f$.

(b) The function φ is Lip(1) and $I(f) := \int \varphi^2(x) f(x)\, dx < \infty$.

(c) φ is differentiable and its derivative $\dot{\varphi}$ is Lip(1).

Furthermore, the design variable X is assumed to have a bounded and positive Lebesgue density g on $\mathbb{R}$ with $E|X|^3 < \infty$.

For $t \in \mathbb{R}$, let $P(t) = P_1(-t)I(t \leqslant 0) + P_2(t)I(t \geqslant 0)$. The two random walks $P_1(t)$ and $P_2(t)$ associated with this process are two independent compound Poisson processes on $[0, \infty)$, both with rate $g(\tau)$ and $P_1(0) = P_2(0) = 0, a.s.$; the distribution of jumps being given by the conditional distribution of $\ln[f(\epsilon + c_0 + c_1 X)/f(\epsilon)]$, given $X = \tau^+$ and the conditional distribution of $\ln[f(\epsilon - c_0 - c_1 X)/f(\epsilon)]$ given $X = \tau^-$, respectively. Furthermore, $P_1(t)$ and $P_2(t)$ tend to $-\infty$, a.s., as $|t| \to \infty$ which implies the existence of the unique random interval $[M_-, M_+)$ on which the process $\{P(t) : t \in \mathbb{R}\}$ attains its global maximum, a. s.

Remark: If the error term is Gaussian, then all conditions (a)-(c) imposed on the error density function are satisfied.

We are now ready to state the asymptotical properties of the estimators which includes the root-n asymptotic normality for the regression parameters and n - consistency of the estimator $\hat{\tau}$ under general error density setting.

Theorem 4.3. (Fixed jump case) *Suppose that conditions* (a)-(c), (4.2) *hold and g is continuous and positive at τ and $E|X|^3 < \infty$. Then, $n(\hat{\tau} - \tau)$ converges weakly to M_- and $n^{1/2}(\hat{\theta}_1 - \theta_1)$ converges weakly to a $N(0, \Gamma^{-1})$ r.v., where $\Gamma = I(f)E\dot{m}_\tau(X)\dot{m}'_\tau(X)$ and*

$$\begin{aligned}\dot{m}_s(x) &\equiv (\partial/\partial\vartheta_1)(m_s(x, \vartheta_1)) \\ &= (I(x \leqslant s),\ xI(x \leqslant s),\ I(x > s),\ xI(x > s))', \quad s \in \bar{\mathbb{R}},\ x \in \mathbb{R},\end{aligned}$$

Notice that

$$E\dot{m}_\tau(X)\dot{m}'_\tau(X) = \begin{bmatrix} G(\tau) & \mu_1(\tau) & 0 & 0 \\ \mu_1(\tau) & \mu_2(\tau) & 0 & 0 \\ 0 & 0 & \bar{G}(\tau) & \bar{\mu}_1(\tau) \\ 0 & 0 & \bar{G}(\tau) & \bar{\mu}_2(\tau) \end{bmatrix} = \Sigma_\tau.$$

Consequently, we conclude that $\hat{\theta}_{11}$ and $\hat{\theta}_{12}$ are asymptotically independent, which is the similar results as in continuous and contiguous cases.

Koul, Qian and Surgailis (KQS, 2003) extended the above results further to M-estimator and obtained same results. Detail proofs can be found in KQS (2003). Ciuperca (2004) derived same results of KQ (2002) when extending the model to nonlinear model.

5 Computational Aspects

Almost all of computational algorithms in the literatures focus on continuous model. Hudson (1966) developed two-stage algorithm using so-called Type-I minimum and Type-II minimum for two phase regression in which the regression function, but not its derivative, is continuous at the change point (see Seber and Wild, p. 456-457). This algorithm was refined by Hinkley (1969, 1971) for two phase linear regression model, and further extended to three phase by Williams (1970). Lerman (1980) noticed the difficulty of solving the problem analytically using Hudson's algorithm and proposed grid search method. Drapper and Smith (1981, Section 5.4) suggested a grid search method based on the smallest residual sum of squares criterion to estimate the join point (Hudson's idea), which is the same algorithm as in section 2.1.

Gallant and Fuller (1973) proposed a different approach and advocated re-parameterizing the regression into a single nonlinear function using indicator functions and recommended using the Gauss-Newton procedure to help estimate the joint point directly. Hawkins (1976) proposed dynamic programming and hierarchical procedure.

Bacon and Watts (BW, 1971) use transition smoothing function *tanh*, also called *sgn* formulation. Tishler and Zang's (TZ, 1981) proposed so-called bent-cable quadratic smoothing around the change point for an arbitrary number of regimes and variables with fast convergence, also named as *max-min* formulation.

Rewrite the continuous two phase linear model into

$$Y_i = \begin{cases} b_0 + b_1 x_i + \epsilon_i, & x_i \leqslant \tau, \\ b_0 + b_1 x_i + b_2(x_i - \tau) + \epsilon_i, & x_i \leqslant \tau. \end{cases}$$

Bacon and Watt (1971)'s transition model is defined as

$$Y_i = b_0 + b_1 x_i + \frac{1}{2}(b_1 + b_2)(x_i - \tau) + \frac{1}{2}(b_2 - b_1)\tanh\frac{(x_i - \tau)}{\gamma} + \epsilon_i$$

where γ is the smoothing parameter and the transition function

$$\tanh(x) = (e^x - e^{-x})/(e^x + e^{-x})$$

satisfies

$$\lim_{\gamma \to 0} \tanh\frac{(x - \tau)}{\gamma} = \operatorname{sgn}(x - \tau).$$

and

$$\operatorname{sgn}(z) = \begin{cases} -1, & z < 0, \\ 0, & z = 0, \\ 1, & z > 0 \end{cases}$$

Tishler and Zang discussed the piecewise linear model:

$$Y_i = \max\{b_1' X_{1i}, b_1' X_{2i}, \cdots, b_p' X_{pi}\} + \epsilon_i$$

where $b_k' = (b_{k1}, \cdots, b_{kp})$ for $k = 1, \cdots, p$. When $p = 2$, it reduces to the two phase linear model which can also be expressed as the following equivalent form:

$$Y_i = b_1' x_{1i} + q(b_2' x_{2i} - b_1' x_{1i}) + \epsilon_i$$

where

$$q(z) = \max(0, z), \ x \in \mathbb{R}.$$

Tishler and Zang approximate $q(z)$ by $\bar{q}_\beta(z)$, which smoothes out the kink and uses a quadratic smoothing function, defined as:

$$\bar{q}_\beta(z) = \begin{cases} 0, & \text{if } z \leqslant -\beta, \\ \frac{(z+\beta)^2}{4\beta}, & \text{if} -\beta < z \leqslant \beta, \\ z, & \text{if } z > \beta. \end{cases}$$

The smoothed model is called TZ's bent-cable model. Figure 5.1 shows the approximation of $q(z)$ by $\bar{q}_\beta(z)$ for small $\beta = .5$.

We analyzed numerous real data sets using three fore-mentioned algorithms (BW, TZ, and Grid search) and concluded that the smoothing

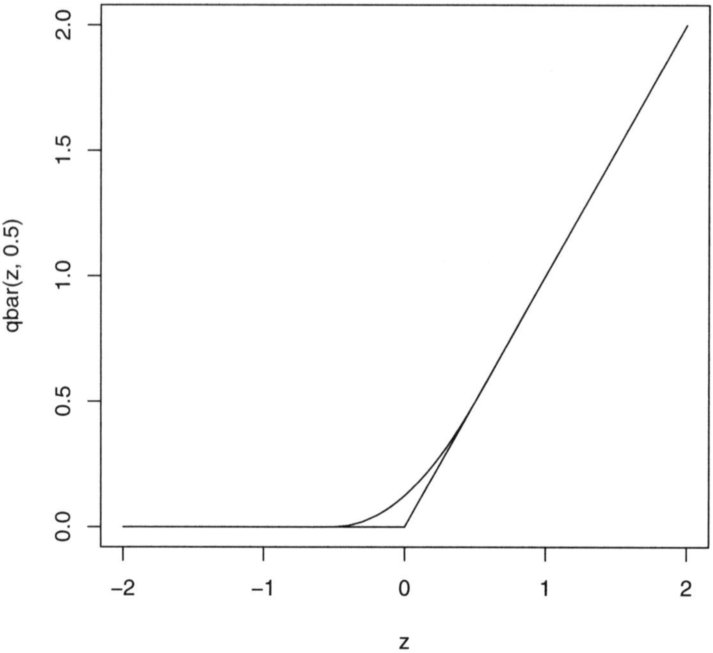

Figure 5.1: The function $q(z)$ and its approximation for $\beta = 0.5$.

algorithms are comparable, but Tishler and Zang's smoothing function is much easier than Bacon and Watt's transition smoothing method to be implemented.

Though the smoothing methods are good for computation using the classical least square method, there are not much asymptotic inference having been made. Recently, Chiu, Lockhart and Routledge (2002) showed that bent-cable asymptotics when the bent is missing using Tishler and Zang's method (1981). They concluded that the bent-cable model converges slowly for large sample and not recommended. Thus, we still recommend two step grid search algorithm.

Among the statistical software packages, there are some ready to use library or procedures. For instance, the *nlin* procedure with method of Marquardt (1963) in SAS provides both change point estimate and coefficient parameter estimates. The *segmented* library in R provided by Muggeo (2003) can estimate regression models with unknown break-points. See Seber and Wild (1989) for more detail discussion.

6 Conclusions and Final Remarks

In this paper, we first motive readers with real life examples using multiple-phase linear regression models. Then, we address hypothesis testing issues through likelihood ratio and informational based approaches. Further we review the asymptotic theorems for the two types of models: Continuous (fixed design and random design) and discontinuous (fixed design and random design) models. At last we summarize the computational aspects for piecewise regression models. All results surveyed in this paper are focused on parametric approaches after 1989. Nonparametric, Bayesian and CUSUM approaches are also used in literatures.

Among the nonparametric methods, kernel type estimation and wavelet methods are the most popular ones. Brodsky and Darkhovsky (1993) edited a book on nonparametric methods in change-point problems. Carlstein et al.'s (1994) monograph on change point problems included twenty eight articles covering different angles of change point problems. Some of the articles involve nonparametric approaches in two phase linear regressions. Wu and Chu (1993), Loader (1996), Müller and Song (1997), Muller and Stadtmüller (1999), Huh and Park (2004) used non-parametric curve estimation methods to construct estimators of the change point in fixed design continuous regression models. Most recently, Gijbels and Goderniaux (2004) studied bootstrap test for an abrupt (fixed jump size) change point in nonparametric regression.

Among Bayesian approaches, Bacon and Watts (1971) utilized Bayesian approach to estimate the joint point of two straight lines and introduced a transition function to represent the model at the intersection of the two lines by a smooth curve. El-Sayyad (1975) gave a Bayesian analysis of the change-point problem and applied it to the cases of normal and exponential distribution. Ferreira (1975) gave a Bayesian analysis of switching regression model with known number of regimes and showed that the mean square error (MSE) of the Bayesian estimates are uniformly smaller than those of the maximum likelihood (ML) estimates. Holbert and Bromelin (1977) derived Bayesian inferences related to shifting sequences and two-phase regression. Piegorsch (1987) considered Bayesian decision procedures with a discretized normal prior density as the prior distribution for the change point. Kezim and Abdelli (2004) conducted a Bayesian analysis of a structural change in the parameters of a time series.

Among the literatures surveyed, most of the researchers focus on the continuous model. On the other hand, these researchers advise caution in the use of the continuity assumption and illustrated, by examples, that it can lead to results different from those when no constraint is imposed. F-test using the extra sum of squares for testing two-phase versus simple linear regression was initially proposed by Hinkley, but the degrees of freedom used is a big mystery.

Furthermore, one notices that the types of models, in terms of fixed jump size versus contiguous or continuous, can result in huge difference for asymptotical theory as indicated by Hawkins (1980), Seber & Wild (1989), Koul and Qian (2002), Koul et al. (2003). To be more specific: we summarize as below:

- In the continuous and contiguous model, the asymptotic distribution of the maximum likelihood estimated change point between the two-phase is related to a linear drifted Brownian process for both fixed and random design cases.
- In discontinuous fixed jump size model, the limiting distribution of the MLE and M-estimation for the change point is related to a compound Poisson process, which shows the distinct difference between the types of discontinuous models: fixed jump size and contiguous cases.

Though in this paper, we only focus on homogeneous variance model, nonhomogeneous error variance is a realistic problem. Kim and Siegmund (KS, 1989) and Kim (K, 1993) studied the likelihood ratio test for a change point in simple linear regression with homogeneous and nonhomogeneous errors, respectively. Brodeau (1999) derived asymptotic results and tests for the choice of approximative models in nonlinear two-phase regression models, under heteroscedastic assumption. Diniz and Brochi (2005) compared the robustness of the tests proposed by Quandt (Q-test), KS-test and K-test through simulation, and concluded that KS-test is superior to Q-test for both models considered while the K-test is more powerful than the other two tests for nonhomogeneous models with known variances.

Recent development indicates that testing the existence of change point is still a hot topic. Jarušková (1998) developed a new test procedure for detecting gradual (contiguous) changes in simple regression models and studied its limit behavior under the null hypothesis 'no change'. Huvšková and her collaborators in their series of papers (1998, 2001, 2002, 2003) proposed a new class of test procedures for testing structural change and studied the limit behaviors both under the null hypothesis (no change) and under some alternatives.

Another closely related topic with application in economics is the topic of "structural change". Hackl and Westlund (1989) presented an annotated bibliography. Benzekri and Brodeau (1991) obtained the $L2$

consistency of least squares estimation in two-phase splined continuously regression models. Neubauer and Zwanzig (1995) and Brodeau (1999) studied the least squared estimators and hypothesis testing of no change using similar techniques of approximation. Hušková and Antoch (2001) studied M-estimator of structural changes and their properties. Hušková and Antoch (2003) constructed permutation test to detect the structural changes in regression and derived the approximated critical values of the test. Huh and Park (2004) studied the change point detection with local polynomial fits for the random design case. Dong (2004) tested homogeneity of error variances via empirical likelihood approach for structural change in regression models. Zeileis (2006) and others developed a class of structural change tests using an econometric computing approach.

Besides the hot topic in change point detection, recent research activities show that the structural change problem in regression models has been extended to more general models such as time series (Qian, 1998; Maekawa, He and Tee, 2004 and references therein), measurement error models (Gbur and Dahm, 1985; Staudenmayer and Spiegelman, 2002), generalized (bi)linear models (Guegan and Pham, 1989; Gabriel, 1998; Pastor and Guallar, 1998; Antoch, Gregoire and Jarušková , 2004) and analyzing longitudinal data (Piepho and Ogutu, 2003).

References

[1] Akaike, H., Information theory and an extension of the maximum likelihood principle. 2nd International Symposium of Information Theory (B.N. Petroc and E. Csaki, Eds), *Akademiai Kiado Budapest,* 1973, 267–281.

[2] Anderson, R. L. and L.A. Nelson, A family of models involving intersecting straight lines and concomitant experimental designs useful in evaluating response to fertilizer nutrients. *Biometrics* 31 (1975), 303–318.

[3] Antoch, Jaromir; Gregoire, Gerard and Jaruvskova, Daniela, Detection of structural changes in generalized linear models. *Stat. Probab. Lett.,* 69 (2004), no. 3, 315–332.

[4] Bacon, Dave and Watts, Don., Estimating the Transition between Two Intersecting Straight Lines. *Biometrika,* 58 (1971) no. 3, 525–534.

[5] Bai, J. and Perron, P., Estimating and testing linear models with multiple structural changes. *Econometrica*, 66 (1998), 47–78.

[6] Beckman, R. J. and Cook, R. D., Testing for two-phase regressions. *Technometrics*, 21 (1979), 65–69.

[7] Benzekri, S. and Brodeau, F., Asymptotic results for parametric estimation in inadequate two phases regression models. *Statistics*, 22 (1991), no. 3, 331–348.

[8] Berman NG, Wong WK, Bhasin S, Ipp E. Applications of segmented regression models for biomedical studies. *Am. J. Physiol.*, 270 (1996), 723–732.

[9] Bhattacharya, G. K. and Johnson, R.A., Non-parametric tests for shift at an unknown time point. *Ann. Math. Stat.*, 39 (1968), 1731–43.

[10] Bhattacharya, P. K., Weak convergence of the log-likelihood process in the two-phase linear regression problem. In Probability, Statistics and Design of Experiments. Wiley eastern, New Delhi. 1990, 145–156.

[11] Bhattacharya, P. K., Some aspects of change-point analysis. IMS Lecture Notes-Monograph Ser., Hayward, CA. (1994), 23, 28–56.

[12] Boehm, B. W., *Software Engineering Economics.* Prentice Hall. 1981.

[13] Blood, E.R., Phillips, J. S., Calhoun, D., and Edwards, D., The role of the Floridan aquifer in depressional wetlands hydrodynamics and hydroperiod. In K. J. Hatcher (ed.), *Proceedings of the Georgia Water Resources Conference*, 1997, pp. 273–279. Athens, GA: Institute of Natural Resources. University of Georgia.

[14] Braña, F. and Ji, X., Influence of incubation temperature on morphology, locomot performance, and early growth of hatchling wallizards (*Podarcias muralis. Journal of Experimental Zoology.* 286, 422–433.)

[15] Brodeau, F., Asymptotic results and tests for the choice of approximative models in nonlinear two-phases regression models, heteroscedastic case. *Metrika*, 49 (1999), no. 2, 85–105.

[16] Brodsky, B.E. and Darkhovsky, B.S., *Nonparametric mehtods in changepoint problems.* Dordrecht, Kluwer Academic Publisherss. 1993.

[17] Brown, R. L; Durbin, J. and Evans, J. M. (1975), Techniques for testing the consistancy of regression relationships over time, *Journal of Royal Statistical Society*, B, 149–192.

[18] Charlstein, E., Müller, H. G. and Siegmund, D., *Change-point Problems.* IMS Lecture Notes, Volume 23, 1994.

[19] Chen, J. and Gupta, A. K., On change point detection and estimation. *Comm. Stat. Simulat. Comput.* 30 (2001) 3, 665–697.

[20] Chiu, Grace; Lockhart, Richard and Routledge, Richard., Bent-cable asymptotics when the bend is missing. *Stat. Probab. Lett.*, 59 (2002), no.1, 9–16.

[21] Chow, G., Tests of equality between two sets of coefficients in two linear regressions. *Econometrica*, (1960) 28, 591–605.

[22] Ciuperca, Gabriela, Maximum likelihood estimator in a two-phase nonlinear random regression model. *Stat. Decisions.*, 22 (2004), no. 4, 335–349.

[23] Cox, C., Threshold dose-response models in toxicology. *Biometrics*, 43 (1987), 544–523.

[24] Csörgő, M. and Horváth, Lajos., Invariance principles for change-point problems. *J. Multivariate Anal.* 27 (1988), no. 1, 151–168.

[25] Csörgő, M. and Horváth, L., *Limit Theorems in Change-point Analysis.* Wiley, New York. 1997.

[26] Davies, R. B., Hypothesis testing when a nuisance parameter is present only under the alternative. *Biometrika*, 64 (1977), 247–254.

[27] Davies, R. B., Hypothesis testing when a nuisance parameter is present only under the alternative. *Biometrika*, 74 (1987) 33–43.

[28] Diniz, C. A. R. and Brochi, L. C. (2005), Robustness of two-phase regression tests. *REVSTAT Statistical Journal*, 3(1), 1–18.

[29] Dong, L. B., Testing change structure in regression: an empirical likelihood approach. *Econometrics Working Paper EWP0405*, ISSN 1485–6441. 2004.

[30] Draper, N. R. and Smith, H. (1981), *Applied regression analysis*, 2nd edn, John Wiley, New York. Section 5.4. with known joint point. 1981.

[31] El-Sayyad, G. M., A Bayesian analysis of the change-point problem. *Egypt. Stat. J.*, 19(1975), 1–13. 20. Section 5.4. with known joint point.

[32] Feder, P. I., On asymptotic distribution theory in segmented regression problems-identified case. *Ann. Stat.*, 3 (1975), 49–83.

[33] Feder, P. I., The log likelihood ratio in segmented regression. *Ann. Stat.*, 3(1975), 84–97.

[34] Ferreira, P. E. (1975), A Bayesian analysis of switching regression model: Known number of regimes. *J. Am. Stat. Assoc.*, 70 (1975), 370–4.

[35] Fiteni, Inmaculada, τ-estimators of regression models with structural change of unknown location. *J. Econometrics.*, 119 (2004), no. 1, 19–44.

[36] Gabriel, K. Ruben, Generalised bilinear regression. *Biometrika*, 85 (1998), no. 3,689–700.

[37] Gallant, A. R. and Fuller, W. A. (1973), Fitting segmented polynomial regression models whose join points have to be estimated. Journal of the American Statistical Association, 68 (1973), 144–147.

[38] Gaylor, D. W., Sheehan, D. M., Young, J. F. and Mattison, D.R. (1988), The threshold dose question in teratogenesis. *Teratology*, 38(1988), 389–391.

[39] Gbur, E. E., Thomas, G. L., and Miller, F. R.(1979), The use of segmented regression models in the determination of the base temperature in heat accumulation models. *Agronomy J.*, 71 (1979), 949–953.

[40] Gbur, E. E. and Dahm, P. F., Estimation of the linear-linear segmented regression model in the presence of measurement error. Commun. *Stat. Theory Methods*, 14 (1985) 809–826.

[41] Gibels, I. and A-C Goderniaux (2004), Bootstrap test for change-points in nonparametric regression. *Journal of Nonparametric Statistics*, 16 (2004), no. 3–4. 591–611.

[42] Chiu, Grace; Lockhart, R. and Routledge, Richard., Bent-cable asymptotics when the bend is missing. *Stat. Probab. Lett.*, 59 (2002), no. 1, 9–16.

[43] Guegan, D. and Pham, D. T. (1989), A note on the estimation of the parameters of the diagonal bilinear model by the method of least squares, *Scandinavian Journal of Statistics*, 16, 129–136.

[44] Hackl, P. and Westlund, A. H. (1989), Statistical Analysis of "structural change": an annotated bibliography. *Journal of Empirical Economics*, 14(2), 167–192.

[45] Hasselblad, V., J. P. Creason, and W.C. Nelson, Regression using "Hockey-stick" functions. Research Traingle Park, NC: US Environmental Protection Agency, 1976.

[46] Hawkins, Douglas M., Point estimation of the parameters of piecewise regression models. *J. Roy. Stat. Soc. Ser. C Appl. Statist.* 25 (1976), no. 1, 51–57.

[47] Hawkins, D. M. (1980), A not on continuous and discontinuous segmented regressions. *Technometrics*, 22, 443–444.

[48] Hinkley, D. V., Inference about the intersection in two-phase regression. *Biometrika*, 56 (1969), 495–504.

[49] Hinkley, D. V., Inference about the change-point in a sequence of random variables. *Biometrika*, 57 (1970), 1–17.

[50] Hinkley, D. V., Inference in two-phase regression. *J. Am. Stat. Assoc.*, 66 (1971), 736–743.

[51] Holbert, Donald; Bromeling,Lyle, Bayesian inferences related to shifting sequences and two-phase regression. *Comm. Stat.-Theory Methods*, A6 (1977), no. 3, 265–275.

[52] Hudson, D. J., Fitting segmented curves whose join points have to be estimated. *J. Am. Stat. Assoc.*, 61 (1966), 1097–1129.

[53] Huh, J. and B. U. Park, Detection of a change point with local polynomial fits for the random design case. *Aust. N. Z. J. Stat.*, 46 (2004), no. 3., 425–441.

[54] Hušková, M., Remarks on test procedures for gradual changes. *Asymptotic methods in probability and statistics* (Ottawa, ON, 1997), 577–583, North-Holland, Amsterdam, 1998.

[55] Hušková, Marie; Antoch, Jaromír, M-estimators of structural changes in regression models. *Probability theory mathematical statistics.* Tatra Mt. Math. Publ. 22 (2001), 197–208.

[56] Huškova, Marie; Antoch, Jaromír, Detection of structural changes in regression. Probastat '02. Part II. Tatra Mt. Math. Publ., 26 (2003), part II, 201–215.

[57] Hušková, M. and Steinebach, J., Asymptotic tests for gradual changes. Statist. *Decisions* 20 (2002), no. 2, 137–151.

[58] Jarušková, Daniela, Change-point estimator in gradually changing sequences. *Comment. Math. Univ. Carolin.*, 39 (1998), no. 3, 551–561.

[59] Kezim, Boualem; Abdelli, Zahia, A Bayesian analysis of a structural change in the parameters of a time series. *Comm. Stat. Theory and Methods*, 33 (2004), no. 8, 1863–1876.

[60] Kim, H. J. and Siegmund, D., The likelihood ratio test for a change point in simple linear regression. *Biometrika*, 76 (1989), 409–423.

[61] Kim, H. J., Two-phase regression with non-homogeneous errors, *Comm. Stat.*, 22 (1993), 647–657.

[62] Kim, H. J. and Cai, L., Robustness of the likelihood ratio for a change in simple linear regression, *J. Am. Stat. Assoc.*, 88 (1993), 864–871.

[63] Koul, H. L., Fitting a two phase linear regression model. *Journal of the Indian Statistical Association*, 38 (2000), 331–353.

[64] Koul, L. H. and Qian, Lianfen, Asymptotics of maximum likelihood estimator in a two-phase linear regression model. *Journal of Statistical Planning and Inference*, 108 (2002), no.1, 99–119.

[65] Koul, L. H., Qian, L. and Surgailis, D., Asymptotics of M-estimators in two-phase linear regression models. *Stochastic Processes and their Applications*, 103 (2003) no.1, 123–154.

[66] Lerman, P. M., Fitting segmented regression models by grid search. *Applied Statistics*, 29 (1980), 77–84.

[67] Loader, Clive R., Change point estimation using nonparametric regression. *Ann. Stat.*, 24 (1996), no. 4, 1667–1678.

[68] Lund, R. and Reeves, J., Detection of undocumented changepoints: A revision of the two-phase regression model. *Journal of Climate.*, 1 (2002), 2547–2554.

[69] Maekawa, Koichi; He, Zonglu; Tee, Kianheng., Estimating break points in a time series regression with structural changes. First special issue: selected papers of the MSSANZ/IMACS 14th Biennial Conference on Modelling and Simulation (Canberra, 2001). *Math. Comput. Simulat.*, 64 (2004), no. 1, 95–101.

[70] Marquardt, D. W., An algorithm for least -squares estimation of nonlinear parameters. *Journal of the Society for Industrial and Applied Mathematics*, 11 (1963), 431–441.

[71] Maronna, R. and Yohai, V. J., A bivariate test for the detection of a systematic change in mean. *J. Am. Stat. Assoc.*, 73 (1978), 640–645.

[72] Muggeo, V. M. R., Estimating regression models with unknown break-points. *Statistics in Medicine.* 22 (2003), 3055–3071.

[73] Müller, Hans-Georg; Stadtmüller, Ulrich. Discontinuous versus smooth regression. *Ann. Stat.*, 27 (1999), no. 1, 299–337.

[74] Neubauer, F. and Zwanzig, S., Test of fit in nonlinear two phases regression models. 21st European Meeting of Statisticians Aarhus, 1995.

[75] Onukogu, I., When is piecewise regression really necessary? *Biometric J.* 26 (1984), 559–566.

[76] Park, S. H., Experimental designs for fitting segmented polynomial regression. *Technometrics* 20(1978), 151–154.

[77] Pastor, R. and Guallar, E., Use of two-segmented logistic regression to estimate change-points in epidemiologic studies. *Am. J. Epidemiol.*, 148 (1998), no 7, 631–642.

[78] Piegorsch, W. W., Discretizing a normal prior for change point estimation in swithcing regressions. *Biometrial Journal*, 29 (1987), 777–782.

[79] Piegorsch, W. W. and Bailer, A., J *Statistics for environmental biology and toxicology.* Chapman and Hall. 1997.

[80] Piegorsch, W. W. and Bailer, A. J., *Analyzing environmental data.* John Wiley & Sons, Ltd. 2005.

[81] Piepho, H. P. and Ogutu, J. O., Inference for the break point in segmented regression with application to longitudinal data. *Biometric J.*, 45(2003) 5, 591–601.

[82] Qian, Lianfen, On maximum likelihood estimation for a threshold autoregression. *Journal of Statistical Planning and Inference.*, 75 (1998) no. 1, 21–46.

[83] Qian, Lianfen and Qingchuan Yao, Software project effort estimation using two-phase linear regression models. *Proceedings of The 15th Annual Motorola Software Engineering Symposium* (SES 2002).

[84] Qian, Lianfen and Soyoung Ryu, Estimating Tree Resin Dose Effect On Termites. *Environmentrics*, 17 (2006), 183–197.

[85] Quandt, R. E., Maximum-likelihood method for estimating the parameters in two separate regression lines that switch at an unknown point. *J. Am. Stat. Assoc.*, 53 (1958). 873–880.

[86] Quandt, R. E., A likelihood ratio test to tesst two separate regression lines as opposed to the null hypothesis that the data follow only one. *J. Am. Stat. Assoc.*, 55 (1960). 324–330.

[87] Seber, G. A. F. and Wild, C. J., *Nonlinear Regression.* Wiley, 1989.

[88] Schulze, Ursula. *Multiphase regression: Stability testing, estimation, hypothesis testing.* Akademie-Verlag, Berlin, 1987. 178 pp. ISBN 3-05-500224-5.

[89] Schwarz, G., Estimating the dimension of a model. *Ann. Stat.*, 6 (1978), 461–464.

[90] Schweder, T., Some "optimal" methods to detect structural shift or outliers in regression. *J. Am. Stat. Assoc.* 71 (1976), 491–501.

[91] Seber, G. A. F. and Wild, C. J., *Nonlinear Regression.* John Wiley. 1989.

[92] Shaban, S. A., Change point problem and two-phase regression: an annotated bibliography. Internat. *Stat. Rev.*, 48 (1980), no. 1, 83–93.

[93] Shepperd, M. and Schofield, C., Estimating software project effort using analogies. *IEEE Transactions on Software Engineering*, 23 (1997), 736–743.

[94] Shin, M. and Goel, A. L., Empirical data modeling in software engineering using radial basis functions. *IEEE Transactions on Software Engineering*, 26 (2000), 567–576.

[95] Smith, A. M. F. and Cook, D. G., Straight lines with a change point: A Bayesian analysis of some renal transplant data. *Appl. Stat.*, 29 (1980), 180–189.

[96] Sprent, P., Some hypotheses concerning two-phase regression lines. *Biometrics*, 17 (1961), 634–645.

[97] Staudenmayer, John and Spiegelman, Donna, Segmented regression in the presence of covariate measurement error in main study/validation study designs. *Biometrics*, 58 (2002), no. 4, 871–877.

[98] Tishler, Asher; Zang, Isreal, A new maximum likelihood algorithm for piecewise regression. *J. Am. Stat. Assoc.*, 76 (1981), no. 376, 980–987.

[99] Tishler, Asher; Zang, Isreal, A maximum likelihood method for piecewise regression models with a continuous dependent variable. *J. Roy. Stat.* Soc. Ser. C (Apllied Statistics) 30 (1981), no. 2, 116–124.

[100] Toms, J. D. and M. L. Lesperance. Piecewise regression: A tool for identifying ecological thresholds. *Ecology*, 84(8) (2003), 2034–2041.

[101] Ulm, K. W., A statistical method for assessing a threshold in epidemiological studies. *Statistics in medicine*, 10 (1991), 341–349.

[102] Van de Geer, S., *Regression analysis and empirical processes.* CWI tract. 45. Amsterdam, NL. 1988.

[103] Vieth, Elisabeth, Fitting piecewise linear regression functions to bilogical resposnes. *Journal of Applied Physiology*, 67 (1989), 390–396.

[104] Wang, X. L., Comments on "Detection of Undocuments Changepoints: A Revision of the Two-Phase Regression Model". *Journal of Climate.* 16 (2003), 3383–3386.

[105] Williams, D. A., Discrimination between regression models to determine the pattern of enzyme syntheisis in synchronous cell cultures. *Biometrics*, 26 (1970), 23–32.

[106] Worsley, K. J., Test for a two-phase multiple regression, *Technometrics*, 25, 35–42.

[107] Worsley, K. J., Confidence regions and tests for a change-point in a sequence of exponential family random variables, *Biometrika*, 73 (1986), 91–104.

[108] Wu, J. S. and Chu, C. K., Modification for boundary effects and jump points in nonparametric regressions. *J. Nonparametr. Stat.*, 2 (1993), no. 4, 341–354.

[109] Wu, Yanhong, Inference for change point and post change means after a CUSUM test. Lecture notes in Statistics, Springer. 2005.

[110] Yohai, V. J. and Zamar, R. H., High breakdown-point estimates of regression by means of the minimization of an efficient scale. *J. Am. Stat. Assoc.*, 83 (1988), no. 402, 406–413.

[111] Zeileis, Achim, Implementing a class of structural change tests: an econometric computing approach. *Comput. Stat. Data Anal.*, 50 (2006), no. 11, 2987–3008.

Estimation in Partially Linear Models With Missing Data: A Review *

Suojin Wang[1], Hua Liang[2] and Lei Jin[1]

[1]*Department of Statistics, Texas A&M University*
College Station, Texas 77845, USA
E-mail: sjwang@stat.tamu.edu, ljin@stat.tamu.edu
and
[2]*Department of Biostatistics and Computational Biology*
University of Rochester Medical Center
Rochester, NY 14642, USA
E-mail: hliang@bst.rochester.edu

Abstract

Partially linear models have a wide range of usage in practice. In this article, we review some recently developed methods when some of the observations in the data are partially missing. Our focus of the discussion is on the estimation of the partial slopes in the models. We also provide a unified framework which includes both the case of the response variable being partially missing and the case of a covariate being partially missing. The empirical likelihood approach to parameter estimation is considered under the unified framework. Some results from simulation experiments are given for an illustration.

Keywords: Confidence region, Empirical likelihood, Measurement error, Missing at random, Missing not at random, Nonparametric regression, Semiparametric estimation.

1 Introduction

Partially linear models have a wide range of usage in various disciplines such as medicine, finance and econometrics. Assume that we have independent and identically distributed observations (Y_i, X_i, Z_i)

---

*This research was partially supported by the Texas A&M Center for Environmental and Rural Health via a grant from the National Institute of Environmental Health Sciences (P30-ES09106), and by the NIAD/NIH grants AI62247 and AI59773. The authors thank the referee and the Editor for their helpful comments.

for $i = 1, \ldots, n$. The partially linear model under our consideration is assumed to link the response variable Y_i to the independent variable X_i in a linear way but is related to the other independent variable Z_i in an unspecified form as follows:

$$Y_i = X_i^{\mathrm{T}}\beta + \nu(Z_i) + \varepsilon_i, \tag{1.1}$$

where X_i is a p-vector covariate, Z_i is a scalar covariate, the function $\nu(\cdot)$ is unknown, and the model error ε_i has mean zero conditional on (X_i, Z_i).

Many authors have studied kernel-based methods for partially linear models; see Engle, Granger, Rice and Weiss (1986), Speckman (1988), Robinson (1988), Severini and Staniswalis (1994), Zeger and Diggle (1994), Opsomer and Ruppert (1999) and Härdle, Liang and Gao (2000), among many others. There have been studies in recent years on the extensions of the standard partially linear model (1.1). For example, Liang, Härdle and Carroll (1999) considered model (1.1) with error-prone X. Another direction of research has been on missing data issues. Liang, Wang, Robins and Carroll (2004) considered the case that X in (1.1) is missing at random in the sense of that the missingness probability of a variable given other observed variables does not depend on the variable; see Little and Rubin (1987) for a review of missing data literature. On the other hand, Wang, Linton and Härdle (2004) considered the case that the response Y is missing at random, while Liang, Wang and Carroll (2006) studied this case when some of the components of X are measured with error. A motivation for such missing data investigations is from AIDS clinical trials, where the response variable, viral load RNA, and the covariates, CD4 measurements, can be missing (or measured with error). To simplify the complicated issue of possibly both of the response variable and the independent variable being partially missing, the missingness patterns have been dealt with separately in the current literature.

In this article, we give a partial review of some innovative methods recently proposed in the literature for estimation of the parameter of interest, β, in partially linear models in the form (1.1) with missing data. Our focus is on (i) the methodology by Liang, et al. (2004) when X in (1.1) is missing at random; and (ii) the methodology by Liang, et al. (2006) when Y in (1.1) is partially missing and X is not missing but possibly measured with error. They are given in Sections 2 and 3 respectively. In Section 4, we provide a unified framework that includes both the case of the response variable being partially missing and the case of a covariate being partially missing. The empirical likelihood approach to parameter estimation is considered under the unified framework. The empirical likelihood method of Liang, et al. (2006) can be viewed as a special case of this unified approach. Some simulation results are illustrated in Section 5. Some concluding remarks are given in Section 6.

2 Partially linear models with missing covariates

In this section we discuss estimation procedures for model (1.1) when the data on X are not fully observed for some study subjects. The problem of missing exposure variables in regression has been treated in great detail in Robins, Rotnitzky and Zhao (1994) who assumed a parametric functional form for $\nu(Z)$. On the other hand, as was argued in Liang, et al. (2004), there is often the need to relax the assumption. As in Robins et al. (1994) we allow the missingness probabilities to depend on both Y and Z but not on the unobserved value of X. This has motivated Liang, et al. (2004) to develop new methods for addressing the issue of estimating β and the nonparametric function $\nu(Z)$ in (1.1) in both the case where the missingness probabilities are known (as in a designed two-stage study) and the case where they are unknown. Their results are built on the work of Wang, Wang, Gutierrez and Carroll (1998), who considered the nonparametric problem (no X) with missing data; see also Cheng (1990, 1994) and Cheng and Chu (1996).

Assume that the X's are missing at random, i.e., given the observable (Z, Y) the missingness of X does not depend on the X value:

$$\pi_x(Y_i, Z_i) = \text{pr}(\delta_{xi} = 1 | X_i, Z_i, Y_i) = \text{pr}(\delta_{xi} = 1 | Z_i, Y_i), \qquad (2.1)$$

where $\delta_{xi} = 1$ if X_i is observed and $\delta_{xi} = 0$ otherwise. If the missing data probability $\pi_x(Y, Z)$ is known, it would be a little easier to carry out the estimation procedure. We will use this setting to describe the procedure below. However, its estimation (denoted by $\widehat{\pi}_x$) with an error of order $o_p(n^{-1/4})$ can be undertaken without affecting the asymptotic properties of the estimator $\widehat{\beta}$. Standard nonparametric estimators such as the kernel based one employed in Liang, et al. (2004) with a proper bandwidth satisfy this condition for the error rate. A further discussion of this is given below Theorem 2.1. We assume that $(Y_i, X_i, Z_i, \delta_{xi})$, $i = 1, \cdots, n$, are independent and identically distributed (i.i.d.). The extension to the case of the longitudinal/clustered data setting has been considered in Liang, et al. (2004, Section 3.4).

The estimation procedure makes use of the Horvitz-Thompson (HT) type weighted local linear kernel method. Let $K(\cdot)$ be a symmetric density function and h be a suitable bandwidth. Then for any response $q(Y, X)$, the local estimate of $\text{E}\{q(Y, X) | Z = z_0\}$ is the intercept α_0

satisfying the following equation in the regression of $q(Y, X)$ on $(Z-z_0)/h$ with weights $K_h(Z - z_0)\delta_x/\pi_x(Y, Z)$ and $K_h(v) = h^{-1}K(v/h)$:

$$\sum_{i=1}^{n} K_h(Z_i - z_0)\frac{\delta_{xi}}{\pi_x(Y_i, Z_i)}\begin{pmatrix} 1 \\ \frac{Z_i - z_0}{h} \end{pmatrix} \left\{q(Y_i, X_i) - \alpha_0 - \alpha_1\left(\frac{Z_i - z_0}{h}\right)\right\} = 0. \quad (2.2)$$

If $\beta = 0$ (no X) in (1.1), the topic becomes a nonparametric problem, for which Wang et al. (1998) developed the HT weighted local linear kernel methods. The estimation procedure for partially linear models with missing covariates can be looked upon as combining the parametric procedure of Robins et al. (1994) with the nonparametric procedure of Wang et al. (1998).

We now describe the simple four-step procedure proposed in Liang, et al. (2004):

(1) For any β, let $\widehat{\nu}(\cdot, \beta, \pi_x)$ be a weighted nonparametric regression of $Y - X^{\mathrm{T}}\beta$ on Z using the HT local linear kernel method, i.e., (2.2) with $q(Y, X) = Y - X^{\mathrm{T}}\beta$.

(2) Form the corresponding HT local linear kernel regression function $\widehat{m}(z, \pi_x)$ for estimating $m(z) = \mathrm{E}(X|z)$ by regressing X on Z, i.e., (2.2) with $q(Y, X) = X$.

(3) Let $\widehat{\phi}(Y, Z, \beta, \widehat{\nu}, \widehat{m})$ be an estimate of $\phi(Y, Z) = \mathrm{E}[\{X - \mathrm{E}(X|Z)\}\{Y - \nu(Z) - X^{\mathrm{T}}\beta\}|Y, Z]$, which is linear in β.

(4) Solve for β in the equation

$$\sum_{i=1}^{n}\{X_i - \widehat{m}(Z_i, \pi_x)\}\left\{Y_i - \widehat{\nu}(Z_i, \beta, \pi_x) - X_i^{\mathrm{T}}\beta\right\}\frac{\delta_{xi}}{\pi_x(Y_i, Z_i)} - \sum_{i=1}^{n}\widehat{\phi}(Y_i, Z_i, \beta, \widehat{\nu}, \widehat{m})\frac{\delta_{xi} - \pi_x(Y_i, Z_i)}{\pi_x(Y_i, Z_i)} = 0. \quad (2.3)$$

Equation (2.3) is easy to solve since it is linear in β. Denote the solution by $\widehat{\beta}_x$. When $\pi_x(Y_i, Z_i)$ is replaced by $\widehat{\pi}_x(Y_i, Z_i)$ in (2.2) and (2.3), we obtain an estimator called $\widehat{\beta}_{\widehat{\pi},x}$.

To carry out Step 3 above, it is recommended to use the HT bivariate estimators $\widehat{\mathrm{E}}(X|y, z)$ and $\widehat{\mathrm{E}}(XX^{\mathrm{T}}|y, z)$ for $\mathrm{E}(X|y, z)$ and $\mathrm{E}(XX^{\mathrm{T}}|y, z)$. For any smooth function g, the HT bivariate local linear estimator of

$\mathrm{E}\{g(X)|y,z\}$ is defined to be α_0 so that $(\alpha_0, \alpha_{1y}, \alpha_{1z})$ minimizes

$$\sum_{i=1}^{n} \{g(X_i) - \alpha_0 - \alpha_{1y}(Y_i - y) - \alpha_{1z}(Z_i - z)\}^2 \times K^*_{\lambda_1,\lambda_2}(Y_i - y, Z_i - z)\frac{\delta_{xi}}{\pi_x(Y_i, Z_i)},$$

where $K^*_{\lambda_1,\lambda_2}(\cdot,\cdot)$ is a two dimensional density function with bandwidths λ_1, λ_2. We can then obtain

$$\begin{aligned}\widehat{\phi}(y,z,\beta,\widehat{\nu},\widehat{m}) &= \widehat{\mathrm{E}}(X|y,z)y - \widehat{\mathrm{E}}(X|y,z)\widehat{\nu}(z,\beta,\pi_x) - \{\widehat{\mathrm{E}}(XX^{\mathrm{T}}|y,z)\}\beta \\ &\quad -\widehat{m}(z,\pi_x)y + \widehat{m}(z,\pi_x)\widehat{\nu}(z,\beta,\pi_x) + \widehat{m}(z,\pi_x)\{\widehat{\mathrm{E}}(X|y,z)\}^{\mathrm{T}}\beta.\end{aligned}$$

Note that there are alternative methods for the regressions in Steps 1–3, including higher degree local polynomial kernel methods, kernel methods with varying bandwidths, smoothing and regression splines, etc. See Liang, et al. (2004) for a further discussion.

We now discuss the asymptotic properties of the resultant point estimators. A set of standard regularity conditions are required to guarantee the asymptotic performance of the estimators. These conditions are listed in Liang, et al. (2004). The main results are as follows:

Theorem 2.1. *Assume that $(Y_i, X_i, Z_i, \delta_{xi})$, $i = 1, \cdots, n$, are independent and identically distributed, and that the regularity conditions mentioned above hold.*

(a) $\sqrt{n}(\widehat{\beta}_x - \beta)$ *is asymptotically normally distributed with mean 0 and covariance matrix* $\Sigma_\beta = \Sigma^{-1}_{X|Z}\Lambda\Sigma^{-1}_{X|Z}$, *where*

$$\Lambda = \mathrm{E}\left(\frac{\varepsilon^2}{\pi_x}\widetilde{X}\widetilde{X}^{\mathrm{T}}\right) - \mathrm{E}\left[\frac{1-\pi_x}{\pi_x}\{\mathrm{E}(\widetilde{X}\varepsilon|Y,Z)\}^{\otimes 2}\right],$$

with $\widetilde{X} = X - \mathrm{E}(X|Z)$, $\Sigma_{X|Z} = \mathrm{Cov}(\widetilde{X})$, *and* $C^{\otimes 2} = C \cdot C^{\mathrm{T}}$ *for any vector* C.

(b) *If* $\inf_{y,z} \pi_x(y,z) > 0$ *and* $\sup_{y,z} |\widehat{\pi}_x(y,z) - \pi_x(y,z)| = o_p(n^{-1/4})$, *then* $\widehat{\beta}_{\widehat{\pi},x}$ *has the same limiting normal distribution as that of* $\widehat{\beta}_x$.

The justifications have been provided in the Appendix of Liang, et al. (2004). They are based on the general asymptotic theory for semiparametric models developed by Newey (1994) and Robins et al. (1994).

Note that the covariance matrix can be estimated consistently with a standard approach. Given the bandwidths of order $n^{-1/6}$, the convergence rate of $\widehat{\pi}_x$ to π_x is $n^{-1/3} = o_p(n^{-1/4})$, satisfying the condition on the convergence in Theorem 2.1(b). It is well known that bandwidth selection plays a key role in the performance of nonparametric kernel estimators in terms of their bias and variance. When implementing the estimation procedure, it is useful to adopt Ruppert, Sheather, and Wand's (1995) approach to search for the bandwidths. In partially linear models, β is of main interest, and $\nu(z)$ is a secondary (sometimes nuisance) function. The numerical performance of the resulting estimators of β appears to be stable around the selected bandwidths both in this context and in the missing response case to be described in Section 3.

While Theorem 2.1 gives the asymptotic properties of the $\widehat{\beta}_x$ and $\widehat{\beta}_{\widehat{\pi},x}$, the asymptotic limit distribution may not work very well in small to moderate sample sizes. This motivates our consideration of the empirical likelihood approach in Section 4.

3 Partially linear models with missing response variables

In this section, we continue to consider the estimation problem in the partially linear model (1.1), but with the response variable Y being sometimes missing. Furthermore, the covariate X is allowed to be measured with error. Similar to the missing covariate case considered in Section 2, we assume that given (X, Z) the missingness of Y does not depend on the Y value:

$$\pi_y(X_i, Z_i) = \text{pr}(\delta_{yi} = 1 | Y_i, X_i, Z_i) = \text{pr}(\delta_{yi} = 1 | X_i, Z_i), \qquad (3.1)$$

where $\delta_{yi} = 1$ if Y_i is observed and $\delta_{yi} = 0$ otherwise. When X is measured without error, then Y is missing at random. However, here we assume that X is generally measured with error. Then the missingness structure is not missing at random. This is a key difference from the usual missing at random assumption employed in most studies in the literature. In the current setting, an i.i.d. sample of $\{(Y_i, W_i, Z_i, \delta_{yi}), i = 1, \ldots, n\}$ is observed, where

$$W_i = X_i + U_i \qquad (3.2)$$

is a surrogate for X_i with $\text{E}(U_i) = 0$, $\text{Cov}(U_i) = \Sigma_{uu}$ and U_i is independent of $(Y_i, X_i, Z_i, \delta_{yi})$.

Local constant smoothers with fixed bandwidths are used for simplicity. The extensions to local polynomial estimation are straightforward, with no change in the limiting distribution of the estimate of β. Define $m_d(z) = \text{E}(\delta_y D | Z = z) / \text{E}(\delta_y | Z = z)$ and $\widetilde{D}_i = D_i - m_d(Z_i)$ for $D = X, W, Y$, etc. Furthermore, denote $\Sigma_{X|Z,\delta_y} = \text{Cov}(\widetilde{X}\delta_y)$.

Note that $\delta_y Y = \delta_y X^{\text{T}}\beta + \delta_y \nu(Z) + \delta_y \varepsilon$. It follows that $\text{E}(\delta_y Y | Z) = \text{E}(\delta_y X^{\text{T}} | Z)\beta + \text{E}(\delta_y | Z)\nu(Z)$ since $\text{E}(\delta_y \varepsilon | Z) = 0$ by (3.1). If X is observed and $m_x(z)$ and $m_y(z)$ are known, one can obtain a least squares type estimator of β as

$$\left[\sum_{i=1}^{n} \delta_{yi}\{X_i - m_x(Z_i)\}^{\otimes 2}\right]^{-1} \left[\sum_{i=1}^{n} \delta_{yi}\{X_i - m_x(Z_i)\}\{Y_i - m_y(Z_i)\}\right].$$

The expectation of the ith term in the right bracket is $\mathrm{E}\{\pi(X,Z)\widetilde{X}\widetilde{Y}\} = \Sigma_{X|Z,\delta_y}\beta$. It is easily shown that this estimator is consistent, asymptotically normal, etc.; see Wang, et al. (2004).

When X is measured with error, and $m_x(z)$ and $m_y(z)$ are unknown, the formula above requires an adjustment for measurement error bias. From (3.2), $\mathrm{E}(\delta_y W|Z) = \mathrm{E}(\delta_y X|Z)$. Using this identity, Liang, et al. (2006) proposed an estimator of β that corrects for attenuation:

$$\widehat{\beta}_y = \left(\sum_{i=1}^{n} \delta_{yi}\left[\{W_i - \widehat{m}_w(Z_i)\}^{\otimes 2} - \Sigma_{uu}\right]\right)^{-1} \times \left[\sum_{i=1}^{n} \delta_{yi}\{W_i - \widehat{m}_w(Z_i)\}\{Y_i - \widehat{m}_y(Z_i)\}\right], \quad (3.3)$$

where $\widehat{m}_w(z)$ and $\widehat{m}_y(z)$ are nonparametric regression estimators, and Σ_{uu} is assumed to be known for now. These estimators take the form

$$\widehat{m}_w(z) = \frac{\sum_{i=1}^{n} K_h(Z_i - z)\delta_{yi}W_i}{\sum_{i=1}^{n} \delta_{yi}K_h(Z_i - z)}, \quad \widehat{m}_y(z) = \frac{\sum_{i=1}^{n} K_h(Z_i - z)\delta_{yi}Y_i}{\sum_{i=1}^{n} \delta_{yi}K_h(Z_i - z)}, \quad (3.4)$$

where $K_h(v) = h^{-1}K(v/h)$ was defined in Section 2.

Note that alternative estimators are readily constructed, but generally suffer from complications. For example, since $Y - \mathrm{E}(Y|Z) = \{X - \mathrm{E}(W|Z)\}^{\mathrm{T}}\beta + \varepsilon$, an obvious approach is to estimate $\mathrm{E}(W|Z)$ and $\mathrm{E}(Y|Z)$ using all the data. The former is easy: any nonparametric regression will do. The latter though is problematic, because of the missing responses, the possibility that missingness depends on X, and the fact that X is unobserved. There does not appear to be an easy way to estimate $\mathrm{E}(Y|Z)$ consistently under the current conditions. Note that one of the most important features of the proposed approach is that, by using the standard measurement error model (3.2), it can handle the not missing at random case with ease and still provide $\sqrt{n}$-consistent estimators, as shown in the theorems that follow.

Theorem 3.1. *Under general regularity conditions, $n^{1/2}(\widehat{\beta}_y - \beta)$ is asymptotically normally distributed with mean 0 and covariance matrix $\Sigma_\beta = \Sigma_{X|Z,\delta_y}^{-1}\Gamma\Sigma_{X|Z,\delta_y}^{-1}$, where*

$$\Gamma = \mathrm{E}[\delta_y\{(\varepsilon - U^{\mathrm{T}}\beta)\widetilde{X}\}^{\otimes 2}] + \mathrm{E}(\delta_y UU^{\mathrm{T}}\varepsilon^2) + \mathrm{E}[\delta_y\{(UU^{\mathrm{T}} - \Sigma_{uu})\beta\}^{\otimes 2}].$$

Proof. The proof and that of Theorem 3.2 given below use a technique similar to that used by Liang, et al. (2004) to prove their Theorems 1 and 2.

We first point out the following fact, which can easily be shown by regularity conditions,

$$\widehat{m}_w(z) - m_w(z) = o_p(n^{-1/4}) \text{ and } \widehat{m}_y(z) - m_y(z) = o_p(n^{-1/4}). \quad (3.5)$$

We use these two equations in what follows. Define

$$\begin{aligned}\Psi&(m_w, m_y, \beta, Y, W, Z, \delta_y)\\ &= \left(\{W - m_w(Z)\}\left[Y - m_y(Z) - \{W - m_w(Z)\}^{\mathrm{T}}\beta\right] + \Sigma_{uu}\beta\right)\delta_y. \quad (3.6)\end{aligned}$$

Our estimator $\widehat{\beta}_y$ solves the estimating equations

$$\sum_{i=1}^{n} \Psi(\widehat{m}_w, \widehat{m}_y, \beta, Y_i, W_i, Z_i, \delta_{yi}) = 0.$$

Let

$$\begin{aligned}\Upsilon&(m_w^* - m_w, m_y^* - m_y, \beta, Y, W, Z, \delta_y)\\ &= \frac{\partial \Psi}{\partial m_w}(m_w^* - m_w) + \frac{\partial \Psi}{\partial m_y}(m_y^* - m_y),\end{aligned}$$

where the partial derivatives are the Frechet partial derivatives. It is easy to obtain that

$$\begin{aligned}\frac{\partial \Psi}{\partial m_w} &= [-Y + m_y(Z) - 2\{W - m_w(Z)\}^{\mathrm{T}}\beta]\delta_y,\\ \frac{\partial \Psi}{\partial m_y} &= -\{W - m_w(Z)\}\delta_y.\end{aligned}$$

A direct calculation yields that $\mathrm{E}\left(\frac{\partial \Psi}{\partial m_w}|Z\right) = 0$ a.e. and $\mathrm{E}\left(\frac{\partial \Psi}{\partial m_y}|Z\right) = 0$ a.e. In addition,

$$\begin{aligned}\|\Psi&(m_w^*, m_y^*, \beta, Y, W, Z, \delta_y) - \Psi(m_w, m_y, \beta, Y, W, Z, \delta_y)\\ &-\Upsilon(m_w^* - m_w, m_y^* - m_y, \beta, Y, W, Z, \delta_y)\|\\ &= O\left(\|m_w^* - m_w\|^2 + \|m_y^* - m_y\|^2\right), \quad (3.7)\end{aligned}$$

where $\|h\|$ denotes a norm for the function h, such as Sobolev norm, i.e, supremum norm for a function and its derivatives. Equation (3.7) is Newey's (1994) assumption 5.1(i). Furthermore, (3.5) ensures assumption 5.1(ii) of Newey. Again his assumption 5.2 holds by the expression of $\Upsilon(\cdot,\beta,Y,W,Z,\delta_y)$. In addition, it follows from the above statements that for any (m_w^*, m_y^*),

$$\mathrm{E}\left\{\Upsilon(m_w^* - m_w, m_y^* - m_y, \beta, Y, W, Z, \delta_y)\right\} = 0,$$

thus verifying Newey's assumption 5.3: his $\alpha(z) = 0$, according to his discussion just above his Lemma 5.1. By that lemma, it follows that $\widehat{\beta}_y$ has the same limit distribution as the solution, call $\widetilde{\beta}_y$, to the equation

$$\sum_{i=1}^{n} \Psi(m_w, m_y, \beta, Y_i, W_i, Z_i, \delta_{yi}) = 0.$$

A direct derivation gives that

$$\widetilde{\beta}_y = \left\{\frac{1}{n}\sum_{i=1}^{n} \widetilde{W}_i \widetilde{W}_i^{\mathrm{T}} \delta_{yi}\right\}^{-1} \frac{1}{n}\sum_{i=1}^{n} \widetilde{W}_i \widetilde{Y}_i \delta_{yi}.$$

It follows that

$$\sqrt{n}(\widetilde{\beta}_y - \beta) = \Delta_n^{-1}\frac{1}{\sqrt{n}}\sum_{i=1}^{n} \delta_{yi}(\varepsilon_i \widetilde{X}_i - \widetilde{X}_i U_i^{\mathrm{T}}\beta + U_i\varepsilon_i - U_i U_i^{\mathrm{T}}\beta + \Sigma_{uu}\beta),$$

where

$$\Delta_n = \frac{1}{n}\sum_{i=1}^{n} \widetilde{X}_i \widetilde{X}_i^{\mathrm{T}} \delta_{yi} + \frac{1}{n}\sum_{i=1}^{n} U_i U_i^{\mathrm{T}} \delta_{yi} - \frac{1}{n}\sum_{i=1}^{n} \delta_{yi}\Sigma_{uu} + o_p(1) \to \Sigma_{X|Z,\delta_y}.$$

Therefore, $\widetilde{\beta}_y$ has the same limit distribution as described in the statement of Theorem 3.1. This completes the proof.

From the proof of Theorem 3.1, it is seen that Σ_β can be estimated via a standard sandwich method as follows. Let

$$\widehat{\Sigma}_{X|Z,\delta_y} = n^{-1}\sum_{i=1}^{n} \delta_{yi}[\{W_i - \widehat{m}_w(Z_i)\}^{\otimes 2} - \Sigma_{uu}],$$

$$\widehat{\Gamma} = n^{-1}\sum_{i=1}^{n} \delta_{yi}\Big(\{W_i - \widehat{m}_w(Z_i)\}[Y_i - \widehat{m}_y(Z_i) - \{W_i - \widehat{m}_w(Z_i)\}^{\mathrm{T}}\widehat{\beta}_y] + \Sigma_{uu}\widehat{\beta}_y\Big)^{\otimes 2}$$

and $\widehat{\Sigma}_\beta = \widehat{\Sigma}_{X|Z,\delta_y}^{-1}\widehat{\Gamma}\widehat{\Sigma}_{X|Z,\delta_y}^{-1}$. Then it is easily shown that $\widehat{\Sigma}_\beta$ is a consistent estimator of Σ_β.

We have so far assumed known covariance matrix Σ_{uu}, while in practice it is generally unknown and needs to be estimated. The usual method of doing so (Carroll, et al., 2006, Chapter 4) is by partial replication, so that we observe $W_{ij} = X_i + U_{ij},\ \ j = 1, \cdots, k_i$. For notational simplicity, we assume that $k_i \equiv k$. Extension to more general settings is straightforward; see Liang, et al. (1999) for a related discussion. Let $\overline{W}_i$ be the sample mean of the replicates W_{ij}. A consistent and unbiased method-of-moments estimate for Σ_{uu} is

$$\widehat{\Sigma}_{uu} = n^{-1}\sum_{i=1}^{n}\sum_{j=1}^{k}(W_{ij} - \overline{W}_i)(W_{ij} - \overline{W}_i)^{\mathrm{T}}.$$

The corresponding estimator of β is

$$\widehat{\beta}_{y,k} = \left(\sum_{i=1}^{n}\delta_{yi}\left[\{\overline{W}_i - \widehat{m}_{\overline{w}}(Z_i)\}^{\otimes 2} - (1/k)\widehat{\Sigma}_{uu}\right]\right)^{-1} \times \left[\sum_{i=1}^{n}\delta_{yi}\{\overline{W}_i - \widehat{m}_{\overline{w}}(Z_i)\}\{Y_i - \widehat{m}_y(Z_i)\}\right], \quad (3.8)$$

where $\widehat{m}_{\overline{w}}(z)$ is the local constant estimate of $m_w(z)$ based on the data $\{(\overline{W}_i, Z_i), i = 1, \cdots, n\}$. We now present the following theorem:

Theorem 3.2. *Under general regularity conditions, the estimator $\widehat{\beta}_{y,k}$ given in (3.8) is consistent for β and asymptotically normal with covariance matrix $\Sigma^{-1}_{X|Z,\delta_y}\Gamma^{*}\Sigma^{-1}_{X|Z,\delta_y}$, where*

$$\Gamma^{*} = \mathrm{E}[\delta_y\{(\varepsilon - \overline{U}^{\mathrm{T}}\beta)\widetilde{X}\}^{\otimes 2}] + \mathrm{E}(\delta_y\overline{U}\,\overline{U}^{\mathrm{T}}\varepsilon^2) + \mathrm{E}[\delta_y\{(\overline{U}\,\overline{U}^{\mathrm{T}} - \Sigma_{uu}/k)\beta\}^{\otimes 2}].$$

By a straightforward but tedious derivation, Theorem 3.2 can be proved in a manner similar to Theorem 3.1. The standard error estimates can be derived. A consistent estimate of $\Sigma_{X|Z,\delta_y}$ in this case is defined as

$$n^{-1}\sum_{i=1}^{n}\delta_{yi}\left[\{\overline{W}_i - \widehat{m}_{\overline{w}}(Z_i)\}^{\otimes 2} - (1/k)\widehat{\Sigma}_{uu}\right].$$

The Γ^{*} can also be estimated. For example, when $k = 2$, let

$$R_i = \{\overline{W}_i - \widehat{m}_{\overline{w}}(Z_i)\}\left[Y_i - \widehat{m}_y(Z_i) - \{\overline{W}_i - \widehat{m}_{\overline{w}}(Z_i)\}^{\mathrm{T}}\widehat{\beta}_{y,k}\right] + (1/2)\left\{(W_{i1} - W_{i2})^{\otimes 2} - \widehat{\Sigma}_{uu}\right\}\widehat{\beta}_{y,k}.$$

Then the sample covariance matrix of the $\delta_{yi}R_i$'s is a consistent estimate of Γ^{*}. See Liang, at al. (1999) for more details.

It is worth mentioning that while our main interest here is the estimation of β, it is often of interest to estimate nonparametric ν in model

(1.1). It is easily achieved once a sensible estimate of β is obtained. For example, in the missing covariate case, $\widehat{\nu}(z, \widehat{\beta}_{\widehat{\pi},x}, \widehat{\pi}_x)$ via Step 1 is a pointwise consistent estimate for $\nu(z)$. It is similar for the missing response case as well. Numerical examples of computing $\widehat{\nu}(z)$ have been illustrated in Liang, et al. (2004) and Liang, et al. (2006).

4 Empirical likelihood based inference

In Sections 2 and 3, a point estimator of β and its asymptotic distributional properties have been given. While the asymptotic results are useful as a guide for large sample cases, the normal approximations may not perform very well for small to moderate samples in constructing a confidence region for β. One potentially promising alternative is the empirical likelihood principle first proposed by Owen (1988) and investigated by many others such as Chen (1994), Qin and Lawless (1994) and Qin (1999); see Owen (2001) for a review. More recently, Liang, et al. (2006) proposed an empirical likelihood approach to the problem of constructing a confidence region in the context of the missing response problem described in Section 3 and showed that the empirical likelihood ratio statistic is asymptotically chi-squared distributed. This line of thought can be extended to a more general estimating equation setting that includes certain other missing data situations such as that considered in Section 2. Here we provide a description of such a unified empirical likelihood approach.

Suppose that we have the following estimating equation that defines an estimator of β:

$$\sum_{i=1}^{n} \Omega(G_i, \beta, \tau) = 0, \tag{4.1}$$

such that $\mathrm{E}\{\Omega(G_i, \beta, \tau)\} = 0$ when β and τ are at their true values, where G_i is the vector of random variables observed at the ith observation, $\Omega(G_i, \beta, \tau)$ is the corresponding estimating function, and τ is the vector of nuisance/secondary parameters and/or functions and is to be properly estimated separately. For example, in the case of missing covariate problem given in Section 2 (Equation (2.3)),

$$\begin{aligned}\Omega = \Omega_x(G, \beta, \tau) = \{X - m(Z)\}\left\{Y - \nu(Z) - X^{\mathrm{T}}\beta\right\} \frac{\delta_x}{\pi_x(Y, Z)} \\ - \phi(Y, Z)\frac{\delta_x - \pi_x(Y, Z)}{\pi_x(Y, Z)} = 0,\end{aligned} \tag{4.2}$$

where $G = (Y, X, Z, \delta_x)^{\mathrm{T}}$ and $\tau = (m, \nu, \phi, \pi_x)^{\mathrm{T}}$. On the other hand, in the case of missing response problem given in Section 3, by (3.6) we have

$$\Omega = \Omega_y(G, \beta, \tau)$$
$$= \Big(\{W - m_w(Z)\}\Big[Y - m_y(Z) - \{W - m_w(Z)\}^{\mathrm{T}}\beta\Big] + \Sigma_{uu}\beta\Big)\delta_y, \quad (4.3)$$

where $G = (Y, W, Z, \delta_y)^{\mathrm{T}}$ and $\tau = (m_y, m_w, \Sigma_{uu})^{\mathrm{T}}$.

Empirical likelihood starts with assigning a probability p_i for observation i, so that $p_i \geqslant 0$ and $\sum_{i=1}^{n} p_i = 1$. If τ are assumed known, then the empirical likelihood ratio function for β is defined as

$$\mathcal{R}(\beta) = \max\left\{\prod_{i=1}^{n} np_i \,\Big|\, \sum_{i=1}^{n} p_i\Omega(G_i, \beta, \tau) = 0, p_i \geqslant 0, \sum_{i=1}^{n} p_i = 1\right\}.$$

In practice, τ is generally unknown. We can then define the semiparametric empirical likelihood ratio as

$$\mathcal{R}_n(\beta) = \max\left\{\prod_{i=1}^{n} np_i \,\Big|\, \sum_{i=1}^{n} p_i\Omega(G_i, \beta, \widehat{\tau}) = 0, p_i \geqslant 0, \sum_{i=1}^{n} p_i = 1\right\}, \quad (4.4)$$

where $\widehat{\tau}$ is properly obtained as is discussed in Sections 2 and 3. We now turn to the discussion of the asymptotic properties of this unified empirical likelihood approach.

Theorem 4.1. *Under general regularity conditions, we have*

(a) $-2\log\{\mathcal{R}(\beta)\}$ *converges to a chi-squared distribution with p degrees of freedom;*

(b) $-2\log\{\mathcal{R}_n(\beta)\}$ *converges to a chi-squared distribution with p degrees of freedom, provided that the estimator $\widehat{\tau}$ has the order of convergence to τ that satisfies*

$$\sum_{i=1}^{n} \Omega(G_i, \beta, \widehat{\tau}) - \sum_{i=1}^{n} \Omega(G_i, \beta, \tau) = o_p(n^{1/2}), \quad (4.5)$$

and

$$\frac{1}{n}\sum_{i=1}^{n} \Omega(G_i, \beta, \widehat{\tau})^{\otimes 2} - \frac{1}{n}\sum_{i=1}^{n} \Omega(G_i, \beta, \tau)^{\otimes 2} = o_p(1). \quad (4.6)$$

Proof. (a) Write $\Omega_i = \Omega(G_i, \beta, \tau)$. Following the argument in Owen (2001, p. 61) it is easy to see that

$$p_i = \frac{1}{n(1 + \boldsymbol{\lambda}^{\mathrm{T}}\Omega_i)} \quad \text{for } i = 1, \ldots, n,$$

where $\boldsymbol{\lambda} = (\lambda_1, \cdots, \lambda_p)^{\mathrm{T}}$ solves

$$n^{-1} \sum_{i=1}^{n} \frac{\Omega_i}{1 + \boldsymbol{\lambda}^{\mathrm{T}} \Omega_i} = 0. \tag{4.7}$$

By (4.7), the constraint $\sum_{i=1}^{n} p_i = 1$ and regularity conditions, some algebra leads to

$$\boldsymbol{\lambda} = \left(\sum_{i=1}^{n} \Omega_i^{\otimes 2} \right)^{-1} \sum_{i=1}^{n} \Omega_i + o_p(n^{-1/2}),$$

and

$$\sum_{i=1}^{n} \boldsymbol{\lambda}^{\mathrm{T}} \Omega_i = \sum_{i=1}^{n} (\boldsymbol{\lambda}^{\mathrm{T}} \Omega_i)^2 + o_p(1).$$

Therefore, with a simple Taylor expansion, we obtain

$$\begin{aligned} -2\log\{\mathcal{R}(\beta)\} &= 2\sum_{i=1}^{n} \log(1 + \boldsymbol{\lambda}^{\mathrm{T}} \Omega_i) \\ &= 2\sum_{i=1}^{n} \left\{ \boldsymbol{\lambda}^{\mathrm{T}} \Omega_i - (1/2)(\boldsymbol{\lambda}^{\mathrm{T}} \Omega_i)^2 \right\} + o_p(1) \\ &= \left(\frac{1}{\sqrt{n}} \sum_{i=1}^{n} \Omega_i \right)^{\mathrm{T}} \left(\frac{1}{n} \sum_{i=1}^{n} \Omega_i^{\otimes 2} \right)^{-1} \left(\frac{1}{\sqrt{n}} \sum_{i=1}^{n} \Omega_i \right) + o_p(1), \end{aligned}$$

which converges to χ_p^2 in distribution.

(b) Let $\widehat{\Omega}_i = \Omega(G_i, \beta, \widehat{\tau})$. Then similar to the proof of part (a), we can obtain that

$$\begin{aligned} -2\log\{\mathcal{R}_n(\beta)\} &= \left(\frac{1}{\sqrt{n}} \sum_{i=1}^{n} \widehat{\Omega}_i \right)^{\mathrm{T}} \left(\frac{1}{n} \sum_{i=1}^{n} \widehat{\Omega}_i^{\otimes 2} \right)^{-1} \left(\frac{1}{\sqrt{n}} \sum_{i=1}^{n} \widehat{\Omega}_i \right) + o_p(1), \\ &= \left(\frac{1}{\sqrt{n}} \sum_{i=1}^{n} \Omega_i \right)^{\mathrm{T}} \left(\frac{1}{n} \sum_{i=1}^{n} \Omega_i^{\otimes 2} \right)^{-1} \left(\frac{1}{\sqrt{n}} \sum_{i=1}^{n} \Omega_i \right) + o_p(1). \end{aligned}$$

Therefore, $-2\log\{\mathcal{R}_n(\beta)\}$ and $-2\log\{\mathcal{R}(\beta)\}$ have the same limiting distribution. The proof is thus complete.

The general result in Theorem 4.1 can be applied to the construction of confidence region in partially linear models with different missing data patterns discussed in Sections 2 and 3. Specifically, in the missing covariate problem we set $\Omega = \Omega_x$ as in (4.2). Then we have

Corollary 4.2. *Under the same general regularity conditions as in Theorem 2.1, $-2\log\{\mathcal{R}_n(\beta)\}$ converges to a chi-squared distribution with p degrees of freedom.*

Similarly, in the missing response problem we set $\Omega = \Omega_y$ as in (4.3). Then we have

Corollary 4.3. *Under the same general regularity conditions as in Theorem 3.1, $-2\log\{\mathcal{R}_n(\beta)\}$ converges to a chi-squared distribution with p degrees of freedom.*

The proofs of these corollaries are easily obtained by checking the conditions (4.5) and (4.6) in Theorem 4.1. Regularity conditions including appropriate rates of convergence of smoothing bandwidths give the order $o_p(n^{-1/4})$ of the uniform convergence of $\widehat{\tau} - \tau$ to 0. This together with the special structure of both Ω_x and Ω_y guarantees the conditions (4.5) and (4.6). In fact, Liang, et al. (2006) used such a direct approximation approach in their initial presentation of the empirical likelihood methodology in partial linear models with missing response and error-prone covariates.

Based on these results, a confidence region of β can be constructed as $\{\beta : -2\log\{\mathcal{R}_n(\beta)\} \leqslant k_\alpha\}$, where k_α is the αth quantile of the chi-squared distribution. In Corollary 4.3 if Σ_{uu} is unknown, we can use replication data in the usual way. Assuming the special case of $k_i \equiv 2$ in Section 3, W_i can then be replaced by $\overline{W}_i$ and $\Sigma_{uu}\beta$ replaced by $\widehat{\Sigma}_{uu}\beta/2$. Clearly, by Theorem 4.1(b), the resulting statistic still has the same asymptotic properties.

The R function `El.s` written by A. B. Owen can be applied in implementing the empirical likelihood method in the current setting.

5 Some numerical demonstrations

Some simulation studies and data analyses regarding the performance of the statistical methods discussed in this article have been illustrated in Liang, et al. (2004) and Liang, et al. (2006). Here we provide some additional simulation results to demonstrate how these methods perform when X and Z are moderately correlated in the context of missing response described in Section 3 as an example. Following the simulation experiment in Liang, et al. (2006), assume that $Y|X,Z \sim N\{\beta_1 X + \nu(Z), \sigma^2(X,Z)\}$ and that the selection probability of Y has the form $\text{pr}(\delta_y = 1|Y,X,Z) = \Phi\{\alpha_0 + \alpha_1 X + \nu_1(Z)\}$, where $\Phi(\cdot)$ is the standard normal distribution function. Moreover, we let $U \sim N(0, 0.09)$, $k_i \equiv 2$, $\beta_1 = 1$, $\alpha_0 = 0$, $\alpha_1 = 2$, $Z \sim \text{Unif}(0,1)$, and $X = 0.5Z + e$ with $e \sim N(0, 0.09)$, so that the correlation coefficient of X and Z is 0.43, and

$\nu(z) = 4\{\exp(-3.25z) - 4\exp(-6.5z) + 3\exp(-9.75z)\}$. The following four cases are under consideration:

(1) $\nu_1(z) = 0.75z$ and $\sigma^2(x,z) = 0.25$;

(2) $\nu_1(z) = \sin(z^2)$ and $\sigma^2(x,z) = 0.25$;

(3) $\nu_1(z) = 0.75z$ and $\sigma^2(x,z) = 0.1\{\sin^2(2\pi x^3) + 0.5z + 0.3\}$;

(4) $\nu_1(z) = 0.75z$ and the error ε follows $0.25^2(\mathcal{X}_2^2 - 2)$, where $\mathcal{X}_2^2$ is a chi-squared variable with 2 degrees of freedom.

Table 1: Point estimates of the naive (NV) and measurement error corrected (ME) methods for $\beta_1 = 1$, together with the 95% confidence intervals based on the empirical likelihood (EL) and normal approximation (NL) methods, and the averages of the lower and upper endpoints of the confidence intervals, their average lengths, and the associated coverage probabilities for the simulated data when the correlation coefficient of X and Z is 0.43. Based on 1,000 replications. "C" stands for "Case".

	Estimate		Confidence Interval		Length		Coverage	
C	NV	ME	EL	NL	EL	NL	EL	NL
			$n = 200$					
1	0.65	1.03	(0.61, 1.56)	(0.52, 1.55)	0.95	1.03	94.0	96.9
2	0.66	1.04	(0.61, 1.56)	(0.53, 1.55)	0.95	1.02	93.3	96.3
3	0.65	1.03	(0.78, 1.36)	(0.67, 1.39)	0.58	0.72	95.3	97.8
4	0.65	1.03	(0.59, 1.55)	(0.52, 1.53)	0.96	1.01	94.8	96.8
			$n = 500$					
1	0.65	1.01	(0.75, 1.32)	(0.71, 1.32)	0.57	0.61	95.6	97.3
2	0.65	1.01	(0.75, 1.32)	(0.71, 1.32)	0.57	0.61	95.9	97.4
3	0.64	0.99	(0.84, 1.19)	(0.79, 1.20)	0.35	0.41	96.7	97.4
4	0.65	1.01	(0.74, 1.32)	(0.71, 1.31)	0.58	0.60	95.0	97.1

The sample sizes of $n = 200$ and 500 are considered. The quartic kernel $K(u) = 15/16(1-u^2)^2 I_{(|u|\leqslant 1)}$ is used in the nonparametric estimation procedure and the bandwidths are selected as suggested in Remark 2 of Liang, et al. (2006). In each of the four cases, 1,000 replications are used. In each case, approximately 35% of the Y's are missing.

The empirical likelihood based confidence intervals for β_1 were computed in the way described below Corollary 4.3, while the normal theory based confidence intervals were computed using the asymptotic normality in Theorem 3.1 as usual. The summary results are given in Table 5. They are similar to those in Table 1 of Liang, et al. (2006), and are between those for the independent X and Z case and those for the case of strongly correlated X and Z, as expected. They appear to confirm the theory. The general messages are that measurement errors can not be ignored in general to avoid possibly significant bias and that the empirical-likelihood based confidence intervals appear to be superior to those based on the normal approximation for moderate sample sizes, especially when the errors in model (1.1) are non-normal and/or they have non-homogeneous variance.

6 Concluding remarks

In this article, we have reviewed some recently proposed easily computable estimators in partially linear models with either missing covariates or missing response variables and error-prone covariates. While the missingness of X is assumed missing at random in the development, the missingness of Y in the second estimation problem is not at random due to measurement error in the X. In both cases, the estimators of the partial slope β are $\sqrt{n}$-consistent and asymptotically normally distributed. We have also provided a unified framework that includes both missing data estimation problems. Empirical likelihood methodology has been considered under this unified framework and shown to be useful for finite samples.

The asymptotic semiparametric efficiency was derived in Liang, et al. (2004) in the case of missing covariates using Robins, et al.'s (1994) results. In general, the simply computable methods discussed in this article appears to have reasonable efficiency. Extensions of these methods to clustered or longitudinal data analysis and other general settings should be possible, but how to take care of correlated data appears to be nontrivial, and is currently under investigation.

References

[1] Carroll, R. J., Knickerbocker, R. K., and Wang, C. Y. (1995), Dimension Reduction in Semiparametric Measurement Error Models, *Annals of Statistics*, 23, 161–181.

[2] Carroll, R. J., Ruppert, D., Stefanski, L. A., and Crainiceanu, C. M. (2006), *Measurement Error in Nonlinear Models: A Modern Perspective*, 2nd Edition. New York: Chapman and Hall.

[3] Chen, S. X. (1994), Empirical Likelihood Confidence Intervals for Linear Regression Coefficients, *Journal of Multivariate Analysis*, 49, 24–40.

[4] Cheng, P. E. (1990), Applications of Kernel Regression Estimation: a Survey, *Communications in Statistics, Theory and Methods*, 19, 4103–4134.

[5] Cheng, P. E. (1994), Nonparametric Estimation of Mean Functionals With Data Missing at Random, *Journal of the American Statistical Association*, 89, 81–87.

[6] Cheng, P. E., and Chu, C. K. (1996), Kernel Estimation of Distribution Functions and Quantiles With Missing Data, *Statistica Sinica*, 6, 63–78.

[7] Engle, R. F., Granger, C. W. J., Rice, J., and Weiss, A. (1986), Semiparametric Estimates of the Relation Between Weather and Electricity Sales, *Journal of the American Statistical Association*, 81, 310–320.

[8] Härdle, W., Liang, H., and Gao, J. (2000), *Partially Linear Models*, Heidelberg: Springer Physica-Verlag.

[9] Liang, H., Härdle, W., and Carroll, R. J. (1999), Estimation in a Semiparametric Partially Linear Errors–in–variables Model, *Annals of Statistics*, 27, 1519–1535.

[10] Liang, H., Wang, S., and Carroll, R. J. (2006), Partially Linear Models with Missing Response Variables and Error-prone Covariates, *Biometrika*, to appear.

[11] Liang, H., Wang, S., Robins, J. M., and Carroll, R. J. (2004), Estimation in Partially Linear Models with Missing Covariates, *Journal of the American Statistical Association*, 99, 357–367.

[12] Little, R. J. A., and Rubin, D. B. (1987), *Statistical Analysis With Missing Data*, New York: John Wiley & Sons.

[13] Newey, W. K. (1994), The Asymptotic Variance of Semiparametric Estimators, *Econometrica*, 62, 1349–1382.

[14] Opsomer, J. D., and Ruppert, D. (1999), A root-n consistent backfitting estimator for semiparametric additive modelling, *Journal of Computational and Graphical Statistics*, 8, 715–732.

[15] Owen, A. B. (1988), Empirical Likelihood Ratio Confidence Intervals for a Single Functional, *Biometrika*, 75, 237–249.

[16] Owen, A. B. (2001), *Empirical Likelihood*, London: Chapman and Hall/CRC.

[17] Qin, J. (1999), Empirical Likelihood Ratio Based Confidence Intervals for Mixture Proportions, *Annals of Statistics*, 27, 1368–1384.

[18] Qin, J., and Lawless, J. (1994), Empirical Likelihood and General Estimating Equations, *Annals of Statistics*, 22, 300–325.

[19] Robins, J. M., Rotnitzky, A., and Zhao, L. P. (1994), Estimation of Regression Coefficients When Some Regressors are not Always Observed, *Journal of the American Statistical Association*, 89, 846–866.

[20] Robinson, P. M. (1988), Root-n-consistent Semiparametric Regression, *Econometrica*, 56, 931–954.

[21] Ruppert, D., Sheather, S. J., and Wand, M. P. (1995), An Effective Bandwidth Selector for Local Least Squares Regression, *Journal of the American Statistical Association*, 90, 1257–1270.

[22] Severini, T. A., and Staniswalis, J. G. (1994), Quasilikelihood Estimation in Semiparametric Models, *Journal of the American Statistical Association*, 89, 501–511.

[23] Speckman, P. (1988), Kernel Smoothing in Partial Linear Models, *Journal of the Royal Statistical Society, Series B*, 50, 413–436.

[24] Wang, C. Y., Wang, S., Gutierrez, R. G., and Carroll, R. J. (1998), Local Linear Regression for Generalized Linear Models with Missing Data, *Annals of Statistics*, 26, 1028–1050.

[25] Wang, Q. H., Linton, O., and Härdle, W. (2004), Semiparametric Regression Analysis With Missing Response at Random, *Journal of the American Statistical Association*, 99, 334–345.

[26] Zeger, S. L., and Diggle, P. J. (1994), Semiparametric Models for Longitudinal Data With Application to CD4 Cell Numbers in HIV Seroconverters, *Biometrics*, 50, 689–699.

Asymptotic Methods in Nonlinear Time Series Models*

Hao Yu

Department of Statistical and Actuarial Sciences
University of Western Ontario
London, Qntario, NGA 5B7, Canada
E-mail: Email: hyu@stats.uwo.ca

Abstract

While a linear stationary ARMA sequence has dynamic conditional mean, its conditional variance remains constant and therefore it is homoskedastic. This means that ARMA models are not suitable to fit many financial time series data exhibiting dynamic conditional variation or conditionally heteroskedastic. On the other hand, the nonlinear time series GARCH sequence is conditionally heteroskedastic but its conditional mean remains constant. Combining two together, we have an ARMA-GARCH model which can capture both dynamic conditional mean and dynamic conditional variation. In this paper, we survey recent developments in the theory of ARMA-GARCH models. First we present some new results on the asymptotic theory of the quasi maximum likelihood estimation (QMLE) for an ARMA-(I)GARCH model. We then consider the high moment partial sum processes based on residuals of an ARMA-(I)GARCH model. Constructing CUSUM statistics for testing ARMA-GARCH model structure changes and using the Jarque-Bera normality test statistic and kernel density function estimation of the unobservable innovation distribution are discussed. Some simulation results of CUSUM statistics and Jarque-Bera normality test are presented.

1 Introduction

A time series sequence $\{Y_t,\ -\infty < t < \infty\}$ is said to follow an autoregressive and moving average model with orders P and Q (ARMA(P, Q)) if it is of the from

$$Y_t - c = \sum_{i=1}^{P} \phi_i (Y_{t-i} - c) + \varepsilon_t + \sum_{j=1}^{Q} \psi_j \varepsilon_{t-j}, \tag{1.1}$$

*Research supported by a grant from the Natural Sciences and Engineering Research Council of Canada.

where $c \in \mathbb{R}, \phi_i \in \mathbb{R}, 1 \leqslant i \leqslant P, \psi_j \in \mathbb{R} = (-\infty, \infty), 1 \leqslant j \leqslant Q$, are the ARMA model parameters and $\{\varepsilon_t, -\infty < t < \infty\}$ is a sequence of iid or white noise random variables (r.v.'s). When $P = 0$ or $Q = 0$, ARMA(P, Q) becomes MA(Q) or AR(P), respectively.

References on ARMA models are quite extensive and abundant. We refer to Brockwell and Davis (1991) and Fuller (1996) and references therein. In general ARMA models cover linear stationary time series and can be expanded to cover homogenous nonstationary sequences (ARIMA models). In this paper we restrict the sequences generated from ARMA models to be stationary.

Let

$$\gamma = (c, \phi_1, \cdots, \phi_P, \psi_1, \cdots, \psi_Q)^{\mathrm{T}} \in \mathbf{\Theta}_\gamma \subset \mathbb{R}^{P+Q+1}$$

and

$$\mathcal{A}_\gamma(z) = 1 - \sum_{i=1}^{P} \phi_i z^i, \quad \mathcal{B}_\gamma = 1 + \sum_{j=1}^{Q} \psi_j z^j.$$

Then the stationarity, invertibility and identifiability conditions of the ARMA model (1.1) are given below.

Assumption 1. *For each $\gamma \in \mathbf{\Theta}_\gamma$, $\mathcal{A}_\gamma(z)\mathcal{B}_\gamma(z) \neq 0$ when $|z| \leqslant 1$, and $\mathcal{A}_\gamma(z)$ and $\mathcal{B}_\gamma(z)$ have no common root with $\phi_P \neq 0$ or $\psi_Q \neq 0$.*

Under Assumption 1, the invertibility implies that there exist constants $\{\pi_i\}$ such that $\sum_{i=1}^{\infty} |\pi_i| < \infty$ and

$$Y_t - c = \sum_{i=1}^{\infty} \pi_i (Y_{t-i} - c) + \varepsilon_t.$$

Then it is easy to see

$$\mathrm{E}[Y_t - c | Y_{t-1}, Y_{t-2}, \cdots] = \sum_{i=1}^{\infty} \pi_i (Y_{t-i} - c).$$

Thus the conditional mean depends linearly on past observations, which implies that past observations can be used to predict future values. This turns out to be one of the most important features of ARMA models. However further computation shows that

$$\mathrm{Var}(Y_t | Y_{t-1}, Y_{t-2}, \cdots) = \mathrm{Var}(\varepsilon_t) = \text{constant},$$

that is, the conditional variance remains constant. Hence ARMA models are homoskedastic and, of course, are not heteroskedastic. Heteroskedasticity means variances or conditional variances are not constant. More specifically, a time series is unconditionally heteroskedastic if unconditional variances are not constant. It is conditionally heteroskedastic if conditional variances are not constant.

Many financial time series data such as stocks, bonds, interest rates, foreign exchanges, etc, often exhibit heteroskedasticity. The heteroskedasticity can be interpreted as volatility clustering: large changes tend to be followed by large changes, and small changes tend to be followed by small changes. Another interpretation is that markets often go through protracted quiet periods interspersed with bursts of turbulence influenced by either "good" or "bad" news. Financial data are also often leptokurtic, or heavy tail distributed. Volatility clustering and heavy tailed distributions are closely related. In general financial data cannot be models by ARMA models alone and this leads naturally to the consideration of nonlinear time series models.

Nonlinear time series models can be introduced in many ways. In this paper we focus on nonlinear models that can be used to model conditional heteroskedasticity.

Let $\{X_t, -\infty < t < \infty\}$ be a time series exhibiting conditional heteroskedasticity. One possibility is that it is of the form

$$X_t = \sigma_t \eta_t, \tag{1.2}$$

where $\{\eta_t, -\infty < t < \infty\}$ is a sequence of i.i.d. r.v.'s with mean zero and variance 1 and σ_t^2 is $\mathcal{F}_{t-1} = \sigma(X_s : s \leqslant t-1)$ adapted. Simple calculations give

$$\mathrm{E}[X_t|\mathcal{F}_{t-1}] = 0, \quad \mathrm{Var}(X_t|\mathcal{F}_{t-1}) = \sigma_t^2,$$

that is, $\{X_t\}$ is a martingale difference and σ_t is the conditional standard deviation. So as long as σ_t is not constant, $\{X_t\}$ is conditionally heteroskedastic. How do we choose a proper functional form of σ_t to link volatility clustering? One simple example is

$$\sigma_t^2 = \alpha_0 + \sum_{i=1}^{p} \alpha_i X_{t-i}^2, \ \alpha_0 > 0, \ \alpha_i \geqslant 0, \ i = 1, \cdots, p, \tag{1.3}$$

that is, conditional variance is determined by square values of past observations, where $\alpha_0, \alpha_1, \cdots, \alpha_p$ are parameters. The model given by (1.2) and (1.3) was first introduced by Engle (1982) and is known as the p-th order autogressive conditional heteroskedastic (ARCH(p)) model. ARCH models have proved to be an extremely popular class of nonlinear time series models for financial data. For surveys of the literature on ARCH models, see Campbell, Lo and MacKinlay (1997) or the volume edited by Rossi (1996) and references therein.

To address some weaknesses of ARCH models, Engle, Lilien, and Robins (1987) introduced the ARCH-M model which extends the ARCH model to allow the conditional variance to affect the mean. Basically (1.2) becomes

$$X_t = c + \delta\, m(\sigma_t) + \sigma_t\, \eta_t,$$

where c and δ are additional parameters with the deterministic function m usually chosen as $m(x) = x, \sqrt{x}$ or $\exp(-x^2)$.

The ARCH model has been generalized to GARCH model by Bollerslev (1986). A GARCH(p, q) sequence $\{X_t, -\infty < t < \infty\}$ is of the form (1.2) with

$$\sigma_t^2 = \alpha_0 + \sum_{i=1}^{p} \alpha_j X_{t-i}^2 + \sum_{j=1}^{q} \beta_j \sigma_{t-j}^2, \tag{1.4}$$

where

$$\alpha_0 > 0,\ \alpha_i \geqslant 0,\ 1 \leqslant i \leqslant p,\ \beta_j \geqslant 0,\ 1 \leqslant j \leqslant q$$

are GARCH parameters.

Obviously, a GARCH(p, q) model reduces to Engle's ARCH(p) model if $q = 0$. When $\sum_{i=1}^{p} \alpha_i + \sum_{j=1}^{q} \beta_j = 1$, model (1.2) and (1.4) is called integrated GARCH (IGARCH), due to the fact: $\mathrm{E}[X_t^2] = \mathrm{E}[\sigma_t^2] = \infty$. Similarly, when $\sum_{i=1}^{p} \alpha_i = 1$ and $q = 0$, ARCH model (1.2) and (1.3) becomes IARCH(p).

Let

$$A_t = \left(\begin{array}{ccc|ccc} \alpha_1\eta_t^2 & \cdots & \alpha_p\eta_t^2 & \beta_1\eta_t^2 & \cdots & \beta_q\eta_t^2 \\ & I_{p-1} & O & & O & \\ \hline \alpha_1 & \cdots & \alpha_p & \beta_1 & \cdots & \beta_q \\ & O & & & I_{q-1} & O \end{array}\right),$$

where I_k is $k \times k$ identity matrix. Nelson (1990) showed that the model (1.2) and (1.4) with $p = q = 1$ has a unique strictly stationary and ergodic solution if and only if $\mathrm{E}\log(\beta_1 + \alpha_1\eta_0) < 0$. The general case was investigated by Bougerol and Picard (1992a, b). They showed that a unique strictly stationary and ergodic solution of the model (1.2) and (1.4) exists if and only if

$$\begin{aligned} \gamma(A_0) &= \inf_{1 \leqslant t < \infty} \frac{1}{t} \mathrm{E}\left(\log \|A_t A_{t-1} \cdots A_1\|\right) \\ &= \lim_{t \to \infty} \frac{1}{t} \log \|A_t A_{t-1} \cdots A_1\| < 0 \quad \text{a.s.}, \end{aligned}$$

where $\|A\| = \sqrt{\mathrm{tr}(AA^{\mathrm{T}})}$. The definition of $\gamma(A_0)$ does not depend on the choice of a norm on the space of the $(p+q) \times (p+q)$ matrices. From now on, $\gamma(A_0) < 0$ is assumed throughout this paper.

In the following, we introduce some basic notations and assumptions so that there is a strictly stationary and ergodic solution $\{X_t, -\infty < t < \infty\}$ of the model (1.2) and (1.4) with certain moments.

Let $\boldsymbol{\delta} = (\alpha_0, \alpha_1, \cdots, \alpha_p, \beta_1, \cdots, \beta_q)^{\mathrm{T}} \in \boldsymbol{\Theta}_\delta \subset \mathbb{R}^+ \times \mathbb{R}_0^{p+q}$, where $\mathbb{R}^+ = (0, +\infty)$ and $\mathbb{R}_0 = [0, +\infty)$ and

$$\mathcal{A}_{\boldsymbol{\delta}}(z) = \sum_{i=1}^{p} \alpha_i z^i, \quad \mathcal{B}_{\boldsymbol{\delta}}(z) = 1 - \sum_{j=1}^{q} \beta_j z^j.$$

The true parameter value is unknown and is denoted as

$$\boldsymbol{\delta}_0 = (\alpha_{00}, \alpha_{01}, \cdots, \alpha_{0p}, \beta_{01}, \cdots, \beta_{0q})^{\mathrm{T}}.$$

Assumption 2. *$\mathcal{A}_{\boldsymbol{\delta}}(z)$ and $\mathcal{B}_{\boldsymbol{\delta}}(z)$ have no common root, $\mathcal{A}_{\boldsymbol{\delta}}(1) \neq 0$, $\alpha_p + \beta_q \neq 0$, and $\sum_{j=1}^{q} \beta_j < 1$ for each $\boldsymbol{\delta} \in \boldsymbol{\Theta}_\delta$.*

Assumption 3. $\mathrm{E}|X_0|^{2\iota} < \infty$ *for some* $\iota > 0$.

Assumption 2 is the identifiability condition for the GARCH model (1.2) and (1.4). When $\iota = 1$, the necessary and sufficient condition for Assumption 3 is

$$\sum_{i=1}^{p} \alpha_{0i} + \sum_{j=1}^{q} \beta_{0j} < 1, \tag{1.5}$$

under which the GARCH model (1.2) and (1.4) has a finite variance. When $0 < \iota \leqslant 1$, a necessary and sufficient condition for Assumption 3 is obtained by Ling (2006). By Proposition 1 of Francq and Zakoïab (2004), Assumption 3 is always true for some $\iota > 0$ if $\gamma(A_0) < 0$.

Theorem 1.1 (Ling, 2006). *Let $0 < \iota \leqslant 1$ and $\underline{\beta} = \min\{\alpha_{0i}, \beta_{0j} : i = 0, 1, \cdots, p, j = 1, \cdots, q\}$. Suppose $\{X_t\}$ is generated by model (1.2) and (1.4).*

(i) *If there exists an integer i_0 such that*

$$\mathrm{E}\|\prod_{k=0}^{i_0-1} A_{0k}\|^{\iota} < 1, \tag{1.6}$$

where A_{0k} is A_k with GARCH parameters replaced their true values, then $\{\varepsilon_t\}$ is strictly stationary and ergodic with $\mathrm{E}|X_t|^{2\iota} < \infty$;

(ii) *If $\underline{\beta} > 0$ and $\{X_t\}$ is strictly stationary with $\mathrm{E}|X_t|^{2\iota} < \infty$, then (1.6) holds;*

(iii) *If $\underline{\beta} > 0$,*

$$\sum_{i=1}^{p} \alpha_{0i} + \sum_{j=1}^{q} \beta_{0j} = 1, \tag{1.7}$$

and η_t has a positive density on $\mathbb{R}$ such that $\mathrm{E}|\eta_t|^{2\tau} < \infty$ for all $\tau < \tau_0$ and $\mathrm{E}|\eta_t|^{2\tau_0} = \infty$ for some $0 < \tau_0 \leqslant \infty$, then $\lim_{x\to\infty} x^2 P(|X_t| > x)$ exists and is positive.

Clearly, when $\iota = 1$, (1.5) and (1.6) are equivalent.

In many cases, the GARCH models provide a reasonably good model for fitting financial time series and estimating conditional volatility. However, there are some aspects of it which can be improved to better capture the characteristics and dynamics of a particular time series. There are many extensions of GARCH models in literature, for example, EGARCH by Nelson (1991), PGARCH by Ding, Granger and Engle (1993), and FIGARCH by Baillie, Bollerslev and Mikkelson (1996). See Gouriéroux (1997) and Mills (1999) for more details.

While a GARCH sequence is heteroskedastic, its conditional mean remains constant due to the fact that it is martingale difference. Thus it cannot handle well many financial data often exhibiting both dynamic conditional mean and dynamic conditional volatility. Hence both ARMA and GARCH models are needed to model such data.

A sequence $\{Y_t, -\infty < t < \infty\}$ generated from an ARMA(P, Q)-GARCH(p, q) process is of the form (1.1) with $\{\varepsilon_t\}$ replaced by

$$\varepsilon_t = \sigma_t \eta_t \text{ and } \sigma_t^2 = \alpha_0 + \sum_{i=1}^{p} \alpha_i \varepsilon_{t-i}^2 + \sum_{j=1}^{q} \beta_j \sigma_{t-j}^2, \tag{1.8}$$

where $\{\eta_t, -\infty < t < \infty\}$ is a sequence of i.i.d. r.v.'s. In other words, an ARMA(P, Q)-GARCH(p, q) model is the ARMA(P, Q) model (1.1) with errors generated from the pure GARCH(p, q) model (1.8). Though it is not necessary, to make model parameters uniquely identifiable, we assume also that

$$\{\eta_t, -\infty < t < \infty\} \text{ is a sequence of i.i.d. (0,1) r.v.'s.} \tag{1.9}$$

When $P = Q = 0$, the model (1.1) and (1.8) reduce to a pure GARCH model with nonzero mean, that is, a pure GARCH model is a special case of an ARMA-GARCH model. Hence, in this paper, we will focus on ARMA-GARCH models only. For recent surveys in the asymptotic theory of pure GARCH models, we refer to Berkes, Horváth, and Kokoszka (2004) and Yu (2004) and references therein.

Notice that the GARCH model (1.8) is the same as the model (1.2) and (1.4) by letting $X_t \equiv \varepsilon_t$. From now on, we will drop the model (1.2) and (1.4) and assume $X_t \equiv \varepsilon_t$. So we can reuse the notations defined for $\{X_t\}$. Let $\boldsymbol{\lambda} = (\boldsymbol{\gamma}^{\mathrm{T}}, \boldsymbol{\delta}^{\mathrm{T}})^{\mathrm{T}}$. Then we assume that the parameter space $\boldsymbol{\Theta} = (\boldsymbol{\Theta}_\gamma, \boldsymbol{\Theta}_\delta) \subset \mathbb{R}^{P+Q+1} \times \mathbb{R}^+ \times \mathbb{R}_0^{p+q}$ is compact. The true parameter value is unknown and is denoted as

$$\boldsymbol{\lambda}_0 = (\boldsymbol{\gamma}_0^{\mathrm{T}}, \boldsymbol{\delta}_0^{\mathrm{T}})^{\mathrm{T}} = \Big((c_0, \phi_{01}, \cdots, \phi_{0P}, \psi_{01}, \cdots, \psi_{0Q})^{\mathrm{T}}, \\ (\alpha_{00}, \alpha_{01}, \cdots, \alpha_{0p}, \beta_{01}, \cdots, \beta_{0q})^{\mathrm{T}} \Big)^{\mathrm{T}}$$

and it is an interior point in $\boldsymbol{\Theta}$. Putting the above together we have the following assumption.

Assumption 4. *$\boldsymbol{\Theta}$ is compact and $\boldsymbol{\lambda}_0$ is an interior point in $\boldsymbol{\Theta}$.*

The rest of this paper is organized as follows. Section 2 investigates the estimation of the ARMA-GARCH parameters. In Section 3, the asymptotic results for high moment partial sum processes are given. Finally, some applications and simulation results are presented in Section 4.

2 Estimation of the true parameter $\boldsymbol{\lambda}_0$

Estimating the true parameter $\boldsymbol{\lambda}_0$ has been investigated by several authors. Weiss (1986) first studied the asymptotic theory for ARMA-ARCH models under the finite 4-th moment on ε_t. More recently, the maximum likelihood estimation (MLE) and QMLE were studied by Ling and Li (1997, 1998, 2003), Ling and McAleer (2003), and Francq and Zakoïab (2004). Since results from Francq and Zakoïab (2004) are under the best possible conditions among cited references, we present their results in details.

Given the observations $\{Y_1, \cdots, Y_n\}$ and the initial values $Y_0, Y_{-1}, \cdots$, we can write the parametric model as

$$\varepsilon_t(\boldsymbol{\gamma}) = (Y_t - c) - \sum_{i=1}^{P} \phi_i (Y_{t-i} - c) - \sum_{j=1}^{Q} \psi_j \varepsilon_{t-j}(\boldsymbol{\gamma}), \tag{2.1}$$

$$\eta_t(\boldsymbol{\lambda}) = \frac{\varepsilon_t(\boldsymbol{\gamma})}{\sigma_t(\boldsymbol{\lambda})}, \; \sigma_t^2(\boldsymbol{\lambda}) = \alpha_0 + \sum_{i=1}^{p} \alpha_i \varepsilon_{t-i}^2(\boldsymbol{\gamma}) + \sum_{j=1}^{q} \beta_j \sigma_{t-j}^2(\boldsymbol{\lambda}). \tag{2.2}$$

Clearly, by (1.1) and (1.8), $\varepsilon_t = \varepsilon_t(\boldsymbol{\gamma}_0)$, $\eta_t = \eta_t(\boldsymbol{\lambda}_0)$ and $\sigma_t^2 = \sigma_t^2(\boldsymbol{\lambda}_0)$.

In practice, we observe only $Y_1, \cdots, Y_n$ which are a realization of length n to the model (1.1) and (1.8). All initial values $Y_0, \cdots, Y_{1-(p-Q)-P}$, $\tilde{\varepsilon}_{-p+Q}, \cdots, \tilde{\varepsilon}_{1-p}, \tilde{\sigma}_0^2, \cdots, \tilde{\sigma}_{1-q}^2$ in the case of $p \geqslant Q$ and $Y_0, \cdots, Y_{1-(p-Q)-P}$, $\tilde{\varepsilon}_0, \cdots, \tilde{\varepsilon}_{1-Q}, \tilde{\sigma}_0^2, \cdots, \tilde{\sigma}_{1-q}^2$ in the case of $p < Q$ will be taken to be fixed (neither random nor functions of the parameters). Then we can rewrite (2.1) and (2.2) as

$$\tilde{\varepsilon}_t(\boldsymbol{\gamma}) = (Y_t - c) - \sum_{i=1}^{P} \phi_i (Y_{t-i} - c) - \sum_{j=1}^{Q} \psi_j \tilde{\varepsilon}_{t-j}(\boldsymbol{\gamma}),$$

$$\tilde{\eta}_t(\boldsymbol{\lambda}) = \frac{\tilde{\varepsilon}_t(\boldsymbol{\gamma})}{\tilde{\sigma}_t(\boldsymbol{\lambda})}, \; \tilde{\sigma}_t^2(\boldsymbol{\lambda}) = \alpha_0 + \sum_{i=1}^{p} \alpha_i \tilde{\varepsilon}_{t-i}^2(\boldsymbol{\gamma}) + \sum_{j=1}^{q} \beta_j \tilde{\sigma}_{t-j}^2(\boldsymbol{\lambda}). \tag{2.3}$$

Define the log-quasi-likelihood function as

$$\tilde{L}_n(\boldsymbol{\lambda}) = \frac{1}{n}\sum_{t=1}^{n}\tilde{l}_t(\boldsymbol{\lambda}), \quad \tilde{l}_t(\boldsymbol{\lambda}) = -\frac{1}{2}\log\tilde{\sigma}_t^2(\boldsymbol{\lambda}) - \frac{1}{2}\tilde{\eta}_t^2(\boldsymbol{\lambda}). \tag{2.4}$$

A QMLE of $\boldsymbol{\lambda}_0$ is any measurable solution of

$$\tilde{\boldsymbol{\lambda}}_n = \arg\max_{\boldsymbol{\lambda}\in\boldsymbol{\Theta}} \tilde{L}_n(\boldsymbol{\lambda}). \tag{2.5}$$

Theorem 2.1 (Francq and Zakoïab, 2004). *Under Assumptions 1, 2, and 4, we have almost surely*

$$\tilde{\boldsymbol{\lambda}}_n \longrightarrow \boldsymbol{\lambda}_0 \text{ as } n \to \infty.$$

Remark 2.2. *The condition that $\boldsymbol{\lambda}_0$ is an interior point in $\boldsymbol{\Theta}$ in Assumption 4 is not required by Francq and Zakoïab (2004). They used the weaker condition $\boldsymbol{\lambda}_0 \in \boldsymbol{\Theta}$. This includes some boundary points such as $\alpha_{01} = 0$. However, due to model restriction, the distribution of $\sqrt{n}(\tilde{\alpha}_1 - \alpha_{01})$ is then concentrated on $[0, \infty]$, so the asymptotic distribution (if it exists) is no longer to be normal.*

Next we discuss the asymptotic normality of $\sqrt{n}(\tilde{\boldsymbol{\lambda}}_n - \boldsymbol{\lambda}_0)$. In the following, for simplicity, we denote

$$\left.\frac{\partial z(\boldsymbol{\lambda})}{\partial \boldsymbol{\lambda}}\right|_{\boldsymbol{\lambda}=\boldsymbol{\lambda}_0} = \frac{\partial z(\boldsymbol{\lambda}_0)}{\partial \boldsymbol{\lambda}}, \quad \left.\frac{\partial^2 z(\boldsymbol{\lambda})}{\partial \boldsymbol{\lambda}\partial \boldsymbol{\lambda}^{\mathrm{T}}}\right|_{\boldsymbol{\lambda}_0} = \frac{\partial^2 z(\boldsymbol{\lambda}_0)}{\partial \boldsymbol{\lambda}\partial \boldsymbol{\lambda}^{\mathrm{T}}}$$

for any function z. Introduce the matrices

$$\mathcal{I} = \mathrm{E}\left(\frac{\partial l_t(\boldsymbol{\lambda}_0)}{\partial \boldsymbol{\lambda}}\frac{\partial l_t(\boldsymbol{\lambda}_0)}{\partial \boldsymbol{\lambda}^{\mathrm{T}}}\right), \quad \mathcal{J} = -\mathrm{E}\left(\frac{\partial^2 l_t(\boldsymbol{\lambda}_0)}{\partial \boldsymbol{\lambda}\partial \boldsymbol{\lambda}^{\mathrm{T}}}\right),$$

where

$$l_t(\boldsymbol{\lambda}) = -\frac{1}{2}\log\sigma_t^2(\boldsymbol{\lambda}) - \frac{1}{2}\eta_t^2(\boldsymbol{\lambda}). \tag{2.6}$$

Theorem 2.3 (Francq and Zakoïab, 2004). *Suppose that Assumptions 1, 2, and 4 hold and Assumption 3 holds for $\iota = 2$. If*

$$\mathrm{E}\eta_t^4 < \infty \tag{2.7}$$

and

$$\mathrm{E}^2\eta_t^3 < \mathrm{E}\eta_t^2\mathrm{E}(\eta_t^2 - 1)^2, \tag{2.8}$$

then

$$\sqrt{n}(\tilde{\boldsymbol{\lambda}}_n - \boldsymbol{\lambda}_0) \xrightarrow{\mathcal{D}} N(0,\ \mathcal{J}^{-1}\mathcal{I}\mathcal{J}^{-1}).$$

If, in addition, the distribution of η_t is symmetric about 0, we have

$$\mathcal{I} = \begin{pmatrix} \mathcal{I}_1 & 0 \\ 0 & \mathcal{I}_2 \end{pmatrix}, \quad \mathcal{J} = \begin{pmatrix} \mathcal{J}_1 & 0 \\ 0 & \mathcal{J}_2 \end{pmatrix},$$

where

$$\mathcal{I}_1 = \frac{1}{2}(\mathrm{E}\eta_t^4 - 1)\mathrm{E}\left(\frac{\partial\sigma_t^2(\boldsymbol{\lambda}_0)/\partial\gamma}{\sqrt{2}\sigma_t^2(\boldsymbol{\lambda}_0)}\frac{\partial\sigma_t^2(\boldsymbol{\lambda}_0)/\partial\gamma^{\mathrm{T}}}{\sqrt{2}\sigma_t^2(\boldsymbol{\lambda}_0)}\right) + \mathrm{E}\left(\frac{\partial\varepsilon_t(\gamma_0)/\partial\gamma}{\sigma_t(\boldsymbol{\lambda}_0)}\frac{\partial\varepsilon_t(\gamma_0)/\partial\gamma^{\mathrm{T}}}{\sigma_t(\boldsymbol{\lambda}_0)}\right),$$

$$\mathcal{I}_2 = \frac{1}{2}(\mathrm{E}\eta_t^4 - 1)\mathrm{E}\left(\frac{\partial\sigma_t^2(\boldsymbol{\lambda}_0)/\partial\boldsymbol{\delta}}{\sqrt{2}\sigma_t^2(\boldsymbol{\lambda}_0)}\frac{\partial\sigma_t^2(\boldsymbol{\lambda}_0)/\partial\boldsymbol{\delta}^{\mathrm{T}}}{\sqrt{2}\sigma_t^2(\boldsymbol{\lambda}_0)}\right),$$

$$\mathcal{J}_1 = \mathrm{E}\left(\frac{\partial\sigma_t^2(\boldsymbol{\lambda}_0)/\partial\gamma}{\sqrt{2}\sigma_t^2(\boldsymbol{\lambda}_0)}\frac{\partial\sigma_t^2(\boldsymbol{\lambda}_0)/\partial\gamma^{\mathrm{T}}}{\sqrt{2}\sigma_t^2(\boldsymbol{\lambda}_0)}\right) + \mathrm{E}\left(\frac{\partial\varepsilon_t(\gamma_0)/\partial\gamma}{\sigma_t(\boldsymbol{\lambda}_0)}\frac{\partial\varepsilon_t(\gamma_0)/\partial\gamma^{\mathrm{T}}}{\sigma_t(\boldsymbol{\lambda}_0)}\right),$$

$$\mathcal{J}_2 = \mathrm{E}\left(\frac{\partial\sigma_t^2(\boldsymbol{\lambda}_0)/\partial\boldsymbol{\delta}}{\sqrt{2}\sigma_t^2(\boldsymbol{\lambda}_0)}\frac{\partial\sigma_t^2(\boldsymbol{\lambda}_0)/\partial\boldsymbol{\delta}^{\mathrm{T}}}{\sqrt{2}\sigma_t^2(\boldsymbol{\lambda}_0)}\right).$$

Remark 2.4. *If we deal only with a pure GARCH(p, q) model, Berkes and Horváth (2003) showed that Theorem 2.3 cannot be true without assumption (2.7).*

Remark 2.5. *After adopting similar notations used in Ling's (2005), we can write*

$$\mathcal{J} = \mathrm{E}[U_t(\boldsymbol{\lambda}_0)U_t^{\mathrm{T}}(\boldsymbol{\lambda}_0)], \quad \mathcal{I} = \mathrm{E}[U_t(\boldsymbol{\lambda}_0)IU_t^{\mathrm{T}}(\boldsymbol{\lambda}_0)],$$

where

$$I = \begin{pmatrix} (\mathrm{E}\eta_t^4 - 1)/2 & -\mathrm{E}\eta_t^3/\sqrt{2} \\ -\mathrm{E}\eta_t^3/\sqrt{2} & 1 \end{pmatrix}, \quad U_t(\boldsymbol{\lambda}) = \left(\frac{\partial\sigma_t^2(\boldsymbol{\lambda})/\partial\boldsymbol{\lambda}}{\sqrt{2}\sigma_t^2(\boldsymbol{\lambda})}, \frac{\partial\varepsilon_t(\gamma)/\partial\boldsymbol{\lambda}}{\sigma_t(\boldsymbol{\lambda})}\right).$$

By Cauchy inequality, it is readily shown that (2.8) holds if and only if $P(\eta_t^2 - 1 = c\eta_t) < 1$ for any constant c, which matches the Assumption (A12) in Francq and Zakoïab (2004). Hence $\mathcal{I}$ and $\mathcal{J}$ are nonsingular. Notice that $I > 0$ if and only if (2.8) holds.

Remark 2.6. *When η_t has a symmetric distribution, $\mathcal{J}^{-1}\mathcal{I}\mathcal{J}^{-1}$ is block-diagonal. This means the ARMA and GARCH parts of $\tilde{\boldsymbol{\lambda}}_n$ are asymptotically independent of each other. So are their asymptotic accuracies.*

Notice that Theorem 2.3 requires $\mathrm{E}\varepsilon_0^4 < \infty$. To understand why $\mathrm{E}\varepsilon_0^4 < \infty$ is required, we use an example given by Ling (2006). Assume that $\{Y_t\}$ is generated from AR(1)-GACH(p, q), i.e., (1.1) is changed to $Y_t = \phi_0 Y_{t-1} + \varepsilon_t$ with $\{\varepsilon_t\}$ from (1.8). Let $\tilde{\phi}_n$ be the LSE of ϕ_0. Then

$$\sqrt{n}(\tilde{\phi}_n - \phi_0) = \frac{\sum_{t=1}^n Y_{t-1}\varepsilon_t/\sqrt{n}}{\sum_{t=1}^n Y_{t-1}^2/n}.$$

The CLT requires $\mathrm{E}(Y_{t-1}\varepsilon_t)^2 = \mathrm{E}(Y_{t-1}^2\sigma_t^2) < \infty$ and hence the condition $\mathrm{E}\varepsilon_0^4 < \infty$ should be minimal for asymptotic normality of $\tilde{\phi}_n$.

However, the condition $\mathrm{E}\varepsilon_0^4 < \infty$ is not well supported by real data. See, for example, Baillie (1996). Hence it is important to weaken the condition $\mathrm{E}\varepsilon_0^4 < \infty$ so that we have the consistency and asymptotical normality of an estimator of $\boldsymbol{\lambda}_0$ under $\mathrm{E}\varepsilon_0^2 < \infty$ or even $\mathrm{E}\varepsilon_0^2 = \infty$ which includes the ARMA(P, Q)-IGARCH(p, q) model.

By introducing a weight into the log-likelihood function $\tilde{L}_n(\boldsymbol{\lambda})$, Ling (2006) obtains the consistency and asymptotical normality only under a fractional moment of ε_0. In the following, we present his results in detail.

Define the weighted log-quasi-likelihood function based on the initial values $Y_0, Y_{-1}, \cdots$ as

$$L_{sn}(\boldsymbol{\lambda}) = \frac{1}{n}\sum_{t=1}^n w_t l_t(\boldsymbol{\lambda}), \tag{2.9}$$

where $l_t(\boldsymbol{\lambda})$ is defined as in (2.6) and the weight w_t satisfies the following assumption:

Assumption 5. $w_t = w(Y_{t-1}, Y_{t-2}, \cdots)$ *and* w *is a measurable and bounded function on* $\mathbb{R}^{Z_0}$ *with* $\mathrm{E}(w_t\xi_{\rho,t-1}^4) < \infty$ *and* $Z_0 = \{0, 1, 2, \cdots\}$ *for any* $0 < \rho < 1$,

where

$$\xi_{\rho,t} = 1 + \sum_{i=0}^{\infty} \rho^i |Y_{t-i}|. \tag{2.10}$$

With finite initial values, the weighted log-quasi-likelihood function (2.9) is modified as

$$\tilde{L}_{sn}(\boldsymbol{\lambda}) = \frac{1}{n}\sum_{t=1}^n \tilde{w}_t \tilde{l}_t(\boldsymbol{\lambda}), \tag{2.11}$$

where $\tilde{l}_t(\boldsymbol{\lambda})$ is defined in (2.4) and $\tilde{w}_t$ is computed from finite initial values. Since the weight w_t is defined by $\{Y_t\}$ itself, the maximizer of $\tilde{L}_{sn}(\boldsymbol{\lambda})$ on $\boldsymbol{\Theta}$ is called the self-weighed QMLE, denoted as $\hat{\boldsymbol{\lambda}}_{sn}$.

Assumption 6. $\mathrm{E}|w_t - \tilde{w}_t|^{\iota_0/4} = O(t^{-2})$, *where* $\iota_0 = \min\{\iota, 1\}$.

Theorem 2.4 (Ling, 2006). *Suppose that Assumptions 1-6 hold. Then,*

(i) $\tilde{\boldsymbol{\lambda}}_{sn} \xrightarrow{\mathcal{P}} \boldsymbol{\lambda}_0$,

(ii) $\mathrm{E}\eta_t^4 < \infty$ *and* $J > 0$ *implies*

$$\sqrt{n}(\tilde{\boldsymbol{\lambda}}_{sn} - \boldsymbol{\lambda}_0) \xrightarrow{\mathcal{D}} N(0,\ \mathcal{J}_w^{-1}\mathcal{I}_w\mathcal{J}_w^{-1}),$$

where $\mathcal{J}_w = \mathrm{E}[w_t U_t(\boldsymbol{\lambda}_0)U_t^{\mathrm{T}}(\boldsymbol{\lambda}_0)]$ *and* $\mathcal{I}_w = \mathrm{E}[w_t^2 U_t(\boldsymbol{\lambda}_0)JU_t^{\mathrm{T}}(\boldsymbol{\lambda}_0)]$.

Remark 2.8. *Since the* $\iota > 0$ *in Assumption 3 can be chosen arbitrarily small, the consistency and asymptotical normality of* $\tilde{\boldsymbol{\lambda}}_{sn}$ *hold only under a fractional moment of* ε_0. *This reduces the moment requirement of* ε_0 *in Theorem 2.3 significantly.*

Remark 2.9. *The inference of* $\tilde{\boldsymbol{\lambda}}_{sn}$ *depends on the selection of the weight* w_t. *We refer to Ling (2006) for choices of* w_t *for different models.*

Remark 2.10. *Theorem 2.4 holds for the ARMA-ARCH and AR-ARCH models by removing the related redundant components.*

Since the self-weighted QMLE $\tilde{\boldsymbol{\lambda}}_{sn}$ is root-n consistent, Ling (2006) uses it as an initial estimator of $\boldsymbol{\lambda}_0$ and obtains the local QMLE for ARMA-GARCH/IGARCH through the following one-step iteration:

$$\tilde{\boldsymbol{\lambda}}_{ln} = \tilde{\boldsymbol{\lambda}}_{sn} - \left[\frac{\partial^2 \tilde{l}_t(\tilde{\boldsymbol{\lambda}}_{sn})}{\partial\boldsymbol{\lambda}\partial\boldsymbol{\lambda}^{\mathrm{T}}}\right]^{-1} \sum_{t=1}^{n} \frac{\partial \tilde{l}_t(\tilde{\boldsymbol{\lambda}}_{sn})}{\partial\boldsymbol{\lambda}}. \tag{2.12}$$

Theorem 2.11 (Ling, 2006). *Suppose that Assumptions 1, 2, and 4-6 hold and that (1.5) or the condition of Theorem 1.1 (iii) is satisfied. If* $\mathrm{E}\eta_t^4 < \infty$ *and* $J > 0$, *then*

$$\sqrt{n}(\tilde{\boldsymbol{\lambda}}_{ln} - \boldsymbol{\lambda}_0) \xrightarrow{\mathcal{D}} N(0,\ \mathcal{J}^{-1}\mathcal{I}\mathcal{J}^{-1}).$$

Remark 2.12. *Notice that the asymptotical normality of the local QMLE* $\tilde{\boldsymbol{\lambda}}_{ln}$ *does not depend on the weight* w_t. *In fact both global QML* $\tilde{\boldsymbol{\lambda}}_n$ *and local QMLE* $\tilde{\boldsymbol{\lambda}}_{ln}$ *have the same asymptotical limit. So they have the same asymptotic efficiency. Of course the moment requirement of* ε_0 *is weaker for the local QMLE than for global QMLE. As pointed by Ling (2006), in general, it is not easy to compare the efficiency of the self-weighted QMLE* $\tilde{\boldsymbol{\lambda}}_{sn}$ *with global QMLE* $\tilde{\boldsymbol{\lambda}}_n$ *or local QMLE* $\tilde{\boldsymbol{\lambda}}_{ln}$. *When* η_t *has a standard normal distribution, then* $\tilde{\boldsymbol{\lambda}}_{ln}$ *or* $\tilde{\boldsymbol{\lambda}}_n$ *is more efficient than* $\tilde{\boldsymbol{\lambda}}_{sn}$.

Remark 2.13. *Theorem 2.11 cannot be applied to the ARMA-IARCH model. However, if the condition (1.5) is used, then Theorem 2.11 holds for ARMA-ARCH, GARCH, and ARCH models only if we remove the redundant parameters and the corresponding components in the variance matrix.*

Though the local QMLE $\tilde{\boldsymbol{\lambda}}_{ln}$ shows promising properties, nevertheless, it is still a local QMLE. Can we find a global QMLE without requiring $\mathrm{E}\varepsilon_0^4 < \infty$? The definition of $l_t(\boldsymbol{\lambda})$ or $\tilde{l}_t(\boldsymbol{\lambda})$ is based on the standard normal density. Berkes and Horváth (2004) showed that other densities can also be used to define $l_t(\boldsymbol{\lambda})$ when dealing with a pure GARCH model. It gives some more efficient estimators and weak moment requirement on η_t. After adopting the same technique to $l_t(\boldsymbol{\lambda})$ for the ARMA-GARCH model, Lu and Yu (2006) find that the estimator not only has similar benefits but it also works under $\mathrm{E}\varepsilon_0^2 < \infty$ or even $\mathrm{E}\varepsilon_0^2 = \infty$ when a student t density is chosen. In the following, we present their results in details.

Assume that h is a density function with third continuous derivatives. Then the log-quasi-likelihood function base on h is defined as:

$$L_{hn}(\boldsymbol{\lambda}) = \frac{1}{n}\sum_{t=1}^{n} l_{ht}(\boldsymbol{\lambda}), \quad l_{ht}(\boldsymbol{\lambda}) = \log \frac{h(\eta_t(\boldsymbol{\lambda}))}{\sigma_t(\boldsymbol{\lambda})}. \tag{2.13}$$

Define function g and its derivatives as:

$$g(x) = \frac{\partial \log h(x)}{\partial x} = \frac{\partial h(x)/\partial x}{h(x)}, \quad g'(x) = \frac{\partial g(x)}{\partial x}, \quad g''(x) = \frac{\partial^2 g(x)}{\partial x^2}.$$

The corresponding $\tilde{L}_{hn}(\boldsymbol{\lambda})$ and $\tilde{l}_{ht}(\boldsymbol{\lambda})$ are defined with finite initial values.

A QMLE of $\boldsymbol{\lambda}_0$ is any measurable solution of

$$\tilde{\boldsymbol{\lambda}}_{hn} = \arg\max_{\boldsymbol{\lambda}\in\boldsymbol{\Theta}} \tilde{L}_{hn}(\boldsymbol{\lambda}). \tag{2.14}$$

Assumption 7. $\mathrm{E}|\log h(\eta_t)| < \infty$.

Assumption 8. *Let* $G(\eta_t, u, v) = \log\big(uh(\eta_t u + v)\big)$, *where* $u > 0, v \in \mathbb{R}$. *Then* $\mathrm{E}G(\eta_t, u, v) < \mathrm{E}G(\eta_t, 1, 0)$ *for any* $u \neq 1$, *and* $v \neq 0$.

Assumption 9. *For some* $\kappa_1 \geqslant 0$ *and some constant* $C > 0$, $|g(x)|$ *is bounded by* $C\big(\max\{|x|,1\}\big)^{2\kappa_1 - 1}$, *and* $|g'(x)|$ *and* $|g''(x)|$ *are bounded by* $C\big(\max\{|x|,1\}\big)^{2(\kappa_1 - 1)}$.

Remark 2.14. *Apparently, Assumption 8 means* $\mathrm{E}G(\eta_t, u, v)$ *has a unique maximum point at* $u = 1$ *and* $v = 0$. *This will ensure that* $\mathrm{E}l_{ht}(\boldsymbol{\lambda})$ *is maximized if and only at true value* $\boldsymbol{\lambda}_0$. *Given Assumption 8, one can obtain easily that* $\mathrm{E}g(\eta_t) = 0$ *and* $1 + \mathrm{E}g(\eta_t)\eta_t = 0$.

Theorem 2.15 (Lu and Yu, 2006). *Suppose that Assumptions 1, 2, 4, and 7-9 hold. Then* $\mathrm{E}|\eta_t|^{2\kappa_1} < \infty$ *implies that almost surely*

$$\tilde{\boldsymbol{\lambda}}_{hn} \longrightarrow \boldsymbol{\lambda}_0 \text{ as } n \to \infty.$$

Remark 2.16. *Just like Theorem 2.1, the condition that* $\boldsymbol{\lambda}_0$ *is an interior point in* $\boldsymbol{\Theta}$ *in Assumption 4 is not required by Lu and Yu (2006). See Remark 2.2.*

Remark 2.17. *When* h *is the standard normal density, then Theorem 2.15 is reduced to Theorem 2.1. As far as consistency is concerned, there is no an apparent advantage of using other density* h*. Only for the asymptotical normality does the advantage with other density* h *show up.*

Introduce the matrices

$$\mathcal{I}_h = \mathrm{E}\left(\frac{\partial l_{ht}(\boldsymbol{\lambda}_0)}{\partial \boldsymbol{\lambda}}\frac{\partial l_{ht}(\boldsymbol{\lambda}_0)}{\partial \boldsymbol{\lambda}^{\mathrm{T}}}\right) \quad \text{and} \quad \mathcal{J}_h = -\mathrm{E}\left(\frac{\partial^2 l_{ht}(\boldsymbol{\lambda}_0)}{\partial \boldsymbol{\lambda}\partial \boldsymbol{\lambda}^{\mathrm{T}}}\right).$$

Theorem 2.18 (Lu and Yu, 2006). *Suppose that Assumptions 1, 2, 4 and 7-9 hold and Assumption 3 holds for* $\iota = (\kappa_1+1)(1-\kappa_2)$ *for some* $0 \leqslant \kappa_2 < 1$*. If*

$$\mathrm{E}|\eta_t|^{4\kappa_1} < \infty, \tag{2.15}$$

$$\mathrm{E}^2\left(g^2(\eta_t)\eta_t\right) < \mathrm{E}g^2(\eta_t)\mathrm{E}(g(\eta_t)\eta_t+1)^2, \tag{2.16}$$

and $\mathcal{J}_h$ *is nonsingular, then*

$$\sqrt{n}(\tilde{\boldsymbol{\lambda}}_{hn} - \boldsymbol{\lambda}_0) \xrightarrow{\mathcal{D}} N(0,\ \mathcal{J}_h^{-1}\mathcal{I}_h\mathcal{J}_h^{-1}).$$

If, in addition, the distribution of η_t *is symmetric about 0,* $\mathrm{E}g'(\eta_t)\eta_t = 0$ *and* $\mathrm{E}g^2(\eta_t)\eta_t = 0$*, we have*

$$\mathcal{I}_h = \begin{pmatrix} \mathcal{I}_{h1} & 0 \\ 0 & \mathcal{I}_{h2} \end{pmatrix}, \quad \mathcal{J}_h = \begin{pmatrix} \mathcal{J}_{h1} & 0 \\ 0 & \mathcal{J}_{h2} \end{pmatrix},$$

where

$$\begin{aligned}\mathcal{I}_{h1} = {} & \frac{1}{2}\mathrm{E}(g(\eta_1)\eta_t+1)^2\mathrm{E}\left(\frac{\partial\sigma_t^2(\boldsymbol{\lambda}_0)/\partial\boldsymbol{\gamma}}{\sqrt{2}\sigma_t^2(\boldsymbol{\lambda}_0)}\frac{\partial\sigma_t^2(\boldsymbol{\lambda}_0)/\partial\boldsymbol{\gamma}^{\mathrm{T}}}{\sqrt{2}\sigma_t^2(\boldsymbol{\lambda}_0)}\right) \\ & +\mathrm{E}g^2(\eta_1)\mathrm{E}\left(\frac{\partial\varepsilon_t(\boldsymbol{\gamma}_0)/\partial\boldsymbol{\gamma}}{\sigma_t(\boldsymbol{\lambda}_0)}\frac{\partial\varepsilon_t(\boldsymbol{\gamma}_0)/\partial\boldsymbol{\gamma}^{\mathrm{T}}}{\sigma_t(\boldsymbol{\lambda}_0)}\right),\end{aligned}$$

$$\mathcal{I}_{h2} = \frac{1}{2}\mathrm{E}(g(\eta_t)\eta_t+1)^2\mathrm{E}\left(\frac{\partial\sigma_t^2(\boldsymbol{\lambda}_0)/\partial\boldsymbol{\delta}}{\sqrt{2}\sigma_t^2(\boldsymbol{\lambda}_0)}\frac{\partial\sigma_t^2(\boldsymbol{\lambda}_0)/\partial\boldsymbol{\delta}^{\mathrm{T}}}{\sqrt{2}\sigma_t^2(\boldsymbol{\lambda}_0)}\right),$$

$$\begin{aligned}\mathcal{J}_{h1} = {} & -\frac{1}{2}\mathrm{E}\Big(g'(\eta_t)\eta_t^2-1\Big)\mathrm{E}\left(\frac{\partial\sigma_t^2(\boldsymbol{\lambda}_0)/\partial\boldsymbol{\gamma}}{\sqrt{2}\sigma_t^2(\boldsymbol{\lambda}_0)}\frac{\partial\sigma_t^2(\boldsymbol{\lambda}_0)/\partial\boldsymbol{\gamma}^{\mathrm{T}}}{\sqrt{2}\sigma_t^2(\boldsymbol{\lambda}_0)}\right) \\ & -\mathrm{E}g'(\eta_t)\mathrm{E}\left(\frac{\partial\varepsilon_t(\boldsymbol{\gamma}_0)/\partial\boldsymbol{\gamma}}{\sigma_t(\boldsymbol{\lambda}_0)}\frac{\partial\varepsilon_t(\boldsymbol{\gamma}_0)/\partial\boldsymbol{\gamma}^{\mathrm{T}}}{\sigma_t(\boldsymbol{\lambda}_0)}\right),\end{aligned}$$

$$\mathcal{J}_{h2} = -\frac{1}{2}\mathrm{E}\Big(g'(\eta_t)\eta_t^2-1\Big)\mathrm{E}\left(\frac{\partial\sigma_t^2(\boldsymbol{\lambda}_0)/\partial\boldsymbol{\delta}}{\sqrt{2}\sigma_t^2(\boldsymbol{\lambda}_0)}\frac{\partial\sigma_t^2(\boldsymbol{\lambda}_0)/\partial\boldsymbol{\delta}^{\mathrm{T}}}{\sqrt{2}\sigma_t^2(\boldsymbol{\lambda}_0)}\right).$$

Remark 2.19. *We have from Remark 2.14 that* $\mathrm{E}g(\eta_t) = 0$ *and* $\mathrm{E}g(\eta_t)\eta_t + 1 = 0$. *This may result in* $\mathrm{E}\eta_t^2 \neq 1$ *unless* h *is chosen as a standard normal density. Since we assume* $\mathrm{E}\eta_t^2 = 1$ *at the beginning to make parameters uniquely identifiable, we have to scale* η_t *to meet those requirements. As a consequence, the ARCH parameters* $\alpha_{00}, \alpha_{01}, \cdots, \alpha_{0p}$ *are scaled accordingly while other parameters remain unchanged.*

Remark 2.20. *Using* $U_t(\boldsymbol{\lambda})$ *defined in Remark 2.5, we can write*

$$\mathcal{J}_h = \mathrm{E}[U_t(\boldsymbol{\lambda}_0) J_h U_t^{\mathrm{T}}(\boldsymbol{\lambda}_0)] \quad \text{and} \quad \mathcal{I}_h = \mathrm{E}[U_t(\boldsymbol{\lambda}_0) I_h U_t^{\mathrm{T}}(\boldsymbol{\lambda}_0)],$$

where

$$J_h = \begin{pmatrix} -\mathrm{E}\big(g'(\eta_t)\eta_t^2 - 1\big)/2 & \mathrm{E}\big(g'(\eta_t)\eta_t\big)/\sqrt{2} \\ \mathrm{E}\big(g'(\eta_t)\eta_t\big)/\sqrt{2} & -\mathrm{E}g'(\eta_t) \end{pmatrix}$$

and

$$I_h = \begin{pmatrix} \mathrm{E}\big(g(\eta_t)\eta_t + 1\big)^2/2 & -\mathrm{E}\big(g^2(\eta_t)\eta_t\big)/\sqrt{2} \\ -\mathrm{E}\big(g^2(\eta_t)\eta_t\big)/\sqrt{2} & \mathrm{E}g^2(\eta_t) \end{pmatrix}.$$

Remark 2.21. *Since* $\mathrm{E}g(\eta_t) = 0$, *by Cauchy inequality, it is readily shown that (2.16) holds if and only if* $P(g(\eta_t)\eta_t + 1 = cg(\eta_t)) < 1$ *for any constant* c. *A simple condition for this is that* η_t *has a positive density on some interval provided that* $g(x)x + 1 = cg(x)$ *has finite roots. Given (2.16), Lu and Yu (2006) prove that* $\mathcal{I}_h$ *is nonsingular. Notice that* $I_h > 0$ *if and only if (2.16) holds. Checking the nonsingularity of* $\mathcal{J}_h$ *can be done through its representation of* $U_t(\boldsymbol{\lambda}_0)$ *and* J_h *in Remark 2.20. Since Francq and Zakoïab (2004) already proved that* $\sigma_t^{-4}(\boldsymbol{\lambda}_0)\left(\partial\sigma_t^2(\boldsymbol{\lambda}_0)/\partial\boldsymbol{\lambda}\right)\left(\partial\sigma_t^2(\boldsymbol{\lambda}_0)/\partial\boldsymbol{\lambda}^{\mathrm{T}}\right)$ *is nonsingular, it is easy to check that* $\mathcal{J}_h$ *is nonsingular if* $\mathrm{E}g'(\eta_t) \leqslant 0$, $\mathrm{E}g'(\eta_t)\eta_t^2 < 1$ *and* $\mathrm{E}g'(\eta_t)\eta_t = 0$.

Remark 2.22. *Which value of* κ_2 *to choose depends on the parameter space* $\boldsymbol{\Theta}$, *in particular,* β_{01}. *In general, we choose* $\kappa_2 = 0$. *If we use the condition* $\beta_{01} > 0$ *specifically, we can find a positive* κ_2 *so the moment of* ε_0 *will be reduced slightly. This enables us to find a global QMLE for an ARMA-IGARCH model but not an ARMA-IARCH model. With* $\kappa_2 = 0$, *Theorem 2.18 holds for ARMA-ARCH, GARCH, and ARCH models only if we remove the reductant parameters and the corresponding components in the covariance matrix.*

In the next, we apply three specific densities: normal, student t and generalized error distribution (GED) densities, to Theorem 2.18.

When h takes the standard normal density as $h(x) = (2\pi)^{-1/2}e^{-x^2/2}$, then we have

$$h'(x) = \partial h(x)/\partial x = -(2\pi)^{-1/2}xe^{-\frac{x^2}{2}},$$

$$g(x) = -x, \quad g'(x) = -1, \quad \text{and} \quad g''(x) = 0.$$

Thus it is easy to check that Assumptions 7-8 hold. Assumption 9 is satisfied with $\kappa_1 = 1$. Hence we establish the asymptotical normality of $\tilde{\boldsymbol{\lambda}}_{hn}$ under $\mathrm{E}|\varepsilon_0|^{4(1-\kappa_2)} < \infty$ and $\mathrm{E}\eta_t^4 < \infty$. In addition, based on Remark 2.21, both $\mathcal{J}_h$ is nonsingular. Essentially, Theorem 2.18 implies Theorem 2.3 when h is chosen as a standard normal density and the condition $\mathrm{E}\varepsilon_0^4 < \infty$ in Theorem 2.3 is reduced to $\mathrm{E}|\varepsilon_0|^{4(1-\kappa_2)} < \infty$ for some $0 < \kappa_2 < 1$, provided that $\beta_{01} > 0$. See Remark 2.22.

When h takes the student t density with degree $d > 1$ as $h(x) = c\left(1 + x^2/d\right)^{-d+1/2}$, where c is a constant (function of d), then we have

$$h'(x) = \frac{\partial h(x)}{\partial x} = -\frac{(d+1)cx}{d}\left(1 + \frac{x^2}{d}\right)^{-\frac{d+1}{2}-1},$$

$$g(x) = -(d+1)\frac{x}{d+x^2},$$

$$g'(x) = -(d+1)\left\{\frac{1}{d+x^2} - \frac{2x^2}{(d+x^2)^2}\right\},$$

$$g''(x) = 2(d+1)\left\{\frac{3x}{(d+x^2)^2} - \frac{4x^3}{(d+x^2)^3}\right\}.$$

Obviously, Assumption 9 is satisfied with $\kappa_1 = 0$. Lu and Yu (2005) verify that Assumption 7 holds if $\mathrm{E}|\eta_t|^\tau < \infty$ for any $\tau > 0$ and Assumption 8 is true if the distribution of η_t is symmetric with monotone decreasing density on its right side. Notice that Assumption 8 implies $\mathrm{E}(1/(d+\eta_t^2)) = 1/(d+1)$ so that $\mathrm{E}\eta_t^2 \neq 1$. Lu and Yu (2006) also show that when η_t is symmetric, $\mathcal{J}_h$ is nonsingular. Thus, based on student t density and η_t being symmetric, we obtain the asymptotical normality of $\tilde{\boldsymbol{\lambda}}_{hn}$ under $\mathrm{E}|\varepsilon_0|^{2(1-\kappa_2)} < \infty$ and $\mathrm{E}|\eta_t|^\tau < \infty$. This last moment condition is surpassed by the model requirement $\mathrm{E}\eta_t^2 < \infty$. In other words, we get the asymptotical normality of a global QMLE under $\mathrm{E}|\varepsilon_0|^{2(1-\kappa_2)} < \infty$ only which covers the ARMA-GARCH and ARMA-IGARCH models.

Density of generalized error distribution (GED) (Nelson, 1991) has the form $f(x) = ue^{-|x|^v}$, where u and v are some constants. This density is not smooth at $x = 0$. Adopting the idea of Hitomi (1997) for double exponential, we smooth the density QED(d) as

$$h(x) = ae^{-(x^2+b^2)^{d/2}}$$

for some positive constant a which is a function of b and d, $b \neq 0$ and $d \geqslant 1$. Then

$$h'(x) = -ad\,e^{-(x^2+b^2)^{d/2}}(x^2+b^2)^{d/2-1}x, \quad g(x) = -d(x^2+b^2)^{d/2-1}x,$$

$$g'(x) = -d(d-2)(x^2+b^2)^{d/2-2}x^2 - d(x^2+b^2)^{d/2-1},$$

$$g''(x) = -d(d-2)(d-4)(x^2+b^2)^{d/2-3}x^3 - 3d(d-2)(x^2+b^2)^{d/2-2}x.$$

Clearly Assumption 9 holds for $\kappa_1 = d/2$. Lu and Yu (2006) show that Assumption 7 is satisfied if $\mathrm{E}|\eta_t|^d < \infty$ and Assumption 8 is true if η_t is symmetric. They also show that when η_t is symmetric, $\mathcal{J}_h$ is nonsingular. Thus, based on QED density and η_t being symmetric, we obtain the asymptotical normality of $\tilde{\boldsymbol{\lambda}}_{hn}$ under $\mathrm{E}|\varepsilon_0|^{(d+2)(1-\kappa_2)} < \infty$ and $\mathrm{E}|\eta_t|^d < \infty$. If we choose $d = 1$, then we get the asymptotical normality of a global QMLE under $\mathrm{E}|\varepsilon_0|^{3(1-\kappa_2)} < \infty$.

In general, it is not easy to compare the efficiency of $\tilde{\boldsymbol{\lambda}}_{hn}$ for different density h. As indicated in Remark 2.19, choosing a density h other than a standard normal density will result in $\mathrm{E}\eta_t^2 \neq 1$. This will make the comparison of the efficiency even hard. However, in case η_t is symmetric, both $\mathcal{J}_h$ and $\mathcal{I}_h$ are block-diagonal matrices and have the forms given in Theorem 2.18. In particular, the covariance matrix $\mathcal{J}_{h2}^{-1}\mathcal{I}_{h2}\mathcal{J}_{h2}^{-1}$ for the GARCH part has the same form as for a pure GARCH model in Berkes and Horváth (2004). They pointed out that if the innovation density is unknown and its tail might be polynomial, the two-sided exponential (QED(1)) method performs better than the QMLE based on standard normal density. The covariance matrix $\mathcal{J}_{h1}^{-1}\mathcal{I}_{h1}\mathcal{J}_{h1}^{-1}$ for the ARMA part is much hard to evaluate due to the presence of a GARCH component. Nevertheless, we believe that ARMA part of QMLE based on student t or QED(d) density will perform better since GARCH errors are likely heavy tailed.

3 High moment partial sum processes of residuals

In this section we will investigate the asymptotic properties of high moment partial sum processes based on residuals of an ARMA-GARCH model. Kulperger and Yu (2005) define the k-th ($k = 1, 2, 3, 4, \cdots$) order high moment partial sum process of residuals is defined as

$$\tilde{S}_n^{(k)}(u) = \sum_{t=1}^{[nu]} \tilde{\eta}_t^k, \ 0 \leqslant u \leqslant 1. \tag{3.1}$$

Its counterpart based on the iid innovations is defined as

$$S_n^{(k)}(u) = \sum_{t=1}^{[nu]} \eta_t^k, \ 0 \leqslant u \leqslant 1. \tag{3.2}$$

They established asymptotical approximation of those high moment partial sum processes based on residuals of a pure GARCH model. Then they used those processes to construct CUSUM statistics for testing GARCH model structure change and the Jarque-Bera omnibus statistic for testing normality of the unobservable innovation distribution of a pure GARCH model. In addition, they verified the use of residuals for constructing a kernel density function estimation of the innovation distribution. Yu and Lu (2006) have extended Kulperger and Yu's (2005) results to ARMA-GARCH/IGARCH models. In this section, we present their results in details.

In the following we assume that $\hat{\boldsymbol{\lambda}}_n = (\hat{\boldsymbol{\gamma}}_n^{\mathrm{T}}, \hat{\boldsymbol{\delta}}_n^{\mathrm{T}})^{\mathrm{T}}$ is an estimator of $\boldsymbol{\lambda}$ based on a sample $Y_1, \cdots, Y_n$, and that it is $\sqrt{n}$ consistent, that is

Assumption 10. $\sqrt{n}|\hat{\boldsymbol{\lambda}}_n - \boldsymbol{\lambda}_0| = O_P(1)$,

where we use $|\cdot|$ to denote the maximum norm of vectors or matrices. Certainly the estimators discussed in Section 2 meet the above assumption under various model assumptions.

Theorem 3.1 (Yu and Lu, 2006). *Suppose that Assumptions 1, 2, 4 and 10 hold and Assumption 3 holds for $\iota = 1-\kappa_2$ for some $0 \leqslant \kappa_2 < 1$. If $\mathrm{E}|\eta_t|^k < \infty$, then*

$$\sup_{0\leqslant u\leqslant 1}\left|\frac{1}{\sqrt{n}}\left(\tilde{S}_n^{(k)}(u) - S_n^{(k)}(u)\right) + \frac{ku\mu_k}{2}\left\langle \Lambda, \sqrt{n}(\hat{\boldsymbol{\lambda}}_n - \boldsymbol{\lambda}_0)\right\rangle \right.$$
$$\left. - ku\mu_{k-1}\left\langle \Gamma, \sqrt{n}(\hat{\gamma}_n - \gamma_0)\right\rangle\right| = o_P(1),$$

$\Lambda = \mathrm{E}\left(\partial \log \sigma_0^2(\boldsymbol{\lambda}_0)/\partial\boldsymbol{\lambda}\right)$, $\Gamma = \mathrm{E}\sigma_0^{-1}(\boldsymbol{\lambda}_0)\left(\partial\varepsilon_0(\gamma_0)/\partial\gamma\right)$, *and $\langle \mathbf{x}, \mathbf{y}\rangle$ is the inner product of the vectors $\mathbf{x}$ and $\mathbf{y}$.*

Remark 3.2. *Theorem 3.1 shows that the asymptotic properties of the high moment partial sum process $\{\tilde{S}_n^{(k)}(u), 0 \leqslant u \leqslant 1\}$ always depend on the parameters of the model for any integer k. This is different from the pure GARCH case where the there is no such a term as $ku\mu_{k-1}\langle\Gamma, \sqrt{n}(\hat{\gamma}_n - \gamma_0)\rangle$ and the approximation for $\{\tilde{S}_n^{(k)}(u), 0 \leqslant u \leqslant 1\}$ can be parameter free if $\mu_k = 0$ for some odd k. See Remark 1.1 in Kulperger and Yu (2005).*

Remark 3.3. *The value of κ_2 can be chosen as the same way as in Remark 2.22. Basically, given $\beta_{01} > 0$, Theorem 3.1 holds for an ARMA-GARCH/IGARCH model but not an ARMA-IARCH model. With $\kappa_2 = 0$, Theorem 3.1 holds for ARMA-ARCH, GARCH, and ARCH models only if we remove the reductant parameters and the corresponding components in the covariance matrix.*

From Theorem 3.1, it is easy to get the following corollary.

Corollary 3.4. *Under the assumptions of Theorem 3.1, we have*

$$\sup_{0\leqslant u\leqslant 1}\left|\left(\tilde{S}_n^{(k)}(u)-u\tilde{S}_n^{(k)}(1)\right)-\left(S_n^{(k)}(u)-uS_n^{(k)}(1)\right)\right|=o_P(1)\ .$$

Remark 3.5. *Corollary 3.4 implies that the CUSUM normalized high moment partial sum process* $\{\tilde{S}_n^{(k)}(u)-u\tilde{S}_n^{(k)}(1), 0\leqslant u\leqslant 1\}$ *behaves as though the residuals* $\{\tilde{\eta}_t, 1\leqslant t\leqslant n\}$ *were asymptotically the same as the unobservable innovations* $\{\eta_t, 1\leqslant t\leqslant n\}$.

The next result follows immediately from Corollary 3.4 based on the invariance principle for partial sums for an iid sequence $\{\eta_t^k\}$ (see for example Billingsley, 1999).

Corollary 3.6. *Under the assumptions of Theorem 3.1 and* $\zeta_k^2=\mathrm{E}(\eta_t^k-\mu_k)^2<\infty$, *we have*

$$\left\{\frac{\tilde{S}_n^{(k)}(u)-u\tilde{S}_n^{(k)}(1)}{\zeta_k\sqrt{n}},\ 0\leqslant u\leqslant 1\right\}$$

converges weakly in the Skorokhod space $D[0,\ 1]$ *with* J_1 *topology to a Brownian bridge* $\{B_0(u), 0\leqslant u\leqslant 1\}$.

Next we define the k-th order moment residual centered partial sum process as

$$\tilde{T}_n^{(k)}(u)=\sum_{t=1}^{[nu]}\left(\tilde{\eta}_t-\bar{\tilde{\eta}}\right)^k,\ 0\leqslant u\leqslant 1\ ,$$

where $\bar{\tilde{\eta}}$ is the sample mean of the residuals. Its counterpart based on the iid innovations is

$$T_n^{(k)}(u)=\sum_{t=1}^{[nu]}\left(\eta_t-\bar{\eta}\right)^k,\ 0\leqslant u\leqslant 1\ ,$$

where $\bar{\eta}$ is the sample mean of innovations.

Theorem 3.7 (Yu and Lu, 2006). *Under the assumptions of Theorem 3.1, we have*

$$\sup_{0\leqslant u\leqslant 1}\left|\frac{1}{\sqrt{n}}\left(\tilde{T}_n^{(k)}(u)-T_n^{(k)}(u)\right)+\frac{ku\mu_k}{2}\left\langle\Lambda,\sqrt{n}(\hat{\boldsymbol{\lambda}}_n-\boldsymbol{\lambda}_0)\right\rangle\right|=o_P(1).$$

Remark 3.8. *Theorem 3.7 shows that the sample mean centering is able to remove ARMA related parameter term* $ku\mu_{k-1}\langle\Gamma,\sqrt{n}(\hat{\gamma}_n-\gamma_0)\rangle$ *in Theorem 3.1.*

Let $\tilde{\sigma}^2_{(n)} = \tilde{T}_n^{(2)}(1)/n$ be the sample moment estimator of μ_2. Just like in pure GARCH case, $\tilde{\sigma}^2_{(n)}$ will play an important role when it is used to self-normalize $\tilde{T}_n^{(k)}(u)$. Denote $\sigma^2_{(n)} = T_n^{(2)}(1)/n$, and note that it is the sample variance of the true innovations, except with divisor n instead of $n-1$.

Theorem 3.9 (Yu and Lu, 2006). *Under the assumptions of Theorem 3.1, we have*

$$\sup_{0\leqslant u\leqslant 1} \frac{1}{\sqrt{n}}\left|\frac{\tilde{T}_n^{(k)}(u)}{\tilde{\sigma}^k_{(n)}} - \frac{T_n^{(k)}(u)}{\sigma^k_{(n)}}\right| = o_P(1).$$

Remark 3.10. *Theorem 3.9 implies that the self-normalized (or estimated scale normalized) high moment centered partial sum process* $\{\tilde{T}_n^{(k)}(u)/\tilde{\sigma}^k_{(n)}, 0\leqslant u\leqslant 1\}$ *behaves as though the residuals* $\{\tilde{\eta}_t, 1\leqslant t\leqslant n\}$ *were asymptotically the same as the unobservable innovations* $\{\eta_t, 1\leqslant t\leqslant n\}$.

Let $\omega_k = \mu_k/\mu_2^{k/2}$ for $k\geqslant 1$ and define $\omega_0 = 1$. For each $k\geqslant 1$, let $\{B^{(k)}(u), 0\leqslant u\leqslant 1\}$ be a zero mean Gaussian process with covariance

$$\begin{aligned} &\mathrm{E}B^{(k)}(u)B^{(k)}(v)\\ &= (\omega_{2k}-\omega_k^2)(u\wedge v) + k\omega_{k-1}(k\omega_{k-1}+k\omega_k\omega_3 - 2\omega_{k+1})uv\\ &\quad + k\omega_k\left((1-k/4)\omega_k + k\omega_k\omega_4/4 - \omega_{k+2}\right)uv \end{aligned} \tag{3.3}$$

for any $0\leqslant u,v\leqslant 1$, where $u\wedge v = \min(u,v)$.

Corollary 3.11 (Yu and Lu, 2006). *Under the assumptions of Theorem 3.1 and* $\mathrm{E}|\eta_0|^{2k}<\infty$ *for some integer* $k\geqslant 1$, *we have*

$$\left\{\frac{1}{\sqrt{n}}\left(\frac{\tilde{T}_n^{(k)}(u)}{\tilde{\sigma}^k_{(n)}} - nu\omega_k\right),\ 0\leqslant u\leqslant 1\right\}$$

converges weakly to the Gaussian process $\{B^{(k)}(u),\ 0\leqslant u\leqslant 1\}$.

Remark 3.12. *When* $k=1$, *(3.3) becomes* $\mathrm{E}\left[B^{(1)}(u)B^{(1)}(v)\right] = u\wedge v - uv$ *for any* $0\leqslant u,v\leqslant 1$, *that is,* $\{B^{(1)}(u),\ 0\leqslant u\leqslant 1\}$ *is a Brownian bridge. For* $k=2$, *(3.3) implies that* $\mathrm{E}\left[B^{(2)}(u)B^{(2)}(v)\right] = (\omega_4-1)(u\wedge v - uv)$ *for any* $0\leqslant u,v\leqslant 1$. *This means* $\{B^{(2)}(u)/\sqrt{\omega_4-1},\ 0\leqslant u\leqslant 1\}$ *is also a Brownian bridge. In general, the Gaussian process* $\{B^{(k)}(u),\ 0\leqslant u\leqslant 1\}$ *for* $k\geqslant 3$ *depends on the moments of the innovation distribution and cannot be identified with a specific classic process such as a Brownian motion or Brownian bridge.*

Remark 3.13. *Both Corollaries 3.6 and 3.11 are used to construct CUSUM statistics given in the next section.*

Corollary 3.14 (Yu and Lu, 2006). *Suppose that Assumptions given in Theorem 3.1 hold. Assume also that* $k \geqslant 1$ *is a odd number and* $\mu_3 = \mu_k = \mu_{k+2} = \mu_{2k+1} = 0$. *Then* $\mathrm{E}|\eta_0|^{2(k+1)} < \infty$ *implies that*

$$\left\{ \frac{1}{\sqrt{n}} \left(\frac{\tilde{T}_n^{(k)}(x)}{\tilde{\sigma}_{(n)}^k} - nx\omega_k, \ \frac{\tilde{T}_n^{(k+1)}(y)}{\tilde{\sigma}_{(n)}^{k+1}} - ny\omega_{k+1} \right), \ 0 \leqslant x, y \leqslant 1 \right\}$$

converges weakly in the Skorokhod space $D^2[0,\ 1]$ *to a two dimensional Gaussian process* $\{(B^{(k)}(x),\ B^{(k+1)}(y))\ 0 \leqslant x, y \leqslant 1\}$, *where* $\{B^{(k)}(x),\ 0 \leqslant x \leqslant 1\}$ *and* $\{B^{(k+1)}(y),\ 0 \leqslant y \leqslant 1\}$ *are two independent zero mean Gaussian processes defined by*

$$\begin{aligned} \mathrm{E}\left[B^{(i)}(x)B^{(i)}(y)\right] &= (\omega_{2i} - \omega_i^2)(x \wedge y) + i\omega_{i-1}(i\omega_{i-1} + i\omega_i\omega_3 - 2\omega_{i+1})xy \\ &\quad + i\omega_i\left((1 - i/4)\omega_i + i\omega_i\omega_4/4 - \omega_{i+2}\right)xy, \ i = k, k+1, \end{aligned}$$

for any $0 \leqslant x, y \leqslant 1$.

Remark 3.15. *Corollary 3.14 will be used to construct the Jarque-Bera test statistic given in the next section.*

4 Applications and simulation

First we consider an application of the CUSUM normalized high moment partial sum process $\left\{\tilde{S}_n^{(k)}(u) - u\tilde{S}_n^{(k)}(1),\ 0 \leqslant u \leqslant 1\right\}$ defined in Corollary 3.4.

Yu and Lu (2006) consider a structural change in the conditional mean c_0 for ARMA-GARCH models. Due to the masking effect of ARMA parameters on c_0, they consider only the GARCH model with nonzero c_0. They formulate it as the following hypothesis test. The null hypothesis is "no-change in the conditional mean"

$$H_0 : \left\{ \begin{array}{l} Y_t = \sigma_t \eta_t + c_0 \\ \sigma_t^2 = \alpha_{00} + \sum_{i=1}^p \alpha_{0i}(Y_{t-i} - c_0)^2 + \sum_{j=1}^q \beta_{0j}\sigma_{t-j}^2 \end{array} \right\}, \ t = 1, \cdots, n$$

and the alternative is "one change in the conditional mean"

$$H_a : \left\{ \begin{array}{l} \left. \begin{array}{l} Y_t = \sigma_t \eta_t + c_{01} \\ \sigma_t^2 = \alpha_{00} + \sum_{i=1}^p \alpha_{0i}(Y_{t-i} - c_{01})^2 \\ \qquad\qquad + \sum_{j=1}^q \beta_{0j}\sigma_{t-j}^2 \end{array} \right\}, \ t = 1, \cdots, [nu^*] \\ \left. \begin{array}{l} X_t = \sigma_t \eta_t + c_{02} \\ \sigma_t^2 = \alpha_{00} + \sum_{i=1}^p \alpha_{0i}(Y_{t-i} - c_{02})^2 \\ \qquad\qquad + \sum_{j=1}^q \beta_{0j}\sigma_{t-j}^2 \end{array} \right\}, \ t = [nu^*] + 1, \cdots, n, \end{array} \right.$$

where $c_{01} \neq c_{02}$ and $0 < u^* < 1$.

To test the above hypothesis, Yu and Lu (2005) use the standard CUSUM test constructed from residuals as

$$CUSUM^{(1)} = \max_{1 \leqslant i < n} \frac{\left|\sum_{t=1}^{i} \tilde{\eta}_t - i\bar{\tilde{\eta}}\right|}{\tilde{\sigma}_{(n)}\sqrt{n}}.$$

By Corollary 3.6, under H_0,

$$CUSUM^{(1)} \xrightarrow{\mathcal{D}} \sup_{0 \leqslant u \leqslant 1} |B_0(u)|,$$

where $\{B_0(u),\ 0 \leqslant u \leqslant 1\}$ is a Brownian bridge. Hence we can reject the H_0 in favor of H_a if $CUSUM^{(1)}$ is large.

Table 4.1 lists the simulation results of GARCH(1,1) with nonzero mean. The parameter in H_0 is $(c_0, \alpha_{00}, \alpha_{01}, \beta_{01}) = (0, 0.5, 0.1, 0.8)$. The break point in H_a is $u^* = 0.5$ and c_0 changes from 0 to 0.5 after $[u^* n]$. Critical values 1.358 and 1.2239 are chosen for significance levels $\alpha = 5\%$ and $\alpha = 10\%$ respectively. Five thousand replications are used.

From the table, it can be seen that sizes are somewhat conservative as sample size n is small. When sample size increases, the sizes are very close to the nominal significance level. The powers increase with the sample size and are bigger than 90% as sample size is bigger than 1000.

Table 4.1: Size and Power of $CUSUM^{(1)}$ Statistic for GARCH(1,1) with Nonzero Mean

$\eta_0 \sim N(0,1)$	n=300	n=600	n=1000	n=3000
Size				
$\alpha = .05$	0.038	0.038	0.043	.050
$\alpha = .1$	0.075	0.079	0.094	.099
Power				
$\alpha = .05$	0.387	0.689	0.902	1
$\alpha = .1$	0.509	0.792	0.944	1

Next structural change hypothesis concerns a change in the conditional variance of an ARMA-GARCH model. Yu and Lu (2006) propose the null hypothesis as for "no-change in the conditional variance"

$$H_{0'}: \left\{ \begin{array}{l} Y_t - c_0 = \sum_{i=1}^{P} \phi_{0i}(Y_{t-i} - c_0) \\ \qquad\qquad + \varepsilon_t + \sum_{j=1}^{Q} \psi_{0j}\varepsilon_{t-j} \\ \varepsilon_t = \sigma_t \eta_t \\ \sigma_t^2 = \alpha_{00} + \sum_{i=1}^{p} \alpha_{0i}\varepsilon_{t-i}^2 + \sum_{j=1}^{q} \beta_{0j}\sigma_{t-j}^2 \end{array} \right\},\ t = 1, \cdots, n$$

against the "one change in the conditional variance" alternative

$$H_{a'}:\begin{cases} Y_t - c_0 = \sum_{i=1}^{P}\phi_{0i}(Y_{t-i}-c_0)+\varepsilon_t+\sum_{j=1}^{Q}\psi_{0j}\varepsilon_{t-j} \\ \varepsilon_t=\sigma_t\eta_t \\ \sigma_t^2=\begin{cases}\alpha_{00}+\sum_{i=1}^{p}\alpha_{0i}\varepsilon_{t-i}^2+\sum_{j=1}^{q}\beta_{0j}\sigma_{t-j}^2 \text{ if } t=1,\cdots,[nu^*] \\ \alpha'_{00}+\sum_{i=1}^{p}\alpha'_{0i}\varepsilon_{t-i}^2+\sum_{j=1}^{q}\beta'_{0j}\sigma_{t-j}^2 \text{ if } t=[nu^*]+1,\cdots,n,\end{cases}\end{cases}$$

where $(\alpha_{00},\alpha_{01},\cdots,\alpha_{0p},\beta_{01},\cdots,\beta_{0q})\neq(\alpha'_{00},\alpha'_{01},\cdots,\alpha'_{0p},\beta'_{01},\cdots,\beta'_{0q})$ and $0<u^*<1$. The statistic is defined as

$$CUSUM^{(2)}=\max_{1\leqslant i<n}\frac{\left|\sum_{t=1}^{i}\left(\tilde{\eta}_t-\bar{\tilde{\eta}}\right)^2-i\tilde{\sigma}_{(n)}^2\right|}{\tilde{\zeta}_2\sqrt{n}},$$

where

$$\tilde{\zeta}_2^2=\frac{1}{n}\sum_{t=1}^{n}\left(\left(\tilde{\eta}_t-\bar{\tilde{\eta}}\right)^2-\tilde{\sigma}_{(n)}^2\right)^2$$

is an estimator of $\zeta_2^2=\mathrm{E}(\eta_0^2-\mu_2)^2=\mu_2^2(\omega_4-1)$. Therefore, by Corollaries 3.11, under H_0,

$$CUSUM^{(2)}\xrightarrow{\mathcal{D}}\sup_{0\leqslant u\leqslant 1}|B_0(u)|,$$

where $\{B_0(u),\ 0\leqslant u\leqslant 1\}$ is a Brownian bridge. Hence we can reject the $H_{0'}$ in favor of $H_{a'}$ whenever $CUSUM^{(2)}$ is large.

Table 4.2 presents the simulation results of ARMA(1,1)-GARCH(1,1) with $\eta_t\sim N(0,1)$ and $\eta_t\sim t(9)$ respectively. The parameter in $H_{0'}$

$$(c_0,\phi_{01},\varphi_{01},\alpha_{00},\alpha_{01},\beta_{01})=(0,0.4,0.6,0.0002,0.1,0.7).$$

Break point 0.5 in $H_{a'}$ is used. Replication is 5000 times. Each time there is only one change in the conditional variance. Critical values 1.358 is chosen for significance level 5%.

From table 4.2, we can see the result for ARMA(1,1)-GARCH(1,1) is similar with that of GARCH(1,1) in Yu (2004). There is size distortion, which is less serious when sample size is bigger. There is power losses when innovation distribution changes from normal to t distribution.

Next we concern the normality test for the GARCH innovation distribution. The Jarque-Bera (JB) (1987) normality test based on residuals of a pure GARCH model has been validated by Kulperger and Yu (2005). By Corollary 3.14, Yu and Lu (2006) show also that JB test works for an

Table 4.2: Size and Power of $CUSUM^{(2)}$ statistic for ARMA(1,1)-GARCH(1,1)

$\eta_0 \sim N(0,1)$	n=500	n=1000	n=1500	n=3000
Null	0.029	0.036	0.037	0.044
$\alpha'_{00} = .0003$	0.392	0.754	0.929	0.999
$\alpha'_{01} = .167$	0.189	0.416	0.621	0.914
$\beta'_{01} = .767$	0.306	0.655	0.851	0.994
$\eta_0 \sim t(9)$	n=500	n=1000	n=1500	n=3000
Null	0.039	0.046	0.048	0.043
$\alpha'_{00} = .0003$	0.191	0.442	0.645	0.930
$\alpha'_{01} = .167$	0.148	0.367	0.551	0.868
$\beta'_{01} = .767$	0.182	0.425	0.636	0.938

ARMA-GARCH/IGARCH model. Basically, the sample skewness and sample kurtosis are defined as

$$\hat{\gamma}_n = \frac{\hat{T}_n^{(3)}(1)/n}{\hat{\sigma}_{(n)}^3}, \quad \hat{\kappa}_n = \frac{\hat{T}_n^{(4)}(1)/n}{\hat{\sigma}_{(n)}^4}.$$

Then JB statistic is defined as

$$JB = \frac{n}{6}\hat{\gamma}_n^2 + \frac{n}{24}\left(\hat{\kappa}_n - 3\right)^2.$$

Assume that the innovation distribution is standard normal. Then, by Corollary 3.14,

$$JB \xrightarrow{\mathcal{D}} \chi^2_{(2)}.$$

The JB is widely used in econometrics and implemented in standard statistical packages such as Splus and R. Jarque and Bera (1987) show that JB is a Lagrange multiplier test statistic of normality against alternatives within the Pearson family of distributions and it is asymptotically equivalent to the likelihood ratio test. Therefore a test based on JB is asymptotically locally most powerful against the Pearson family. The hypothesis of normality is rejected for large sample size, if the computed value of JB is greater than the appropriate critical value of a $\chi^2_{(2)}$. Lu (2001) has used Monte Carlo simulation to obtain critical values for several different n. Based on this a finite sample size correction can also be used to improve the choice of critical value. Lu (2001) obtained the finite sample size correction for the sizes 0.05 and 0.10 critical values of the JB test as

$$JB_{0.05} = 5.991645 - 15.17n^{-1/2} + 345.9n^{-1} - 3110.8n^{-3/2}, \quad n \geqslant 200$$

and

$$JB_{0.10} = 4.60517 - 9.78n^{-1/2} + 132.25n^{-1} - 1696n^{-3/2}, \quad n \geqslant 200,$$

respectively.

A test based on a vector of certain weighted residual empirical processes has been proposed by Koul and Ling (2006). It can be used to test the normality of the GARCH innovation distribution. To compare the performance of Koul and Ling's (2006) test with JB test, we use the same model AR(1)-GARCH(1,1) with parameters of $(\phi_{01}, \alpha_{00}, \alpha_{01}, \beta_{01}) = (0.5, 0.025, 0.25, 0.5)$. We take sample size 200 and 400 with 1000 replications. The Null distribution of η_t is N(0,1) and its alternatives are set as:

$$H_{a1} : \eta_t \sim \sqrt{3/5}\, t(5); \quad H_{a2} : \eta_t \sim \sqrt{1/2}\, t(4); \quad H_{a3} : \eta_t \sim \sqrt{1/3}\, t(3);$$

$$H_{a4} : \eta_t \sim \text{double exponential} \;; \quad H_{a5} : \eta_t \sim [0.5N(-3,1){+}0.5N(3,1)]/\sqrt{10}\,.$$

Table 4.3 lists the JB test results based on corrected critical values by Lu (2001). Table 4.4 lists the JB test results based on χ_2 critical values. It can be seen that JB test based on both corrected critical values and χ_2 critical values are much more powerful than Koul and

Table 4.3: JB statistic for AR(1)-GARCH(1,1) with corrected critical value

	n=200		n=400	
	$\alpha = 0.1$	$\alpha = 0.05$	$\alpha = 0.1$	$\alpha = 0.05$
Null	0.088	0.043	0.096	0.052
H_{a1}	0.842	0.804	0.990	0.982
H_{a2}	0.960	0.933	0.996	0.993
H_{a3}	0.988	0.985	1	1
H_{a4}	0.977	0.963	0.999	0.998
H_{a5}	1	1	1	1

Ling's (2006) test. In particular, JB test has substantial power gains under H_{a1} and H_{a2}. The sizes are conservative when χ_2 critical values are used. While applying corrected critical values, the sizes are very close to the nominal significance level for $n = 400$, though the sizes are still somewhat conservative for $n = 200$.

Table 4.4: *JB* statistic for AR(1)-GARCH(1,1) with χ_2 critical value

		n=200			n=400	
	$\alpha = 0.1$	$\alpha = 0.05$	$\alpha = 0.01$	$\alpha = 0.1$	$\alpha = 0.05$	$\alpha = 0.01$
Null	0.064	0.032	0.015	0.063	0.034	0.008
H_{a1}	0.842	0.789	0.710	0.989	0.978	0.958
H_{a2}	0.929	0.908	0.875	0.997	0.996	0.991
H_{a3}	0.982	0.977	0.963	1	1	1
H_{a4}	0.942	0.915	0.856	0.999	0.997	0.994
H_{a5}	1	1	1	1	1	1

Assume that the innovation distribution has a uniformly continuous density function $f(x)$ which is unknown. Let h_n be a sequence of positive numbers and K be a probability density function (kernel) with mean 0 and variance 1. Then the kernel density estimation of $f(x)$ based on the residuals is defined as

$$\tilde{f}_n(x) = \frac{1}{nh_n} \sum_{t=1}^{n} K\left(\frac{x - \tilde{\eta}_t}{h_n}\right), \quad x \in \mathbb{R} .$$

Its counterpart based on iid innovations is defined as

$$f_n(x) = \frac{1}{nh_n} \sum_{t=1}^{n} K\left(\frac{x - \eta_t}{h_n}\right), \quad x \in \mathbb{R}.$$

Under some regular conditions on the kernel density K and the bandwidth h_n, Yu and Lu (2005) obtain that

$$\sup_{x\in\mathbb{R}} |\tilde{f}_n(x) - f_n(x)| = o_P(1)$$

which implies also

$$\sup_{x\in\mathbb{R}} |\tilde{f}_n(x) - f(x)| = o_P(1).$$

References

[1] R. T. Baillie, Long memory processes and fractional integration in econometrics. J. Econometrics **73** (1996), 5–59.

[2] R. T. Baillie, T. Bollerslev, and H.O. Mikkelsen, Fractionally Integrated Generalized Autoregressive Conditional Heteroskedasticity. J. Econometrics **74** (1996), 3–30.

[3] I. Berkes and L. Horváth, The rate of consistency of the quasi-maximum likelihood estimator. Statistics and Probability Letters **61** (2003), 133-143.

[4] I. Berkes and L. Horváth, The efficiency of the estimatiors of the parameters in GARCH processes. Ann. Stat. **32** (2004), 633–655.

[5] I. Berkes, L. Horváth, and P. Kokoszka, Probabilistic and statistical properties of GARCH processes. Fields Institute Communications **44** (2004), 409–429.

[6] P. Billingsley, Convergence of probability measures. 2nd Edition. Wiley, New York 1999.

[7] T. Bollerslev, Generalized autoregressive conditional heteroskedasticity. J. Econometrics **31** (1986), 307–327.

[8] P. Bougerol and N. Picard, Strict stationarity of generalized autoregressive processes. Ann. Probab. **20** (1992a), 1714–1730.

[9] P. Bougerol and N. Picard, Stationarity of GARCH processes and of some nonnegative time series. J. Econometrics **52** (1992b), 115–127.

[10] P. Brockwell and R. Davis, Time Series: Theory and Methods. 2nd Edition. Springer-Verlag, New York 1991.

[11] J. Y. Campbell, A. W. Lo, and A. C. MacKinlay, The Econometrics of Financial Markets. Princeton University Press 1997.

[12] Z. Ding, C. W. J. Granger, and R. F. Engle, A Long Memory Property of Stock Market Returns and a New Model. Journal of Empirical Finance **1** (1993), 83–106.

[13] R. F. Engle, Autoregressive conditional heteroscedasticity with estimates of the variance of U.K. inflation. Econometrica **50** (1982), 987–1008.

[14] R. F. Engle, D. M.Lilien, and R. P. Robins, Estimating Time Varying Risk Premia in the Term-Structure: the ARCH-M Model, Econometrica **55** (1987), 391–407.

[15] C. Francq and J. M. Zakoïan, Maximum likelihood estimation of pure GARCH and ARMA-GARCH processes. Bernoulli **10** (2004), 605–637.

[16] W. Fuller, Introduction to Statistical Time Series. 2nd Edition. John Wiley, New York 1996.

[17] C. Gourieroux, ARCH models and financial applications. Springer-Verlag, New York 1997.

[18] K. Hitomi, (1997). Smoothed least absolute deviation estimator and Smoothed censored least absolute deviation esitmators. *mimeo.*

[19] C. M. Jarque and A. K. Bera, A test for normality of observations and regression residuals. International Statistical Review **55** (1987), 163–172.

[20] H. L. Koul and S. Ling, Fitting an error distribution in some heteroscedastic time series models. Ann. Stat. **34** (2006), 994–1012

[21] R. J. Kulperger and H. Yu, High moment partial sum processes of residuals in GARCH models and their applications. Ann. Stat. **33** (2005), 2395-2422.

[22] S. Ling, Self-weighted and local quasi-maximum likelihood estimators for ARMA-GARCH/IGARCH models. To appear in J. Econometrics (2006).

[23] S. Ling and W. K. Li, On fractionally integrated autoregressive moving-average time series models with conditional heteroscedasticity. J. Am. Stat. Assoc. **92** (1997), 1184–1194.

[24] S. Ling and W. K. Li, Limiting distributions of maximum likelihood estimators for unstalbe ARMA models with GARCH errors. Ann. Stat. **26** (1998), 84–125.

[25] S. Ling and W. K. Li, Asymptotic inference for unit root processes with GARCH(1,1) errors. Econometric Theory **19** (2003), 541–564.

[26] S. Ling and McAleer Asymptotic theory for a vector ARMA-GARCH model. Econometric Theory **19** (2003), 280–310.

[27] P. G. Lu, Identification of distribution for ARCH innovation. Master's Project. University of Western Ontario 2001.

[28] P. G. Lu and H. Yu, Asymptotic properties of quasi-maximum likelihood estimators of an ARMA-(I)GARCH model. Preprint 2006.

[29] T. C. Mills, The cconometric Modelling of financial Time Series. Cambridge University Press, Cambridge 1999.

[30] D. B. Nelson, Stationarity and Persistence in the GARCH(1,1) Model. Econometric Theory **6** (1990), 318–334.

[31] D. B. Nelson, Conditional Heteroscedasticity in Asset Returns: A New Approach, Econometrica **59** (1991), 347-370.

[32] P. Rossi, Modelling Stock Market Volatility: bridging the gap to continuous time. Academic Press, Toronto 1996.

[33] A. A. Weiss, Asymptotical theory for ARCH models: estimation and testing. Econometrics Theory **2** (1986), 107–131

[34] H. Yu, Analyzing residual processes of (G)ARCH time series models. Fields Institute Communications **44** (2004), 469–486.

[35] H. Yu and P. G.Lu, High moment partial sum processes of residuals in ARMA-(I)GARCH models. Preprint. 2006.

Nonparametric Classification and Probabilistic Classifier with Environmental and Remote Sensing Applications

Jun Yu and Bo Ranneby

Centre of Biostochastics
Swedish University of Agricultural Sciences
SE-901 83, Umeå, Sweden
E-mail: jun.yu@sekon.slu.se, bo.ranneby@sekon.slu.se

Abstract

National and international policies today require environmental monitoring and follow-up systems that detect, in a quality assured way, changes over time in land use and landscape indicators. Remote sensing of satellite images offers great potential to assess wall-to-wall changes in the health of ecosystems and identify risks. Questions related to environmental health and spatial patterns call for new statistical tools. We present in this chapter some new developments on the classification of land use and spatial indicators using multispectral and multitemporal satellite images. They are developed under non-standard conditions - conditions under which many statistical methods do not work properly but frequently appear in environmental and remote sensing applications. Error rates of traditional remote sensing classification methods are usually quite high but can be improved by (1) denoising the images using the wavelet transform, (2) reclassification using Markov random field approaches, or (3) applying new classification algorithms based on nonparametric classifiers. In order to assess quality of classification at pixel level, a new concept, the probabilistic classifiers, is introduced. These classifiers are useful for measuring uncertainty at pixel level and obtaining reliable area estimates locally. Results from both simulation studies and real applications are presented.

1 Introduction

Environmental pressures on ecosystems and biodiversity, the continuous loss of suitable habitats for different species and the threat of irreversible damages jeopardize the ambitions of a sustainable development. Environmental work needs to be enforced at a global, national, regional and local level. This includes the implementation of different measures as well as policymaking and the formulation of long-term strategies at all levels.

The environmental strategies also include the formulation of environmental quality objectives at all levels which in turn accentuates the need for accurate, rapid, comprehensive and cost-effective monitoring and follow-up systems.

Spatial patterns at different scales are important characteristics of environmental quality. Spatial elements that indicate high environmental quality are usually quite rare and important changes are sometimes very subtle. Existing monitoring programs linked to landscape surveillance, e.g. the Swedish National Forest Inventory (NFI), do not permit evaluation of data at a local level or of smaller regions. The landscape perspective requires wall-to-wall analyses that are scale-independent, i.e. the results can be used at different levels and the accuracy of the measurements can be clearly defined.

Remote sensing of satellite images offers great potential to assess wall-to-wall changes in the health of ecosystems and identify risks. So far, however, little of this potential has been realized.

Questions related to environmental health, biodiversity, and spatial patterns in general require new statistical methods. Methods and models must be developed under so called non-standard conditions - conditions under which many statistical methods do not work properly but frequently appear in environmental and remote sensing applications.

In this chapter the focus will be on the classification of land use and spatial indicators using multispectral and multitemporal satellite images. Section 2 is devoted to a brief review of classification methods, both noncontextual and contextual, and also simulation studies on classifiers using the wavelet transform and Markov random field approaches. New developments of nonparametric classifiers will be presented in Sections 3 and 4. A real application of the new methodologies is shown in Section 5. Some final remarks are given in Section 6.

2 Classification

2.1 Background

Supervised classification of remote-sensing images has been widely used as a powerful means to extract various kinds of information concerning the earth environment. The objective of supervised classification in remote sensing is to identify and partition the pixels comprising the noisy image of an area according to its class, with the parameters in the model for pixel values estimated from training samples (ground truths).

Usually, the spectral signature is the main aspect of the classes used to classify the pixels. For multispectral image data acquired by remote sensing devices, one should be aware that they are very complex entities which have not only spectral attributes (with correlated bands) but also spatial attributes. Proper utilization of this spatial contextual information, in addition to spectral information, can improve the classification performance significantly in many applications compared to the conventional noncontextual rules such as discriminant analysis and the k-nearest neighbour classifier. See [33, 46] for a comprehensive review of these methods.

In noncontextual rules, each pixel is classified solely on its spectral intensities and it does not account for spatial dependence. These approaches effectively assume the spectral intensities in neighbouring pixels to be independent, therefore important information from neighbouring pixels is neglected. Such approaches might be reasonable if the pixel-sizes are large or when the densities of the spectral intensities are well separated for different classes. In the classification of the states of forests, for example, the densities of the spectral intensities are seldom well separated. Also spatial dependency grows more important as the spatial resolution increases. Therefore, various methods for contextual classification of multi-spectral scanner data have been developed in order to take the neighbouring pixel information into account. In particular, they have been introduced to cope with segmentations and classifications of remotely sensed data (see [13, 23, 38, 48, 54] and therein). They tend, however, to be computationally intensive, due to either estimation of autocorrelation parameters and transition probabilities in pixelwise classification (such as Haslett's method studied in [23]) or iterations in simultaneous classification of all pixels (such as the iterated conditional modes (ICM) and Gibbs sampler algorithms studied in [13] and [53]. Moreover, contextual methods usually have special requirements on the ground truths that provides information about the neighboring pixels, such as homogeneous crosses of pixels. This is not easily accessible in practice, often due to the limitation of resources (such as financial, time, well-trained surveyors) and also the inventory season.

Despite the computational effort they require, contextual classifiers that rely on macro models for the behaviour of the classes, such as Markov random fields (MRF), are of great interest as a smoothing process over an initial classification image. Previous studies [12, 13, 38] have shown that the ICM method (introduced by Besag [4]) has a promising performance on achieving high accuracy classifications, with less cost on computational effort compared to other Bayesian based contextual classifiers.

Another problem with the parametric classification methods is the assumption of (approximative) normality for the feature vector, which is often not fulfilled in remote sensing applications. For non-normal densities it is nevertheless possible to use maximum likelihood (ML) classifiers but with other models for the class distributions. For example, in Taxt et al. [51] mixtures of normal distributions are used for classification of images. Their results indicate a substantial increase in correct classification rates compared to classification under normal densities. It is worth noting that mixture models are examples of models when traditional estimation methods such as ML have a tendency to fail. It has been observed that in remotely sensed data feature vectors quite frequently possess bimodal or multimodal empirical distributions. If the training sets are objectively selected empirical distributions of this kind will become even more common. Therefore general estimation methods, which give efficient and robust estimates also when traditional methods break down, are of fundamental importance for parametric classification. For recent achievements in this direction we refer to Ranneby et al. [43], Jimenez & Yukich [29] and Teterukovskiy & Ranneby [52].

Recently, the wavelet transform has been attracting attention in diverse areas such as medical imaging, pattern recognition, data compression, numerical analysis, and signal processing, especially for non-stationary signal analysis applications, since it provides information in both spatial and frequency domains due to its inherent nature of space-frequency analysis (see e.g. [1, 2, 9, 39]). In wavelet representations, the images are decomposed using basis functions localised in spatial position, orientation, and spatial frequency (scale). Because satellite images consist of thousands of pixels pertaining to the energy of light reflected from the ground, it displays non-stationary signal characteristics. This specific characteristic of satellite images and the characteristics of the wavelet transform motivate the investigation of use of the wavelet transform in target classification of scenes. Related studies have been done in e.g. [30, 50, 56].

In a simulation study, Yu & Ekström [57] evaluated various traditional classifiers by using the wavelet transform on multispectral images, and found that no matter which type of spatial correlation between pixels the image has and which kind of conventional classification method one uses, the classification by using the wavelet transform (specially the wavelet shrinkage) yields a much lower classification error than that working directly on the original image data, as soon as we have a class image with some structure in it. It is worth noting that this study is based on a model for feature vector assuming normality of class distributions.

Concerning the conventional nonparametric classification methods such as the k-nearest neighbor classifier, its successful combination with the wavelet transform and information theory (Ranneby & Yu [44, 45]) has shown a great potential in remote sensing applications. The methodology and current results will be shown in the following sections.

By a simulation study (also based on the assumption of normal distributed feature vector), Teterukovskiy & Yu [53] has shown that the combination of the spectral-contextual classifiers (using ICM and Gibbs sampler) can improve to a great extent the accuracy of conventional non-contextual classification methods. The use of the Gibbs sampler for reclassification is particularly justified if the initial classification performed poorly.

In the next section we will introduce the conventional classification methods (noncontextual and contextual), the wavelet transform, and then provide a simulation study for investigating the performance of wavelet-based classifiers and MRF-based reclassification.

2.2 Methods

We assume that a given pixel from the scene belongs to one of a fixed number of classes, say, $1, \cdots, K$. The proportion of pixels belonging to class c in the population under study is denoted by π_c, which is usually unknown. Each pixel gives rise to certain measurements (e.g., spectral signatures in remote sensing images, or their various transformations), which form the feature vector $\mathbf{X} \in \mathbb{R}^d$. Our task is to allocate each pixel into one of the classes in $\{1, \cdots, K\}$, on the basis of the observed value $\mathbf{X} = \mathbf{x}$ and/or prior information about the classes.

2.2.1 Conventional classification methods

Three conventional classification methods were used in our simulation studies: linear discriminant analysis (LDA), quadratic discriminant analysis (QDA), and the k-nearest neighbour rule (k-NN), where the first two are known as parametric classification rules and the last one as a nonparametric rule.

In parametric models, the feature vectors from class c are assumed to be distributed according to the density $p_c(\mathbf{x})$. Then the posterior distribution of the classes after observing $\mathbf{x}$ is $p(c|\mathbf{x}) \propto \pi_c p_c(\mathbf{x})$.

If we assume the probability model in which the observations for class c are (d-dimensional) multivariate normal with mean $\boldsymbol{\mu}_c$ and covariance matrix Σ_c, the Bayes rule is to allocate a future observation $\mathbf{x}$ to the class that minimises

$$Q_c = -2\log p(c|\mathbf{x}) = (\mathbf{x} - \boldsymbol{\mu}_c)^{\mathrm{T}}\Sigma_c^{-1}(\mathbf{x} - \boldsymbol{\mu}_c) + \log|\Sigma_c| - 2\log\pi_c.$$

This method is known as *quadratic discriminant analysis* due to the quadratic form of $\mathbf{x}$ it involves.

When the classes have a common covariance matrix Σ, Q_c becomes a linear function of $\mathbf{x}$ plus a quadratic term which does not depend on the class. So minimizing Q_c is equivalent to maximizing the linear terms

$$L_c = 2\mathbf{x}^{\mathrm{T}}\Sigma^{-1}\boldsymbol{\mu}_c - \boldsymbol{\mu}_c^{\mathrm{T}}\Sigma^{-1}\boldsymbol{\mu}_c + 2\log\pi_c,$$

which leads to the *linear discriminant analysis.*

In practice, we replace $\boldsymbol{\mu}_c$, Σ_c or Σ, and π_c with their estimates (e.g., maximum likelihood) by using a training sample.

The assumption of the normality reflects the following situation [33]: there is an "ideal prototype" (a point in $\mathbb{R}^d$) of class c and all class members are distorted versions of it. Slight distortions are more likely to occur than big distortions, causing more objects to be located in the close vicinity of the ideal prototype than far away from it. The prototype is represented by the mean $\boldsymbol{\mu}_c$ and the scatter of the points around it is associated with the covariance matrix Σ_c.

B. Nonparametric model

In nonparametric designs we do not assume any model of the probability distributions. There is a number of nonparametric classifiers based on nonparametric estimates of the class densities or of the log posterior, such as kernel methods, orthogonal expansion, projection pursuit, and others. Here we use the simple adaptive kernel method which gives a classifier known as the *k-nearest neighbor rule.* The k-NN rule is introduced by Fix and Hodges [22], which is one of the most theoretically elegant and yet simple classification techniques. It is based on finding the k nearest (usually in the Euclidean distance) pixels from the training samples (prototypes), and taking a majority vote amongst the classes of these k samples, or equivalently, estimating the posterior distributions $p(c|\mathbf{x})$ by the proportions of the classes amongst the k samples. The nearest neighbor rule (when k is 1) is of special interest in our later applications.

Nonparametric classification is often associated with the notion "prototype". A prototype is a representative point of certain class in the feature space $\mathbb{R}^d$ and is attached with both field and satellite information. Different classes can have different number of prototypes, depending on the property of field data or ground truth we have. Prototype selection plays a key roll in the nonparametric classification. k-NN is a prototype classifier.

Specifically, assuming the number of prototypes for class c is N_c and $N = \sum_1^K N_c$ is the number of all prototypes. For an arbitrary pixel with observed feature value $\mathbf{x}$, let $B(\mathbf{x})$ denote the hypersphere with centre $\mathbf{x}$ and containing exactly k prototypes and $||B(\mathbf{x})||$ the volume of $B(\mathbf{x})$. The number of prototypes in $B(\mathbf{x})$ from class c is denoted by k_c. The unconditional probability density of $\mathbf{x}$ and the class-conditional probability density for class c can be approximated as

$$p(\mathbf{x}) \approx \frac{k}{N||B(\mathbf{x})||} \quad \text{and} \quad p_c(\mathbf{x}) \approx \frac{k_c}{N_c||B(\mathbf{x})||},$$

respectively. Then the posterior probabilities are obtained as

$$p(c|\mathbf{x}) = \frac{p_c(\mathbf{x})\pi_c}{p(\mathbf{x})} \approx \frac{\frac{k_c}{N_c||B(\mathbf{x})||} \cdot \frac{N_c}{N}}{\frac{k}{N||B(\mathbf{x})||}} = \frac{k_c}{k},$$

which implies that the k-NN classification rule assigns the class label independent of the volume $||B(\mathbf{x})||$.

When we are dealing with sparse or rarely occurring objects, however, it is unrealistic to assume that the number of prototypes for these classes is large so the approximation above may be violated. Thus there is an obvious risk that such classes will be underestimated. To overcome this problem alternatives of deriving the probabilities are needed.

The k-NN classifier uses the whole prototype set of any class to choose the neighborhood, which differs from using the k-nearest neighbor density estimate for each class. Ties in distances can occur with finite-precision data. One solution to this problem is to include in the vote all distances equal to the kth largest.

Notice that "nearest" implies a certain metric over the feature space $\mathbb{R}^d$. No matter which metric is used, the k-NN rule is applied in the same way. However, this metric has to be defined properly because the closeness in the feature space is no longer so simple as in Euclidean case. This is connected to the similarity in spatial statistics, which means that the smaller the distance, the higher the similarity between input data and the prototype.

2.2.2 Contextual Bayesian classification

By contextual classification it is understood that the class of a given pixel is based both on the spectral value of this pixel and the spectral values and/or classes of its spatial neighbours. In most classification tasks where some sort of structure can be observed in the image, this additional information proves to be very important for the overall quality of classification. In general, the neighbours are not only spatial located but can also be temporally connected. In order to include information from the neighbors it is necessary to specify a joint model for the behaviour of all classes. One common choice for such model is a Markov random field. In this model one supposes that the class of pixel $\mathbf{s}$, given the classes in the rest of the image, depends only on the classes in the neighbors $\partial\mathbf{s}$ of $\mathbf{s}$. Naturally, by extending the neighborhood to the whole image, one can consider *any* spatial random model as a Markov random field. If the neighborhood is taken small enough, and the dependence between neighbors is chosen wisely, the inference on the image using MRF model becomes possible. Classification results obtained by these methods effectively outperform the results of non-contextual classifiers.

We have chosen the following form of dependence between neighboring classes [4]. The second-order neighborhood $\partial\mathbf{s}$ is considered, and the probability

$$P(C(\mathbf{s}) = c) \propto \exp\left(\beta \times \sum_{\mathbf{t} \in \partial\mathbf{s}} I(C(\mathbf{t}) = C(\mathbf{s}))\right), \tag{2.1}$$

where $I(\cdot)$ is the indicator function. Using this representation for the prior distribution of the classes we can write the posterior distribution as

$$p\left(c|\mathbf{x}\right) \propto \Phi_c(\mathbf{x}) \times \exp\left(\beta \times \sum_{\mathbf{t} \in \partial\mathbf{s}} I(C(\mathbf{t}) = C(\mathbf{s}))\right).$$

Here $\Phi_c(\mathbf{x})$ denotes the Gaussian distribution with mean and covariance matrix depending on the classes. The parameter β indicates the strength of attraction within classes.

Maximization of the posterior distribution gives us the maximum a posteriori estimate (MAPE). Unfortunately, direct maximization of the distribution $p(c|\mathbf{x})$ requires unrealistic computational effort. Therefore, several algorithms for approximation of MAPE were suggested (see [55]). Two approaches are considered in this paper: one quickly giving the *local* maximum of the posterior distribution, and another - approaching the *global* maximum but on the expense of computing time.

A. Iterated conditional modes

The ICM method was described by Besag [4], and is, in fact, a statistical reformulation of the coordinate-wise maximal descent method well-known in optimization. It is often used in image analysis applications since it is fast and easy to implement. Instead of maximizing $p(c|\mathbf{x})$ the maximization of the marginal distributions $p\left(C(\mathbf{s})|\mathbf{x}\right)$ is considered. For each pixel $\mathbf{s}$ such marginal distribution is concentrated in the number of atoms equal to the number of classes. Choosing the mode of this distribution, and then repeating the procedure for all pixels (possibly, several times) will soon produce a configuration which does not have neighbors with higher posterior probability. Such configurations are by definition the local maxima of the posterior distribution.

Since this method searches for the local maximum, it strongly depends on the initial configuration. This is a drawback on one hand, but can also be seen as a guarantee that the classification will not deviate strongly from the true one if the prior distribution is chosen somewhat incorrectly. In cases where we want to obtain the global maxima of the posterior distribution, the ICM is inappropriate. A simple modification of this algorithm provides (at least, theoretically) means to obtain a global maximum.

B. The Gibbs sampler and simulated annealing

Instead of choosing the mode of the marginal distribution, Geman and Geman [26] suggested sampling from it. They called their algorithm the Gibbs sampler since it has an obvious connection with the Gibbs fields in statistical physics. They have proved that provided all pixels are visited frequently enough, the stationary distribution of the Gibbs sampler is the posterior distribution $p(c|\mathbf{x})$. For details about the Gibbs sampler we refer to the original paper and a book by Winkler [55].

We shall use the modification of the Gibbs sampler called *simulated annealing*. The idea of this method is to emphasise the modes of the posterior distribution by modifying the conditional distributions which are used in the Gibbs sampler in the following way. Let us write out the targeted posterior distribution in the form

$$p(c|\mathbf{x}) = \frac{1}{Z}\exp(-H(c,\mathbf{x})),$$

where Z is the normalizing constant. Then the distribution

$$p^{\mathrm{T}}(c|\mathbf{x}) = \frac{1}{Z^{\mathrm{T}}}\exp(-\frac{H(c,\mathbf{x})}{T}), \quad T > 0$$

will for small T have the same maxima as $p(c|\mathbf{x})$ but easier detectable. This observation is the basis of the simulated annealing. The parameter T is called the *temperature*. If, in the course of running the Gibbs sampler, the temperature is lowered according to a certain *cooling schedule*, then the limit distribution of the sampler will be concentrated on the set of maximizers of the posterior distribution. When we in the section devoted to results refer to the Gibbs sampler, it is always understood that the sampler is coupled with simulated annealing with a linear cooling schedule.

2.2.3 Wavelet transform

In our simulation study, we adopt the two-dimensional discrete wavelet transform (2-D DWT) due to its fast computational algorithm for both decomposition and reconstruction of the image and also the discrete nature of the image pixels. The discrete wavelet transform results from multi-resolution analysis [36], which involves decomposition of an image in frequency channels of constant bandwidth on a logarithmic scale. It has advantages such as similarity of data structure with respect to the resolution (i.e., the higher resolution image includes the lower resolution images) and available decomposition at any level (scale). These properties are beneficial to remotely sensed data analysis where one may wish to see macroscopic characteristics of the scene at first stage using a low resolution level and then analyse an area of interest in detail using an appropriate higher resolution level [56].

"Pixels" in wavelet transformed sub-images (i.e., wavelet coefficients) represent the characteristics in sub-frequency bands, which also represent the characteristics of the pixels in the original image. Since these new "pixels" bear information also on the neighbouring pixels, the use of conventional classification methods on the wavelet images rather than on the original image seems appropriate. Furthermore, a repeated wavelet transform to a specific image may serve to classification of targets efficiently. This is an advantage compared to contextual methods which require extensive computation time on correlation between adjacent pixels.

Alternatively, by using wavelet shrinkage we can effectively remove the noise from images and therefore improve the classification results.

The wavelet transform has also been used to achieve image compression and noise removal, where the efficiency of the representation is important for the clarity of the analysis. The technique is based on the principle of *wavelet shrinkage* [17], which refers to taking the wavelet transform of the image, removing noise by shrinking wavelet coefficients towards zero, and inverting the thresholded wavelet transform coefficients. Much study has been made concerning the choice of the shrinkage function and the threshold value.

There are two basic choices of the shrinkage function, that is, *soft* shrinkage:

$$\delta_\lambda^S(x) = \begin{cases} \operatorname{sign}(x)(|x| - \lambda), & \text{if } |x| > \lambda, \\ 0, & \text{if } |x| \leqslant \lambda, \end{cases}$$

and *hard* shrinkage:

$$\delta_\lambda^H(x) = \begin{cases} x, & \text{if } |x| > \lambda, \\ 0, & \text{if } |x| \leqslant \lambda, \end{cases}$$

where $\lambda > 0$ is the threshold. While at first sight the hard shrinkage may seem a more natural approach, the continuity of the soft shrinkage has important advantages [28]. Several other functions which preserve the highest coefficients and have more smooth transition from noisy to important coefficients have been proposed by Gao and Bruce [25] and Gao [24]. In general, these sophisticated shrinking schemes are computationally more intensive. Soft shrinkage is a trade-off between a fast and straightforward method and a continuous approach [28].

The simplest scheme for threshold assessment uses a single threshold for all detail wavelet coefficients. Donoho and Johnstone [16] gave two ways to set this threshold: a "minimax" threshold and the "universal" threshold. The minimax threshold minimizes a bound on the asymptotic risk while the universal threshold ensures that, asymptotically, all detail coefficients are annihilated. They also developed a technique, called SUREShrink, to select a threshold by minimizing the Stein Unbiased Risk Estimator at each resolution level. Gao and Bruce [25] studied the threshold selection for the so-called firm shrinkage. More references can be found in Jansen [28].

In general, wavelet shrinkage methods have a number of theoretical advantages, including near-optimal mean square error and near-ideal spatial adaptation. The process of shrinking coefficients is much like the process of keeping only few important coefficients in image compression algorithms, the aim of which is to form a representation of an image using as little storage as possible and subject to a specified error of image quality. See [7, 10, 15, 17] and the references therein for details. The wavelet shrinkage methodology has been extended to a broader class of models and situations, including classification and discriminant analysis (see e.g. [8, 47]).

2.3 Simulation study

Simulation studies [53, 57] with the following objectives have been performed

- to compare conventional classification methods with wavelet based methods;
- to evaluate the ability of contextual classifiers (the ICM and the Gibbs sampler) to improve the initial classifications based on the chosen conventional classification methods.

The studies were based on different types of scenes (ground truths) together with multi-spectral remotely sensed data with correlated frequency bands. The observed vectors of spectral reflectance in pixels, given the classes, have varying degrees of spatial dependency, from independence to strong autocorrelation. The data are generated from bivariate-normal class distributions. Moreover, the data are generated in such a way that the spectral intensities are highly overlapped. Therefore, the training sets will be highly overlapping in the representation space.

2.3.1 Autocorrelation model and scenes

In this subsection we describe how the data that we used to illustrate the classification procedures introduced in the previous section, were simulated.

The stochastic model for the d-dimensional spectral intensity vector, $\{X(\mathbf{s}), \mathbf{s} \in \mathbb{R}^2\}$, is assumed to be spatial stationary and isotropic, with the following structure:

$$X(\mathbf{s}) = Y(\mathbf{s}) + \varepsilon(\mathbf{s}), \tag{2.2}$$

where $\varepsilon(\mathbf{s})$ is the autocorrelated noise process which accounts for the class independent variations, and $\{Y(\mathbf{s}), \mathbf{s} \in \mathbb{R}^2\}$ are the "true" intensities, conditionally independent given their classes, and also independent of the noise process. The distribution of $Y(\mathbf{s})$ given its class at pixel $\mathbf{s}$, $C(\mathbf{s}) = c$, is multivariate normally distributed with mean vector $\boldsymbol{\mu}_c$ and covariance matrix $(1-\theta)\lambda_c\Gamma$, where θ and λ_c's are proportion parameters and Γ is a fixed $d \times d$ positive definite matrix. Thus, the covariance matrices of Y are allowed to be different for different classes. The assumption of the distribution for the noise vectors $\varepsilon(\mathbf{s})$ is as follows: $\varepsilon(\mathbf{s}) \sim N(\mathbf{0}, \theta\Gamma), Cov(\varepsilon(\mathbf{s}), \varepsilon(\mathbf{t})) = \rho^{|\mathbf{s}-\mathbf{t}|}\theta\Gamma$, where $|\mathbf{s}-\mathbf{t}|$ is the Euclidean distance between pixels $\mathbf{s}$ and $\mathbf{t}$, and ρ is the spatial autocorrelation between the noise vectors.

Hence, model (2.2) implies that the spectral intensity X is multivariate normal with

$$\begin{aligned} E(X(\mathbf{s}) \mid C(\mathbf{s}) = c) &= \boldsymbol{\mu}_c, \\ Var(X(\mathbf{s}) \mid C(\mathbf{s}) = c) &= \Sigma_c = [(1-\theta)\lambda_c + \theta]\Gamma, \\ Cov(X(\mathbf{s}), X(\mathbf{t}) \mid C(\mathbf{s}), C(\mathbf{t})) &= Cov(\varepsilon(\mathbf{s}), \varepsilon(\mathbf{t})) = \rho^{|\mathbf{s}-\mathbf{t}|}\theta\Gamma. \end{aligned} \tag{2.3}$$

Three different true maps, each of size 40×40 pixels, were considered to represent the ground truth of spatial distributions of the classes. They are denoted by RANDOM, STRIPE, and TRUE, as shown in Figure 2.1. The number of classes is $K = 3$. The priors are the same on all scenes: $\pi_1 = 0.2, \pi_2 = 0.3$, and $\pi_3 = 0.5$. At the RANDOM scene the classes of the pixels were determined by independent random drawings from the prior distribution, giving a completely chaotic pattern. The STRIPE scene is at the other extreme: all pixels from a given row have the same class. The number of rows of the three classes are 8, 12, and 20, respectively, according to the prior distribution, and the rows are arranged in two stripes for each class. The TRUE scene is a small selection from a real scene and is of most interest for our simulation study.

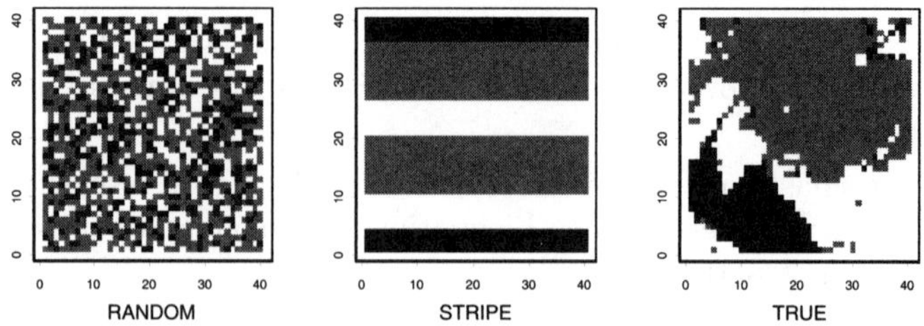

Figure 2.1: Three true maps. Legends: Class 1 = black; Class 2 = white, Class 3 = grey

Here we consider an image with two correlated bands, that is, $d = 2$. We choose the values of λ_c's as 0.4, 1.4, and 1.0, respectively, which gives different covariance matrices for the three classes. The class expectations and the matrix Γ are chosen as

$$\boldsymbol{\mu}_1 = \begin{pmatrix} 125 \\ 128 \end{pmatrix}, \ \boldsymbol{\mu}_2 = \begin{pmatrix} 130 \\ 135 \end{pmatrix}, \ \boldsymbol{\mu}_3 = \begin{pmatrix} 127 \\ 110 \end{pmatrix}, \ \Gamma = \begin{pmatrix} 4 & 8 \\ 8 & 81 \end{pmatrix}.$$

As a measure of the spatial dependency of the noise term we use the nearest neighbor correlation for class 3, $\alpha = \rho\theta$, as specified in the following table.

α	0	0.25	0.72
ρ	0	0.5	0.9
θ	0.5	0.5	0.8

Thus, $\alpha = 0$ implies spatial independency between pixels, which is the case most appropriate for the noncontextual classification methods. The value of the autocorrelation parameter ρ depends on the pixel size. The larger the pixels, the smaller the value. We refer to [27] for an illustration of this relationship.

2.3.2 Results

In the Monte Carlo study, we generated 100 replicates of a noisy image with two correlated bands, for each particular choice of scene and α-value. For training data, we follow the approach of Klein and Press [32], by taking observations from all the pixels in some columns. Specifically, columns 3, 8, 13, 18, 23, 28, 33, and 38 were chosen, which consists of 1/5 of the scene. For each replicate, the parameters $\boldsymbol{\mu}_c$ and Σ_c in the model were estimated (using maximum likelihood) and plugged into the classification rules, and all pixels, except those in the training set, were classified.

A. Wavelet

The total misclassification rates for a given scene are estimated by averaging the relative number of misclassified pixels in each replicate. Tables A.1-A.5 in Appendix show the mean values of misclassification percentages for the TRUE scene and those five different structures of spatial dependency. In all these tables, each column represents one classification method mentioned in Section 2.2.1, where the prior distribution in LDA and QDA was chosen to be the class proportion coming from the training data, and the number of neighbours in k-NN was chosen to have three different alternatives: $k = 1, 3$ or 10, in order to check the neighbouring effect. The rows present the image data types, from the original one to the wavelet images with different levels (from $n = 1$ to 3) as well as denoised images based on the wavelet shrinkage at different degrees. In the wavelet shrinkage, the soft shrinkage was chosen and the threshold was selected as the 50%, 60%, or 70% quantile of the absolute values of the wavelet coefficients, and then the coefficients under this threshold, except those at the coarsest level, were removed. Note that this type of wavelet shrinkage corresponds to the TOP method, with which one will have a good control [6]. The wavelet function in the 2D-DWT was chosen to be the *haar* wavelet because of the simple class structure of the true map. Thus, as an example, "shrink.n3.60" stands for the reconstructed image by using 2D-DWT up to level 3 and then shrinking the wavelet coefficients with the threshold equal to the 60% quantile. This notation for the row name is also valid for Table A.6, though the quantile level was different there. The best results among classification methods for each image data type (row wise) are indicated with bold face, and the total best in every table is even underlined.

What we can observe and then conclude from Tables A.1-A.5 is as follows:

- Classification by using the wavelet shrinkage results in much lower mean classification error than that working directly on the original image data (the reduction can be up to 56%), and level 3 always performed better than the lower levels, except for the case when $\alpha = 0$. Therefore the wavelet transform should be executed at least twice in wavelet shrinkage.
- Classification based on wavelet sub-images gives us slightly better results only at level 1 and for small α-values than working directly on the original image data. This is somewhat unexpected, by considering our previous study [58] where the classification based on

wavelet sub-images was very convincing. One probable reason is that the underlying true scene there was more simple and regular, which is more comparable to the STRIPE scene in this study.

- In general, the mean misclassification rate increases as the spatial dependency between pixels strengthens. There are exceptions, however, in particular when $\alpha = 0.40$. This is probably caused by the fact that the empirical variation in this case is clearly higher than the theoretical one.
- QDA and LDA work better than the k-NN classifier, especially in the case when $k = 1$. This is quite natural because our model is multivariate normal which fits the basic requirements in LDA and QDA.
- The threshold setting in the range of 50% to 70% in wavelet shrinkage does not play an important role in changing the misclassification rate. As we also tested different threshold levels outside of the above range, a general impression has been that for this kind of scenes the threshold should be set neither too high nor too low. The results can be much worse if the threshold is poorly chosen. Higher and lower thresholds are more suitable to those extreme scenes such as STRIPE and RANDOM, respectively.

Note that only the mean values are reported in these tables though the corresponding standard deviations were also calculated. The reason was that they performed similarly with respect to the mean values and no special outliers could be observed.

For the RANDOM and STRIPE scenes, the same type of simulation study as for the TRUE has been carried out. On the RANDOM scene, where there is no structure at all, there is generally no gain in using the wavelet methods. Using QDA on the original image data performed best for all α cases, where the misclassification rates vary between 11.92% and 17.68%.

On the STRIPE scene, which represents the other extreme, the results are contrary. By using the wavelet methods the mean misclassification rates could be reduced drastically, except one case with "haar.n3". We summarize the corresponding five tables as on the TRUE scene into one table for the STRIPE scene, shown in Table A.6, where the best result for each α-value is presented and the classification rules used are indicated with legends. The 1-NN classifier based on wavelet shrinkage with threshold 95% at level 3 is a clear winner in all cases of spatial dependency, with a reduction (from only using the original image data) of a factor from 8 when $\alpha = 0.72$ up to 87 when $\alpha = 0$!

It is interesting to observe that, in the case of STRIPE, the threshold setting (from 85% to 95%) in wavelet shrinkage does make sense for reducing the misclassification rate. One needs to use only 5% of the wavelet coefficients for obtaining a high classification precision. There is one exception where the wavelet shrinkage with threshold 0.95 at level 2 seems to work not so well.

It is also worth noticing that for the STRIPE scene, in general, the mean misclassification rate increases as the spatial dependency of the noise term increases, and the relative increase is also faster than that in the case of TRUE. Here again the case when $\alpha = 0.40$ is an exception due to the same reason as we explained before in this subsection.

Summarizing our simulation study, it showed that the wavelet based method usually reduces the misclassification rates considerably and performs best among the methods under study, which agrees with a previous study [58] of wavelet classification for remotely sensed data with a single frequency band. Our studies have convinced us that the wavelet method is useful in the subject of statistical classification of remotely sensed imagery.

Note that results can be even better by optimising wavelet shrinkage via varying different choices of thresholds and wavelet functions. Concerning the threshold assessment, in general, a threshold is a trade-off between closeness of fit and smoothness. A small threshold yields a result close to the input and a larger one produces an output with a simple and smooth representation in the chosen basis. Paying too much attention to smoothness, however, destroys some of the signal singularity, in image processing it may cause blur and artifacts [28]. This has been reflected in our simulation results: the threshold level should be set quite high when the original image is very regular and simple structured such as the STRIPE scene, while it should be very low when the original image is of chaotic pattern such as the RANDOM scene. Otherwise, the threshold should be of moderate size as shown in the case of TRUE scene. In real application, however, the situation is more complicated. Often one has no replicates of the image under investigation so that the performance of different methods can not be evaluated as in our Monte Carlo study. Alternatively, we can make our selection through estimating the misclassification rate by using cross-validation or bootstrap methods and the ground reference data. Examples can be found in Steele et al. [49] and Yu et al. [58].

Compared to contextual classification methods which require extensive computation time on correlation estimation between adjacent pixels, the discrete wavelet transform provides a very efficient tool (due to the fast pyramid algorithm) for applications using satellite images with a large amount of pixels. It should be mentioned that there was an idea in this study to compare the contextual classification methods such as those studied in Mohn et al. [38] with the wavelet-based classification methods. But the comparison could not be conducted due to the limitations of the contextual methods. Specifically, the spatial correlations ρ and θ could not be properly estimated no matter what estimation methods including that used in [38]. We found that the estimates were very unstable, often giving values outside the interval $[0, 1]$ of ρ.

One more disadvantage of the contextual methods is the requirement on the training data that provides information about the neighboring pixels, e.g., field plots are of the form of (homogeneous) crossings. This is not easily accessible in practice.

There are possibilities of further development associated with the use of the wavelet transform in image classification problem. For example, how to further utilise the inherent properties of wavelet coefficients such as dependency within and between the resolution levels (an application of hidden Markov chain to this problem has been investigated for one-dimensional signal analysis [14]).

B. Markov random field

Here only 50 replicates were generated due to the time consuming computation. The results of initial classifications by k-NN and QDA are then used as the inputs for the ICM and the Gibbs sampler. The parameter β in (2.1) is chosen to be fixed at 0.75 over all iterations. The choice of this parameter is somewhat of an expert knowledge. The model for classes that we adopted is in fact similar to Potts model, which is in its turn a generalization of the Ising model in statistical physics. By comparing our true scene with realizations of the Ising field we were able to make a realistic choice of β. In real life applications, where the true scene is unknown, this problem is not so easily solved. The number of iterations is fixed in advance at eight for the ICM reclassification. The implementation of the simulated annealing with the fast cooling schedule allowed us to stop the Gibbs sampler after only 100 iterations.

The total misclassification rates are estimated by averaging the relative number of misclassified pixels in each replicate. Table 2.1 shows the mean values of misclassification percentages. The first column indicates the three different structures of spatial dependency ($\alpha = 0, 0.25$, and 0.72, respectively). The last three columns represent the initial classification methods, where the prior distribution used in QDA was chosen to be the class proportion coming from the training data, and the number of neighbours in k-NN was chosen to have two different alternatives: $k = 1$ or 3, in order to check the neighboring effect. In each cell the first and second rows show the misclassification rates of the initial classification and of the ICM reclassification, respectively. The best results at each spatial dependence level are indicated with bold face.

Table 2.1: Mean misclassification rates(%)

Spatial dependency		1-NN	3-NN	QDA
$\alpha = 0$	Original	19.36	16.44	13.30
	ICM	3.86	3.84	**3.80**
$\alpha = 0.25$	Original	18.95	16.20	13.12
	ICM	4.58	**4.56**	**4.56**
$\alpha = 0.72$	Original	20.82	17.87	14.67
	ICM	**6.98**	7.10	7.15

It can be seen from Table 2.1 that:

- The ICM reclassification can improve to a great extent the accuracy of k-NN and QDA classifiers. The mean misclassification rate from the initial classification based on these non-contextual classifiers can be reduced up to 80%.
- For the initial classification, QDA worked better than the k-NN classifier, especially in the case when $k = 1$. This is quite natural because our model is multivariate normal which fits the basic requirements in QDA. But the performance of the initial classification seems not to affect our final results. This is explained by the fact that the ICM classifier seeks the local maximum of the posterior distribution, starting from all initial classifications not far from this local maximum, however distinct they may be.
- In general, the mean misclassification rate increases as the spatial dependency between pixels strengthens.

We have also investigated the ability of the Gibbs sampler to maximize the posterior distribution and obtained the results of the implementation of the Gibbs sampler for reclassification of the images with strongest autocorrelation (0.72), whose initial classification was obtained by QDA - the best noncontextual method among those that we considered. For illustrative purpose, we display in Figure 2.2 the result that the Gibbs sampler attained in the worst case. That is, a collection of three replicates that proved most difficult for QDA, which means that the initial classification by QDA of those replicates had the highest misclassification rate (26.28% in average).

The obtained results are as follows. For images with poor initial classification, the reclassification by the Gibbs sampler is preferable to that by the ICM. This is explained by the fact that the bad initial classification is located far (in the image space) from the global maximum of the posterior distribution, and therefore such algorithms as the ICM that searches for a local maximum can not be expected to perform well. For images with average and good initial classifications, the implementation

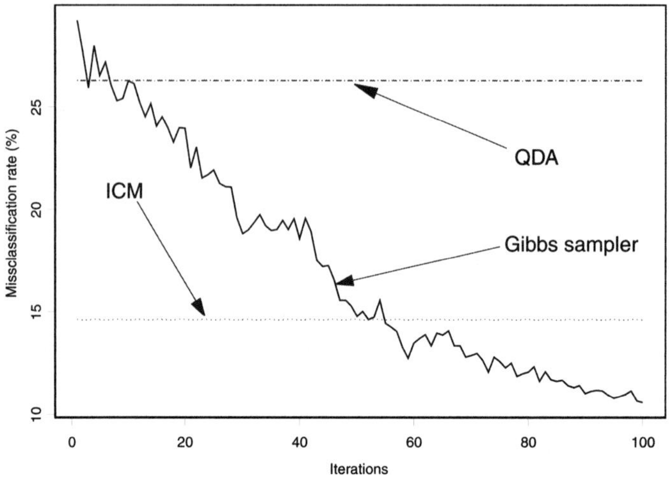

Figure 2.2: Misclassification rates for QDA, QDA+ICM and QDA+Gibbs sampler for bad initial classification

of the Gibbs sampler leads to unnecessary loss of CPU time. For images with the best initial classification, the Gibbs sampler has not attained even the local maximum after a 100 iterations.

This study indicates that the combination of the spectral-contextual classifiers can improve to a great extent the accuracy of conventional non-contextual classification methods. It is also shown how the most complicated cases can be handled by the Gibbs sampler.

3 Nonparametric classification — new developments

In this section we will present some new developments in methodology for nonparametric classification using satellite image data, with classification of forest land as an example.

3.1 Background

In Sweden 16 national environmental quality objectives (EQO) were adopted by the Parliament in 1999. In June 2000 the Committee on Environmental Objectives (CEO) presented its final report [37] in which it suggests goals, targets and action strategies to be attained and implemented in order for the environmental quality objectives to be reached. The report formed a basis for the Government Bill presented in 2001 where the Government further proposed sub-goals and interim targets to be attained within one generation. For the EQO "Sustainable Forests", the interim targets set is:

- A further 900,000 hectares of forest land of high conservation value will be excluded from forest production by the year 2010.
- By 2010 the amount of dead wood, the area of mature forest with a large deciduous element and the area of old forest will be maintained and increased by
  - increasing the quantity of hard dead wood by at least 40% throughout the country and considerably more in areas where biological diversity is particularly at risk;
  - increasing the area of mature forest with a large deciduous element by at least 10%;
  - increasing the area of old forest by at least 5%;
  - increasing the area regenerated with deciduous forest.

To be able to follow the achievement of the interim targets it is necessary to estimate areas for different forest classes (e.g. old deciduous forest). In this application the forest is defined into 24 different classes with respect to species composition and age, supplemented with the class "low basal area", when species composition and age do not have any meaning. Table 3.1 presents the definition of those 25 forest classes to be used:

Table 3.1: Definition of forest classes

Proportion Deciduous		Age			
		10-39	40-69	70-109	110+
0-19%	Pine $\geqslant$ 70%	1	7	13	19
	Spruce $\geqslant$ 70%	2	8	14	20
	"Remaining"	3	9	15	21
20-49%		4	10	16	22
50-69%		5	11	17	23
$\geqslant$ 70%		6	12	18	24
Class 25 = basal area $\leqslant 1$ m^2 per hectare					

Remote sensing classification of sparse and rarely occurring objects in the landscape is complicated and reported classification rates are usually poor. For several reasons traditional methods will not give satisfactory results:

- The number of field plots in different classes varies too much leading to an unfavorable treatment of sparsely occurring objects.

- Data editing methods, like the multiedit algorithm [46], often drop whole classes when applied to moderately sized training sets with more dimensions and classes.
- Noncontextual methods neglect information from neighboring plots and the feature vector is not normally distributed.
- Parametric contextual classification methods do utilize information from neighboring plots but the demands for field data increases, and estimation of some parameters in the model is unreliable and in many situations impossible.
- The feature vector is not normally distributed and the different class distributions are often highly overlapping.

To overcome these problems nonparametric approaches have been used. Information from neighboring pixels will be utilized by using the wavelet transform to denoise the original images by removing small wavelet coefficients and inverting the wavelet transform. Hereafter a suitable classification algorithm is applied.

For previous work in this direction see Choi and Baraniuk [11], Ranneby & Yu [45], Teterukovsky and Yu [53], Yamaguchi et al. [56], Yu et al. [58], Yu and Ekström [57], and Yu and Ranneby [59].

3.2 The classification method

The remote sensing methods will be based on a new approach for classification of multitemporal satellite data sets, combining multispectral and change detection techniques. The classification method has to be adapted for handling a mixture of reference data from different sources in order to compensate for the lack of ground samples from rare classes within the main reference data source.

The classification algorithm will be based on the nearest neighbor method. The main procedures are as follows.

- Define the target function. The probabilities of correct classification for each class are used, so that each class is of equal importance. Note that it differs from the overall correct classification, which is often used in various applications.
- Denoise the feature vector. The wavelet shrinkage method based on 2D-wavelet transform is used to denoise the images. The feature vector consists of components that are pixel values from different spectral bands of different temporal images over both years and seasons.
- Remove outliers from the reference data. When the feature vector has high dimension and the number of classes is large, new data

editing methods are needed in order to remove outliers due to poor quality of field data, and to find out the prototypes for each class in the case of nearest neighbor classifiers. This must be done so that poor quality data is removed at the same time as most of the natural variations within the different classes are kept. This is achieved by working with a further subdivision of the classes with respect to other information, as e.g. basal area, and for the new classes (now 109 classes) suitable techniques for outlier removal are applied.

- Calculate the information values in the components in the feature vector. Before the feature vector can be used the components have to be rescaled. The size of the rescaling factor is a function of the dependence between the studied objects and the component. Since the classes are defined by more than one variable, there will not exist any natural order relation between the classes so conventional dependence measures, such as the correlation coefficient, cannot be used. Instead measures with the origin in information theory are used ([41, 42]).
- Determine a proper metric. We prefer to use the following metric:

$$\sqrt{\sum_{i=1}^{d} w_i^2 |x_i(\mathbf{s}) - x_i(\mathbf{t})|^2}, \tag{3.1}$$

where $\mathbf{x}(\mathbf{s}) = [x_1(\mathbf{s}), x_2(\mathbf{s}), \cdots, x_d(\mathbf{s})]$, $\mathbf{x}(\mathbf{t}) = [x_1(\mathbf{t}), x_2(\mathbf{t}), \cdots, x_d(\mathbf{t})]$ are two pixels at location $\mathbf{s}$ and $\mathbf{t}$, with attribute of the spectral components, d is the dimension of feature vector, $w_i = \frac{\alpha_i^p / q_i}{\sum \alpha_i^p / q_i}$ is the weighting factor, q_i and α_i are the inter-quartile range and the information value of the ith component, respectively, and p is determined so the target function is maximized. Usually p equals to 1.

- Determine prototypes for the classes. Here the information about the occurrence of different classes in the reference data is used.
- Run a nonparametric classification. Here the nearest neighbor classifier is used.
- Declare the quality of classification result by using probability matrices. The probability matrix is based on the confusion matrix and defined as $\mathbf{P} = [P_{ij}]$ where P_{ij} is the probability that class i is classified as class j.

The classification algorithm is derived in order to optimize the average probability for correct classification, i.e. each class is equally important. If we want to have good classification performance also for small classes they have to be overestimated (unless the classification is 100% perfect). Traditional methods as maximum likelihood classification or LDA with or without prior information fail completely (Ranneby & Yu [45]).

3.3 Field data and remotely sensed data

The study area consists of the Counties of Dalarna and Gävleborg covering a total area of about 5.1 million hectare. Landsat-7 ETM+ data and Landsat-5 TM data were used. The field data used were collected from the Swedish NFI, years 1996 - 2000.

Satellite data from three different dates were used representing:

a) The best available cloud-free satellite scenes (primarily Landsat-7 ETM+) from one of the last three years for obtaining optimal coverage over the study area from the vegetation period (June - August);

b) Old satellite data (5 - 10 years old) from vegetation period for mapping of clear-cut areas and for classification;

c) One satellite scene from the spring (before leafing) for enhanced classification of tree species.

The images used were registered by the Landsat satellites with the TM sensor (Thematic Mapper) as the principal instrument. The TM sensor has seven spectral bands in visible, near infrared, mid infrared and thermal infrared with 30 meter ground resolution. Most of the satellite data were geometric corrected with ortho-correction methods, using the digital elevation model to remove image parallaxes. The geometric accuracy requirement for multi-temporal data classification is to aim for less than one pixel root mean square error. This quality was not reached for all scenes used, leading to local misalignments between scenes in some cases. The main reason for this has been the deteriorated geometric quality of the ageing Landsat-5 TM sensor, which was launched already in 1984 and could not be replaced completely until 2000 by Landsat-7. The TM data are resampled to 25 meter pixel size for the continued processing. The image data is stored as 8 bit integer data in each band, corresponding to digital numbers from 0 - 255.

Each spectral band has its wavelength and nominal spectral location which determine the intended principal applications. A study at the USGS EROS Data Center [40] showed an interpreter preference for several specific band-color combinations for various features. For the mapping of water sediment patterns, a normal color composite of bands 1, 2, and 3 (displayed as blue, green, and red) was preferred. For most other applications, such as mapping urban features and vegetation types, the combinations of (a) bands 2, 3, and 4 (color infrared composite), (b) bands 3, 4, and 7, and (c) bands 3, 4, and 5 were preferred. In general, vegetation discrimination is enhanced through the incorporation of data from one of the mid-IR bands (5 or 7). Combinations of any one visible (bands 1 to 3), the near-IR (band 4), and one mid-IR (band 5 or 7) band are also very useful. See Lillesand & Kiefer [34] for more reading in remote sensing and image interpretation. In our classification of forest, spectral information from bands 3, 4, and 5 are used. This gives a 12-dimensional feature vector.

The classification into the 25 different classes is performed outside clear-cut areas (which are treated as class No. 26) and areas covered by clouds. To reduce the amount of unclassified areas due to cloud coverage those gaps are filled using only one of the recent images.

3.4 Quality assessment

Quality of remote sensing classification is usually evaluated by splitting the field data in two pieces or using some cross-validation technique, e.g. leave-one-out cross-validation. As the number of field plots usually is limited the precision in the estimated confusion matrix will be poor. Efron and Tibrishani [18] tried to reduce the uncertainty by using a bootstrap approach. Their assumptions are not general enough to make the method appropriate in our applications. Monte Carlo techniques have been used to obtain reliable estimates of the confusion matrix under general assumptions. Unbiased and high-precision estimates of the confusion matrix are obtained. Even if the classified image gives incorrect proportions of the different classes it is possible to obtain unbiased area estimates (for the whole satellite scene), as illustrated in the next subsection.

3.5 Results

The probability for correct classification varies between 30% and 40% for most of the pure classes although it is considerably lower for some other classes, especially those with different types of mixed forest. The largest class is C_1 (younger pine forest) and the smallest is C_{24} (older deciduous forest). We have compared our classification rates with traditional methods as maximum likelihood classification or LDA with or without prior information. Note that it was impossible to apply QDA here because the field plots for small classes were too few to be able to estimate the covariances. LDA with prior information is the most commonly used method. With this method the probability for correct classification is 0% for 14 out of the 25 classes. The reason is that all small classes essentially are classified as the larger classes, particularly C_1, which gets a rate of 70% for correct classification. If we do not use any prior information then the rates increases for the small classes but drops dramatically (to 12%) for class C_1. A comparison for the largest and smallest classes is given in Table 3.2 below.

Table 3.2: Probabilities of correct classification for C_1 and C_{24}

Class	LDA with prior	LDA without prior	Our method
C_1	70%	12%	39%
C_{24}	0%	34%	37%

As we want to have high classification accuracy also for small classes they have to be overestimated. To overcome this problem we calculated matrices for area correction and the precision in the correction matrices. This was done by using Monte Carlo techniques. From this it is possible to obtain unbiased estimates for the different classes and standard errors of the area estimates. Example of estimated areas and standard errors are shown in Table 3.3.

It is worth to notice that the estimation property is reliable on the scene level and it is inappropriate to apply it to the local area or pixel level. To obtain area estimation of good quality (such as unbiasness), a new approach is needed, and we will take it up in the coming section.

As illustrated by Table 3.3 the smallest classes (C_{23} and C_{24}, older deciduous) are overestimated. Before correction the proportions are 1.79% and 0.99%, respectively and after application of the area correction matrices the estimates are 0.16% and 0.07%, respectively. The largest class C_1 is underestimated having a proportion of 11.98% before correction and 15.51% after correction. However, large classes do not have to be overestimated, c.f. class C_7 with estimates 9.54% before correction and 8.40% after, An eventual over- or underestimation depends on the pattern for misclassifications. Class 7 (pine forest 40-70 years) is contaminated with C_1 (younger pine forest) and C_{13} (older pine forest), which both cover larger areas than C_7. The standard errors are (in absolute value) small for all classes. These standard errors should be compared with what can be obtained using NFI-data. For the largest class C_1 the standard error would be around 3% and for the second largest, C_{13}, around 3.7%. For small classes the standard errors are considerably larger and for rare classes, such as C_{24}, the NFI will not give any reliable estimates at all.

Table 3.3: Observed and estimated areas

Class	Observed areas	Estimated areas	Standard errors
C_1	11.98%	15.51%	0.14%
C_2	4.18%	3.96%	0.06%
C_3	2.23%	3.33%	0.05%
C_4	7.91%	7.39%	0.08%
C_5	1.26%	1.49%	0.04%
C_6	1.21%	2.08%	0.05%
C_7	9.54%	8.40%	0.06%
C_8	2.85%	3.29%	0.04%
C_9	2.58%	3.61%	0.04%
C_{10}	4.02%	4.74%	0.07%
C_{11}	2.71%	1.54%	0.04%
C_{12}	2.72%	1.61%	0.05%
C_{13}	8.40%	10.20%	0.08%
C_{14}	5.93%	5.49%	0.05%
C_{15}	2.96%	3.80%	0.03%
C_{16}	4.95%	4.04%	0.06%
C_{17}	1.51%	0.84%	0.03%
C_{18}	0.56%	0.50%	0.03%
C_{19}	9.35%	8.16%	0.08%
C_{20}	1.89%	1.89%	0.03%
C_{21}	1.76%	1.61%	0.03%
C_{22}	2.20%	1.28%	0.02%
C_{23}	1.79%	0.16%	0.03%
C_{24}	0.99%	0.07%	0.01%
C_{25}	4.51%	5.00%	0.06%

4 Probabilistic classifier

For several applications, e.g. habitat suitability models, it is necessary to use pixelwise information. However, as the distributions for different classes are highly overlapping it is not possible to get satisfactory accuracy at pixel level. Instead it is necessary to introduce a new concept, pixelwise probabilistic classifiers. This concept is closely related to fuzzy classification, see Bezdek et al. [5] for different definitions. Instead of classifying each pixel to a specific class, each pixel is given a probability distribution describing how likely the different classes are. The drawback with fuzzy classifiers is that until now the methods are not based on any rigorous theory and the literature is sparse where the probabilistic classifier definition is used.

In principle, probabilistic classifiers can be derived both from parametric and nonparametric classification methods. However, deriving probabilistic classifiers from parametric methods is a complicated task since for several classes the distributional assumptions are not satisfied. Furthermore, it is of key importance to have consistent parameter estimates. To avoid the problem of misspecification, nonparametric approaches are preferred. The probabilistic classifier derived here is based on the k-NN method with $k = 1$, defined in Subsection 2.2.1.

There exist articles about fuzzy classification and k-NN, see e.g. Bereau and Dubuisson [3] and Kissiov and Hadjitodorov [31] but with a different scope. To be useful for e.g. area estimation the probabilities must be extracted from proper class distributions. When the classifier uses distances in feature space to derive the probabilities it is intuitively obvious that the probabilities will be inversely proportional to the distances raised to some power. However, there is only one value that is correct and that value depends on the number of components in the feature vector (i.e., the dimension of the feature space). Actually, by using the nonparametric density estimation (Loftsgaarden & Quesenberry [35]), the density for class c at $\mathbf{x}(\mathbf{s})$ will be inversely proportional to the volume of the nearest neighbor ball with center $\mathbf{x}(\mathbf{s})$. Note that the volume is proportional to the radius, based on the metric defined in (3.1), raised to the dimension of the feature space.

Specifically, let the class labels be $1, 2, \cdots, K$. For each pixel with location $\mathbf{s}$ and feature vector $\mathbf{x}(\mathbf{s}) = [x_1(\mathbf{s}), x_2(\mathbf{s}), \cdots, x_d(\mathbf{s})]$, the probability that this pixel belongs to class c is defined as

$$p_c(\mathbf{s}) = \frac{\operatorname{dist}(\mathbf{x}(\mathbf{s}), c)^{-d}}{\sum_{c=1}^{K} \operatorname{dist}(\mathbf{x}(\mathbf{s}), c)^{-d}}, \tag{4.1}$$

where $\operatorname{dist}(\mathbf{x}(\mathbf{s}), c)$ = minimum distance in the feature space (using the metric defined in (3.1)) from $\mathbf{x}(\mathbf{s})$ to the prototypes of class c. Thus, the probability vector at pixel $\mathbf{s}$ is obtained as $\mathbf{p}(\mathbf{s}) = [p_1(\mathbf{s}), p_2(\mathbf{s}), \cdots, p_K(\mathbf{s})]$.

The probabilistic classifiers can be used in different ways. For instance, it can be used to measure uncertainty at pixel level, or to obtain reliable area estimates locally. It can also be applied to correct the erroneous mask of maps (see comments in the last section).

A traditional classification is obtained by choosing the class with the highest probability for each pixel. With the probabilistic classifier, it is possible to give quality declaration at pixel level. The pixel-wise vectors of probabilities can be used to judge how reliable a traditional classification is and to derive measures of the uncertainty (entropy) for the individual pixels. Based on the probability vector, the entropy at pixel $\mathbf{s}$ is defined as $-\sum_{c=1}^{K} p_c(\mathbf{s}) \log p_c(\mathbf{s})$. It is extremely important that proper probability distributions allowing frequency interpretation are derived; otherwise misleading results are obtained.

In application, in addition to the classified map a raster map expressing the quality (entropy) for each pixel is produced. Areas where the precision is unsatisfactory will be identified. In these areas it is possible to take additional field plots and perform an improved classification. Figure 4 illustrates our concept of probabilistic classifiers operated on a small site of size 16×16 pixels. The area consists of mixed forest, which implies difficulty to classify. Red colored pixels indicate higher entropy values or unsatisfactory quality. The diagram on top right shows the probability distributions of 16 pixels (one row) picked from the entropy map on the left side. For a given pixel, the more even distributed probabilities among classes, the high entropy. If one class dominates the probability vector (i.e., having high probability), then the pixel has very low entropy and is reliably classified.

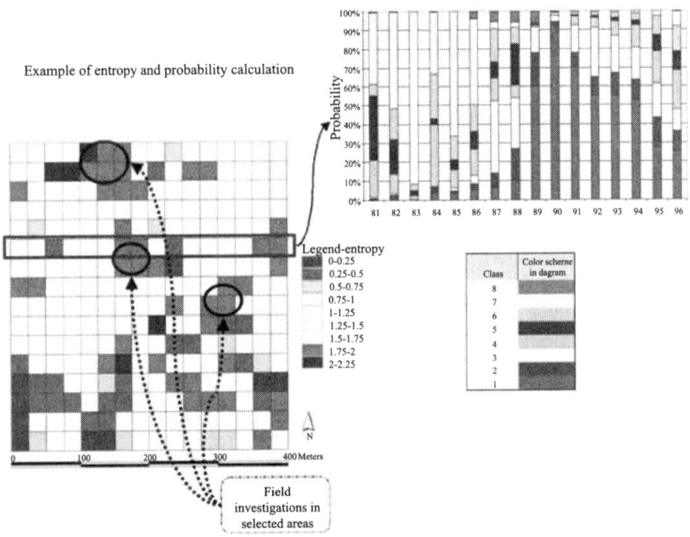

Figure 4.1: The concept of probabilistic classifiers. Test area with mixed forest, 8 classes. Field plots can be added in areas where entropy is high (i.e. quality at pixel level is unsatisfactory) to produce additional reference data as input to a renewed (improved) classification

Although it is possible to obtain satisfactory area estimates on scene level using area correction matrices derived from the confusion matrices as we discussed in the previous section, the precision is not good enough for smaller areas or on pixel level. The pixelwise vectors of probabilities can be used to get unbiased area estimates also for quite small areas and for deviating areas, such as potential areas for nature reserves. The area estimates is obtained simply by summation of the "class probabilities" over the selected subregion.

5 Application to classification of agricultural crops

In this section we will present a case study for classification of agricultural crops using multitemporal satellite images. The nonparametric classification algorithm and the probabilistic classifier introduced in Section 3 and 4 will be applied. We begin with a study area where satellite images (SPOT and Landsat TM) from different seasons over a year are available. Then the application extends to a larger scale on the Dalälven's catchment area, which is covered by five Landsat TM scenes.

Section 5.1 presents the study area and input data including masks, satellite images, and field data. Results of classification for the study area and discussion are given in Section 5.2. A real application of the new approach to a large scale is presented in Section 5.3. More detailed class definitions and classification results are referred to Appendix. Further reading can be found in [59].

5.1 Study area and input data

The study area is located in the eastern part of Dalarna in Sweden. Characteristic for agriculture fields in that area is that they are long and narrow, which is a complication for the classification. As a consequence, the existing remote sensing classification methods will not give satisfactory results. Input data includes various masks, satellite data (Landsat TM and SPOT images), difference images, as well as field data ("block database").

- Map masks. Masks from the topographic maps will be used for stratification before classification. Agricultural area was defined by the agricultural mask from the 1:100000 scale "blue map" (Blå Kartan).
- Cloud cover masks. All scenes have been manually interpreted for cloud cover and other image data errors. Areas covered by clouds, haze and shadows from clouds were digitized on the screen. Other pixels having erroneous measurements for different other reasons were also marked and included in the "cloud mask", which in reality is a mask defining pixels not to be used for classification. Examples of this are data dropouts, resampling effects at the scene edges and "no data" pixels outside the imaged area.
- Satellite data. The images to be used are either Landsat TM images or SPOT images. The satellite data were geometric corrected with ortho-correction methods. The data are resampled to 25 m

pixel size. Table 5.1 presents the satellite images used in the study area.

Table 5.1: SPOT and Landsat images used in the study area

	Scene code	Date	Spectral bands used
SPOT-2	054-226/8	1998-10-24	XS 1, 2, 3
Landsat-5	194-018	1999-05-07	TM 2, 3, 4, 5
Landsat-5	194-018	1999-07-10	TM 2, 3, 4, 5
SPOT-4	054-226/227	1999-07-30	XS 1, 2, 3, 4
Landsat-7	195-018	1999-09-11	TM 2, 3, 4, 5

- Field data. The "block database", i.e. a GIS database with polygons grouping the parcels of agricultural units ("blocks") that are registered in the IAKS database (the administrative database for agricultural aid in Sweden).

5.2 Results and discussion

The crops and agricultural land were predefined into 25 classes, including cereals (autumn-sown and spring-sown), oil seed crops, potatoes, grassland on arable land, energy forest (salix), and so on. Details on the definition of crop classes and different thematic levels can be found in Tables A.7 and A.8 in Appendix. Because the spectral signature differs between the center and the boundary of the arable units, and from the one field units to multi-field units, the pixels along the edges and transition zones were grouped into their own classes.

In field data, only few were used as reference data for classification, i.e. build-up of prototypes for each class. The major part was used for validation.

Table 5.2: Probability matrices at level 1

	5 seasonal scenes			1 single scene		
	C_1	C_2	C_3	C_1	C_2	C_3
C_1	.90	.09	.01	.84	.14	.02
C_2	.36	.62	.02	.49	.49	.03

For the purpose of comparison, classification was done using a single image and five images in different seasons, respectively. The probability matrices at level 1 (consisting of three classes: C_1 - arable land, C_2 - grazing land, and C_3 - boundary pixels) are shown in Table 5.2, and the corresponding results at level 2 are presented in Tables 5.3 and 5.4. It is evident that the use of multitemporal images over seasons can substantially improve the classification result compared to a single image. With one single image, however, spring-sown cereals and spring-sown oil seed

Table 5.3: Probability matrix using five seasonal scenes at level 2

	C_1	C_2	C_3	C_4	C_5	C_6	C_7	C_8
C_1	.49	.35	.00	.00	.03	.00	.14	.00
C_2	.04	.78	.01	.00	.03	.00	.14	.01
C_3	.00	.07	.72	.00	.01	.00	.20	.00
C_4	.01	.09	.01	.65	.04	.00	.20	.00
C_5	.02	.02	.02	.00	.63	.01	.27	.02
C_6	.00	.04	.00	.00	.11	.56	.27	.01
C_7	.01	.04	.00	.00	.20	.01	.71	.02

Table 5.4: Probability matrix using one single scene at level 2

	C_1	C_2	C_3	C_4	C_5	C_6	C_7	C_8
C_1	.19	.30	.01	.03	.11	.00	.34	.01
C_2	.04	.53	.01	.01	.10	.01	.27	.02
C_3	.01	.07	.70	.00	.05	.00	.17	.00
C_4	.02	.02	.01	.17	.37	.00	.39	.01
C_5	.02	.04	.00	.05	.47	.02	.38	.02
C_6	.01	.14	.00	.00	.12	.28	.42	.03
C_7	.02	.06	.01	.02	.28	.04	.54	.03

crops (C_2 and C_3 in Table 5.4) have quite high accuracy (53% and 70% respectively). At the thematic level 1 we have very good classification accuracy for arable land (84%), but much lower for grazing land (50%). This is mainly depending on confusion with pasture or grass for hay or silage on arable land.

Looking further into the classification accuracy at level 3 in Table A.9 (five seasonal images) and A.10 (one single image) in Appendix, it can be found that even with images from five occasions it is difficult to identify winter barley and winter wheat. One possible reason is that the image from 7th of May is too early. This is confirmed by the information value for spectral band 5. The value is only half of that from the image taken in July. Probability of correct classification for these two classes are slightly below 40%. Both crops occupy, however, less than 1% each of the cultivated area in the study area. The other classes of grains have considerably higher probabilities of correct classification (55% – 93%). For other crops and agricultural landscapes, the probability of correct classification is above 50%, except meadow, fallow (> 1 year) and other land use, which hardly surprisingly have low accuracy.

Even though the classification accuracy at pixel level by using one single image is low, the results can be useful in some applications such as input to source apportionment mdels. Here it is, however, important to develop sensible area estimates for different catchments. As mentioned before, one problem in area estimation based on the classified image is that small classes will be overestimated whereas large classes often underestimated. Therefore, the observed area has to be corrected. By using information from the block database and the Monte Carlo method, unbiased estimate of the confusion matrix can be derived. Then it is possible to calculate an "area correction matrix", which can be used to obtain unbiased area estimates for the whole satellite scene.

The area estimation in the study area before and after correction is summarized in Table A.11 for all types of units. Although the observed proportions, based on one image and five images respectively, may differ significantly, the differences after corrections become small. For the one image case, one can see that the observed (classified) proportion for spring barley (C_2), pasture on arable land (C_{12}) and meadow (C_{15}) was 13.2%, 19.5% and 2.16%, respectively. After correction the proportions became 29%, 34% and 0.27% respectively.

It is worth noticing that the above probabilities can only be used for evaluation of quality at the scene level. In practice, often the quality of classification at pixel level is required. The probabilistic classifier introduced in Section 4 is run on two small test sites. These two sites were selected with different characteristics of spatial distributions (homogeneous versus heterogeneous), each of size 16×16 pixels.

The probability for each class at pixel level and the entropy for each pixel were calculated. Figure 5.1 present the map of entropy for test site 1 and test site 2. Note that they are based on five scenes and at thematic level 2. Higher values of entropy indicate uncertain classification at that pixel. By setting some threshold, we are able to identify areas, where the classification accuracy is unsatisfactory. In our test site 1, most of the entropies are very low, except a few pixels with medium size entropies. This indicates that the quality of classification at this site is high. As also expected, the accuracy of classification at site 2 (heterogeneous area) became lower, so that one obtains a number of pixels with high entropies (red colored).

In areas having an unsatisfactory classification it is possible to take additional field plots and perform an improved classification. Here it is important that the classification method can handle different types of reference data. The final product will be two maps, one showing the pixel wise classification and the other giving the classification accuracy at pixel level.

5.3 Large scale applications

The classification method developed in the study area was later applied to the Dalälven's catchment area. This region is of size 29000 km^2 and requires five Landsat TM scenes to be completely covered. Unfortunately, it was not possible to find cloud free satellite images over the

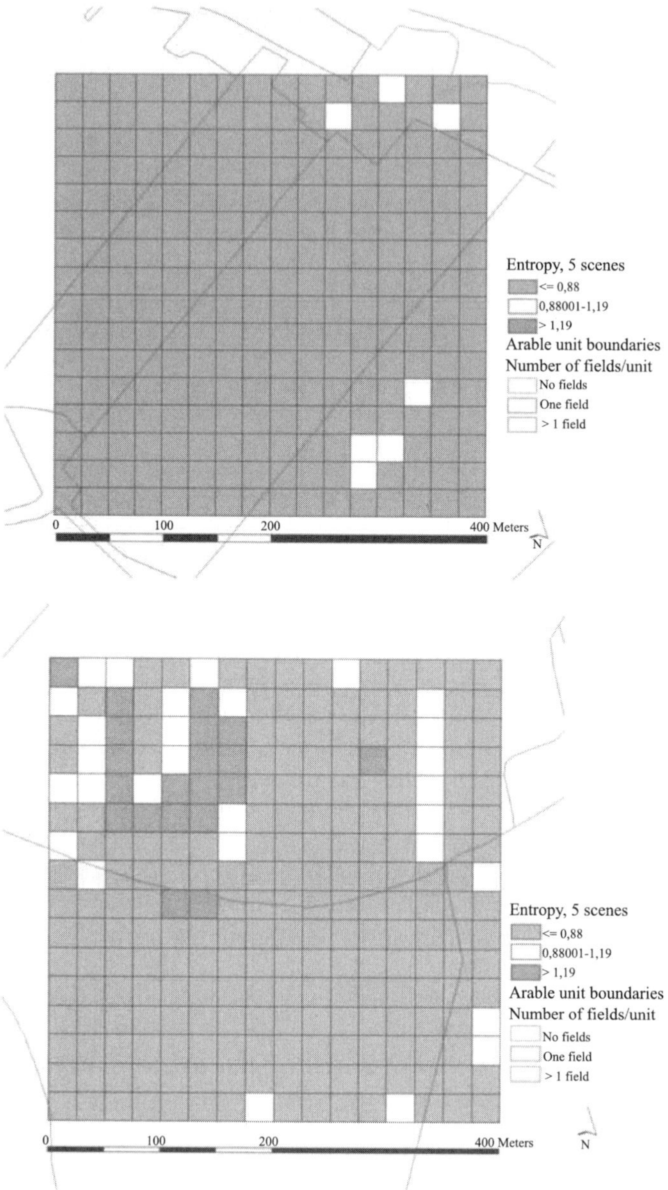

Figure 5.1: Entropy map of test site 1 (above) and test site 2 (below)

whole catchment area from the same season and year. The two scenes 194-018 and 196-017 were used here, since they cover the large part of the region (see Table 5.5). The time points at July 1999 and August 1999 respectively were chosen so that the cloud cover is minimized in both scenes.

Table 5.5: Landsat images used in classification for Dalälven's catchment area

	Scene	Date	Spectral bands used
Landsat-5	194-018	1999-07-10	TM 2,3,4,5
Landsat-5	196-017	1999-08-01	TM 2,3,4,5

According to results from the study area, it would be desirable if we could use several seasonal scenes also for the whole catchment. It was, however, impossible to get Landsat TM images to fulfill such requirement. Alternatively, one can consider using either SPOT images (10 meter resolution) or MERIS images (200 meter resolution) to handle the agricultural fields. The former demands much more work in preprocessing to make a useful mosaic spatially over the whole catchment as well as temporally over different seasons, whereas the latter has too low spatial resolution to distinguish different agricultural units. Ideas with SPOT mosaic are very interesting and have shown to be useful, and successfully applied in other studies.

The crops and agricultural land were predefined in the same way as in the study area. They are presented in Tables A.12 and A.13 in Appendix for scene 194-018 and scene 196-017, respectively. Note that the number of classes differs from that in the study area. This is simply because of the lack of field data in different scenes. Summarized results of our classification are shown in Table 5.6 for the thematic level 1 and Table 5.7-5.8 for the thematic level 2.

From Table 5.6, one can see that classification accuracies at thematic level 1 are quite similar and satisfactory for both scenes. Looking at Tables 5.7 and 5.8 for thematic level 2, two things are obvious and common in both scenes, that is, C_1 (autumn-sown cereals) is confused with C_2 (spring-sown cereals) and C_7 (other crops) is confused with all other classes. One of the reasons is that C_2 and C_7 are both dominating classes in arable land and grazing land respectively.

For the sake of space limitation, details of probability matrices at thematic level 3 and the area estimation for these two scenes are omitted here but they are available upon request.

Table 5.6: Probability matrices at level 1

	scene 194-018		scene 196-017	
	C_1	C_2	C_1	C_2
C_1	0.82	0.18	0.81	0.19
C_2	0.35	0.65	0.25	0.75

Table 5.7: Probability matrix for scene 194-018 at level 2

	C_1	C_2	C_3	C_4	C_5	C_6	C_7	C_8
C_1	0.20	0.26	0.07	0.05	0.05	0.01	0.34	0.03
C_2	0.08	0.48	0.05	0.02	0.03	0.02	0.26	0.04
C_3	0.02	0.08	0.61	0.01	0.03	0.01	0.19	0.04
C_4	0.03	0.02	0.24	0.21	0.07	0.00	0.38	0.04
C_5	0.02	0.05	0.12	0.05	0.24	0.02	0.44	0.05
C_6	0.02	0.19	0.01	0.00	0.06	0.36	0.32	0.04
C_7	0.02	0.06	0.07	0.02	0.14	0.04	0.58	0.05

Table 5.8: Probability matrix for scene 196-017 at level 2

	C_1	C_2	C_3	C_4	C_5	C_6	C_7	C_8
C_1	0.09	0.38		0.09	0.14		0.28	0.00
C_2	0.02	0.49		0.09	0.09		0.29	0.02
C_3								
C_4	0.01	0.26		0.34	0.08		0.27	0.04
C_5	0.02	0.14		0.11	0.33		0.35	0.05
C_6								
C_7	0.01	0.09		0.01	0.12		0.75	0.02

6 Final remarks

Remote sensing classification of sparse and rarely occurring objects in the landscape is complicated. A new approach, based on nonparametric statistical methods, is developed to overcome the problems arisen with traditional classification methods. It utilises information from multitemporal and multispectral satellite data and change detection techniques. The classification method is also adapted for handling a mixture of reference data from different sources in order to compensate for the lack of ground samples from rare classes within the main reference data source. The new algorithm is derived in order to optimize the average probability for correct classification, i.e. each class is equally important. As we want to have high classification accuracy also for small classes they have to be overestimated. By using Monte Carlo techniques, it is possible to obtain unbiased estimates for the different classes and standard errors of the area estimates, at the scene level.

As we have seen the probabilistic classifiers are useful for measuring uncertainty at pixel level and obtaining reliable area estimates locally. It can also be applied to correct the masks of maps. The masks from the topographic maps are usually used to define different land use of vegetation, e.g. forest land, peat land, or agricultural land. It has, however, been found that the errors in masks can be significant, and thus development of better masks is needed. An application of the probabilistic classifiers to obtain new masks for forest land and peatland has been conducted successfully in other studies of our group. The idea is as follows: a) use the probabilistic classifier together with information from digital elevation model (wetness index) and neighboring pixels to get the probability vector for classes (both forest and peatland) at pixel level and consider it as a prior probability distribution for each pixel; b) apply the probabilistic classifier by using remotely sensed data to get the probability vector for classes at pixel level; c) join a) with b) to get the a posterior probability vector for each pixel; d) aggregate the classes to two categories: Forest and Peatland; e) calculate entropies for each pixel and determine the threshold value according to the entropies so that pixels belong to one of the following zones: Forest, Peatland, or Transition Zone; f) recalculate the probability vectors for pixels in Forest and Peatland zones. In such a way, two masks with higher quality are obtained for forest and peatland, respectively. A third one, the Transition Zone, is given for pixels with a certain degree of uncertainty.

As mentioned in the introduction, in order to make ecosystem health assessment effective, information must be captured rapidly, comprehensively, and economically. High quality monitoring systems are already operational for e.g. the aquatic and marine environments. Concerning the terrestrial environment and the temporal and spatial variation of land use and landscape parameters more development and new approaches are needed. Especially if we want access to dynamic and scale-independent systems, i.e. systems that can detect environmental trends with sufficient accuracy on a wall-to-wall basis at local, regional and national level. The new classification approaches introduced in this chapter can form a basis for cost-effective environmental monitoring.

Field data have to be combined with remote sensing information obtained from different sensors and varying time intervals. The size, shape and location of the field plots must be determined so that the information value of each plot is high. This can be achieved using a grid of existing plots supplemented with new plots at selected locations, but still obtained by objective algorithms. Using the probabilistic classifier it is possible to derive cost-efficient monitoring systems. Start with the in situ measurements that are available, e.g. from the NFI. Take away outliers and perform an initial classification. Use the pixelwise quality assessment obtained from the probabilistic classifier to identify areas where the quality is unsatisfactory, add plots in those areas, and reclassify. Estimates are obtained by combining field and remote sensing information, and it is possible to get a wall-to-wall coverage with pixelwise estimates. In these situations, standard error estimation is not straightforward and requires careful handling.

Resampling and subsampling methods have been suggested in the literature to nonparametrically estimate different characteristics, such as the distribution and the variance of statistics computed from spatial lattice data. The advantage of such methods is that statistical inference can be done without knowledge of the underlying spatial dependence mechanism and marginal distributions that generated the data. Furthermore, for the user no explicit theoretical derivation is necessary but instead intensive computing. Most of the proposed methods in the literature assume stationary data. However, when modeling real-life data, the hypothesis of stationarity often must be rejected. Especially for nonstationary spatial data, there are very few available methods to do inference. Our research group has proposed a number of resampling and subsampling methods for nonstationary spatial lattice data, see. e.g., [19, 20, 21].

From our simulation studies, it has been shown that by using MRF approaches, such as ICM and Gibbs sampler, for reclassification the accuracy of classification can be improved. However, besides their high demands of computational capacity, the MRF approaches suffer from the common drawback of smoothing techniques, that is, rarely occurring objects or small classes tend to be "smoothed" away because they are often sparsely located pixels. Thus, the usefulness of the MRF approaches in real applications remains to be approved. We believe, therefore, that today nonparametric methods are most appropriate.

References

[1] A. Aldroubi and M. Unser, editors. *Wavelets in Medicine and Biology.* CRC Press, Boca Raton, FL, 1996.

[2] A. Antoniadis and G. Oppenheim, editors. *Wavelet and Statistics,* volume 103 of *Lecture Notes in Statistics.* Springer-Verlag, New York, 1995.

[3] M. Bereau and B. Dubuisson. A fuzzy extended k-nearest neighbor rule. *Fuzzy Sets and Systems*, 44:17–32, 1991.

[4] J. Besag. On the statistical analysis of dirty pictures. *Royal Statistical Society, Ser. B*, 48:259–302, 1986.

[5] J. C. Bezdek, J. M. Keller, R. Krishnapuram, and N. R. Pal. *Fuzzy Models and Algorithms for Pattern Recognition and Image Processing*. Kluwer Academic Publishers, 1999.

[6] A. Bruce and H. Y. Gao. *Applied wavelet Analysis with S-PLUS*. Springer, New York, 1996.

[7] R. W. Buccigrossi and E. P. Simoncelli. Image compression via joint statistical characterization in the wavelet domain. *IEEE Trans. Image Processing*, 8(12):1688–1701, December 1999.

[8] J. B. Buckheit and D. L. Donoho. Improved linear discriminant analysis using time-frequency dictionaries. Technical report, Stanford University, 1995.

[9] J. S. Byrnes, J. L. Byres, K. A. Hargreaves, and K. Berry, editors. *Wavelets and Their Applications*. Kluwer Academic Publishers, Dordrecht, 1994.

[10] A. Chambolle, R. A. DeVore, N-Y. Lee, and B. J. Lucier. Nonlinear wavelet image processing: variational problems, compression, and noise removal through wavelet shrinkage. *IEEE Trans. Image Processing*, 7:319–335, 1998.

[11] H. Choi and R. G. Baraniuk. Image segmentation using waveletdomain classification. *SPIE Proceedings*, 3816, 1999.

[12] F. J. Cortijo and N. Perez De La Blanca. A comparative study of multispectral image classifiers: applications to classification problems with high-dimensional data and high-overlapping spectral signatures. simulation of high-dimensional images. Technical Report DECSAI-97-03-04, University of Granada, Spain, 1997.

[13] F. J. Cortijo and N. Perez De La Blanca. Improving classical contextual classifications. *Int. J. Remote Sensing*, 19(8):1591-1613.

[14] M. S. Crouse, R. D. Nowak, K. Mhirsi, and R. G. Baraniuk. Waveletdomain hidden Markov models for signal detection and classification. *SPIE Proceedings*, 3162:36–47, 1997.

[15] R. A. DeVore, B. Jawerth, and B. J. Lucier. Image compression through wavelet transform coding. *IEEE Trans. Inform. Theory*, 38:719–746, 1992.

[16] D. L. Donoho and I. M. Johnstone. Ideal spatial adaptation by wavelet shrinkage. *Biometrika*, 81:425–455, 1994.

[17] D. L. Donoho, I. M. Johnstone, G. Kerkyacharian, and D. Picard. Wavelet shrinkage: asymptopia? *J. Roy. Stat. Soc. Ser. B*, 57:301–369, 1995.

[18] B. Efron and R. Tibshirani. Improvement on cross-validation: the .632+ bootstrap method. *J. Am. Stat. Assoc.*, 92(438):548–560, 1997.

[19] M. Ekström. Nonparametric estimation of the variance of sample means based on nonstationary spatial data. *Comm. Stat. Theor. Meth*, 31:1743–1775, 2002.

[20] M. Ekström. Subsampling variance estimation for statistics computed from nonstationary spatial lattice data. Reseach Report 7, Centre of Biostochastics, SLU, Umeå, 2005.

[21] M. Ekström and S. Sjöstedt-de Luna. Subsampling methods to estimate the variance of sample means based on nonstationary spatial data with varying expected values. *J. Am. Stat. Assoc.*, 99:82–95, 2004.

[22] E. Fix and J. L. Hodges. Discriminatory analysis-nonparametric discrimination: consistency properties. Report 4, US Air Force School of Aviation Medicine, Random Field, Texas, 1951.

[23] A-M. Flygare. *Classification of Remotely Sensed Data Utilising the Autocorrelation between Spatio- Temporal Neighbours.* PhD thesis, Department of Mathematical Statistics, UmeåUniversity, 1997.

[24] H. Y. Gao. Wavelet shrinkage denoising using the non-negative garrots. *J. Comput. Graph. Stat.*, 7(4):469–488, December 1998.

[25] H. Y. Gao and A. G. Bruce. Waveshrink with firm shrinkage. *Statistica Sinica*, 7(4):855–874, 1997.

[26] S. Geman and D. Geman. Stochastic relaxation, gibbs distributions, and the bayesian restoration of images. it IEEE Trans. Pattern Analysis and Machine Intelligence, 6(6):721-742, November 1984.

[27] N. L. Hjort, E. Mohn, and G. Storvik. Contextual classification of remotely sensed data, based on an auto-correlation model. *In Sbfi & al.*, 1985.

[28] M. Jansen. *Noise Reduction by Wavelet Thresholding*, volume 161 of *Lecture Notes in Statistics.* Springer, New York, 2001.

[29] R. Jimenez and J. Yukich. Asymptotics for statistical distances based on voronoi tessellation. *J. Theor. Probab.*, 15:503–541, 2002.

[30] K. J. Jones. 3-d wavelet image processing for spatial and spectral resolution of landsat images. *SPIE Proceedings*, 3391:218–225, 1998.

[31] V. T. Kissiov and S. T. Hadjitodorov. A fuzzy version of the k-nn method. *Fuzzy Sets and Systems*, 49:323-329, 1992.

[32] R. Klein and S. J. Press. Contextual bayesian classification of remotely sensed data. *Comm. Stat. - Theor. Meth.*, 18(9):3177–3202, 1989.

[33] L. I. Kuncheva. *Fuzzy Classifier Design*, volume 49 of *Studies in Fuzzness and Soft Computing.* Springer, Heidelberg, Germany, 2000.

[34] T. M. Lillesand and R. W. Kiefer. *Remote Sensing and Image Intepretation.* John Wiley & Sons, Newe York, 1994.

[35] D. O. Loftsgaarden and C. P. Quesenberry. A nonparametric estimate of a multivariate density function. *Annals of Mathematical Statistics*, 36:1049–1051, 1965.

[36] S. Mallat. The theory for multiresolution signal decomposition: the wavelet representation. *IEEE Trans. Pattern Analysis and Machine Intelligence*, 11(7):654–693, 1989.

[37] Miljödepartementet. Framtidens miljö-allas vårt ansvar. Slutbetänkande från miljömålskommittén. *SOU*, (52), 2000.

[38] E. Mohn, N. L. Hjort, and G. O. Storvik. A simulation study of some contextual classification methods for remotely sensed data. *IEEE Trans. Geosci. Remote Sensing*, GE-25(6):796–804, November 1987.

[39] P. Müller and B. Vidakovic. *Bayesian Inference in Wavelet-Based Models.* Number 141 in Lecture Notes in Statistics. Springer, New York, 1999.

[40] NOAA. Visual interpretation of tm band combinations being studied. *Landsat Data Users Notes*, (30), March 1984.

[41] C. Rajski. A metric space of discrete probability distributions, *Information and Control*, 4:373–377, 1961.

[42] C. Rajski. On the normed information rate of discrete random variables. *Trans. Third. Prague Conference*, pages 583–585, 1964.

[43] B.Ranneby, S. R. Jammalamadaka, and A Teterukovskiy. The maximum spacing estimate for multivariate observation. *J. Stat. Plann. Infer.*, 129:427–446, 2005.

[44] B. Ranneby and J. Yu. Classification of agricultural crops and quality assessment using multispectral and multitemporal images. Research Report 7, Centre of Biostochastics, SLU, Umeå, Sweden, 2003.

[45] B. Ranneby and J. Yu. Remote sensing classification of environmental indicators. *Proceedings of 30th International Symposium on Remote Sensing of Environment-Information for Risk Management and Sustainable Development, Hawaii*, December 2003.

[46] B. D. Ripley. *Pattern Recognition and Neural Networks.* Cambridge University Press, Cambridge UK, 1996.

[47] N. Saito and R. Coifman. Local discriminant bases. *SPIE Proceedings-Wavelet Applications*, 2303:2–14, July 1994.

[48] K. M. S. Sharma and A. Sarkar. A modified contextual classification technique for remote sensing data. *Photogrammetric Engineering & Remote Sensing*, 64(4):273–280, April 1998.

[49] B. M. Steele, J. C. Winne, and R. L. Redmond. Estimation and mapping of misclassification probabilities for thematic land cover maps. *Remote Sens. Environ.*, 66:192–202, 1998.

[50] H. H. Szu, J. L. Moigne, N. S. Netanyahu, and C. C. Hsu. Integration of local texture information in the automatic classification of Landsat images. *SPIE Proceedings*, 3078:116–127, 1997.

[51] T. Taxt, N. Lid Hjort, and L. Eikvil. Statistical classification using a linear nixture of two multinormal probability densities. *Pattern Recognition Letters*, 12:731–737, 1991.

[52] A. Teterukovsky and B. Ranneby. The maximum spacing estimation based on dirichlet tessellation. 2003. in Doctoral thesis Silvestria 277, Swedish University of Agricultural Sciences.

[53] A. Teterukovsky and J. Yu. Contextual reclassification of multispectral images: A markov random field approach. *Information Processes*, 2:12–21, 2002.

[54] T. Watanabe, H. Suzuki, S. Tanba, and R. Yokoyama. Improved contextual classifiers of multispectral image data. *IEICE Trans. Fundamentals*, E77-A(9):1445–1450, September 1994.

[55] G. Winkler. *Image Analysis, Random Fields and Markov Chain Monte Carlo Methods.* Springer, Berlin, second edition, 2006.

[56] Y. Yamaguchi, T. Nagai, and H. Yamada. JERS-1 SAR image analysis by wavelet transform. *IEICE Trans. Commun.*, E78B(12):1617–1621, December 1995.

[57] J. Yu and M. Ekström. Multispectral image classification using wavelets: a simulation study. *Pattern Recognition*, 36:889–898, 2002.

[58] J. Yu, M. Ekström, and M. Nilsson. Image classification using wavelets with application to forestry. *Proceedings of the 4th International Conference on Methodological Issues in Official Statistics, Stockholm*, October 12–13 2000.

[59] J. Yu and B. Ranneby. Nonparametric and probabilistic classification of agricultural crops using multitemporal images. *Journal of Remote Sensing*, 5, 2007. To appear.

A Appendix

Table A.1: Mean misclassification rates (%) for $\alpha = 0$

	1-NN	3-NN	10-NN	LDA	QDA
orig	19.36	16.44	14.88	**13.20**	13.30
haar.n1	15.85	15.23	13.66	**9.03**	9.99
haar.n2	17.30	16.84	15.94	**13.22**	14.64
haar.n3	18.45	18.48	18.94	**18.06**	19.15
shrink.n1.50	9.50	7.64	7.05	**6.21**	6.24
shrink.n2.50	8.64	7.09	6.63	5.80	**5.78**
shrink.n3.50	8.71	7.29	6.80	5.94	**5.88**
shrink.n1.60	10.26	7.82	7.11	**6.31**	6.38
shrink.n2.60	8.60	7.05	6.50	<u>**5.77**</u>	<u>**5.77**</u>
shrink.n3.60	8.51	7.12	6.60	5.89	**5.87**
shrink.n1.70	11.97	8.98	8.09	**7.33**	7.37
shrink.n2.70	9.32	7.68	6.91	**6.27**	6.37
shrink.n3.70	8.97	7.70	6.90	**6.30**	6.35

Table A.2: Mean misclassification rates (%) for $\alpha = 0.25$

	1-NN	3-NN	10-NN	LDA	QDA
orig	18.95	16.20	14.88	13.15	**13.12**
haar.n1	18.05	17.58	15.97	**11.51**	12.60
haar.n2	18.03	17.64	17.16	**14.12**	16.22
haar.n3	18.34	**18.27**	18.94	18.36	19.52
shrink.n1.50	12.09	10.25	9.65	**8.24**	8.31
shrink.n2.50	10.42	8.70	8.17	7.07	**7.04**
shrink.n3.50	9.95	8.24	7.70	6.86	**6.76**
shrink.n1.60	12.98	10.65	9.91	**8.55**	8.68
shrink.n2.60	10.43	8.65	7.97	**6.92**	6.96
shrink.n3.60	9.88	8.16	7.45	6.66	<u>**6.56**</u>
shrink.n1.70	14.40	11.83	10.93	**9.51**	9.56
shrink.n2.70	10.88	9.02	8.31	**7.33**	7.43
shrink.n3.70	10.13	8.49	7.72	6.99	**6.93**

Table A.3: Mean misclassification rates (%) for $\alpha = 0.40$

	1-NN	3-NN	10-NN	LDA	QDA
orig	24.65	21.21	19.25	17.31	**17.27**
haar.n1	21.86	22.10	19.69	**14.57**	15.83
haar.n2	19.46	19.22	18.67	**15.53**	18.11
haar.n3	18.40	**18.24**	18.78	18.81	19.67
shrink.n1.50	17.19	15.09	13.77	**11.90**	11.94
shrink.n2.50	15.10	12.71	11.67	10.24	**10.23**
shrink.n3.50	14.27	12.10	10.94	9.64	**9.60**
shrink.n1.60	17.21	15.06	13.90	**11.96**	12.05
shrink.n2.60	14.43	12.17	11.09	9.81	**9.77**
shrink.n3.60	13.50	11.17	10.14	9.05	**9.03**
shrink.n1.70	18.16	15.92	14.75	**12.72**	12.89
shrink.n2.70	14.20	12.13	10.86	9.82	**9.77**
shrink.n3.70	12.88	10.95	9.93	**8.91**	8.95

Table A.4: Mean misclassification rates (%) for $\alpha = 0.45$

	1-NN	3-NN	10-NN	LDA	QDA
orig	16.70	14.32	13.07	11.68	**11.51**
haar.n1	17.21	17.07	15.31	**12.82**	13.30
haar.n2	18.21	18.17	18.06	**16.52**	17.54
haar.n3	**18.04**	18.19	18.83	19.87	19.33
shrink.n1.50	12.48	10.56	10.00	8.94	**8.90**
shrink.n2.50	11.62	9.75	9.07	8.37	**8.17**
shrink.n3.50	11.06	9.44	8.81	8.20	**7.95**
shrink.n1.60	13.51	11.32	10.45	9.34	**9.31**
shrink.n2.60	11.58	9.80	9.19	8.57	**8.36**
shrink.n3.60	11.01	9.32	8.77	8.33	**8.06**
shrink.n1.70	15.30	12.80	11.75	10.61	**10.59**
shrink.n2.70	11.85	10.17	9.71	9.21	**9.01**
shrink.n3.70	11.33	9.76	9.02	8.71	**8.58**

Table A.5: Mean misclassification rates (%) for $\alpha = 0.72$

	1-NN	3-NN	10-NN	LDA	QDA
orig	20.82	17.87	16.22	14.95	**14.67**
haar.n1	20.14	20.55	18.57	**16.15**	16.66
haar.n2	19.50	19.88	20.41	**19.21**	19.93
haar.n3	**18.13**	18.20	18.86	21.19	20.37
shrink.n1.50	16.47	14.71	13.59	12.46	**12.28**
shrink.n2.50	15.53	13.55	12.69	11.86	**11.56**
shrink.n3.50	15.30	13.20	12.28	11.60	$\underline{\mathbf{11.24}}$
shrink.n1.60	17.25	15.24	14.10	12.83	**12.74**
shrink.n2.60	15.38	13.52	12.70	11.97	**11.73**
shrink.n3.60	14.74	12.89	12.01	11.70	**11.34**
shrink.n1.70	18.64	16.69	15.38	14.01	**13.89**
shrink.n2.70	15.51	13.84	13.29	12.59	**12.31**
shrink.n3.70	14.79	13.01	12.13	12.15	**11.88**

Table A.6: Mean misclassification rates (%) for STRIPE: the lowest for each α-value. *Legends:* $*$ *QDA*, $\star$ *LDA*, $\dagger$ *1-NN*, $\ddagger$ *3-NN.*

	$\alpha = 0$	$\alpha = 0.25$	$\alpha = 0.40$	$\alpha = 0.45$	$\alpha = 0.72$
orig	13.03^{*}	12.98^{*}	17.16^{*}	11.99^{*}	15.46^{*}
haar.n1	$0.66^{\star}$	$3.17^{\star}$	$6.59^{\star}$	$4.68^{\star}$	8.29^{*}
haar.n2	$7.86^{\dagger}$	$8.59^{\dagger}$	$9.60^{\dagger}$	$9.26^{\dagger}$	$10.79^{\dagger}$
haar.n3	$19.79^{\dagger}$	$17.66^{\dagger}$	$17.34^{\dagger}$	$17.95^{\dagger}$	$17.98^{\dagger}$
shrink.n1.85	0.53^{*}	$2.91^{\star}$	6.19^{*}	4.33^{*}	8.10^{*}
shrink.n2.85	0.27^{*}	1.21^{*}	$2.76^{\dagger}$	$2.58^{\dagger}$	$4.69^{\dagger}$
shrink.n3.85	0.34^{*}	1.00^{*}	$2.64^{\dagger}$	$2.59^{\dagger}$	$4.82^{\dagger}$
shrink.n1.90	0.58^{*}	$2.91^{\star}$	6.19^{*}	4.33^{*}	8.10^{*}
shrink.n2.90	$0.20^{\dagger}$	$1.07^{\dagger}$	$2.42^{\dagger}$	$1.89^{\dagger}$	$3.44^{\dagger}$
shrink.n3.90	0.19^{*}	$0.78^{\dagger}$	$2.05^{\dagger}$	$1.84^{\dagger}$	$3.48^{\dagger}$
shrink.n1.95	0.58^{*}	$2.91^{\star}$	6.19^{*}	4.33^{*}	8.10^{*}
shrink.n2.95	10.00^{*}	$10.16^{\star}$	$10.91^{\star}$	$11.61^{\ddagger}$	$12.46^{\dagger}$
shrink.n3.95	$\underline{0.15}^{\dagger}$	$\underline{0.65}^{\dagger}$	$\underline{1.57}^{\dagger}$	$\underline{1.23}^{\dagger}$	$\underline{1.87}^{\dagger}$

Table A.7: Class definition and thematic levels for the study area

Crops	Level 3	Level 2	Level 1
Winter barley	1	1	1
Spring barley	2	2	1
Oats	3	2	1
Winter wheat	4	1	1
Spring wheat	5	2	1
Mixed grain	6	2	1
Rye	7	1	1
Spring turnip rape	8	3	1
Peas	9	7	1
Linseed	10	3	1
Potatoes	11	4	1
Pasture or grass for hay or silage on arable land	12	5	1
Seed ley	13	5	1
Grazing land	14	7	2
Meadow	15	5	2
Woodland pasture	16	7	2
Fallow (one year)	17	7	1
Fallow (> one year)	18	7	1
Energy forest (salix)	19	6	1
Reed canary grass (1999)	20	5	1
Strawberries	21	7	1
Other cultivation of berries	22	7	1
Forest plantation on arable land	23	7	1
Other land use	24	7	2
Boundary pixels	25	8	3

Table A.8: Definition of the thematic levels

Code	Level 2	Level 1
1	Autumn-sown cereals	Arable land
2	Spring-sown cereals	Grazing land
3	Spring-sown oil seed crops	Boundary pixels
4	Potatoes	
5	Grass land on arable land for hay or silage	
6	Energy forest (salix)	
7	Other crops	
8	Boundary pixels	

Table A.9: Probability matrix using five seasonal scenes at level 3

	C_1	C_2	C_3	C_4	C_5	C_6	C_7	C_8	C_9	C_{10}	C_{11}	C_{12}	C_{13}	C_{14}	C_{15}	C_{16}	C_{17}	C_{18}	C_{19}	C_{20}	C_{21}	C_{22}	C_{23}	C_{24}	C_{25}
C_1	.38	.35	.14	.04	0	0	.01	0	0	0	0	.01	0	.01	0	0	.05	0	0	0	0	0	0	0	0
C_2	.02	.65	.11	.01	0	.02	.01	0	0	0	0	.03	.01	.03	0	.01	.08	.01	0	0	0	0	0	0	.01
C_3	.02	.22	.55	0	0	0	.01	0	0	.01	0	.02	.01	.03	0	.01	.10	0	0	0	0	0	0	.01	.01
C_4	.08	.22	.03	.36	0	0	.07	0	0	0	0	.04	0	.03	0	0	.15	.02	0	0	0	0	0	0	0
C_5	0	0	0	0	.93	0	0	0	0	0	0	0	0	0	0	0	.07	0	0	0	0	0	0	0	0
C_6	0	.03	0	.02	0	.90	0	0	0	0	0	.02	0	0	0	0	.03	0	0	0	0	0	0	0	0
C_7	.03	.15	0	.07	0	0	.67	0	0	0	0	0	0	0	0	0	.04	.03	0	0	0	0	0	0	0
C_8	0	.04	.01	0	0	0	0	.77	.12	0	.01	0	.01	0	0	.01	.02	0	0	0	0	0	0	0	0
C_9	0	.07	.04	0	0	0	0	0	.82	0	0	.01	0	.03	0	0	.02	0	0	0	0	0	0	0	0
C_{10}	.01	.11	.02	0	0	0	.01	0	0	.46	0	.02	0	0	0	0	.36	0	0	0	0	0	0	0	.01
C_{11}	0	.06	.03	0	0	0	.01	0	0	.01	.65	.01	.02	.03	.01	.01	.14	.02	0	0	.01	0	0	0	0
C_{12}	0	.01	0	.01	0	0	.01	0	0	.02	0	.62	0	.10	0	.02	.13	.01	.01	.01	.01	0	0	0	.02
C_{13}	.02	.13	.27	0	0	0	0	0	0	0	0	.06	.49	.02	0	0	.01	0	0	0	0	0	0	0	0
C_{14}	0	.02	.01	0	0	0	0	0	0	0	0	.25	0	.48	.02	.11	.04	.02	.01	0	0	.01	0	.01	.03
C_{15}	0	0	0	0	0	0	0	0	0	.02	0	.12	0	.68	.13	.04	0	0	0	0	0	0	0	0	0
C_{16}	0	.01	0	0	0	0	0	0	0	.01	.01	.04	0	.23	0	.65	.01	.01	0	0	0	0	0	.01	.01
C_{17}	0	.05	.01	.01	0	.01	.01	0	0	0	0	.07	0	.08	0	.02	.63	.02	.01	.02	0	0	0	.02	0
C_{18}	0	.02	.01	0	0	0	.01	0	0	.02	0	.25	0	.17	0	.03	.16	.21	.01	.02	0	0	0	.04	.03
C_{19}	0	.02	.03	0	0	0	0	0	0	0	0	.11	0	.12	0	.03	.05	.05	.56	0	0	0	.01	.01	.01
C_{20}	0	0	0	0	0	0	0	0	0	.02	0	0	0	.02	0	.10	.15	0	0	.71	0	0	0	0	0
C_{21}	0	.03	0	0	0	0	0	0	0	0	.06	.03	0	0	0	0	.17	0	0	0	.72	0	0	0	0
C_{22}	0	.02	0	0	0	0	0	0	0	0	0	.10	.02	.42	0	.02	.07	0	0	0	0	.35	0	0	.02
C_{23}	0	0	.02	0	0	0	0	0	0	0	0	.08	0	.19	0	0	0	.02	.10	0	0	0	.59	.02	0
C_{24}	0	.05	.01	0	0	0	.01	0	0	0	0	.13	0	.32	0	.04	.17	.06	.03	0	0	0	0	.16	.01

Table A.10: Probability matrix using one single scene at level 3

	C_1	C_2	C_3	C_4	C_5	C_6	C_7	C_8	C_9	C_{10}	C_{11}	C_{12}	C_{13}	C_{14}	C_{15}	C_{16}	C_{17}	C_{18}	C_{19}	C_{20}	C_{21}	C_{22}	C_{23}	C_{24}	C_{25}
C_1	.08	.30	.12	.06	.01	0	.01	0	.01	0	.01	.07	0	.05	0	.03	.19	.01	.01	0	0	0	0	.01	.02
C_2	.01	.28	.25	.03	.01	0	0	0	.04	.01	.01	.05	.04	.04	0	.04	.11	.01	.01	0	0	0	.01	0	.02
C_3	.01	.19	.30	.03	.02	0	0	0	.04	.01	.01	.06	.04	.05	0	.04	.13	.01	.01	0	0	0	.01	0	.02
C_4	.01	.16	.06	.19	0	0	.02	0	0	.02	.03	.13	0	.08	.01	.04	.24	.01	0	0	0	0	0	0	0
C_5	0	.04	.14	0	.21	0	0	.07	.04	0	.07	.11	0	.14	0	0	.18	0	0	0	0	0	0	0	0
C_6	.02	.13	.12	0	0	.05	0	0	0	0	0	.23	0	.17	0	0	.05	.13	0	0	0	0	0	.03	.07
C_7	0	0	0	.09	0	0	.21	0	0	.03	.04	.19	0	.03	0	0	.39	0	0	0	0	0	0	0	.01
C_8	0	.01	.02	0	.02	0	0	.82	.08	.01	0	0	.01	0	0	0	.02	0	0	0	0	0	0	0	0
C_9	0	0	.01	0	.01	0	0	.08	.84	0	0	.02	0	0	0	.03	0	0	0	0	0	0	0	0	0
C_{10}	0	.07	.05	.03	.03	0	0	.01	0	.14	.01	.07	0	.01	0	0	.35	.01	0	.15	.07	0	0	0	0
C_{11}	0	0	.01	.02	.01	0	0	0	0	.01	.17	.18	.02	.01	.02	.05	.25	0	0	.16	.08	0	0	0	.01
C_{12}	0	.02	.02	.01	0	0	0	0	0	0	.05	.35	.05	.10	.05	.03	.14	.01	.02	.03	.06	0	.02	.01	.02
C_{13}	0	.21	.04	.02	.01	0	0	0	0	0	.02	.06	.41	.11	0	.01	.04	0	.01	0	0	0	0	.01	.04
C_{14}	0	.03	.02	.01	0	0	0	0	0	0	.01	.23	.06	.31	.01	.15	.04	.01	.05	0	0	0	.01	.02	.03
C_{15}	0	0	0	0	0	0	0	0	0	0	.01	.16	.08	.34	.10	.05	.04	0	.15	0	.02	0	.04	0	0
C_{16}	0	.01	0	.01	0	0	0	0	0	0	0	.09	0	.18	.01	.67	.01	0	.01	0	0	0	0	.01	0
C_{17}	0	.04	.03	.03	.01	0	.01	0	0	.02	.03	.14	.02	.10	.02	.04	.30	.01	.02	.09	.06	0	0	0	.02
C_{18}	0	.04	.03	.03	.01	0	0	0	0	.01	.03	.21	.04	.13	.01	.07	.14	.07	.02	.03	.03	0	.07	.01	.03
C_{19}	.01	.05	.09	0	0	0	0	0	0	0	0	.07	.04	.27	0	.05	.05	.01	.28	0	0	0	.02	.02	.03
C_{20}	0	0	0	0	0	0	0	0	0	0	.15	0	0	0	0	.05	.22	0	0	.54	.05	0	0	0	0
C_{21}	0	0	0	0	.03	0	0	0	0	.03	0	.06	0	0	.11	0	.17	0	0	.03	.58	0	0	0	0
C_{22}	0	0	0	0	0	0	0	0	0	0	.02	.37	0	.42	.02	.02	0	0	0	0	0	.10	0	0	.07
C_{23}	0	.05	0	0	0	0	0	0	0	0	0	.14	.06	.10	0	.02	.02	0	.06	0	0	0	.54	.02	0
C_{24}	0	.04	.03	.01	.01	0	0	0	0	0	.01	.17	.03	.26	.01	.05	.10	.01	.11	.01	.01	0	.02	.07	.02

Table A.11: Area estimation for all types of units

	5 seasonal scenes		one single scene	
Class	Observed	Corrected	Observed	Corrected
C_1	1.41%	0.89%	0.45%	0.81%
C_2	25.89%	30.96%	13.20%	29.05%
C_3	11.06%	11.60%	13.05%	10.64%
C_4	0.99%	0.98%	1.93%	0.92%
C_5	0.17%	0.06%	0.70%	0.04%
C_6	0.42%	0.06%	0.02%	0.08%
C_7	0.95%	0.09%	0.33%	0.08%
C_8	0.84%	0.95%	1.07%	0.94%
C_9	0.37%	0.24%	2.32%	0.31%
C_{10}	1.39%	0.22%	0.55%	0.21%
C_{11}	1.47%	1.30%	2.60%	1.05%
C_{12}	25.82%	32.50%	19.51%	34.00%
C_{13}	0.61%	0.32%	4.43%	0.32%
C_{14}	6.70%	8.82%	7.22%	10.21%
C_{15}	0.33%	0.21%	2.16%	0.27%
C_{16}	1.60%	0.29%	3.44%	0.34%
C_{17}	14.86%	7.35%	13.98%	7.35%
C_{18}	1.05%	1.21%	0.76%	1.33%
C_{19}	1.25%	1.24%	1.66%	1.28%
C_{20}	0.58%	0.05%	2.48%	0.06%
C_{21}	0.49%	0.07%	3.66%	0.05%
C_{22}	0.11%	0.05%	0.15%	0.07%
C_{23}	0.12%	0.05%	1.45%	0.07%
C_{24}	0.59%	0.48%	0.52%	0.52%
C_{25}	0.93%		2.38%	

Table A.12: Class definition and thematic levels for scene 194-018

Crops	Level 3	Level 2	Level 1
Winter barley	1	1	1
Spring barley	2	2	1
Oats	3	2	1
Winter wheat	4	1	1
Spring wheat	5	2	1
Mixed grain	6	2	1
Triticale	7	1	1
Rye	8	1	1
Spring turnip rape	9	3	1
Peas	10	7	1
Linseed	11	3	1
Potatoes	12	4	1
Pasture or grass for hay or silage on arable land	13	5	1
Grazing land	14	7	2
Meadow	15	5	2
Woodland pasture	16	7	2
Fallow (one year)	17	7	1
Fallow ($>$ one year)	18	7	1
Energy forest (salix)	19	6	1
Strawberries	20	7	1
Other cultivation of berries	21	7	1
Other land use	22	7	2
Boundary pixels	23	8	3

Table A.13: Class definition and thematic levels for scene 196-017

Crops	Level 3	Level 2	Level 1
Winter barley	1	1	1
Spring barley	2	2	1
Oats	3	2	1
Mixed grain	4	2	1
Potatoes	5	4	1
Pasture or grass for hay or silage on arable land	6	5	1
Grazing land	7	7	2
Meadow	8	5	2
Woodland pasture	9	7	2
Fallow (one year)	10	7	1
Strawberries	11	7	1
Other land use	12	7	2
Boundary pixels	13	8	3

Mixed Linear Model Approaches for Complex Trait Analysis*

Jian Yang and Jun Zhu
Institute of Bioinformatics, Zhejiang University
Hangzhou, 310029, China
E-mail: jzhu@zju.edu.cn

1 Introduction

In contrast to the Mendelian traits controlled by single genes, there are complex traits of which their phenotypic variations result from the segregation of alleles at multiple quantitative trait loci with effects sensitive to the genetic, sexual, and environmental factors (Mackay 2001). Many experiments and surveys for complex trait analysis often contain random sources of variation, such as the NC (North Carolina) designs I and II (Comstock and Robinson, 1952; Hallauer and Miranda, 1981), diallel mating designs (Griffing 1956; Gardner and Ebarhert 1966), and multiple-environment trials (METs) in plant and animal genetics, and the cross-over trial and multi-centre trail in medical researches (Brown and Prescott 1999). Systematic statistical methods have been developed for partitioning the variation of experimental output into the attribution of several variance components, including experimental (or biological) factors, controllable systematic errors and random noises. The majority of these methods focus on analysis of variance (ANOVA) by testing the significance of the variance components based on the reduction of sum of square (Searle et al. 1992). However, in some cases, researchers desired to know the values of the variance and effects for each explanatory factor, which is unable to be solved in the ANOVA methods. To meet these requirements, some other methods were developed, such as maximum likelihood (ML, Hartly and Rao 1967), restricted maximum likelihood (REML, Patterson and Thompson 1971) and minimum norm quadratic unbiased estimation (Rao 1971) for variance estimation, as well as the best linear unbiased predictor (BLUP, Henderson 1963) and linear

*The preparation of this manuscript was partially supported by the National Basic Research Program of China (973 Program) and the National High Technology Research and Development Program of China (863 Program).

unbiased predictor (LUP, Zhu and Weir 1996) for random effect prediction. These methods have been greatly achieved in plant and animal genetics (Zhu and Weir 1994; Tempelman 1998) and clinical researches (Putt and Chinchilli 1999; Rashid 2003), however, they presented poor computational efficiency because of requiring to calculate the inverse of an matrix (where n is the number of observations). With the advance of biotechnology, the commonly used mixed linear model approaches encountered great challenges, especially in mapping quantitative trait loci (QTLs) and analysis of microarray data. In QTL mapping analysis with data from METs, likelihood ratio (LR) tests of mixed linear models were conducted to scan genome for putative QTLs by data from multiple environments (Wang et al. 1999; Piepho 2000). Since LR test also requires calculating the inverse of an matrix, genome scan procedure conducted by LR tests is intensively time-consuming, especially when performing two-dimensional genome scan for digenic epistasis. Microarray techniques are powerful laboratory tools to simultaneously monitor the genome-wide expression in different conditions. A common microarray experiment usually involves more than thousands of genes measured in several chips under multiple biological or technical treatments. In the analysis of differentially expressed genes (DEGs) by mixed linear model, the overall data from such a huge experiment is unable to be handled by a single model, therefore, each gene is separately tested for differential expression under different treatment levels, and consequently results in large number of tests for the overall data of one experiment. More efficient mixed linear approach is desired to handle these kinds of hug datasets. In this chapter, we purport to introduce the mixed linear model approaches used for complex trait analysis. We start with the commonly used approaches such as REML, MINQUE, and BLUP etc. Then we propose an efficient approach combined with Henderson method III for testing the significance of a group of effects (Searle 1992) and Bayesian method via Gibbs sampling for estimating variance components and fixed effects, predicting random effects and performing statistical inference of them (Wang et al. 1994). This approach is much more efficient than the conventional approaches, and can meet the requirement for computational efficiency in QTL mapping and microarray data analysis. Finally, we introduce the applications of this approach for QTL mapping and microarray data analysis.

2 Commonly Used Approaches for Parameter Estimation and Statistical Inference of Mixed Linear Model

A general form of mixed linear model can be expressed as

$$\begin{aligned} y &= Xb + U_1e_1 + U_2e_2 + \cdots + U_re_r + e_\epsilon \\ &= Xb + \sum_{u=1}^{r+1} U_ue_u \\ &\sim N(Xb, V = \sum_{u=1}^{r+1} \sigma_u^2 U_u R_u U_u' = \sum_{u=1}^{r+1} \sigma_u^2 Z_u), \end{aligned} \tag{2.1}$$

where y is an $(n \times 1)$ vector of trait values; b is an $(1 \times p)$ vector of fixed effects with the known coefficient matrix X; e_u $(u = 1, 2, \cdots, r)$ is an $(1 \times q_u)$ vector of the u-th random effect with the known coefficient matrix U_u; $e_u \sim N(0, \sigma_u^2 R_u)$, where R_u is Wright's relationship matrix describing the relationship among the elements of e_u; e_ϵ is an $(n \times 1)$ vector of residual effects; $e_{r+1} = e_\epsilon \sim N(0, \sigma_\epsilon^2 I)$, where I (U_{r+1}) is an $(n \times n)$ identity matrix.

2.1 Variance Component Estimation and Significance Test

The methods for estimating variance components e.g. maximum likelihood (ML; Hartley and Rao 1967), restricted maximum likelihood (REML; Patterson and Thompson 1971), or minimum norm quadratic unbiased estimation (MINQUE; Rao 1971) were developed in 1970's and were firstly applied in animal genetics and breeding. Under the normal assumption of trait data, the ML estimators can be solved based on Hartley-Rao (HR) equation via an iterative process. For the h-th iterative step, the HR equation can be written as

$$[tr(\hat{V}_{[h]}^{-1} Z_u)\hat{V}_{[h]}^{-1} Z_v][\hat{\sigma}^2_{u[h+1]}] = [y'\hat{Q}_{[h]} Z_u \hat{Q}_{[h]} y], \tag{2.2}$$

where $\hat{Q}_{[h]} = \hat{V}_{[h]}^{-1} - \hat{V}_{[h]}^{-1} X (X'\hat{V}_{[h]}^{-1} X)^{-} X' \hat{V}_{[h]}^{-1}$, $\hat{V}_{[h]} = \sum_u^{r+1} \hat{\sigma}^2_{u[h]} Z_u$, and $u, v = 1, 2, \cdots, r+1$.

At the beginning of the iterative process, arbitrary positive value is assigned to each variance component. Then, the h-th estimators are given as the initiators for the $[h+1]$-th iterative step until the estimators obtained from the h-th and $[h+1]$-th steps converged. If there are any negative values of the solution to equation (2), they can be replaced by zeros for $u, v = 1, 2, \cdots, r$ and a small positive value for $u, v = r+1$

(Giesbrecht and Burns 1985). In the presence of fixed effects in the HR equation, ML estimates are influenced by the fixed effects and thus rarely used for estimation of variance components. The REML method can overcome the influence of the fixed effects. Unlike ML method, REML method only maximizes the portion of the likelihood that does not depend on the fixed effects, and gets the following estimation equation

$$[tr(\hat{Q}_{[h]}Z_u\hat{Q}_{[h]}Z_v)][\hat{\sigma}^2_{u[h+1]}] = [y'\hat{Q}_{[h]}Z_u\hat{Q}_{[h]}y]. \tag{2.3}$$

As compared to ML, REML is a preferable method for analyzing large data sets with complex structure, which provides the estimates of variance components that are identical to those obtained by ANOVA method for completely balanced data (Lynch and Walsh 1997). The MINQUE method does not need the assumption of normal distribution for the observed data, and can give estimated variances without iteration. MINQUE method is used to consider the effects of minimizing the Euclidian norm rather than minimizing a variance. It uses the following equation for estimating the variance components

$$[tr(Q_\alpha Z_u Q_\alpha Z_v)][\hat{\sigma}^2_u] = [y'Q_\alpha Z_u Q_\alpha y], \tag{2.4}$$

where $Q_\alpha = V_\alpha^{-1} - V_\alpha^{-1}X(X'V_\alpha^{-1}X)^- X'V_\alpha^{-1}$ with $V_\alpha = \sum_u^{r+1} \alpha_u Z_u$; $[\alpha_u]$ are prior values.

As compared to ML and REML methods, MINQUE method has the advantages of simple computation, and no requirement for normal distribution. It has been proved that if the prior values are chosen not based on the observation data, MINQUE estimates are unbiased. MINQUE(1), which has 1 for all the prior values, is widely used in most applications of MINQUE method (Giesbrecht 1983).

For large sample size, REML and MINQUE estimates have asymptotic multivariate normal distribution,

$$[\hat{\sigma}^2_u] \sim N([\sigma^2_u], 2H^{-1}), \tag{2.5}$$

where $H = [tr(QZ_uQZ_v)]$ with $Q = V^{-1} - V^{-1}X(X'V^{-1}X)^- X'V^{-1}$ for REML (Searle 1970); $H = [tr(Q_\alpha Z_u Q_\alpha Z_v)]$ with $Q_\alpha = V_\alpha^{-1} - V_\alpha^{-1}X(X'V_\alpha^{-1}X)^- X'V_\alpha^{-1}$ for MINQUE (Rao and Kleffe 1980). Under the null hypothesis H_0: $\sum_{u=1}^{r+1} c_u\sigma_u = \rho$, we can have the following χ^2-statistic

$$\chi^2_{df=1} = \frac{(\sum_{u=1}^{r+1} c_u\hat{\sigma}^2_u - \rho)^2}{2(\sum_{u=1}^{r+1} c_u^2 \hat{H}^{-1}_{uu} + 2\sum_{u=1}^{r}\sum_{u<v}^{r+1} c_u c_v \hat{H}^{-1}_{uv})}. \tag{2.6}$$

Zhu (1989) conducted Monte Carlo simulation to investigate the powers of testing the significance of variance components by equation (2.6), and pointed out that using χ^2-test based on the asymptotic multivariate normal distribution, the powers of detecting significant variance components are awfully low. Zhu and Weir (1996) suggested using the Jackknife technique (Miller 1974) to construct a t-statistic by resampling observations, which could significantly improve the powers.

In addition, with simple extension, the MINQUE method can be used to estimate the covariance components between two correlated traits (Zhu and Weir 1994)

$$[tr(Q_\alpha Z_u Q_\alpha Z_v)][\hat{\sigma}_{u/u}] = [y_1' Q_\alpha Z_u Q_\alpha y_2]. \tag{2.7}$$

When using MINQUE method for estimating the variance-covariance components of two related traits, the prior values of variance components are given as 1 and these of covariance components are given as 0, which is therefore named as MINQUE(0/1) (Zhu and Weir, 1994). Thus, we can have the following general equation for estimating variance and covariance components for one trait ($y_1 = y_2$), and covariance components between two traits ($y_1 = y_2$)

$$[tr(Q_{(0/1)} Z_u Q_{(0/1)} Z_v)][\hat{\sigma}_{u/u}] = [y_1' Q_{(0/1)} Z_u Q_{(0/1)} y_2], \tag{2.8}$$

where $Q_{(0/1)} = V_{(0/1)}^{-1} - V_{(0/1)}^{-1} X (X' V_{(0/1)}^{-1} X)^{-} X' V_{(0/1)}^{-1}$ with $V_{(0/1)} = \sum_u^{r+1} Z_u$.

2.2 Analysis of Fixed and Random Effects

Two of the mostly used approaches for fixed effect estimation in mixed linear model are the generalized least square (GLS) and ordinary least square (OLS). The general formulae for GLS and OLS estimates can be expressed as

$$\hat{b} = (X'V^{-1}X)^{-1}X'V^{-1}y \sim N(b, (X'V^{-1}X)^{-1})(\text{GLS}), \tag{2.9}$$

$$\hat{b} = (X'X)^{-1}X'y \sim N(b, (X'X)^{-1}X'VX(X'X)^{-1})(\text{OLS}), \tag{2.10}$$

Under the null hypothesis $H_0 : c'b = m$, a t-statistic can be constructed with degrees of freedom $df = rank(X)$ by $t^* = |c'(\hat{b}-b)/\sqrt{c'(X'V^{-1}X)^{-1}c}|$ for GLS and $t^* = |c'(\hat{b}-b)/\sqrt{c'(X'X)^{-1}X'VX(X'X)^{-1}c}|$ for OLS, respectively. By replacing the variance components in V by their unbiased estimates, the above t-statistic can be used to test the difference between any effect and zero, as well as to do pair-wise comparison and any specific linear contrast of several effects. As regarded to the random effects, Henderson (1963) described a prediction procedure known as BLUP (best linear unbiased prediction) by the formula as

$$\hat{e}_u = \sigma_u^2 R_u U_u' Q y \sim N(0, \sigma_u^2 R_u U_u' Q U_u' R_u) \tag{2.11}$$

where $Q = V^{-1} - V^{-1}X(X'V^{-1}X)^{-}X'V^{-1}$. In this formula, the real parameters of all the variance components are required. In practices, each σ_u^2 is replaced by its unbiased estimate, and thus BLUP losses the property of "best" and "linear". Based on the MINQUE approach, Zhu and Weir (1996) proposed a method named as LUP (linear unbiased estimation) for random effect prediction

$$\hat{e}_u = \alpha_u R_u U_u' Q_\alpha y \sim N(0, \alpha_u^2 R_u U_u' Q_\alpha V Q_\alpha U_u' R_u), \tag{2.12}$$

where Q_α has the same definition as that in equation (2.4). Analog to that of MINQUE method, if the prior values are chosen not based on the observation data, the LUP predictors are unbiased. Hypothesis tests of random effects can be conducted by constructing t-statistics according to the distributions of the BLUP and LUP predictors with some approximations. Alternatively, we also can use the Jackknife resampling technique which was thought to be more powerful (Zhu and Weir 1996).

3 HAB (Henderson and Bayesian) Method for Analysis of Mixed Linear Model

All the aforementioned approaches require to calculate the inverse of variance-covariance matrix V with the size of $n \times n$, which has the computational complexity approximate to $O(n^3)$. Moreover, if the hypothesis test is based on resampling techniques such as Jackknife or Bootstrap, the whole procedure will be rather time-consuming. In the following two sections, we introduce an efficient approach combined with Henderson method III for significance test of the effects of interest and Bayesian method via Gibbs sampling for parameter estimation. This approach can completely avoid of calculating the inverses of $n \times n$ matrices.

3.1 Test Significant of Multiple Effects by Henderson method III

Henderson method III borrows the idea from analysis of fixed effect models based on sums of squares. The sums of squares used are the reductions in sums of squares due to fitting full model and various reduced models (Searle et al. 1992). In equation (1), we clearly distinguish the fixed effects and random effects. Here, we remove this distinction and combine b and all the e_u into θ, and write the model equation as

$$y = W\theta + \epsilon, \tag{3.1}$$

where $W = [X \vdots U_1 \vdots U_2 \vdots \cdots \vdots U_r]$. Analog to that in fixed model analysis, the reduction sum of square due to fitting the model (3.1) can be expressed as

$$R(\theta) = y'W(W'W)^{-}W'y. \tag{3.2}$$

Consider a sub-vector of θ, denoted as θ_T , and denote the remaining elements of θ in vector form as θ_C. Thus, model (3.1) can be re-written as

$$y = W_C\theta_C + W_T\theta_T + \epsilon. \tag{3.3}$$

When θ_T is excluded from model (3.3), the reduction sum of squares will decrease by

$$\begin{aligned} R(\theta_T|\theta_C) &= y'W(W'W)^{-}W'y - y'W_C(W_C'W_C)^{-}W_C'y \\ &= y'M_CW_T(W_T'M_CW_T)^{-}W_T'M_Cy, \end{aligned} \tag{3.4}$$

where $M_C = I - W_C(W_C'W_C)^{-1}W_C' = M_C' = M_C^2$ with $M_CW_C = 0$. Without knowing which elements of θ are fixed effects and which are random, we can have the variance of θ as

$$Var(\theta) = E(\theta\theta') - E(\theta)E(\theta'), \tag{3.5}$$

For any quadric form $y'Ay$, with equation (3.5) we have the expectation of which as

$$\begin{aligned} E(y'Ay) &= tr[AWVar(\theta)W' + A\sigma_\epsilon^2 I] + E(\theta')W'AWE(\theta) \\ &= tr[W'AWVar(\theta) + W'AWE(\theta)E(\theta')] + \sigma_\epsilon^2 tr(A) \quad (3.6) \\ &= tr[W'AWE(\theta\theta')] + \sigma_\epsilon^2 tr(A). \end{aligned}$$

Thus the expectation of $E[R(\theta_T|\theta_C)]$ can be obtained as

$$\begin{aligned} E[R(\theta_T|\theta_C)] &= tr\left\{\begin{bmatrix} W_T' \\ W_C' \end{bmatrix}[M_CW_T(W_T'M_CW_T)^{-}W_T'M_C][W_T \vdots W_C]E(\theta\theta')\right\} \\ &\quad +\sigma_\epsilon^2 tr[M_CW_T(W_T'M_CW_T)^{-}W_T'M_C] \qquad (3.7) \\ &= tr[W_T'M_CW_TE(\theta_T\theta_T')] + \sigma_\epsilon^2(r_W - r_{W_C}), \qquad (3.8) \end{aligned}$$

where r_W and r_{W_C} are the ranks of W and W_C, respectively.

Under the null hypothesis $H_0 : \theta_T = 0$, we could have the following F-statistic

$$F = \frac{R(\theta_T|\theta_C)/(r_W - r_{W_C})}{SSE/(n - r_W)}. \tag{3.9}$$

Since there is no distinguish between fixed and random effects in this method, can comprise of several fixed or random factors, or both fixed and random factors. In the whole procedure of this method, it only need to calculate the inverse of an $s \times s$ matrix (s is the size of θ), and thus can greatly improve the computational efficiency.

3.2 Bayesian Analysis of Mixed Linear Model via Gibbs Sampling

For a general mixed model equation, the estimates of all the effects can be obtained by solving the following normal equation (Henderson 1988)

$$\begin{bmatrix} X'X & X'U_1 & X'U_2 & \cdots & X'U_r \\ U_1'X & U_1'U_1 + R_1^{-1}\frac{\sigma_{r+1}^2}{\sigma_1^2} & U_1'U_1 & \cdots & U_1'U_r \\ U_2'X & U_2'U_1 & U_2'U_2 + R_2^{-1}\frac{\sigma_{r+1}^2}{\sigma_2^2} & \cdots & U_2'U_r \\ \vdots & \vdots & \vdots & \vdots & \vdots \\ U_r'X & U_r'U_1 & U_r'U_2 & \cdots & U_r'U_r + R_r^{-1}\frac{\sigma_{r+1}^2}{\sigma_r^2} \end{bmatrix} \times \begin{bmatrix} \hat{b} \\ \hat{e}_1 \\ \hat{e}_2 \\ \vdots \\ \hat{e}_r \end{bmatrix} = \begin{bmatrix} X'y \\ U_1'y \\ U_2'y \\ \vdots \\ U_r'y \end{bmatrix}. \quad (3.10)$$

Based on this equation, Wang et al. (1994) developed a Bayesian method for making inference in a Gaussian mixed linear model via Gibbs sampling. In Bayesian analysis of mixed linear model, prior distributions of all of the unknown parameters are given as

$$p(b) \propto constant, \quad (3.11)$$

$$e_u|R_u, \sigma_u^2 \sim N_{q_u}(0, R_u\sigma_u^2), \quad (3.12)$$

$$p(\sigma_u^2) \propto (\sigma_u^2)^{-\nu_u/2-1}\exp(-\frac{1}{2}\nu_u s_u^2/\sigma_u^2), \quad (3.13)$$

where q_u is the rank of U_u; ν_u is a "degree of belief"; s_u^2 is a prior value of σ_u^2. The joint posterior distribution density functions of all the parameters are

$$p(\theta, \sigma|y, s, v) \propto \Pi_{i=1}^{r+1}\left[(\sigma_i^2)^{-(q_i+\nu_i+2)/2}\exp\left(-\frac{\lambda_i}{2\sigma_i^2}\right)\right], \quad (3.14)$$

where $\theta' = [b', e', \cdots, e_r']$, $\sigma' = [\sigma_1^2, \sigma_2^2, \cdots, \sigma_{r+1}^2]$, $v' = [\nu_1, \nu_2, \cdots, \nu_{r+1}]$ and $s' = [s_1^2, s_2^2, \cdots, s_{r+1}^2]$; $\lambda_i = e_i'R^{-1}e_i$ if $i \leqslant r$, and $\lambda_{r+1} = (y - Xb - \sum_i^r U_ie_i)'(y - Xb - \sum_i^r U_ie_i) + \nu_{r+1}s_{r+1}^2$. The full conditional posterior distributions of the parameters are

$$\theta_i|y, \theta_{-i}, \sigma \sim N(\tilde{\theta}_i, \tilde{\nu}_i), \quad (3.15)$$

$$\sigma_i^2|y, \theta, \sigma_{-i}, s, v \sim \tilde{\nu}_i\tilde{s}_i^2\chi_{\tilde{\nu}_i}^{-2}, \quad (3.16)$$

where θ_{-i} and σ_{-i} denote θ and σ without the ith element; $\tilde{\theta}_i = (\pi_\tau - \sum_{j=1,j\neq i}^{N} \omega_{ij}\theta_j)\omega_{ii}$, $\tilde{\nu}_i = \sigma_{r+1}^2/\omega_{ii}$ and ω_{ij} is the ijth element of the first matrix of (3.10); $\tilde{s}_i^2 = \lambda_i/\tilde{\nu}_i$.

The full conditional posterior distributions, equation (3.15) and equation (3.16), are called as Gibbs samplers. Our objective is to generate random samples from the joint posterior distribution, through updating and drawing samples from the Gibbs samplers. The Gibbs sampling procedure is formally performed as

(I) set arbitrary initial values for θ and σ;

(II) generate θ from equation (3.15) and update θ;

(III) generate σ from equation (3.16) and update σ;

(IV) repeat (II) and (III) for k times.

The initial iterations are usually not collected as samples for that the chain may not have reached the equilibrium distribution yet. We run the chains under different specifications (initial values, chain length and number of samplers saved). If these different specifications result in similar results, it is assumed that the Gibbs samplers have converged to equilibrium distribution. This procedure is called burn-in. After burn-in, samples are stored every d iterations, where d is a small positive integer. Parameter estimations and statistical inferences of the parameters can be made by summarizing the Gibbs samplers.

4 Application of HAB Method in Mapping QTL for Complex Traits (HAB-QTL)

Understanding of the genetic architecture of complex trait is a major challenge in the post-genomics area, especially for epistasis and QTL-environment interaction (QE), and the other more complicated high order interactions. With the advancement of molecular marker technique, fine-scale genetic maps of various organisms are now available. Using such maps, methods of mapping QTLs responsible for the quantitative traits in experimental populations have been established (Haley and Knott, 1992; Jansen, 1994; Lander and Botstein, 1989; Zeng, 1994). More methods were proposed for detecting QE interactions (Piepho, 2000) and epistatic interactions among QTLs (Kao, et al., 1999; Yi, et al., 2003; Ljungberg, et al., 2004), some of which are based on simultaneous scans to detect epistatic QTLs that do not have individual effects (Ljungberg, et al., 2004). However, none of these methods has integrated the QE interaction and epistasis into one mapping system. Wang et al. (1999) proposed a two-locus model for data from multiple-environment trials (METs), which could simultaneously analyze QTL main effects, epistatic effects as well as their interactions with environments. But it has some drawbacks in cofactor selection, false positive rate control and computational efficiency. In this section, we introduce a new method which applies the HAB method for mapping the genetic architecture of complex traits under multiple environments (HAB-QTL) (Yang et al. 2007). The HAB-QTL method starts with a full-QTL model, which integrates the effects of multiple QTLs with epistasis, and QE interactions into one mapping system. Estimation of genetic effects is performed on the full-QTL model using the Bayesian method described in section 3.2. A systematic mapping strategy is developed to search for QTLs and epistasis by F-test based on Henderson method III (see section 3.1). In the whole mapping process, permutation test approach is adopted to control the genome-wise false positive rate.

4.1 Modeling genetic architecture for data from METs

Consider a population consisting of N recombinant inbred lines (RILs) or double-haploid lines (DHLs) derived from a cross between two homozygous inbred lines (P_1 and P_2). The experiments are conducted in p different environments. Suppose there are s segregating QTLs $(Q_1, Q_2, \cdots, Q_s)$ each with two genotypes QQ and qq, in which t pairs of QTLs are involved in epistatic interactions. Let a random variable ξ_{ki} be the genotype of Q_k from the i-th line taking $\xi_{ki} = 1$ if the genotype of Q_k is Q_kQ_k and $\xi_{ki} = -1$ if the genotype of Q_k is q_kq_k. Let $x_{ki} = E(\xi_{ki}|$ the genotypes of flanking markers) which can be achieved by a general algorithm proposed by Jiang and Zeng (1997) with dominant, co-dominant or missing markers. Thus, the phenotypic value of the i-th line in the j-th environment (y_{ij}) can be expressed by the following mixed linear model

$$y_{ij} = \mu + \sum_k^s a_k x_{ki} + \sum_{k,h \in (1,2,\cdots,s), k \neq h}^t aa_{kh} x_{ki} x_{hi} + e_j + \sum_k^s ae_{kj} x_{ki} + \sum_{k,h \in (1,2,\cdots,s), k \neq h}^t aae_{khj} x_{ki} x_{hi} + \epsilon_{ij}, \quad (4.1)$$

where μ is the population mean; a_k is the additive effect of Q_k , fixed effect; aa_{kh} is the additive additive epistatic effect between Q_k and Q_h, fixed effect; e_j is the main effect of the j-th environment, random effect; ae_{kj} is additive environment interaction effect between Q_k and environment j, random effect; aae_{khj} is the interaction effect between

aa_{kh} and environment j, random effect; ϵ_{ij} is the residual effect. In such a mixed linear model, there is no constraint that any pair of effects involved in interaction must have individual effects. We can express model (4.1) in a matrix form as

$$\begin{aligned} y &= \mathbf{1}\mu + X_A b_A + X_{AA} b_{AA} + U_E e_E + \sum_{k=1}^{s} U_{A_k E} e_{A_k E} \\ &\quad + \sum_{h=1}^{t} U_{AA_h E} e_{AA_h E} + e_\epsilon \\ &= [\mathbf{1} \vdots X_A \vdots X_{AA}][\mu \vdots b'_A \vdots b'_{AA}]' + \sum_{u=1}^{r} U_u e_u + I e_\epsilon \\ &= Xb + \sum_{u=1}^{r+1} U_u e_u, \end{aligned} \tag{4.2}$$

where y and e_ϵ have the same definition as those in model (2.1); $\mathbf{1}$ is a $n \times 1$ vector with all the elements are 1. $b'_A = [a_1, a_2, \cdots, a_s]$ and $b_{AA} = [aa_1 aa_2 \cdots aa_s]'$ with the coefficient matrix X_A and X_{AA}; $e'_E = [e_1, e_2, \cdots, e_p]$, $e'_{A_k E} = [ae_{k1}, ae_{k2}, \cdots, ae_{kp}]$ and

$$e'_{AA_h E} = [aae_{h1}, aae_{h2}, \cdots, aae_{hp}]$$

with the coefficient matrices U_E, $U_{A_k E}$ and $U_{AA_h E}$, respectively.

If the locations of all the QTLs and the epistatic interactions among them are known, all the genetic effects in model (4.2) can be estimated by the Bayesian method described in section 3.2. In reality, however, such information is actually unavailable before mapping. In the following section, we will introduce a systematic mapping strategy, consisting of the procedures of marker interval selection, detection of marker interval interaction, one- and two-dimensional genome scans, to search for the locations of all the QTLs and the interactions among them.

4.2 Scanning genome for QTL and epistasis

4.2.1 Mapping QTL by one-dimensional (1D) genome scan

It is a multiple dimensional search problem to search for the multiple loci in the whole genome. To address this problem, Zeng (1994) proposed an approach called composite interval mapping (CIM) which could simplify the process of mapping multiple QTLs from multiple dimensional to one dimensional search problem. It can be accomplished by testing for a QTL at a particular genomic region conditioned on the selected markers controlling the genetic variance of other QTLs located outside the genomic region to be tested. Let (M_l^-, M_l^+) be interval I_l and M_l^- and M_l^+ be its flanking markers. Let M_lM_l and m_lm_l be the two genotypes of M_l . The QTL mapping model for testing a locus k within a particular genomic region can be written as

$$y_{ij} = \mu_j + a_{kj}x_{ki} + \sum_{l=1}^{c}(\alpha_{jl}^-\xi_{il}^- + \alpha_{jl}^+\xi_{il}^+) + \epsilon_{ij}, \tag{4.3}$$

where μ_j is the population mean of the j-th environment; α_{jl} is the effects of the l-th marker in the j-th environment; ξ_{il} takes 1 or -1 depending on whether the genotype of M_l is M_lM_l or m_lm_l; the remaining variables and parameters have the same definition as those in model (4.1).

Model (4.3) can be written in matrix form as

$$y = W_Q b_Q + W_M b_M + \epsilon, \tag{4.4}$$

where $b'_Q = [a_1, a_2, \cdots, a_p]$; $b'_M = [\mu', \alpha'_1, \alpha'_2, \cdots, \alpha'_p]$ with $\mu' = [\mu_1, \mu_2, \cdots, \mu_p]$ and $\alpha'_j = [\alpha_{j1}^-, \alpha_{j1}^+, \alpha_{j2}^-, \alpha_{j2}^+, \cdots, \alpha_{jc}^-, \alpha_{jc}^+]$; W_Q and W_M are the coefficient matrices corresponding to b_Q and b_M, respectively; y and ϵ have the same definition as those in model (4.2).

Under the null hypothesis $H_0 : a_1 = a_2 = \cdots = a_p = 0$, we have the following F-statistic for any locus in the genome

$$F = \frac{SSR(b_Q|b_M)/(r_W - r_{W_M})}{SSE/(n - r_W)}. \tag{4.5}$$

We can perform the F-test throughout the whole genome step by step separated by 1 cM (1D genome scan). When the F-values at a region exceed a pre-defined critical threshold, a QTL is indicated at the position with the regional maximum F-value. Before scanning the genome for putative QTLs, marker interval analysis is required to select the candidate marker intervals as cofactors for model (31). Piepho and Gauch (2001) proposed a marker pair selection (MPS) approach to selected cofactors for CIM. The MPS approach has two advantages: 1) markers enter the model in adjacent pairs, which reduces the number of models to be considered, thus alleviating the problem of over-fitting and increasing the chances of detecting QTL; 2) an exhaustive search for all the marker pairs per chromosome is used instead of simple forward selection, which maximizes the chance of finding the best-fitting models. For a trait value of the i-th line in the j-th environment, a single-interval model of the l-th marker interval is written as

$$y_{ij} = \mu_j + \alpha_{jl}^-\xi_{il}^- + \alpha_{jl}^+\xi_{il}^+ + \epsilon_{ij}, \tag{4.6}$$

where all the parameters and variables have the same definition as those in model (4.3). We can conduct the F-tests based on the Henderson method III for all the marker intervals, and plot the F-values along the whole genome. When the F-values at a region exceed the threshold value, a candidate marker interval is selected at the position with the maximum F-value.

4.2.2 Mapping epistasis by two-dimensional (2D) genome scan

Suppose that p QTLs have been mapped by the 1D genome scan. Obviously, interactions between these QTL can be tested by model selection technique analogous to the strategy of MIM (multiple interval mapping; Kao and Zeng, 1999). However, there may be epistasis involving some QTL not detected by the 1D genome scan due to small or even no individual effects. In order to find all possible epistasis, we adopt the 2D genome scan procedure conditioned on the effects of the p QTLs mapped by the 1D genome scan as well as a group of marker interval pairs selected by marker interval interaction analysis. For any pair of marker intervals (I^A and I^B), we can use the following two-interval model to test the interaction effect between them

$$y_{ij} = \mu_j + \alpha\alpha_j^{A^-B^-}\xi_i^{A^-}\xi_i^{B^-} + \alpha\alpha_j^{A^+B^+}\xi_i^{A^+}\xi_i^{B^+} + \sum_{k=1}^{c}(\alpha_{jk}^{-}\xi_{ik}^{-} + \alpha_{jk}^{+}\xi_{ik}^{+}) + \epsilon_{ij}, \tag{4.7}$$

where $\alpha\alpha_j^{A^-B^-}$ and $\alpha\alpha_j^{A^+B^+}$ are the interaction effects between the flanking markers of I^A and I^B, (M^{A^-}, M^{B^-}) and (M^{A^+}, M^{B^+}). Under the null hypothesis $H_0 : \alpha\alpha_1^{A^-B^-} = \alpha\alpha_1^{A^+B^+} = \cdots = \alpha\alpha_p^{A^-B^-} = \alpha\alpha_p^{A^+B^+} = 0$, Henderson method III based F-tests can be performed for all possible pair-wise marker intervals. We plot these F-values to get a 2D F-statistic profile. When the F-values at a region exceed the pre-defined threshold value, a candidate marker interval interaction is selected at the position with the maximum F-value.

Let $(I_1^A, I_1^B), (I_2^A, I_2^B), \cdots, (I_f^A, I_f^B)$ be the candidate marker interval interactions selected above. The model for testing the significance of epistatic interaction between loci k and h can be written as

$$\begin{aligned} y_{ij} &= \mu_j + aa_{khj}x_{ki}x_{hi} + \sum_{s}^{p} a_{sj}x_{si} \\ &+ \sum_{l}^{f}\left[\alpha\alpha_{jl}^{A^-B^-}\xi_{il}^{A^-}\xi_{il}^{B^-} + \alpha\alpha_{jl}^{A^+B^+}\xi_{il}^{A^+}\xi_{il}^{B^+}\right] + \epsilon_{ij}, \end{aligned} \tag{4.8}$$

where all the parameters and variables have the similar definition as those described above. Under the null hypothesis of $H_0 : \alpha\alpha_{kh1} = \cdots = \alpha\alpha_{khp} = 0$, we can use the Henderson method III based F-statistic to test any pair of loci in the genome (2D whole-genome scan). However, this

2D whole-genome scan strategy is very time-consuming especially for the data from METs. In order to reduce the computational complexity, we skip the genomic regions, in which no significant marker interval interaction is detected by the interval interaction analysis.

4.2.3 Threshold determination and model selection

As described above, the present mapping strategy consists of the procedures of marker interval selection, detection of marker interval interaction, one- and two-dimensional genome scans. In each of these procedures, multiple hypothesis tests are performed across the entire genome. Thus, it is necessary to adjust for the critical threshold value of the F-statistic to control the experiment-wise false positive rate. The permutation testing (Doerge and Churchill, 1996) is employed to determine an empirical threshold value of the F-statistic. However, models (4.3), (4.6), (4.7) and (4.8) are complex models, which contain only not the variables to be tested, but also the variables for background variance control. If the observation values are directly permutated, the relationship between trait and background-control variables will be destroyed, and thus permutation testing will give artificially low threshold. Therefore, some adjustments are needed to apply the permutation testing for complex models. Without loss of generality, we express the models (4.3), (4.6), (4.7) and (4.8) in a general matrix form as

$$y = W_T b_T + W_C b_C + \epsilon, \tag{4.9}$$

where b_T represents the effects to be tested with coefficient matrix W_T; b_C represents the effects for background variance control with coefficient matrix W_C. For each permutation step, we randomly shuffle shuffling the line order of the matrix w_T to destroy the relationship between the variables to be tested and the trait values, but keep the matrix W_C unchanged. This method is distribution-free, applicable in different population structures and especially simple to be conducted. However, a significant disadvantage of it is the computational burden. At least 1,000 and 10,000 permutations are suggested to obtain reasonably accurate estimates of the threshold value for type I error rates of 0.05 and 0.01, respectively.

Furthermore, at the end of each mapping procedure, the peaks which exceed the critical F-value calculated by permutation testing are selected from the F-statistic profile. However, some of the peaks are ghost peaks due to the high correlation of closely linked markers and random noise etc. Thus, we perform a stepwise selection on all the significant peaks selected from the F-statistic profile after each mapping procedure using the F-statistic as criterion. The stepwise selection strategy is comprised of two steps i.e. the forward selection step and backward selection step.

Firstly, all of the selected peaks are ranked by F-values, and the one with the maximum F-value is picked up into model (4.9) as the effects for background control (b_C). In forward selection step, the remaining peaks are put into model (4.9) as tested effects (b_T) for F-test one by one, and the most significant one is retained in the model. In backward selection step, the peaks retained in the model are tested by F-statistic to examine if some retained peaks will return to non-significance due to the new peak selected by forward selection. These two selection step are iteratively performed until the forward selection step is unable to select any peak into the model and the backward selection step is unable to drop any retained peak out of the model.

5 Application of HAB Method in Microarray Data Analysis (HAB-Array)

Gene expression microarrays are new biotechnology which open a new area on genome research achieved by simultaneously investigate hundred thousands of genes in one experiment. The essential problem of gene expression microarray data analysis is to identify differentially expressed genes (DEGs) under certain treatments. Microarray experiments commonly involve a large number of complex procedures, including RNA preparation, hybridization, and scanning, which bring many random variations into the raw intensities. In addition, microarray experiments generate large and complex multivariate datasets-thousands of interrelated variables (genes), small sample sizes (sometimes less than 10 arrays) from unbalanced design, with little or no replication and missing values because of the failed spots. All of these present challenge of statistical methods for identifying genes with differential expression. Various statistical methods are being proposed and applied in the selection of genes with differential expression, such as the Significance Analysis of Microarray (SAM) (Tusher, et al., 2001), fixed ANOVA method (Kerr, et al., 2000; Kerr and Churchill, 2001) and mixed linear model method (Jin et al., 2001; Lu et al., 2005). In this section, we introduce a method that applies the HAB method for identification of differentially expressed genes (HAB-array) as biomarkers for classifying treatment subtypes (Zou et al. 2006).

5.1 Mixed linear model for cDNA microarray data

Currently, there are many different types of expression microarray systems, where spotted cDNA microarrays (DeRisi et al., 1996) and Affymetrix oligonucleotide microarrays (Fodor et al., 1993) are most commonly used. Since cDNA microarrays have two different dyes and relatively large variability between arrays, four basic experimental effects: treatment of primary interest, genes, dyes, and arrays, need to be considered (Kerr et al., 2001). For Affymetrix oligonucleotide microarrays, as to only one dye designed in technology scenario, and high reproducibility of different arrays as well, the effects due to arrays and dyes can be ignored. We take cDNA microarray experiment of loop design as example to illustrate the HAB-array method. Let y_{ijkl} be the logarithm of measurement from the i-th array, the j-th treatment, the k-th dye and the l-th gene. The model can be expressed as

$$y_{ijkl} = \mu + G_l + A_i + T_j + D_k + GA_{li} + GD_{lk} + GT_{lj} + \epsilon_{ijkl}, \qquad (5.1)$$

where μ is the overall expression level; G_l is the effect of the l-th gene; A_i is the effect of the i-th array, $A_i \sim N(0, \sigma_A^2)$; T_j is the effect of the j-th treatment, $T_j \sim N(0, \sigma_T^2)$; D_k is the effect of the k-th dye, $D_k \sim N(0, \sigma_D^2)$; GA_{li} is the interaction effect between the l-th gene and the i-th array, $GA_{li} \sim N(0, \sigma_{GA}^2)$; GT_{lj} is the interaction between the l-th gene and the j-th treatment, $GT_{lj} \sim N(0, \sigma_{GT}^2)$; GD_{lk} is the interaction between gene l and dye k, $GD_{lk} \sim N(0, \sigma_{GD}^2)$; ϵ_{ijkl} is the residual effect accounting for the uncontrollable random errors, $\epsilon_{ijkl} \sim N(0, \sigma_\epsilon^2)$. All the terms in model (38) have been clearly interpreted by Kerr et al. (2000). The gene effects G_l account for the expression level of genes averaged over the other factors. The array effects A_i account for overall differences among arrays because arrays are not hybridized under exactly identical experimental condition. The dye effects D_k account for differences between the average signals from each dye. Dye bias can be obviously observed in an experiment where two identical mRNA samples are labeled with different dye and the red intensities are usually higher for the green dye. The treatment effects T_j account for overall differences among treatments. Such differences could arise if some treatments have more transcription activity in general, or simply because of differential concentration of mRNA in the labeled sample. The GA_{li} account for the differences between the average expression of gene l on array i and G_l. They mainly attribute to the fact that there is not complete control over the amount and concentration of cDNA immobileized from one array to another. The GD_{lk} terms are gene-specific dye effects, which occur when subsets of genes labeled with one dye exhibit higher fluorescent signal than that with the other dye. All of these effects are generally not of interest, but account for sources of variation in microarray data. The effects of interest in model (5.1) are the interactions between genes and treatments. These effects capture deviations from the overall averages that are attributed to the combination of a treatment j and a gene l. Non-zero differences in gene-treatment interactions across treatments for a given gene indicate differential expression.

5.2 Identification of DEGs by HAB-Array method

First of all, a normalization procedure is performed to minimize the global systematical variations involved in the experiments from the raw fluorescence measurements. The raw fluorescence measurements are normalized as $\gamma_{ijkl} = y_{ijkl} - (\bar{y}_{i...} + \bar{y}_{.j..} + \bar{y}_{..k.} - 2\bar{y}_{....})$. The normalized data ($\gamma_{ijkl}$) can be viewed as a variation for each gene after removing systematic experimental errors and are the input data for the following single-gene model,

$$\gamma_{ijkl} = \mu_l + A_{li} + T_{lj} + D_{lk} + \epsilon_{ijkl}, \tag{5.2}$$

where μ_l is the mean expression level of gene l; A_{li}, T_{lj} and D_{lk} are equivalent to GA_{li}, GT_{lj} and GD_{lk} in model (5.1), and are assumed to be normally distributed random variables with zero means and variance components σ^2_{GA}, σ^2_{GT} and σ^2_{GD}, respectively; ϵ_{ijkl} is the residual error of gene l, $\epsilon_{ijkl} \sim N(0, \sigma^2_\epsilon)$.

Model (5.2) can also be written in matrix form as

$$\gamma = W_{BV}b_{BV} + W_{GT}b_{GT} + \epsilon, \tag{5.3}$$

where $b'_{GT} = [T_{l1}, T_{l2}, \cdots]$; $b'_{BV} = [\mu_l, A_{l1}, A_{l2}, \cdots, D_{l1}, D_{l2}, \cdots]$; W_{BV} and W_{GT} are the coefficient matrices corresponding to b_{BV} and b_{GT}, respectively; γ is the vector of the inputted data; ϵ is the vector of residual effects.

Under the null hypothesis $H_0 : T_{l1} = T_{l2} = \cdots = 0$, we could have a F-statistic to test the significance of the differential expression for each gene based on Henderson method III (See section 3.1).

$$F = \frac{SSR(b_{GT}|b_{BV})/(r_W - r_{W_{BV}})}{SSE/(n - r_W)}, \tag{5.4}$$

where $W = [W_{BV} \vdots W_{GT}]$; r_W and $r_{W_{BV}}$ are the ranks of W and W_{BV}, respectively.

One important problem arises from the analysis of gene expression microarray data is the variation in expression under constant conditions is not stable from gene to gene, which is known as variance heterogeneity. To address this problem, F-statistic in equation (5.4) is proposed. However, it has low power in typical microarray experiments because of small sample sizes, and it is sensitive to variations in the estimation of residual variance. Recently, variance stabilizing transformations have been developed that can remove the systematic dependence of the variance on the mean, so that inference techniques that assume a common error variance across genes can be applied. But some studies indicate that variance stabilizing transformation also appears that there is still considerable variance heterogeneity that can interfere with global analysis of expression data. A compromise between power and bias may be obtained by adding global variance to the denominators of the F-statistic in equation (5.4),

$$F = \frac{SSR(b_{GT}|b_{BV})/(r_W - r_{W_{BV}})}{MSE_{global} + SSE/(N - r_W)}, \tag{5.5}$$

where $MSE_{global} = \frac{1}{N}\sum_{g=1}^{N} SSE/(N - r_W)$; N is the number of genes to be identified.

With the assumption of normal distribution of observed data, the F-statistics obtained in both (5.4) and (5.5) follow standard F distribution. Since F-tests are performed for all the genes, false discovery rate (FDR) (Benjamini and Hochberg, 1995) is used to control the experimental-wise type I error. Since microarray data often displays dramatically non-normal distribution, researchers prefer to use randomization methods, like permutation and bootstrapping, to establish distribution free significance levels for statistical tests. Ideally, sample shuffling is applied to simulate the null distribution of the test statistic, and a minimum of about 6 replicates per condition (yielding a total of 924 distinct permutations) is recommended for a two-sample comparison. However, microarray experiment is expensive and most time few replicates are available. In this condition, the number of possible permutations will not be sufficient, and residual shuffling will be alternatively conducted across genes under the assumption of homogeneous variance of all the genes.

6 Conclusions

In conventional mixed linear model analysis, datasets collected from agricultural, biological and medical studies and surveys are usually not too large, and thus are able to be handled by the commonly used approaches such as REML and BLUP etc. However, with the advance of biotechnology, data from biology experiments exponentially increased. In quantitative genetics, researches did not encounter so much computational challenge until they met the QTL mapping study, especially for those cases with the aim of analyzing epistasis by data from multiple environment trials (METs). While searching for QTLs, the genome scan strategy is widely used to fit model throughout the whole genome step by step with each step separated by 1cM. Nowadays, most genetic maps are larger than 1000 cM, so there will be at least one thousand tests to be performed. If those methods, which require calculating the inverse of an $n \times n$ matrix, are used in the model fitting process, and if the data sets are collected from METs with large number environments, it will be unable to be implemented. Another challenge comes from the analysis of mciroarray data, which also requires huge number of test points. These two problems can be reasonably, if not completely, solved by HAB method. However, the HAB method also has some drawbacks. Since the whole derivation is based on the assumption of normal distribution for the observation values, any traits that are not following normal distribution are unable to be solved in HAB framework. In such situation, we can use MINQUE and LUP that have no requirement for normal distribution of the observation data. As regard to the initiation of Gibbs sampling process, there is trade off in whether using the estimates (or predictors) from the commonly used mixed linear model approaches as the initiators for Gibbs sampling or not. Obviously, this kind of initiation can accelerate the convergence of Markov Chain, but it will bring extra computational spending because of the involving of calculating the inverse of the $n \times n$ matrix when using the commonly used mixed model approaches. In addition, in the Bayesian analysis of mixed linear model, the chosen of degree of belief is a tough problem. In our practices, we found that there was no unique mode of choosing the degree of belief, which needs to be determined for each model specially. Therefore, further studies are required to find a general approach to determine the degree of belief for various kinds of mixed linear models.

References

[1] Brown, H. K., and Prescott, R. J., 1999. Applied mixed models in medicine. John Wiley & Sons Inc. New York.

[2] Benjamini, Y., Hochberg, Y., 1995. Controlling the False Discovery Rate: A practical and powerful approach to multiple testing. J. Roy. Stat. Soc. Ser. B (Methodological), 57: 289–300.

[3] Comstock, R. E., and Robinson, H. F., 1952. Estimation of average dominance of genes. In: Heterosis. Gowan, J. W. (ed.), Iowa State University Press, Ames, Iowa.

[4] DeRisi, J., Penland, L., Brown, P. O., Bittner, M. L., Meltzer, P. S., Ray, M., Chen, Y., Su, Y. A., Trent, J. M., 1996. Use of a cDNA microarray to analyse gene expression patterns in human cancer. Nat. Genet., 14: 457–460.

[5] Doerge, R. W., and Churchill, G. A., 1996. Permutation tests for multiple loci affecting a quantitative character. Genetics, 142: 285–294.

[6] Fodor, S. P. A., Rava, R. P., Huang, X. C., Pease, A. C., Holmes, C. P., Adams, C. L., 1993. Mutliplexed biochemical assays with biological chips. Nature, 364: 555–556.

[7] Gardner, C. O., and Ebarhert, S. A., 1966. Analysis and interpretation of the variety cross diallel and related populations. Biometrics, 22: 439–452.

[8] Giesbrecht, F. G., 1983. An efficient procedure for computing MINQUE of variance components and generalized least squares estimates of fixed effects. Comm. Stat. Theor. Meth., 12: 2169–2177.

[9] Giesbrecht, F. G., and Burns, J. C., 1985. Two-stage analysis based on mixed model: Large sample asymptotic theory and small-sample simulation results. Biometrics, 41: 477–486.

[10] Griffing, B., 1956. Concept of general and combining ability in relation to diallel crossing systems. Aust. J. Biol. Sci., 9: 463–493.

[11] Haley, C. S., and Knott, S. A., 1992. A simple regression method for mapping quantitative trait loci in line crosses using flanking markers. Heredity, 69: 315–324.

[12] Hallauer, A. R., and Miranda, J. B., 1981. Quantitative genetics in maize breeding. Iowa State University Press, Ames, Iowa.

[13] Hartley, H. D. and Rao, J. N. K., 1967. Maximum-likelihood estimation for the mixed analysis of variance model. Biometrika, 54: 93–108.

[14] Henderson, C. R., 1963. Selection index and expected genetic advance. In: Statistical Genetics and Plant Breeding. Hanson, W. D., and Robinson, H. F., (eds.). Washington DC. National Academy of Science, National Research Council Publication, 982: 141–163.

[15] Henderson, C. R., 1988. Progress in statistical methods applied to quantitative genetics since 1976. In: Proceedings of the Second International Conference on Quantitative Genetics. Weir, B. S., Eisen, E. J., Goodman, M. M., et al., (eds.), Sinouer Assoc., Sunderland, MA.

[16] Jansen, R. C., 1994. Controlling the type I and type II errors in mapping quantitative trait loci. Genetics, 138: 871–881.

[17] Jiang, C., and Zeng, Z. B., 1997. Mapping quantitative trait loci with dominant and missing markers in various crosses from two inbred lines. Genetica, 101 :47–58.

[18] Jin, W., Riley, R. M., Wolfinger, R. D., White, K. P., Passador-Gurgel, G., Gibson, G., 2001. The contributions of sex, genotype and age to transcriptional variance in drosophila melanogaster. Nat. Genet., 29: 389–395.

[19] Kao, C. H., Zeng, Z. B., and Teasdale, R. D., 1999. Multiple interval mapping for quantitative trait loci. Genetics, 152: 1203–1216. Kerr,

M. K., Martin, M., Churchill, G. A., 2000. Analysis of variance for gene expression microarray data. J. Comp. Biol., 7: 819–837.

[20] Kerr, M. K., Churchill, G. A., 2001. Experimental design for gene expression microarrays. Biostatistics, 2: 183–201.

[21] Lander, E. S., and Botstein, S. 1989. Mapping Mendelian factors underlying quantitative traits using RFLP linkage maps. Genetics, 121: 185–199.

[22] Ljungberg, K., Holmgren, S., and Carlborg, O., 2004. Simultaneous search for multiple QTL using the global optimization algorithm DIRECT. Bioinformatics, 20: 1887–1895.

[23] Lynch, M., and Walsh, B., 1997. Genetics and Analysis of Quantitative Traits. Sinauer Associates, Inc. Publishers Sunderlad, Massachusetts, USA.

[24] Lu, Y., Zhu, J., Liu, P., 2005. A two-step strategy for detecting differential gene expression of cDNA microarray data. Physiological Genomics. Cur. Genet., 47: 121–131.

[25] Mackay, T. F., 2001. The genetic architecture of quantitative traits. Annu. Rev. Genet., 35: 303–339.

[26] Patterson, A. H., and Thompson, R., 1971. Recovery of inter-block information when block sizes are unequal. Biometrika, 58: 545–554.

[27] Piepho, H. P., 2000. A mixed-model approach to mapping quantitative trait loci in barley on the basis of multiple environment data. Genetics, 156: 2043–2050.

[28] Piepho, H. P., and Gauch H. G., 2001. Marker pair selection for mapping quantitative trait loci. Genetics, 157: 433–444.

[29] Putt, M., and Chinchilli, V. M., 1999. A mixed effects model for the analysis of repeated measures cross-over studies. Statistics in Medicine, 18: 3037–3058.

[30] Rao, C. R., 1971. Estimation of variance and covariance components MINQUE theory. J. Multivar. Anal., 1: 257–275.

[31] Rao, C. R., and Kleffe, J., 1980. Estimation of variance components. In: Handbook of statistics. Krishnaiah, P. R. (ed.). North-Holland, New York.

[32] Rashid, M. M., 2003. Rank-based tests for non-inferiority and equivalence hypotheses in multi-center clinical trials using mixed models. Stat. Med., 22: 291–311.

[33] Searle, S. R., 1970. Large sample variances of maximum likelihood estimators f variance components using unbalanced data. Biometrics, 26: 505–524.

[34] Searle, S. R., Casella, G., and McCulloch, C. E., 1992. Variance components. John Wiley & Sons Inc. New York.

[35] Tempelman, R. J., 1998. Generalized linear mixed models in dairy cattle breeding. J. Dairy Sci., 81: 1428–1444.

[36] Tusher, V.G, Tibshirani, R., Chu, G., 2001. Significance analysis of microarrays applied to the ionizing radiation response. Proc. Natl. Acad. Sci. USA, 98: 5116–5121.

[37] Wang, C. S., Rutledge, J. J., Gianola, D., 1994. Bayesian analysis of mixed linear models via Gibbs sampling with an application to litter size in Iberian pigs. Genetics Selection Evolution, 26: 91–115.

[38] Wang, D. L., Zhu, J., Li, Z. K., and Paterson, A. H., 1999. Mapping QTLs with epistatic effects and QTL environment interactions by mixed linear model approaches. Theor. Appl. Genet., 99: 1255–1264.

[39] Yang, J., Zhu, J., and Williams, R. W., 2007. Mapping the genetic architecture of complex traits in experimental populations. Bioinformatics (submitted).

[40] Yi, N., Xu, S., and Allison, D. B., 2003. Bayesian model choice and search strategies for mapping interacting quantitative trait Loci. Genetics, 165: 867–883.

[41] Zeng, Z. B., 1994. Precision mapping of quantitative trait loci. Genetics, 136: 1457–1468.

[42] Zhu, J., 1989. Estimation of genetic variance components in the general mixed model. Ph. D. Dissertation, North Carolina State Univ. Raleigh, N. C.

[43] Zhu, J., and Weir, B. S., 1994. Analysis of cytoplasmic and maternal effects: I. a genetic model for diploid plant seeds and animals. Theor. Appl. Genet., 89: 153–159.

[44] Zhu, J., and Weir, B. S., 1996. Mixed model approaches for diallel analysis based on a bio-model. Genet. Res., 68: 233–240.

[45] Zou, Y. Y., Yang, J., and Zhu, J., 2006. A robust statistical procedure to discover expression biomarkers using microarray genomic expression data. J. Zhejiang Univ SCIENCE B, 7: 603–607.

Part III: Mathematical Finance and Insurance

Inference and Computation for Stochastic Volatility Models Related to Option Pricing

Chuanshu Ji

Department of Statistics and Operations Research
University of North Carolina, Chapel Hill, NC 27599, USA
E-mail: cji@email.unc.edu

Abstract

This paper reviews the Black-Scholes theory and studies a number of extensions. In particular, we focus on logarithmic stochastic volatility (SV) models and square-root SV models with jumps. Inference on those models, using Markov chain Monte Carlo methods, is summarized. Generalized Black-Scholes (GBS) pricing formulas are presented as an important step in calibration of SV models using both return and option data. A Gaussian approximation scheme is proposed to significantly reduce the dimensionality of certain stochastic integrals involved in the GBS formulas, which shows a great promise in improving computational efficiency.

1 Introduction

The celebrated Black-Scholes (BS) theory [2] in financial economics lays a foundation for three important aspects: a probability model, an option pricing formula and a statistical inference procedure for volatilities. Although the past few decades have seen a great deal of progress in various extensions of the BS theory, a brief recap still provides a clear theme for our review and ongoing research.

- *The geometric Brownian motion* [*BS Stochastic Differential Equation* (SDE)]:

$$dS_t/S_t = \mu\ dt + \sigma\ dW_t, \tag{1.1}$$

where S_t represents a stock price at time t, the constants μ and σ represent the expected rate of return and volatility respectively, and $\{W_t\}$ is a standard Wiener process. The BS model consists of only two primary assets, i.e. the stock $\{S_t\}$ and a bank account $\{B_t\}$ with constant interest rate r:

$$B_t = B_0 \, e^{rt}. \tag{1.2}$$

- *Pricing formulas for derivative securities*

Let $t \leqslant T$. A contingent claim $g(S_T)$ written on the stock price S_T can be given a time t "fair" price $P(t, S_t)$ in two equivalent ways. The analytical approach yields $P(t,s)$, $t \in [0,T]$, $s \geqslant 0$ as the solution to the boundary value problem

$$\frac{\partial P(t,s)}{\partial t} + rs \, \frac{\partial P(t,s)}{\partial s} + \frac{1}{2} \, s^2 \sigma^2 \, \frac{\partial^2 P(t,s)}{\partial s^2} \; - \; r \, P(t,s) \; = \; 0$$
$$\text{with} \quad P(T,s) = g(s). \tag{1.3}$$

The probabilistic approach expresses $P(t, S_t)$ as a conditional expectation given $\mathcal{F}_t$ under a risk-neutral probability measure Q:

$$P(t, S_t) = E^Q \left[e^{-r(T-t)} g(S_T) | \, \mathcal{F}_t \right], \tag{1.4}$$

where Q is defined through its Radon-Nikodym derivative with respect to the real-world probability measure P (also referred to as the physical measure or objective measure) that governs the BS SDE (1.1)

$$\left. \frac{dQ}{dP} \right|_{\mathcal{F}_t} = \exp \left[- \, \frac{\mu - r}{\sigma} \, W_t \, - \, \frac{t}{2} \, \left(\frac{\mu - r}{\sigma} \right)^2 \right]; \tag{1.5}$$

and $\{\mathcal{F}_t, \; t \geqslant 0\}$ is the filtration generated by the Wiener process $\{W_t\}$. Note in this case, the information up to time t, i.e. $\mathcal{F}_t$, is summarized by the current state S_t. In the special case of an European call option, we have $g(S_T) = \max\{S_T - K, \; 0\}$, where the positive constant K is called the strike price at the maturity time T. The time t price $P(t, S_t)$, denoted by a function $C^{BS}(S_t, \sigma^2, r, K, T-t)$, enjoys a closed form solution by using either approach:

$$C^{BS}(S_t, \sigma^2, r, K, T-t) = S_t \; \Phi(d_1) - \; e^{-r(T-t)} K \; \Phi(d_2), \tag{1.6}$$

where $\Phi(\cdot)$ is the cdf of $N(0,1)$ distribution, and

$$d_1 = \frac{1}{\sigma \, \sqrt{T-t}} \left[\log \frac{S_t}{K} + (r + \sigma^2/2)(T-t) \right], \tag{1.7}$$

$$d_2 = d_1 - \sigma \, \sqrt{T-t}. \tag{1.8}$$

Remarks. (i) *The equivalence between (1.3) and (1.4) was established via the Feymann-Kac formula (cf. [46], [56]). In what follows, we will follow the probabilistic approach and refer to [23] for the analytical approach.*

(ii) *The fundamental risk-neutral valuation principle for pricing derivatives originated from the seminal papers of [2], [53]. A rigorous probability framework was due to [35] and [36]. Several excellent books were written in mathematical finance at different levels, such as [38], [1], [13], [15], to name a few. For a systematic treatment of stochastic calculus, including the Girsanov theorem, see [46] and [56].*

- *Estimation of volatilities*

To put the pricing formula (1.6) in action, numerical values of t, T, S_t, K, r, σ need to be specified. Among them, t, T, S_t, K are known or observable; while r and σ require more attention. In a simple case, it may not be hard to estimate r. However, some other situations go beyond the scope of this work, in which r is not only changing over time, but also considered to be stochastic. This is a rich research area of interest rate modelling. We refer to [44] and [58] for more detailed accounts.
Focusing on estimation of volatility σ, there are two substantially different methods:

(i) *Historical volatility*

This is just to estimate a population variance by a sample variance. Suppose we observe $S_{t_0}, S_{t_1}, \cdots, S_{t_n}$ at $t_0 < t_1 < \cdots < t_n$. Let

$$R_i = \log S_{t_i} - \log S_{t_{i-1}} \tag{1.9}$$

be the return incurred in $(t_{i-1}, t_i]$. Assume $t_i - t_{i-1} = \Delta, \forall\ i = 1, \cdots, n$. Then $R_1, \cdots, R_n$ are i.i.d. normal random variables with mean $(\mu - \sigma^2/2)\Delta$ and variance $\sigma^2\Delta$. Naturally, a historical volatility estimate for σ is given by

$$\widehat{\sigma} = \sqrt{\frac{\sum_{i=1}^n (R_i - \bar{R})^2}{(n-1)\Delta}} \quad \text{with} \quad \bar{R} = \frac{1}{n}\sum_{i=1}^n R_i.$$

Note that a common sense expression of return R_i, given by $(S_{t_i} - S_{t_{i-1}})/S_{t_{i-1}}$, is a first-order approximation of the definition in (1.9) with small Δ.

(ii) *Implied volatility*

An argument against the use of historical volatility is based on (a) volatility is not constant, but changes over time; (b) historical volatility only gives an estimate for the volatility over a past time period, but what is more relevant concerns the *market expectation* of the volatility in the near future. More specifically, if we want to determine the time t price of an option with maturity time T, then the market expectation of the volatility in the life time $(t, T]$ of the option should be incorporated consistently in the volatility estimation. One way to realize the market expectation is the following: Note that $C^{BS}(\cdot)$ in (1.6) is an invertible map of σ with other variables held fixed. Get a market price c_t from a "benchmark" call option written on the same stock $\{S_t\}$. Then find σ by inverting $C^{BS}(\cdot)$ from c_t. Such a value of σ is called the implied volatility.

Mathematically, implied volatility does not follow a traditional statistical inference guideline as historical volatility does. It is similar to a dynamical programming strategy where at each step a conditional expectation of future reward given the current state is compared to the instantaneous reward, in order to determine an optimal strategy. The idea is far reaching. We will revisit it in later sections when studying calibration of stochastic volatility (SV) models using both return and option data.

2 SV models and Markov chain Monte Carlo (MCMC) calibration

2.1 ARCH/GARCH models

Several drawbacks of the BS model were widely recognized, particularly the constant volatility σ. Since then, many efforts have been made to model volatility as a time-varying stochastic quantity. Along this line, the ARCH/GARCH (*autoregressive conditional heteroscedasticity/ generalized* $\sim$) family is an important development in modelling volatility.

ARCH(p):

$$X_t = \sigma_t \; \epsilon_t, \tag{2.1}$$

$$\sigma_t^2 = \alpha_0 + \sum_{i=1}^{p} \alpha_i \; X_{t-i}^2, \tag{2.2}$$

where $\epsilon_1, \epsilon_2, \cdots$ are i.i.d. $N(0,1)$ random variables, and $\alpha_0 > 0$, $\alpha_i \geqslant 0$, $i = 1, \cdots, p$ are parameters. Note that X_t is considered to be the return in $(t-1, t]$, thus $\sigma_t^2 = E(X_t^2|\mathcal{F}_{t-1}) = Var(X_t|\mathcal{F}_{t-1})$ is the conditional variance or second moment of X_t. In particular, α_0 represents the long-run volatility.

GARCH(p, q):

$$X_t = \sigma_t \, \epsilon_t, \tag{2.3}$$

$$\sigma_t^2 = \alpha_0 + \sum_{i=1}^{p} \alpha_i \, X_{t-i}^2 + \sum_{j=1}^{q} \beta_j \, \sigma_{t-j}^2 \tag{2.4}$$

with the additional parameters $\beta_j \geqslant 0$, $j = 1, \cdots, q$. Apparently, ARCH(p) is a special case of GARCH(p, q) with $q = 0$.

Since the original works of [18] and [3], ARCH/GARCH models have enjoyed great popularity. The likelihood functions have simple forms which make inference and prediction very convenient. There are many extended versions in the ARCH/GARCH family, among which [55] is noteworthy in relating GARCH and SV models. The books [22] and [65] also devote a couple of chapters to ARCH/GARCH models.

In ARCH/GARCH models, the sequence $\{\sigma_t^2\}$ represents time-varying volatilities, and they are stochastic. However, the term "stochastic volatility" usually is not meant to include ARCH/GARCH models, but reserved for a different model class we are to discuss. The main distinction is that there is a single source of uncertainty from asset returns, and each σ_t^2 is defined as a deterministic function of returns. In contrast, volatility in a SV model is defined as a stochastic process generated from an uncertainty source (often termed "factor") in addition to that for asset returns. Sometimes more than one factors are introduced. The necessity for SV models lies in the modelling flexibility. Certain interesting behaviors in financial markets, such as the famous "volatility smile", can be modelled by SV models, but not by ARCH/GARCH models. Here is a brief account: A call option is said to be "in the money" (ITM) at time t if $S_t > K$ and "out of the money" (OTM) if $S_t < K$. For put option with the contingent claim $g(S_T) = \max\{K - S_T, \, 0\}$, the inequalities are reversed. If $S_t = K$, the option is said to be "at the money" (ATM). Having observed the market prices of a number of European call options with the same maturity date T on the same underlying stock, the plot of implied volatility as a function of strike price K should be a horizontal straight line, if the assumption of constant volatility σ in the BS model is correct. Contrary to this, empirical evidences show that options far OTM or deep ITM are traded at higher implied volatilities than ATM options. The graph of the observed implied volatility function often looks like a smile shape, which gives rise to the term "volatility smile".

2.2 SV models

Let $S = \{S_t\}$ denote a continuous-time process for an asset price under a physical measure P. The following SV model was considered in [6]:

$$y_t = \log S_t, \tag{2.5}$$

$$dy_t = \mu\, dt + e^{h_t/2}\left(\sqrt{1-\rho^2}\, dW_{1t} + \rho\, dW_{2t}\right), \tag{2.6}$$

$$dh_t = (\alpha + \beta h_t)\, dt + \sigma\, dW_{2t}, \tag{2.7}$$

where $W = \{W_t\} = \{(W_{1t}, W_{2t})\}$ is a standard 2D Wiener process defined on a probability space $(\Omega, \mathcal{F}, P)$. Let $\{\mathcal{F}_t : 0 \leqslant t \leqslant T\}$ denote the filtration generated by W. The parameter ρ is the correlation between the asset return and the volatility factor.

This model consists of an observation equation (2.6) and a state evolution equation (2.7). It can be viewed as an example in state-space models or in filtering problems. The logarithmic volatility formulation guarantees that the resulting volatility $e^{h_t/2}$ is always positive. Although other forms of volatility evolution are available, log-SV models were introduced earlier and considered in [10], [64], [62], [51], [25], [41], [14], [31], [60], [61], [47], [23], [30], [9], [42], and [8].

To price an option on S, a standard approach is to assume no arbitrage and specify two risk premia processes. See [57] and [20]. In doing so, the SDEs describing the returns process have the same functional form under the physical measure P and the risk neutral measure Q. These two risk premia processes are

$$\tilde{\nu}_t = \nu_1 + \nu_2\, h_t, \tag{2.8}$$

$$\lambda_t = \frac{1}{\sqrt{1-\rho^2}}\left\{e^{-h_t/2}\left[(\mu - r) + e^{h_t}/2\right] - \rho\, \tilde{\nu}_t\right\}, \tag{2.9}$$

where r is the short rate, presumed constant. Assume the L_2 norms of the risk premia are bounded and $E\left[\exp\left(\int_0^T (\lambda_u^2 + \tilde{\nu}_u^2)\, du\right)\right] < \infty$. As shown in [35] and [36], there exists an equivalent martingale measure Q such that the time t price of a contingent claim $g(S_{t'})$ (with $t' > t$) can be expressed as the conditional expectation $E^Q[e^{-r(t'-t)} g(S_{t'}) \mid \mathcal{F}_t]$. For the purpose of computing the expectation, the measure Q can be expressed either as a Radon-Nikodým derivative with respect to the measure P,

$$\left.\frac{dQ}{dP}\right|_{\mathcal{F}_t} \tag{2.10}$$

$$= \exp\left(-\int_0^t \lambda_u\, dW_{1u} - \int_0^t \tilde{\nu}_u\, dW_{2u} - \frac{1}{2}\int_0^t \lambda_u^2\, du - \frac{1}{2}\int_0^t \tilde{\nu}_u^2\, du\right),$$

or as a system of SDEs in terms of the Wiener processes

$$W_{1t}^Q = W_{1t} + \int_0^t \lambda_u du, \tag{2.11}$$

$$W_{2t}^Q = W_{2t} + \int_0^t \tilde{\nu}_u du, \tag{2.12}$$

which is

$$dS_t/S_t = r\ dt + e^{h_t/2}[\sqrt{1-\rho^2} dW_{1t}^Q + \rho\ dW_{2t}^Q], \tag{2.13}$$

$$dh_t = [\alpha - \nu_1\sigma + (\beta - \nu_2\sigma)h_t]\ dt + \sigma\ dW_{2t}^Q. \tag{2.14}$$

SV models enjoy tremendous modelling flexibility. They can model various important market behaviors, such as the volatility smile. A key idea, termed generalized BS (GBS) pricing formulas, was hinted in [39] but advocated explicitly in [60], [61], [59], [27] and [29]. Here is the GBS formula for the log-SV model (2.6)-(2.7) and (2.13)–(2.14):

$$P(S_t, h_t, \theta, \nu) = E^Q\left[C^{BS}\left(S_t e^{Z_{t,T^*}}, \overline{V}_{t,T^*}, r, K, T^* - t\right)|\mathcal{F}_t\right], \tag{2.15}$$

$$Z_{t,T^*} = \rho \int_t^{T^*} e^{h_u/2}\ dW_{2u}^Q - \frac{\rho^2}{2}\int_t^{T^*} e^{h_u}\ du, \tag{2.16}$$

$$\overline{V}_{t,T^*} = \frac{1-\rho^2}{T^*-t}\int_t^{T^*} e^{h_u}\ du, \tag{2.17}$$

where $\theta = \{\mu, \alpha, \beta, \sigma, \rho\}$ and $\nu = \{\nu_1, \nu_2\}$ denote the parameters in (2.6)–(2.7) and (2.13)–(2.14); Q is a risk-neutral measure [Note: Q is not unique, which is specified by the choice of risk premia (2.8) and (2.9)]; $C^{BS}(\cdot)$ is the BS call option pricing formula given by (1.6) with life span $(t,\ T^*]$. Hence the GBS pricing formula is interpreted as a conditional expectation of the original BS option price with certain adjusted parameters, taken over the unobservable log-volatilities $h_u,\ t < u \leqslant T^*$ in the life time of the option. The special case with constant volatility and $\rho = 0$ will turn (2.15) to the original BS call option price. A put option can be priced using the parity relation between a put (P_t) and a call (C_t), which is $P_t = C_t - S_t + K\exp\left[-r\left(T^* - t\right)\right]$. Other derivatives can also be priced in a similar way.

Later we will elaborate on how to compute GBS formulas efficiently by using certain probability approximations.

2.3 MCMC calibration

(i) *Estimation based on return data only*

Statistical inference for the SV model (2.6) – (2.7) concerns the model parameter θ (parameter estimation) and latent volatilities $h = \{h_{t+i\Delta},\ i = 0, 1, .., n_y - 1\}$ (state estimation), based on a time series of return data $Y = \{y_{t+i\Delta},\ i = 0, 1, .., n_y - 1\}$ generated from a physical measure P. Note that Δ is the sampling window of observations; the initial time t is arbitrary to maintain flexibility; the risk-neutral measure Q, risk premia parameter ν are irrelevant because no option data are used for the time being (inference with option data to be studied later).

Discretization of the diffusion model (2.6) – (2.7) via an Euler scheme (see [48]) yields a likelihood. The presence of latent volatility sequence h poses an indirect inference problem, which can be tackled by several methods: EM algorithms, generalized method of moments (GMM), efficient method of moments (EMM), MCMC, etc. As for references, see [52] for EM algorithms, [33] and [5] for GMM, [24] for EMM, and [50] for MCMC. When applying to calibration of SV models, each method has its own pros and cons. No attempts are made in this paper for comparisons among different methods. Here we only mention [26] and [8] as examples of using GMM or EMM, and [41] as the earliest contribution to MCMC calibration of SV models. More references will be given when discussing inference on SV models using both return and option data in case (ii).

In the current context, MCMC methods generate a discrete-time Markov chain whose stationary distribution is the joint posterior distribution of model parameter θ and latent variables h, hence samples from the posterior distribution of any marginal are easy to obtain by selecting elements from the joint series.

Discretization can introduce bias. This bias may be eliminated by using any one of the fill-in methods proposed by [19] or [17]. But as verified by [20], for the type of data and class of models considered here, accuracy is adequate for a Δ of one day so there is no need to use fill-in methods if data sampled at a daily frequency are available.

The MCMC method we adopt is a type of "Metropolis within Gibbs". In the spirit of Gibbs, the unknowns θ and h are classified in several blocks. Each block is sampled conditioning on the rest and one cycles through different blocks. Because the normalizing factors for the conditional distributions involved are not known, a Metropolis sampler is used to generate runs within each block. Implementation of a Metropolis step consists of generating a candidate move from a proposal density and deciding whether to accept the proposed move based on a Metropolis ratio. The art is that the proposal density is usually specified in a problem-dependent manner. In other words, a few guidelines are available, but no rigid rules.

Denote the posterior distribution for parameters and latent variables by $P(h, \theta \mid Y)$. Holding Y fixed, this density is proportional to the likelihood times the prior

$$P(h, \theta \mid Y) \propto P(Y, h \mid \theta) P(\theta) = P(Y \mid h, \theta) P(h \mid \theta) P(\theta) \tag{2.18}$$

with

$$\begin{aligned} & P(Y|h, \theta)\ P(h|\theta) \\ &= \prod_{i=0}^{n-1} p(y_{t+(i+1)\Delta}|y_{t+i\Delta}, h_{t+i\Delta}, \theta)\ p(h_{t+(i+1)\Delta}|h_{t+i\Delta}, \theta), \end{aligned} \tag{2.19}$$

where $p(y_{t+(i+1)\Delta}|y_{t+i\Delta}, h_{t+i\Delta}, \theta)$ is the normal density of $y_{t+(i+1)\Delta}$ with the conditional mean $\mu\Delta + y_{t+i\Delta}$ and conditional variance $e^{h_{t+i\Delta}}\Delta$, and $p(h_{t+(i+1)\Delta}|h_{t+i\Delta}, \theta)$ is the normal density of $h_{t+(i+1)\Delta}$ with the conditional mean $\alpha\Delta + (1+\beta\Delta)h_{t+i\Delta}$ and conditional variance $\sigma^2\Delta$, following the discretization of (2.6)–(2.7).

The Gibbs blocking is as follows. The elements within the block of h are moved one at a time. The coordinate variable $h_{t+i\Delta}$ at each i is generated from the conditional density

$$p(h_{t+i\Delta} \mid h_{t+(i-1)\Delta}, h_{t+(i+1)\Delta}; y_{t+(i-1)\Delta}, y_{t+i\Delta}, y_{t+(i+1)\Delta}; \theta)$$

due to the Markovian structure in the dynamics of (2.6)–(2.7). The drift parameters μ and $\{\alpha, \beta\}$ are moved as two separate blocks and the diffusion parameters $\{\sigma, \rho\}$ as another. The priors are uniform on their supports with the exception for μ which has a conjugate normal prior, allowing μ to be updated directly without Metropolis-Hastings acceptance/rejection intermediation. All priors are proper with boundary parameters set in the experiments to be effectively noninformative.

The single-move Gibbs sampler is not as efficient as some other Bayesian MCMC inference methods for calibrating SV models when using only returns data Y. In particular, the multi-move Gibbs sampler with Kalman filters introduced in [63], [47] and [9] is more efficient computationally. Note in case (ii) we still use the single-move Gibbs sampler for calibrating SV models based on both return and option data because, to the best of our knowledge, no multi-move MCMC methods are available for that task. Given that we must use it for that purpose, we also use it for returns alone to reduce our coding and debugging burden.

(ii) *Estimation based on both return and option data*

The MCMC (Bayesian) inference on SV models is essentially an extension of the (frequentist) BS historic volatility approach in a broad sense that it only uses return data. Many financial econometricians argue that option data are at least as relevant as return data, if not more, in volatility calibration. Technically, the idea of BS implied volatility cannot be extended to SV models naively because the pricing formula, as a map from latent volatilities to an option price, is not invertible, i.e. it is an ill-posed inverse problem due to the high dimensionality of h. Nevertheless, we could and should still incorporate observed option data in volatility estimation by checking the discrepancy (called a pricing error) between an observed option price and one calculated via a GBS formula.

Consider the additional information provided by options written on Y. The option prices C we consider are on homogeneous instruments sampled at the same time as $y_{t+i\Delta}$ so that C has elements $c_{t+j\Delta}$ for $j = 0, 1, \cdots, n_c - 1$, where $c_{t+j\Delta}$ is the logarithm of the observed option price. (Note: Usually $n_c < n_y$ and the time t where C begins is larger than that of Y. But we do not consider those complications here.)

As is common in derivative pricing applications (cf. [20]), the distribution of C is determined by both an option pricing formula, which is (2.15) in our case, and an assumption of pricing errors. Pricing errors are not only required to permit application of Bayesian and similar likelihood based methods but are also plausible reflections of bid-ask bounce and similar micro structure considerations (cf. [59], [43]). We assume that errors are additive in log-scale,

$$c_{t+j\Delta} = f_{t+j\Delta} + \delta\widetilde{\epsilon}_j, \quad j = 0, 1, \cdots, n_c - 1, \tag{2.20}$$

where, from (2.15),

$$f_{t+j\Delta} = \log\left[P\left(S_{t+j\Delta}, h_{t+j\Delta}, \theta, \nu\right)\right], \tag{2.21}$$

and the errors $\widetilde{\epsilon}_j$, $j = 0, 1, \cdots, n_c - 1$ are i.i.d. $N(0, 1)$ random variables, independent of the processes S and h. Note that (2.20) treats observed option data as a noisy version of the option prices predicted based on a GBS pricing formula. In [45], a different but somewhat equivalent approach is taken by using the implied volatility data (VIX) to be a noisy version of the latent volatility variables.

There are several unknowns now, including $\theta = \{\mu, \alpha, \beta, \rho, \sigma\}$ and h that appear in the returns dynamics (2.6)–(2.7), the risk premia parameters $\nu = \{\nu_1, \nu_2\}$ that appear in the risk-neutral dynamics (2.13)–(2.14), and the pricing error standard deviation δ from (2.20).

We assume the three blocks $\{Y, h, \theta\}$, ν, and δ are mutually independent under the prior distribution. Under this assumption, the joint posterior distribution of interest becomes

$$P\left(h, \theta, \nu, \delta | C, Y\right) \propto P\left(C | Y, h, \theta, \nu, \delta\right) P\left(Y | h, \theta\right) P\left(h | \theta\right) P\left(\theta\right) P\left(\nu\right) P\left(\delta\right). \tag{2.22}$$

We have expressions for $P(Y|h,\theta)$ and $P(h|\theta)$ from case (i). Following (2.20), the likelihood of the option prices is

$$P(C|Y,h,\theta,\nu,\delta) \propto \prod_{j=0}^{n_c-1} \delta^{-1} \exp\left(-\frac{\left(c_{t+j\Delta} - f_{t+j\Delta}\right)^2}{2\,\delta^2}\right). \tag{2.23}$$

The Gibbs blocking is as follows. As in case (i) of this subsection, the elements $h_{t+i\Delta}$ of h are moved one at a time, the drift parameters μ and $\{\alpha,\beta\}$ are moved as two separate blocks, and the diffusion parameters $\{\sigma,\rho\}$ as another. The risk premium parameters ν and the pricing error parameter δ are moved as two separate blocks. The prior on ν is uniform. The prior for δ is the conjugate, so that δ can be directly sampled from a product of inverse Gamma distributions without Metropolis-Hastings acceptance/rejection intermediation. All priors are proper with parameters set to be effectively noninformative.

3 GBS formulas and probability approximations for numerical integration

A price that comes with the modelling capability by using SV models is the computational intensity required for calibrating SV models using financial data from two sources: underlying asset returns and option prices. Adding option data to this task creates a significant challenge for the following reason. In a SV model, a volatility time series consists of latent variables h_t, $t=0,1,\cdots,T$. Calibration of parameter θ (including the risk-neutral parameter ν as some components) and volatilities $\{h_t\}$ is an indirect inference problem, inviting certain Monte Carlo based algorithms, such as EMM or MCMC. There is an extensive literature in calibration of SV models using both return and option data. Here we only mention a few. See [7], [57] for EMM, and [30], [45], [20], [6] for MCMC. Those algorithms yield updated values $\theta^{(m)}$ and $h_t^{(m)}$, $t=0,1,\cdots,T$ in the mth iteration, $m=1,\cdots,M$. In particular, as an important part of the mth iteration in fitting a SV model, an option price $f_t^{(m)}$ is calculated by using a pricing formula based on the SV model with the current parameter value $\theta^{(m-1)}$ and volatility value $h_t^{(m-1)}$, and compared to the observed option price C_t. The comparison will lead to an adjustment of $h_t^{(m-1)}$ to $h_t^{(m)}$, and also updating of $\theta^{(m-1)}$ to $\theta^{(m)}$ based on $h_t^{(m)}$, $t=0,1,\cdots,T$. Notice that a general method of computing $f_t^{(m)}$ is to perform high-dimensional numerical integration — treated as a conditional expectation over the space of sample paths of future volatilities, and this has to be done for *every* t and *every* m in the algorithm. The required computational time quickly adds up with large values of T

(several hundred days) and M (typically more than $100,000$ iterations). So far, there have been basically two ways to handle the computation in practice. A basic strategy is to use "brute force" simulation, i.e. to generate a large number of "additional" sample paths of future volatilities based on which the Monte Carlo integration yields an approximation to $f_t^{(m)}$. An alternative approach relies on the availability of closed-form option pricing formulas which can avoid the aforementioned brute force numerical integration. However, Heston's model (cf. [37]) appears to be the only case beyond the original BS setting that enjoys a closed-form solution (still, certain low-dimensional numerical integration needed for a Fourier inversion strategy). The intractability of the brute force calculation and the limitation with closed-form pricing formulas present an urgent need for developing new efficient computational methodology.

In this section, we present a Gaussian approximation scheme for numerical computation of the option price and summarize the related work. The proposed Gaussian approximations promote a significant dimension reduction for numerical integration, from the space of volatility sample paths (with dimensionality of several hundreds) to the sample space of bivariate Gaussian vectors. See [6] (http://www.stat.unc.edu/faculty/cji/research.html) for detailed analytical arguments and numerical results. We will also include some ongoing work in [66] for Gaussian approximations applied to a SV model with jumps.

3.1 Gaussian approximations for logarithmic SV models

The obvious (and usual) approach to computing the integrals in (2.15)–(2.17) is to discretize (2.13)–(2.14), generate sample paths according to the recursions implied by the discretization, evaluate (2.15)–(2.17) over each sample path, and take an average. Indeed, this is how we check the accuracy of our proposed Gaussian approximations by creating a benchmark value of the option price. A number of useful tips with solid theoretical justifications are given in [48], [16], [32] regarding Monte Carlo simulation of SDEs. For the purpose of calibrating SV models via MCMC considered in this work, the brute force simulation is too computationally intensive to be practicable. Our proposal is as follows.

With a small time increment Δ, the Euler discretization of the volatility process under the risk-neutral measure Q, i.e. (2.14) yields the recursion

$$h_{t+\Delta} = a + b\, h_t + c\, \epsilon_{t+\Delta}, \tag{3.1}$$

where $\epsilon_{t+\Delta}$ is a $N(0,1)$ random variable, which is independent of the past, i.e. independent of the spot volatility h_t, of the drift terms

$a = (\alpha - \nu_1\sigma)\Delta$ and $b = 1 + (\beta - \nu_2\sigma)\Delta$, and of the diffusion term $c = \sqrt{\Delta}\sigma$. Notice that our notation departs from convention because we have incorporated Δ into the expressions for a, b, and c in order to simplify later formulas. Let $n = (T^* - t)/\Delta$. Our task is to derive the asymptotic distribution of the two sums

$$U_n = \sum_{j=0}^{n-1} e^{h_{t+j\Delta}}, \tag{3.2}$$

$$V_n = \sum_{j=0}^{n-1} e^{h_{t+j\Delta}/2} \epsilon_{t+(j+1)\Delta}, \tag{3.3}$$

which appear in the discretization of (2.16)–(2.17) using (3.1), i.e.

$$U_n\, \Delta \approx \int_t^{T^*} e^{h_u}\, du \tag{3.4}$$

$$V_n\, \sqrt{\Delta} \approx \int_t^{T^*} e^{h_u/2}\, dW_{2u}^Q. \tag{3.5}$$

For fixed t, let $E_t(\cdot)$, $Var_t(\cdot)$ and $Cov_t(\cdot)$ denote the conditional expectation, variance and covariance operators respectively, given h_t.

Theorem 3.1. *Assume $|b| < 1$. Fix t, Δ and an initial state h_t. As $n \to \infty$, the limiting distribution of $n^{-1/2}(U_n - E_tU_n,\ V_n - E_tV_n)$ conditioning on h_t is a bivariate normal distribution with mean $(0,0)$ and covariance matrix $A = \begin{pmatrix} a_{11} & a_{12} \\ a_{21} & a_{22} \end{pmatrix}$ provided $a_{11}a_{22} > a_{12}^2$, with*

$$\begin{aligned} a_{11} &= \lim_{n\to\infty} n^{-1} Var_t(U_n), \\ a_{12} = a_{21} &= \lim_{n\to\infty} n^{-1} Cov_t(U_n,\ V_n), \\ a_{22} &= \lim_{n\to\infty} n^{-1} Var_t(V_n); \end{aligned}$$

where

$$E_tU_n = \sum_{i=0}^{n-1} \exp\left[\frac{a(1-b^i)}{1-b} + b^i h_t + \frac{c^2(1-b^{2i})}{2(1-b^2)}\right]; \tag{3.6}$$

$$E_tV_n = 0; \tag{3.7}$$

$$
\begin{aligned}
&Var_t(U_n) \qquad (3.8)\\
&= \sum_{i=0}^{n-1} Var_t(e^{h_{t+i\Delta}}) + 2\sum_{i=0}^{n-2}\sum_{j=i+1}^{n-1} Cov_t(e^{h_{t+i\Delta}}, e^{h_{t+j\Delta}}),\\
&Cov_t(U_n,\ V_n) \qquad (3.9)\\
&= \sum_{i=1}^{n-1}\sum_{j=0}^{i-1} c\, b^{i-j-1}\ \exp\Bigg[(b^i + b^j/2)\,h_t + \frac{a\,(3/2 - b^i - b^j/2)}{1-b}\\
&\qquad + \frac{c^2\,(5/4 - b^{2i} - b^{2j}/4 + b^{i-j} - b^{i+j})}{2(1-b^2)}\Bigg]\\
&Var_t(V_n) = E_t U_n; \qquad (3.10)
\end{aligned}
$$

with

$$
\begin{aligned}
&Var_t(e^{h_{t+i\Delta}})\\
&= \exp\left(2b^i h_t + \frac{2a(1-b^i)}{1-b} + \frac{c^2(1-b^{2i})}{1-b^2}\right)\left[\exp\left(\frac{c^2(1-b^{2i})}{1-b^2}\right) - 1\right],
\end{aligned}
$$

$$
\begin{aligned}
&Cov_t(e^{h_{t+i\Delta}}, e^{h_{t+j\Delta}})\\
&= \exp\left((b^i b^j)h_t + \frac{a(2-b^i-b^j)}{1-b} + \frac{c^2(2-b^{2i}-b^{2j})}{2(1-b^2)}\right)\\
&\quad \times \left[\exp\left(\frac{c^2(b^{j-i}-b^{j+i})}{1-b^2}\right) - 1\right].
\end{aligned}
$$

Theorem 3.1 follows from a CLT for mixing sequences (cf. [40], [4]). The detailed proof, implementation of MCMC, and numerical results in simulation studies and an application to real foreign exchange data can be found in [6]. In particular, the assessment of computational efficiency (speed/accuracy) shows great promises of the proposed Gaussian approximations.

3.2 Gaussian approximations for SV models with jumps

Among several important recent contributions to calibration of SV models in recent years, SV models with jumps (simply denoted by SVJ in what follows) have received much attention. See [57], [21], [20], [8]. Jumps and SV are two ways to incorporate additional sources for risk and uncertainty into a model (SDE) for asset returns. Each of them has its own risk premium. Merton was the first one to introduce a jump-diffusion model for stock returns in [54]. An encyclopedic treatment of financial models with jumps is given in the excellent book [12]. Here we present a SVJ model, a GBS pricing formula, and related Gaussian approximations. The notation in this subsection, following [66], is kept "local" and need not be consistent with those used in previous sections.

(i) *SVJ models*

Consider the SVJ model

$$d(\log S_t) = (r_t + \eta_s V_t - V_t/2)\,dt + \sqrt{V_t}\,dW_t^s + U_{N_t} dN_t - \mu_1 \lambda\, V_t dt, \quad (3.11)$$

$$dV_t = \kappa(\theta - V_t)\,dt + \sigma_v \sqrt{V_t}\,dW_t^v, \quad (3.12)$$

where for $t \geqslant 0$, $S = \{S_t\}$ and $V = \{V_t\}$ represent continuous-time asset price and SV processes respectively. r_t is the interest rate at time t (assumed to be known). Both $\{W_t^s\}$ and $\{W_t^v\}$ are standard 1D Wiener processes, which satisfy $Cov(W_t^s, W_t^v) = \rho\, t$ with the correlation coefficient ρ to incorporate the leverage effect. The SV dynamic (3.12) corresponds to a square-root diffusion with a mean-reversion level θ and a reversion rate κ.

Price jumps are captured via a pure jump process that contains two components: a jump time process and a jump size process. Successive jumps at $0 < \tau_1 < \tau_2 < \cdots$ form a point process with a state-dependent intensity function $\{\lambda\, V_t : t \geqslant 0\}$ for some constant $\lambda > 0$, which allows jumps to occur more frequently in high volatility regimes. Given the arrival of the ith jump, the price jumps from $S_{\tau_i -}$ to $S_{\tau_i -}\exp(U_i^s)$, where $U_i^s, i = 1, 2, \cdots$ are iid normal random variables with mean μ_J and variance σ_J^2. Assume $\{U_i^s\}$ is independent of processes $\{\tau_i\}$, $\{W_t^s\}$ and $\{W_t^v\}$. Denote the mean relative jump size, conditioning on a jump event, by $\mu_1 = E(\exp(U_1^s) - 1) = \exp(\mu_J + \sigma_J^2/2) - 1$. It follows that the last term in (3.11) compensates the instantaneous price increment introduced by the pure jump process. The model parameters are $\eta_s, \kappa, \theta, \sigma_v, \lambda, \mu_J, \sigma_J$.

(ii) *Candidate pricing kernels*

A pricing kernel, introduced in [34], is basically the Radon-Nikodym derivative $\frac{dP}{dQ}$ of a risk-neutral measure Q with respect to a physical measure P.

The unhedgeable risk factors introduced by the SV and jump processes make an incomplete market in our study. Consequently, a pricing kernel is not unique. To price those risk factors (the diffusive price shocks, the price jumps, and the diffusive volatility shocks), we define a candidate pricing kernel π_t by

$$d(\log \pi_t) = h(V_t)\,dt + a(V_t)\,dW_t^v + b(V_t)\,dW_t^{v\perp} + (\log \pi_t - \log \pi_{t-})\,dN_t, \quad (3.13)$$

where $W = (W^v,\ W^{v\perp})$ is a standard 2D Wiener process. N_t represents the number of price jumps up to time t, i.e. we assume that $\{\pi_t\}$ and $\{S_t\}$ share the same jump time process. Let $\log\pi_t - \log\pi_{t-} = U_i^\pi$ if $\tau_i = t$. Assume $U_i^\pi, i = 1, 2, \cdots$ are iid normal random variables with mean μ_π and variance σ_π^2, and $\{U_i^\pi\}$ is independent of W and $\{N_t\}$. However, we allow the correlation $Corr(U_1^\pi, U_1^s) = \rho_\pi$.

The assumption of no arbitrage requires that the pricing kernel defined in (3.13) satisfies the following constraints:

$$r_t + h(V_t) + \frac{1}{2}a^2(V_t) + \frac{1}{2}b^2(V_t) + \mu_2\lambda\ V_t = 0, \tag{3.14}$$

$$\begin{aligned}&\eta_s V_t - V_t/2 + \sqrt{V_t}\ [\rho\ a(V_t)\ + \sqrt{1-\rho^2}\ b(V_t)] \\ =& -\ (\mu_3 - \mu_2 - \mu_1)\lambda\ V_t,\end{aligned} \tag{3.15}$$

where

$$\begin{aligned}\mu_2 &= E[\exp(U_1^\pi) - 1] = \exp(\mu_\pi + \sigma_\pi^2/2) - 1,\\ \mu_3 &= E[\exp(U_1^\pi + U_1^s) - 1] = \exp(\mu_\pi + \sigma_\pi^2/2 + \rho_\pi\sigma_\pi\sigma_J + \mu_J + \sigma_J^2/2) - 1.\end{aligned}$$

(iii) *GBS pricing formulas*

Having introduced a pricing kernel $\{\pi_t\}$, all conditional expectations in what follows are taken under the physical measure P, and no risk-neutral measure Q is necessary. Let F_{t,T^*} denote the σ-field generated by $\{V_u, N_u : t < u \leqslant T^*\}$. Define

$$B^*(t,T^*) = E_t\left(\frac{\pi_{T^*}}{\pi_t}\middle| F_{t,T^*}\right), \tag{3.16}$$

$$\xi_{t,T^*} = E_t\left(\frac{\pi_{T^*}}{\pi_t}\frac{S_{T^*}}{S_t}\middle| F_{t,T^*}\right). \tag{3.17}$$

The time t price $P(S_t, V_t, t, T^*)$ of an European call option with maturity time T^* follows the GBS pricing formula:

$$\begin{aligned}&P(S_t, V_t, t, T^*) \\ =\ & E_t\left[\ \frac{\pi_{T^*}}{\pi_t}\ (S_{T^*} - K)^+\right] \\ =\ & E_t\left[\ C^{BS}\left(S_t\xi_{t,T^*},\ (\overline{\sigma_{t,T^*}})^2, \{r_u,\ t\leqslant u\leqslant T^*\}, K, T^*-t\right)\right] \\ =\ & E_t\left[\ S_t\xi_{t,T^*}\ \Phi(d_1) - KB^*(t,T^*)\ \Phi(d_2)\right],\end{aligned} \tag{3.18}$$

where

$$\begin{aligned}d_1 &= \frac{1}{\sqrt{T^*-t}\ \overline{\sigma_{t,T^*}}}\ \left[\log\left(\frac{S_t\xi_{t,T^*}}{KB^*(t,T^*)}\right) + \frac{1}{2}\left(\overline{\sigma_{t,T^*}}\right)^2\ (T^*-t)\right],\\ d_2 &= d_1 - \overline{\sigma_{t,T^*}}\ \sqrt{T^*-t},\end{aligned}$$

$$(\overline{\sigma_{t,T^*}})^2 = \frac{(1-\rho^2)\int_t^{T^*} V_u \, du + \sigma_J^2 \, (N_{T^*} - N_t)}{T^* - t}, \tag{3.19}$$

$$\begin{aligned} B^*(t,T^*) = \Bigg\{ \exp \Bigg[\int_t^{T^*} \left(-r_u - \frac{1}{2} a^2(V_u) - \mu_2 \lambda \, V_u \right) du \\ + \int_t^{T^*} a(V_u) \, dW_u^v \Bigg] \Bigg\} \, (\mu_2 + 1)^{N_{T^*} - N_t}, \end{aligned} \tag{3.20}$$

$$\begin{aligned} &\xi_{t,T^*} \\ &= \left[\exp\left(- \int_t^{T^*} \mu_3 \lambda \, V_u \, du \right) \right] \Bigg\{ \exp \Bigg[\int_t^{T^*} \left(a(V_u) + \rho \, \sqrt{V_u} \right) \, dW_u^v \\ &- \frac{1}{2} \int_t^{T^*} \left(a(V_u) + \rho \, \sqrt{V_u} \right)^2 du \Bigg] \Bigg\} \, (\mu_3 + 1)^{N_{T^*} - N_t}. \end{aligned} \tag{3.21}$$

(3.18) clearly demonstrates the idea of GBS. The BS option price in this case is a function measurable on F_{t,T^*}, i.e. a function of unobserved latent variables $\{V_u, N_u, t < u \leqslant T^*\}$. Taking another conditional expectation over $\{V_u, N_u, t < u \leqslant T^*\}$ then yields the option price $P(S_t, V_t, t, T^*)$ for the SVJ model. Note that in this GBS formula, $B^*(t,T^*)$, ξ_{t,T^*} and $\overline{\sigma_{t,T^*}}$ are interpreted as the bond price (or the discount factor), the stock price and the (integrated) volatility respectively in a *virtual* world where the information about future volatilities and jumps in F_{t,T^*} were known at time t.

(iv) *Gaussian approximations in computation of* $P(S_t, V_t, t, T^*)$

To compute $P(S_t, V_t, t, T^*)$ numerically, we propose a similar Gaussian approximation scheme. There are a number of integrals in (3.19)-(3.21): $\int_t^{T^*} V_u \, du$, $\int_t^{T^*} a^2(V_u) du$, $\int_t^{T^*} \sqrt{V_u} \, a(V_u) \, du$; $\int_t^{T^*} \sqrt{V_u} \, dW_u^v$, $\int_t^{T^*} a(V_u) \, dW_u^v$. The specification $a(V_u)$ = constant $\sqrt{V_u}$ will enables us to focus on the joint distribution of two integrals: $\int_t^{T^*} V_u \, du$ and $\int_t^{T^*} \sqrt{V_u} \, dW_u^v$. Once again, let $n = (T^* - t)/\Delta$ and define

$$U_{1n} = \sum_{j=0}^{n-1} V_{t+j\Delta} \approx \Delta^{-1} \int_t^{T^*} V_u \, du,$$

$$U_{2n} = \sum_{j=0}^{n-1} \sqrt{V_{t+j\Delta}} \, \epsilon_{t+(j+1)\Delta} \approx \Delta^{-1/2} \int_t^{T^*} \sqrt{V_u} \, dW_u^v,$$

where $\epsilon_{t+\Delta}$, $\epsilon_{t+2\Delta}, \cdots$ are iid $N(0,1)$ random variables. For fixed Δ and large n, a Gaussian approximation scheme is proposed for the joint distribution of $(U_{1n},\ U_{2n})$. The theoretical justification will be provided by a CLT. Here we present the first and second moments of $(U_{1n},\ U_{2n})$.

An Euler discretization scheme yields the SV dynamics

$$V_{t+\Delta} = a + b\ V_t + c\ \sqrt{V_t}\ \epsilon_{t+\Delta}, \tag{3.22}$$

where

$$\begin{aligned} a &= \kappa\ \theta\ \Delta \\ b &= 1 - \kappa\ \Delta \\ c &= \sigma_v \sqrt{\Delta} \end{aligned}$$

are transformations of the original parameters κ, θ and σ_v.

For fixed t, let $E_t(\ \cdot\ )$, $Var_t(\ \cdot\ )$ and $Cov_t(\ \cdot\ )$ denote the conditional expectation, variance and covariance operators respectively, given V_t.

Theorem 3.2.

$$E_t U_{1n} = \frac{a}{1-b}\left(n - \frac{1-b^n}{1-b}\right) + \frac{1-b^n}{1-b}\ V_t \tag{3.23}$$

$$E_t U_{2n} = 0 \tag{3.24}$$

$$Var_t U_{1n} = B_0 + B_1\ V_t + B_2\ V_t^2 \tag{3.25}$$

$$Var_t U_{2n} = E_t U_{1n} \tag{3.26}$$

$$\begin{aligned} Cov_t(U_{1n},\ U_{2n}) &= \frac{ac}{(1-b)^3}\ [n - 2 - nb + nb^{n-1} - (n-2)b^n] \\ &\quad + \frac{c\ [1 - nb^{n-1} + (n-1)b^n]}{(1-b)^2}\ V_t \end{aligned} \tag{3.27}$$

where the coefficients B_0, B_1 and B_2 in (3.25) enjoy the closed-form expressions

$$\begin{aligned} B_0 &= \frac{ac^2(n-2)}{(1-b)^2(1+b)} - \frac{abc^2(1-b^{n-2})(1+2b-b^n)}{(1-b)^3(1+b)^2} \\ &\quad + \frac{2abc^2}{(1-b)^3(1+b)}\Big[(n-3)(1+b^{n-2}+b^{n-1}) \\ &\quad - \frac{(1-b^{n-3})(2b+2b^2+b^{n+1})}{(1-b)(1+b)}\Big] + \frac{2a^2b^n(1+b)}{(1-b)^2}\left(n - 1 - \frac{1-b^{n-1}}{1-b}\right) \end{aligned} \tag{3.28}$$

$$B_1 = \frac{1-b^{n-1}}{(1-b)^3}\ [c^2(1+b^n) + 2ab^n(1-b^2)] - \frac{2(n-1)\ b^{n-1}c^2}{(1-b)^2} \tag{3.29}$$

$$B_2 = 0 \tag{3.30}$$

4 Conclusion

Following the spirit of the Black-Scholes theory, several SV models are studied along with the related GBS pricing formulas and MCMC inference methods. As a new contribution, a proposed Gaussian approximation scheme significantly reduces the dimensionality of certain stochastic integrals involved in the GBS formulas, which makes the computation on Monte Carlo based inference much more efficiently.

There are several related areas not presented in this paper, including the PDE approach for option pricing (cf. [23]), realized volatilities and high frequency data (cf. [12]), and the utility-based approach for option pricing with equilibrium models (cf. [49], [11], [28]). The cited references will provide more detailed information and treatments.

References

[1] Björk, T. (2004). *Arbitrage Theory in Continuous Time* (2nd edition). Oxford Univ. Press.

[2] Black, F. and Scholes, M. (1973). The pricing of options and corporate liabilities. *Journal of Political Economy* **81**, 637–659.

[3] Bollerslev, T. (1986). Generalized autoregressive conditional heteroscedasticity. *Journal of Econometrics* **31**, 307–327.

[4] Bradley, R.C. (1986). Basic properties of strong mixing conditions. *Dependence in Probability and Statistics* (edited by E. Eberlein and M.S. Taqqu), 165–192. Birkhäuser.

[5] Campbell, J. Y., Lo, A. W. and MacKinlay, A. C. (1997). *The Econometrics of Financial Markets.* Princeton Univ. Press.

[6] Cheng, A., Gallant, A. R., Ji, C., and Lee, B. (2005). A central limit theorem for computation of option prices for stochastic volatility models. Under revision for *Journal of Econometrics.*

[7] Chernov, M. and Ghysels, E. (2000). A study towards a unified approach to the joint estimation of objective and risk neutral measures for the purpose of options valuation. *Journal of Financial Economics* **56**, 407–458.

[8] Chernov, M., Gallant, A. R., Ghysels, E., and Tauchen, G. (2003). Alternative models for stock price dynamics. *Journal of Econometrics* **116**, 225–257.

[9] Chib, S., Nardari, F. and Shephard, N. (2002). Markov chain Monte Carlo methods for stochastic volatility models. *Journal of Econometrics* **108**, 281–316.

[10] Clark, P. K. (1973). A subordinated stochastic process model with finite variances for speculative prices. *Econometrica* **41**, 135–156.

[11] Cochrane, J. H. (2005). *Asset Pricing* (revised edition). Princeton Univ. Press.

[12] Cont, R. and Tankov, P. (2004). *Financial Modelling with Jump Processes.* Chapman & Hall.

[13] Cvitanić, J. and Zapatero, F. (2004). *Introduction to the Economics and Mathematics of Financial Markets.* The MIT Press.

[14] Danielson, J. (1994). Stochastic volatility in asset prices: estimation with simulated maximum likelihood. *Journal of Econometrics* **63**, 375–400.

[15] Duffie, D. (1996). *Dynamic Asset Pricing Theory* (2nd edition). Princeton Univ. Press.

[16] Duffie, D. and Glynn, P. (1995). Efficient Monte Carlo simulation of security prices. *Annals of Applied Probability* **5**, 897–905.

[17] Elarian, O., Chib, S. and Shephard, N. (2001). Likelihood inference for discretely observed non-linear diffusions. *Econometrica* **69**, 959–994.

[18] Engle, R. F. (1982). Autoregressive conditional heteroscedasticity with estimates of the variance of the United Kingdom inflation. *Econometrica* **50**, 987–1007.

[19] Eraker, B. (2001). MCMC analysis of diffusion models with applications to finance. *Journal of Business and Economic Statistics* **19**, 177–191.

[20] Eraker, B. (2004). Do stock prices and volatility jump? Reconciling evidence from spot and option prices. *Journal of Finance* **54**, 1367–1404.

[21] Eraker, E., Johannes, M., and Polson, N. (2003). The impact of jumps in returns and volatility. *Journal of Finance* **53**, 1269–1300.

[22] Fan, J. and Yao, Q. (2003). *Nonlinear Time Series.* Springer-Verlag.

[23] Fouque, J. P., Papanicoulaou, G. and Sircar, R. (2000). *Derivatives in Financial Markets with Stochastic Volatility.* Cambridge Univ. Press.

[24] Gallant, A. R. and Tauchen, G.E. (1996). Which moments to match? *Econometric Theory* **12**, 657–681.

[25] Gallant, A. R., Hsieh, D. A., and Tauchen, G. E. (1991). On fitting a recalcitrant series: the Pound/Dollar exchange rate, 1974–1983, nonparametric and semiparametric methods in econometrics and statistics. *Proceedings of the Fifth International Symposium in Economic Theory and Econometrics* (edited by Barnett, W. A., Powell, J., and Tauchen, G.E.), 199–240. Cambridge Univ. Press.

[26] Gallant, A. R., Hsu, C.-T. and Tauchen, G. (1999). Using daily range data to calibrate volatility diffusions and extract the forward

integrated variance. *The Review of Economics and Statistics* **81**, 617–631.

[27] Garcia, R., Ghysels, E. and Renault, E. (2003). The econometrics of option pricing. Preprint.

[28] Garcia, R., Luger, R. and Renault, E. (2003). Empirical assessment of an intertemporal option pricing model with latent variables. *Journal of Econometrics* **116**, 49–83.

[29] Garcia, R., Luger, R. and Renault, E. (2005). Option prices, preferences, and state variables. Preprint.

[30] Ge, X. (2000). Bayesian calibration of stochastic volatility models. Ph. D. dissertation. Univ. of North Carolina at Chapel Hill.

[31] Ghysels, E., A. Harvey and E. Renault (1996): Stochastic volatility. *Handbook of Statistics* **14** (edited by Maddala, G.S. and Rao, C. R.). Elsevier Science B. V.

[32] Glasserman, P. (2000). *Monte Carlo Methods in Financial Engineering*. Springer-Verlag.

[33] Hansen, L. (1982). Large sample properties of generalized method of moments estimators. *Econometrica* **50**, 1029–1054.

[34] Hansen, L. and Richard, S. (1987). The role of conditioning information in deducing testable restrictions implied by dynamic asset pricing models. *Econometrica* **55**, 587–613.

[35] Harrison, M. and Kreps, D. (1979). Martingales and arbitrage in multiperiod security markets. *Journal of Economic Theory* **20**, 381–408.

[36] Harrison, M. and Pliska, S. (1981). Martingales and stochastic integrals in the theory of continuous trading. *Stochastic Process and Their Application* **11**, 215–260.

[37] Heston, S. L. (1993). A closed-form solution for options with stochastic volatility with application to bond and currency options. *Review of Financial Studies* **6**, 327–343.

[38] Hull, J. C. (2000). *Options, Futures, and Other Derivatives* (4th edition). Prentice Hall.

[39] Hull, J. and White, A. (1987). The pricing of options on assets with stochastic volatility. *Journal of Finance* **42**, 281–300.

[40] Ibragimov, I. A. and Linnik, Yu. V. (1971). *Independent and Stationary Sequences of Random Variables*. Wolters-Noordhoff Publishing, Groningen.

[41] Jacquier, E., Polson, N., and Rossi, P. (1994). Bayesian analysis of stochastic volatility models (with discussion). *Journal of Business and Economic Statistics* **12**, 371–417.

[42] Jacquier, E., Polson, N., and Rossi, P. (2004). A Bayesian analysis fat-tailed stochastic volatility models with correlated errors. *Journal of Econometrics* **122**, 185–212.

[43] Jacquier, E. and Jarrow, R. (2000). Bayesian analysis of contingent claim model error. *Journal of Econometrics* **94**, 145–180.

[44] James, J. and Webber, N. (2000). *Interest Rate Modelling.* John Wiley & Sons.

[45] Jones, C. S. (2003). The dynamics of stochastic volatility: evidence from underlying and option markets. *Journal of Econometrics.* **116**, 181–224.

[46] Karatzas, I. and Shreve, S. E. (1991). *Brownian Motion and Stochastic Calculus* (2nd edition). Springer-Verlag.

[47] Kim, S., Shephard, N., and Chib, S. (1998). Stochastic volatility: likelihood inference and comparison with ARCH models. *Review of Economic Studies* **65**, 361–393.

[48] Kloeden, P. E., and Platen, E. (1992). *Numerical Solution of Stochastic Differential Equations.* Springer-Verlag.

[49] Kou, S. G. (2002). A jump-diffusion model for option pricing. *Management Science* **48**, 1086–1101.

[50] Liu, J. (2001). *Monte Carlo Strategies in Scientific Computing.* Springer-Verlag.

[51] Melino, A. and Turnbull, S. (1990). Pricing foreign currency options with stochastic volatility. *Journal of Econometrics* **45**, 239–265.

[52] Meng, X. -L. and van Dyk, D. A. (1997). The EM algorithm — an old folk song sung to a fast new tune. *Journal of Royal Statistical Society, Series B* **59**, 511–567.

[53] Merton, R. (1973). The theory of rational option pricing. *Bell Journal of Economics and Management* **7**, 141–183.

[54] Merton, R. (1976).Option pricing when underlying stock returns are discontinuous. *Journal of Financial Economics* **3**, 125–144.

[55] Nelson, D. B. (1991). Conditional heteroscedasticity in asset returns: a new approach. *Econometrica* **59**, 347–370.

[56] Øksendal, B. (1995). *Stochastic Differential Equations* (5th edition). Springer-Verlag.

[57] Pan, J. (2002). The jump-risk premia implicit in options: evidence from an integrated time-series study. *Journal of Financial Economics* **63**, 3–50.

[58] Rebonato, R. (1996). *Interest-rate Option Models* (2nd edition). John Wiley & Sons.

[59] Renault, E. (1997). Econometric models of option pricing errors. *Advances in Economics and Econometrics* (edited by D. Kreps and K. Wallis), 223–278. Cambridge Univ. Press.

[60] Renault, E. and Touzi, N. (1996). Option hedging and implied volatilities in a stochastic volatility model. *Mathematical Finance* **6**, 279–302.

[61] Romano, M. and Touzi, N. (1997). Contingent claims and market completeness in a stochastic volatility model. *Mathematical Finance* **7**, 399–412.

[62] Scott, L. (1987). Option pricing when the variance changes randomly: theory, estimation and an application. *Journal of Financial and Quantitative Analysis* **22**, 419–438.

[63] Shephard, N. and Pitt, M. (1997). Likelihood analysis of non-Gaussian measurement time series. *Biometrika* **84**, 653–667.

[64] Taylor, S. J. (1986). *Modelling Financial Time Series.* John Wiley & Sons.

[65] Tsay, R. S. (2002). *Analysis of Financial Time Series.* John Wiley & Sons.

[66] Yoon, J., Renault, E. and Ji, C. (2007). Calibration of stochastic volatility models with jumps and computation using generalized Black-Scholes pricing formulas. Work in preparation.

A Selective Overview of Applications of Choquet Integrals*

Zengwu Wang and Jia-An Yan †
Institute of Applied Mathematics
Academy of Mathematics and Systems Science
Chinese Academy of Sciences, Beijing 100080, China

Abstract

In this survey paper, after presenting basic results about capacities and Choquet integrals, we review some applications of Choquet integrals to decision making under risk and uncertainty, finance, economics and insurance.

1 Introduction

Capacities and Choquet integrals, introduced by Choquet (1953), are reasonable generalizations of probability measures and mathematical expectations. Since 1953 many researchers tried to extend the classical results of measure theory to the case of capacities and Choquet integrals. Some theoretical results about Choquet integrals were obtained, such as integral representation theorem (Greco, 1982 and Zhou, 1998), convergence theorem (Denneberg, 1994), Radon-Nikodym theorem (Graf, 1980), Fubini theorem (Denneberg, 1994 and Ghirardato, 1997). Besides, Bayes' theorem of capacity (Wasserman and Kadane, 1990), entropies for discrete capacity (Marichal, 2002 and Kojadinovic et al., 2005), and law of large numbers for capacities (Maccheroni and Marinacci, 2005) were established. Several definitions of conditional Chouquet integral were also proposed, see Dempster (1967), Shafer (1976), Gilboa and Schmeidler (1993), Cohen et al. (1997), Chateauneuf et al. (2001), Denneberg (1994, 2002), Lehrer (2005) and Lapied and Kast (2005). However, all

*Yan's work was supported by the National Natural Science Foundation of China (No. 10571167).

†Corresponding author: Jia-An Yan (E-mail: jayan@amt.ac.cn.)

of them have some limitations. A reasonable definition of conditional Choquet integral is still a challenging open problem until now.

The original goal of Choquet was to use capacities and Choquet integrals in statistical mechanics and potential theory. Capacities and Choquet integrals started to attract economists' attentions after the seminal contribution of Shapely (1953) in the study of cooperative games. However, the connection with decision theory was not explicitly recognized at that time. It was only with the work of Dempster (1967), later developed by Shafer (1976), that applications to uncertainty and the representation of beliefs are considered. Some years later the interest of statistics for the subject was aroused by the understanding of its implications of robust Bayesian inference, explicitly presented by Huber (1973). For applications to statistics, see a review paper by Wasserman (1990).

It has been observed that non-additive uncertainty can arise in a lot of circumstance with many different reasons. For example, it can be due to the psychological attitude called "uncertainty aversion", made famous by celebrated Ellsberg's paradox. Decision theorists rediscovered capacities and Choquet integrals in 1989, when David Schmeidler first put forward an axiomatic model of choice with non-additive beliefs. Stimulated by subsequent contributions of Gilboa and Schmeidler (1989), the array of decision theories with non-additive beliefs became substantial, see Grabisch(1996), Marichal (2000), Narukawa and Murofushi (2004), and Kojadinovic (2005) for recent developments and references therein. In these works the authors proposed a kind of integral w.r.t. a non-additive set function (i.e. the Choquet integral) as a representation of the decision maker's preference. For applications to Artificial Intelligence and Automated Reasoning, see for instance Fagin et al. (1995).

There are many applications in finance and asset pricing (Chateauneuf et al., 1996 and Castagnoli et al., 2004), insurance (Wang, et al., 1997 and Waegenaere et al., 2003), agency theory (Ghirardato, 1994) and game theory (Eichberger and Kelsey, 1994).

The remainder of this survey is arranged as follows: In Section 2 we introduce basic results of capacities and Choquet integrals. From Section 3 to Section 6 we present some applications of Choquet integrals to decision making under risk and uncertainty, finance, economics and insurance. The multi-criteria decision making problem is presented in Section 3. The decision making under uncertainty (CEU model) and risk (RDEU model) are presented in Section 4. Section 5 devotes to present Choquet pricing and portfolio inertia. The relations of Choquet pricing with linear pricing and Maximal (Minimal) pricing are also presented in this section. Axiomatic approaches to insurance pricing under risk and uncertainty and to general risk measures in terms of Choquet integrals are presented in Section 6, the relations between risk measures and the equivalent utility principle are also presented.

2 Capacities and Choquet Integrals

Throughout this survey, for the ease of expositions, we will work on a nonempty set Ω with an algebra $\mathcal{A}$ of subsets of Ω. We denote by $L^\infty(\mathcal{A})$ the set of all bounded, real-valued, $\mathcal{A}$-measurable functions on Ω.

2.1 Definitions and Notations

Definition 2.1. *A set function $\mu : \mathcal{A} \to \mathbb{R}$ is called a (normalized) capacity if $\mu(\emptyset) = 0, \mu(\Omega) = 1$ and $\mu(A) \leqslant \mu(B)$ for all $A \subset B$* and $A, B \in$ *$\mathcal{A}$;*

(2) *A capacity μ is called concave (or submodular) if*

$$\mu(A \cup B) + \mu(A \cap B) \leqslant \mu(A) + \mu(B), \quad A, B \in \mathcal{A}, \tag{2.1}$$

If the inequality in (2.1) is reversed, μ is called convex (or supermodular);

(3) *A capacity μ is called superadditive (resp. subadditive), if*

$$\mu(A \cup B) \geqslant (\text{resp.} \leq) \mu(A) + \mu(B), \quad A, B \in \mathcal{A} \text{ and } A \cap B = \emptyset;$$

(4) *The conjugate capacity $\bar{\mu}$ of capacity μ is defined as $\bar{\mu}(A) = 1 - \mu(A^c)$, $A \in \mathcal{A}$;*

(5) *A capacity μ is called continuous from below (resp. continuous from above) if $A_n \in \mathcal{A}, A_n \subset$ (resp. $\supset$)A_{n+1} for all $n \in \mathbb{N}$ with $A = \cup_{n=1}^\infty A_n$(resp.$\cap_{n=1}^\infty A_n) \in \mathcal{A}$ implies $\lim_{n\to\infty} \mu(A_n) = \mu(A)$. If a capacity μ is continuous from below as well as from above, we call it continuous.*

Let $\mathbb{P}$ be a probability measure on $(\Omega, \mathcal{A})$ and $\gamma : [0,1] \to [0,1]$ an increasing function with $\gamma(0) = 0$ and $\gamma(1) = 1$. Then $\mu = \gamma \circ \mathbb{P}$ is a capacity. μ is called a *distorted probability* and γ the corresponding *distortion.* Corresponding to the dual capacity, the *dual distorted probability* of $\gamma \circ \mathbb{P}$ is $\bar{\mu} = \bar{\gamma} \circ \mathbb{P}$, where $\bar{\gamma}(x) = 1 - \gamma(1-x)$ is the *dual distortion.* If γ is a concave (resp. convex) function then $\gamma \circ \mathbb{P}$ is a concave (resp. convex).

The notion of integral w.r.t. a capacity, due originally to Choquet (1953), was rediscovered and extended by Schmeidler (1986, 1989).

Definition 2.2. *Given $X \in L^\infty(\mathcal{A})$, we define the Choquet integral of X w.r.t. μ by*

$$\mu(X) := \int_\Omega X d\mu = \int_0^\infty \mu([X \geqslant x])dx + \int_{-\infty}^0 (\mu([X \geqslant x]) - 1)dx,$$

where the integral is taken in the sense of Riemann.

Remark 2.1. (1) *If μ is a probability measure, the Choquet integral of X w.r.t. μ coincides with its mathematical expectation.*

(2) *If X is a simple function of the form $X = \sum_{i=1}^{n} x_i I_{A_i}$, where $A_1, \cdots, A_n$ are disjoint and enumerated so that (x_i) are in descending order, i.e. $x_1 \geqslant \cdots \geqslant x_n$, then*

$$\mu(X) = \sum_{i=1}^{n}(x_i - x_{i+1})\mu(S_i) = \sum_{i=1}^{n} x_i(\mu(S_i) - \mu(S_{i-1})),$$

where $S_i = A_1 \cup \cdots \cup A_i, i = 1, \cdots, n, S_0 = \emptyset$, and $x_{n+1} = 0$.

(3) *If $\mu = \gamma \circ \mathbb{P}$ is a distorted probability, then*

$$\mu(X) = \int_0^{\infty} \gamma(\mathbb{P}(X \geqslant x))dx + \int_{-\infty}^{0} (\gamma(\mathbb{P}(X \geqslant x)) - 1)dx, \quad \text{for } X \in L^{\infty}(\mathcal{A}).$$

2.2 Basic Results of Choquet Integrals

In this subsection we will introduce basic results of Choquet integrals, which proofs can be found in Denneberg (1994) and/or in Yan (2005).

2.2.1 Properties of the Choquet Integral

Definition 2.3. *Two functions X* and *Y in $L^{\infty}(\mathcal{A})$ are called comonotonic, if there is no pair $\omega_1, \omega_2 \in \Omega$ such that $X(\omega_1) < X(\omega_2)$ and $Y(\omega_1) > Y(\omega_2)$.*

The following theorem gives equivalent conditions for a pair of functions to be comonotonic (see Yan (2005) for a proof).

Theorem 2.1. *Given two functions X* and *Y in $L^{\infty}(\mathcal{A})$, the following statements are equivalent:*

(i) *X and Y are comonotonic;*

(ii) *There exists a function $Z : \Omega \to \mathbb{R}$ in $L^{\infty}(\mathcal{A})$ and two increasing functions f and g on $\mathbb{R}$ such that $X = f(Z), Y = g(Z)$;*

(iii) *There exist two continuous, increasing functions f and g on $\mathbb{R}$ satisfying $f(z) + g(z) = z(z \in \mathbb{R})$ and $X = f(X+Y), Y = g(X+Y)$.*

The following theorem summarizes the basic properties of the Choquet integral, cf. Denneberg (1994) and Yan (2005).

Theorem 2.2. *If μ is a capacity on $\mathcal{A}$ and $X, Y \in L^{\infty}(\mathcal{A})$, then*

(i) *$\mu(I_A) = \mu(A), A \in \mathcal{A}$;*

(ii)[1] (**positive homogeneity**) *$\mu(\lambda X) = \lambda\mu(X)$, for all $\lambda \in \mathbb{R}_+$;*

[1]This property is equivalent to the following property: $\mu(\lambda X) = \lambda^+\mu(X) + \lambda^-\mu(-X)$ for any $\lambda \in \mathbb{R}$, which and (*iv*) also imply that $\mu(\lambda X) = \lambda^+\mu(X) - \lambda^-\bar{\mu}(X)$ for any $\lambda \in \mathbb{R}$.

(iii) (**superhomogeneity** (*resp.* **subhomogeneity**)) *If μ is concave (resp. convex), then*

$$\mu(\lambda X) \geqslant \text{ (resp. } \leqslant) \ \lambda\mu(X), \ \text{for all } \lambda \in \mathbb{R};$$

(iv) (**asymmetry**) $\mu(X) = -\bar{\mu}(-X)$;
(v) (**monotonicity**) *If $X \leqslant Y$, then $\mu(X) \leqslant \mu(Y)$;*
(vi) (**translation invariance**) $\mu(X+c) = \mu(X) + c$ *for all* $c \in \mathbb{R}$;
(vii) (**comonotonic additivity**) $\mu(X+Y) = \mu(X)+\mu(Y)$ *if X and Y are comonotonic;*

2.2.2 The Integral Representation Theorem

The following representation theorem is due to Greco (1982). Its proof and some extended results can be found in Schmeidler (1986).

Theorem 2.3. *A mapping $\Gamma : L^\infty(\mathcal{A}) \to \mathbb{R}$ with $\Gamma(I_\Omega) = 1$ can be represented as a Choquet integral*

$$\Gamma(X) = \mu(X), \ X \in L^\infty(\mathcal{A}),$$

where $\mu(A) = \Gamma(I_A)$ on $\mathcal{A}$, if and only if it satisfies the following conditions:
(i) *(comonotonic additivity) If X and Y are comonotonic, then $\Gamma(X+Y) = \Gamma(X) + \Gamma(Y)$;*
(ii) *(monotonicity) $X \geqslant Y$ implies $\Gamma(X) \geqslant \Gamma(Y)$.*

Let $\mathcal{F}$ be a collection of bounded real-valued functions on Ω. $\mathcal{F}$ is called a *Stone vector lattice* if : (i) $\mathcal{F}$ is a vector space; (ii) $\mathcal{F}$ is a lattice, i.e., $X \vee Y, X \wedge Y \in \mathcal{F}$ for all $X, Y \in \mathcal{F}$; and (iii) $\mathcal{F}$ contains all constant functions on Ω. Let I be a (real-valued) function from $\mathcal{F}$ to $\mathbb{R}$. I is called a *quasi-integral* if: I is comonotonically additive, monotonic and continuous in the sense that

$$\lim_{n\to\infty} I(X_n) = I(X), \ \text{if } X, X_n \in \mathcal{F}, n \geqslant 1,$$

where (X_n) is decreasing and tends to X. A quasi-integral I is always positive homogeneous.

If a capacity is continuous from above we call it a *upper-continuous capacity.* The following theorem (due to Zhou (1998)) establishes a one-to-one correspondence between upper-continuous capacities and quasi-integrals.

Theorem 2.4. *Assume that I is a quasi-integral on a Stone lattice $\mathcal{F}$ on Ω and $I(1) = 1$. Then there exists a unique upper-continuous capacity μ on $\mathcal{S} := \{[X \geqslant c], X \in \mathcal{F}, c \in \mathbb{R}\}$ which represents I. On the other hand, for any upper-continuous capacity μ, the functional I defined on $\mathcal{F}$ by the Choquet integral $I(X) := \int X d\mu$ is a quasi-integral.*

2.2.3 Jensen's Inequality of Choquet Integrals

To the best of our knowledge, Jensen's inequality of Choquet integrals is first exploited by Asano (2003). However, a necessary and sufficient condition for Jensen's inequality to hold in the nonlinear case is presented by Hu (2005). The subsequent results are due to Asano (2003) and Hu (2005), respectively. Theorem 2.10 is a special case of Hu (2005), however, it can be proved in a different manner.

Theorem 2.5. *For any non-decreasing convex (resp. concave) function* $u : \mathbb{R} \to \mathbb{R}$, *if* $u(X) \in L^\infty(\mathcal{A})$, *then*

$$\mu(u(X)) \geqslant (\text{resp.} \leqslant) u(\mu(X)). \tag{2.2}$$

Theorem 2.6. *Given a function* $u : \mathbb{R} \to \mathbb{R}$, *if* $u(X) \in L^\infty(\mathcal{A})$, *then* (2.2) *holds for any convex (resp. concave) function* u *if and only if* $\mu \geqslant (\text{resp.} \leq) \bar{\mu}$ [2].

Corollary 2.1. *Assume that* $\mu \geqslant \bar{\mu}$. *Then for any* $X \in L^\infty(\mathcal{A})$ *we have*

$$(\mu(|X|^\alpha))^{\frac{1}{\alpha}} \leqslant (\mu(|X|^\beta))^{\frac{1}{\beta}}, \ \forall 0 < \alpha \leqslant \beta;$$

$$1/\mu(|X|^\lambda) \leqslant \mu(1/|X|^\lambda), \ \forall \lambda > 0.$$

2.2.4 Other Results

- **Newman-Pearson Lemma of Capacities**: Huber and Strassen (1973);
- **Radon-Nikodym Theorem of Capacities**: Graf (1980);
- **Bayes' Theorem of Capacities**: Wasserman and Kadane (1990);
- **Convergence Theorem of Choquet Integrals**: Denneberg (1994);
- **Fubini Theorem of Choquet Integrals**: Denneberg (1994) and Ghirardato (1997);
- **Law of Large Numbers of Capacities**: Maccheroni and Marinacci (2005);
- **Conditional Capacities and Conditional Choquet integrals**: Lapied and Kast (2005) and references therein.

[2] Economically, condition $\bar{\mu} \geqslant \mu$ can be explained as the measure $u(\mu, A)$ of uncertainty aversion (cf. Dow and Werlang, 1992) is non-negative . That is, Jensen's inequality of Choquet integrals holds for any concave function if and only if the measure $u(\mu, A)$ of uncertainty aversion is non-negative. The other case $\bar{\mu} \leqslant \mu$ can be explained in economics by the same manner.

3 Multi-criteria Decision Making

In multi-criteria decision making one aims at ordering multidimensional alternatives. For simplicity, we consider a finite set of alternatives $\mathcal{A} = \{a, b, c, \dots\}$ and a finite set of criteria $N = \{1, \cdots, n\}$. Each alternative $a \in \mathcal{A}$ is associated with a profile $(a_1, \dots, a_n) \in E^n$, where E is an interval of $\mathbb{R}$, and a_i represents the partial score of a related to criterion i. A traditional approach for this problem is to use the weighted sum $V(a) = \sum_{i \in N} w_i a_i$, where weight w_i represents the (subjective) importance given by a decision maker to criterion i. Despite its simplicity, this approach has a shortcoming that it implicitly assumes the independence of criteria, which is however rarely verified, since the criteria often interact. A substitute to the weighted sum is the Choquet integral w.r.t. a capacity μ. This approach was proposed by Grabisch(1996), Marichal (2000), and further developed by Kojadinovic (2005).

3.1 An Example of Multi-criteria Decision Making

Before presenting Kojadinovic's work, we study an example which is borrowed from Grabisch(1996).

Let us consider a problem of evaluating students in a high school w.r.t. three subjects: Mathematics (M), Physics (P) and Literature (L), whose marks are given on a scale from 0 to 20. Usually, this is done by a weighted sum. Suppose that the school is more science-oriented than literature-oriented. Then the weights could be, for example, $w_M = \frac{3}{8}, w_P = \frac{3}{8}$ and $w_L = \frac{2}{8}$, respectively. The scores of three students a, b and c are listed in Table 3.1. Then the weighted sum will give the following results: student a has the highest rank, and student b has the lowest rank.

If the school wants to favor well equilibrated students without weak points, then student c should be considered better than student a. However, since students good at mathematics are also good at physics (and vice versa), too much importance is given to the mathematics and physics in the weighted sum. To overcome this shortcoming, we are going to construct a suitable capacity and use the Choquet integral to replace the weighted sum. In doing so, we keep the initial ratio of weights $(3, 3, 2)$, and attribute the weights to the set $\{M, P\}$ not too higher than that to $\{M\}$ or $\{P\}$. Besides, in order to favor students equally good at scientific subjects and literature, the weights attributed to the sets $\{L, M\}$ and $\{L, P\}$ should be greater than the sum of individual weights. Thus, we propose a capacity μ as follows:

$$\mu(\{M\}) = \mu(\{P\}) = 0.45, \quad \mu(\{L\}) = 0.3,$$
$$\mu(\{M, P\}) = 0.5, \quad \mu(\{M, L\}) = \mu(\{P, L\}) = 0.9,$$

and $\mu(\{M, P, L\}) = 1$. Applying the Choquet integral w.r.t. the above capacities leads to the following desired results in Table 3.1: student c has the highest rank. Also, note that student b has still the lowest rank, as requested by the scientific tendency of this high school.

Table 3.1

Student	M	P	L	Weighted sum	Choquet integral
a	18	16	10	15.25	13.9
b	10	12	18	12.75	13.6
c	14	15	15	14.625	14.9

3.2 Entropies of Discrete Capacities

In order to be able to present Kojadinovic's work on multi-criteria decision making, we need two concepts of entropy for discrete capacities, which were introduced by Marichal (2002) and Kojadinovic (2005), respectively.

Definition 3.1. *Let μ be a capacity on $\Omega = \{1, 2, \cdots, n\}$. The entropy of μ on Ω is defined by*

$$H_M(\mu) = \Sigma_{i=1}^n \Sigma_{A \subset \Omega \setminus \{i\}} \gamma_{|A|}(n) h\Big(\mu(A \cup \{i\}) - \mu(A)\Big),$$

where $\gamma_{|A|}(n) = \frac{(n-|A|-1)!|A|!}{n!}$, $h(x) = -x \ln x$ if $x > 0$ and 0 if $x = 0$, and $|A|$ represents the number of elements in A.

The following theorem summaries main properties of $H_M(\mu)$.

Theorem 3.1. *Let μ be a capacity on Ω.*

(i) (**symmetry**) *For any permutation π on Ω, we have $H_M(\pi\mu) = H_M(\mu)$, where $\pi\mu$ is defined by $\pi\mu[(X_1, \cdots, X_n)] = \mu[(X_{\pi(1)}, \cdots, X_{\pi(n)})]$;*

(ii)(**maximality**) *We have $0 \leqslant H_M(\mu) \leqslant \ln n$. Moreover, $H_M(\mu) = \ln n$ if and only if $\mu(A) = \frac{|A|}{n}$ for all $A \subset \Omega$;*

(iii)(**decisivity**) *$H_M(\mu) = 0$ if and only if μ is a binary-valued capacity, that is, $\mu(A) \in \{0, 1\}$ for all $A \subset \Omega$;*

(iv)(**expansibility**) *If $k \in \Omega$ is a null element, i.e. $\mu(A \cup \{k\}) = \mu(A)$ for all $A \subset \Omega \setminus \{k\}$, then we have $H_M(\mu) = H_M(\mu_{-k})$, where μ_{-k} denotes the restriction of μ to $\Omega \setminus \{k\}$;*

(v)(**strictly monotonicity**) *Define capacities $\mu^*(A) = \frac{|A|}{n}$ for all $A \subset \Omega$ and $\mu_\lambda = \mu + \lambda(\mu^* - \mu)$ with $\lambda \in [0, 1]$ and $\mu \neq \mu^*$, then*

$$0 \leqslant \lambda_1 < \lambda_2 \leqslant 1 \Longrightarrow H_M(\mu_{\lambda_1}) < H_M(\mu_{\lambda_2}).$$

The entropy $H_M(\mu)$ of the capacity μ is merely a measure of the uniformity of μ, which is a generalization of Shannon entropy for discrete probability measures. In the same manner, Kojadinovic (2005) extended the entropy of Havrda and Charvat (1967) to the capacity case as

$$H_{HC}^{\beta}(\mu)=\frac{1}{1-\beta}\left[\sum_{i\in N}\sum_{A\subset N\backslash\{i\}}\gamma_{|A|}(n)[\mu(A\cup\{i\})-\mu(A)]^{\beta}-1\right], \quad (3.1)$$

where $\beta > 0$ and $\beta \neq 1$.

It is easy to verify that $H_{HC}^{\beta}(\mu)$ possesses the five properties listed in Theorem 3.1. Another straightforward way to measure the uniformity of a capacity is to compute its variance

$$\begin{aligned} V(\mu) :&= \frac{1}{n}\sum_{i\in N}\sum_{A\subset N\backslash\{i\}}\gamma_{|A|}(n)[\mu(A\cup\{i\})-\mu(A)-\frac{1}{n}]^2 \\ &= \frac{1}{n}\sum_{i\in N}\sum_{A\subset N\backslash\{i\}}\gamma_{|A|}(n)[\mu(A\cup\{i\})-\mu(A)]^2-\frac{1}{n^2}, \end{aligned}$$

which together with (3.1) implies that

$$H_{HC}^{2}(\mu)=\frac{n-1}{n}-nV(\mu).$$

Recently, an axiomatic approach to the definition of the entropy of a discrete capacity was proposed by Kojadinovic et al. (2005).

3.3 Identification of Capacity by Minimum Variance Principle

The use of Choquet integrals as an aggregation operator clearly requires the prior identification of the underlying capacity μ. In above example, although the capacity μ proposed is consistent with the decision maker's preference, but its construction didn't follow some principle. A natural question arises: can we identify a suitable capacity μ following some principle compatible with the initial preference of the decision maker? In what follows, we will present a " Minimum Variance Principle" (introduced by Kojadinovic (2005)) to give a positive answer to this question.

The global importance of a criteria $i \in N$ can be measured by means of its *Sharpley value*, cf. Shapley (1953), which is defined by

$$\phi_{\mu}(\{i\})=\sum_{A\subset N\backslash\{i\}}\gamma_{|A|}(n)[\mu(A\cup\{i\})-\mu(A)].$$

The average interaction between two criteria i and j can be measured by means of their *Shapley interaction index*, see Grabisch (1997), which is defined as

$$I_{\mu}(\{i,j\})=\sum_{A\subset N\backslash\{i,j\}}\gamma_{|A|}(n-1)[\mu(A\cup\{i,j\})-\mu(A\cup\{i\})-\mu(A\cup\{j\})+\mu(A)].$$

For probability distributions, the strict concavity of the Shannon entropy and its naturalness as a measure of uncertainty gave rise to the *maximum entropy principle*, which was pointed out by Jaynes (1957). This principle states that, among all the probability distributions of the possible outcomes of a random variable that are in accordance with the available prior knowledge (i.e. a set of constraints), one should choose the one that has maximum uncertainty. The strict concavity of H_M and H_{HC}^{β} suggests to extend such an inference principle to capacities. In such a context, $H_M(\mu)$ (or $H_{HC}^{\beta}(\mu)$) can be interpreted as a measure of the average value over all $a \in E^n$ of the degree to which the arguments $a_1, \cdots, a_n$ of a profile contribute the calculation of the aggregation value $\mu(a)$. The best studied such situation is probably the case that the objective function is a quadratic form. This means that the most interesting objective function from a practical perspective is $H_{HC}^2(\mu)$, maximizing which is equivalent to minimizing $V(\mu)$. The constraints that we shall consider are naturally based on the indices presented in the last subsection, which are all linear w.r.t. to the underlying capacity. We can state the above suggested approach as the following optimization problem:

$$\min V(\mu)$$

$$\text{subject to} \begin{cases} \mu(a) - \mu(b) \geqslant \delta_\mu, \\ -\delta_\mu \leqslant \mu(a) - \mu(b) \leqslant \delta_\mu, \dots, \\ \phi_\mu(\{i\}) - \phi_\mu(\{j\}) \geqslant \delta_{\phi_\mu}, \\ -\delta_{\phi_\mu} \leq \phi_\mu(\{i\}) - \phi_\mu(\{j\}) \leq \delta_{\phi_\mu}, \dots, \\ I_\mu(\{i,j\}) - I_\mu(\{k,l\}) \geqslant \delta_{I_\mu}, \\ \dots . \end{cases}$$

where δ_μ, δ_{ϕ_μ} and δ_{I_μ} are fixed preference thresholds to be defined by the decision maker. The constraints are mathematical expressions of prior information or initial preferences. For example, $\phi_\mu(\{i\}) - \phi_\mu(\{j\}) \geqslant \delta_{\phi_\mu}$ means to what extent criterion i is more important than criterion j; $-\delta_{\phi_\mu} \leq \phi_\mu(\{i\}) - \phi_\mu(\{j\}) \leq \delta_{\phi_\mu}$ means to what extent criterion i and criterion j are considered as equally important.

In terms of Möbius representation of a capacity μ, μ can be represented as

$$\mu(T) = \sum_{S \subset T} m_\mu(S),$$

where

$$m_\mu(S) = \sum_{H \subset S} (-1)^{(|S|-|H|)} \mu(H).$$

So a solution to the above optimization problem, if feasible, concerns with $2^n - 1$ coefficients, and needs much constraints. However, if we are limited to the case where μ is *2-additive*, in the sense that $m_\mu(S) = 0$, for all S with $|S| \geqslant 3$, and $m_\mu(S) \neq 0$ for at least one S with $|S| = 2$, then we will only require the decision maker to provide Sharpley value and interaction indices which are sufficient to define preferences over the alternatives.

For computations in solving this optimization problem, we refer the reader to Kojadinovic (2005), where some interesting examples are also presented.

4 Choquet Expected Utility (CEU)

In this section we follow closely Chaueauneuf and Cohen (2000) to present the Choquet expected utility model, introduced by Schmeidler (1989).

4.1 Axiomatic Characterization of CEU

A *(simple) lottery* $(x_1, p_1; \cdots; x_n, p_n)$ is a finite probability distribution over the set $\mathcal{M} \subset \mathbb{R}$ of *outcomes*, assigning probability p_i to outcome $x_i \in \mathcal{M}, i = 1, \cdots, n$, where $\sum_{i=1}^{n} p_i = 1$ for $0 \leqslant p_i \leq 1$. A lottery $(x_1, p_1; \cdots; x_n, p_n)$ can be regarded as a map p from $\mathcal{M}$ to $[0, 1]$ such that the set $\{x : p(x) \geqslant 0\}$ is finite and $\sum_x p(x) = 1$. The set of all lotteries is denoted by $\mathcal{X}$. A compound lottery of the form "with probability α you'll get lottery p, and with probability $1 - \alpha$ you'll get lottery q" is still a lottery, denoted by $\alpha p + (1 - \alpha)q$. By $\succeq$ we denote the preference relation of a decision maker on $\mathcal{X}$, with $\succ, \sim, \preceq$ and $\prec$ are defined as usual.

We introduce the following axioms:

(i)(**Weak Order**) (a) (Completeness) For all p and q in $\mathcal{X}$, $p \succeq q$ or $q \succeq p$; (b) (Transitivity) If $p \succeq q \succeq r$, then $p \succeq r$;

(ii)(**Independence**) For all p, q and r in $\mathcal{X}$, and for all α in $(0, 1)$, $p \succ q$ implies $\alpha p + (1 - \alpha)r \succ \alpha q + (1 - \alpha)r$;

(iii) (**Continuity**) For all p, q and r in $\mathcal{X}$, if $p \succ q$ and $q \succ r$, then there exist α and β in $(0, 1)$ such that $\alpha p + (1 - \alpha)r \succ q$ and $q \succ \beta p + (1 - \beta)r$;

The well-known **von-Neumann-Morgenstern's Expected Utility Theorem** states that a necessary and sufficient condition for the agent's preferences over $\mathcal{X}$ satisfy the above three axioms is the existence of a function $u : \mathcal{M} \to \mathbb{R}$ (called the *vNM utility function*) such that $p \succeq q$ iff $\sum_x u(x)p(x) \geqslant \sum_x u(x)q(x)$. This utility function u is unique up to an affine transformation.

In Anscombe and Aumann (1963), they considered the reference relation $\succeq$ over the set of *acts*, instead of the set of lotteries. Here, for simplicity, we assume that the state space (i.e. the set of states of the world) Ω is finite, and the algebra of events is $\mathcal{A} = 2^{\Omega}$. By an act of the decision maker, we mean a mapping X from Ω to the set of lotteries $\mathcal{X}$. We denote by L_0 the set of all acts.

In the Anscombe and Aumann's model, the independence axiom (ii) reads as follows:

(ii)'(**Independence**) For all X, Y and Z in L_0, and for all α in $(0,1)$, $X \succ Y$ implies $\alpha X + (1-\alpha)Z \succeq \alpha Y + (1-\alpha)Z$.

Schmeidler (1989) introduces an axiom of *comonotonic independence* to relax the von Neumann-Morgenstern axiom of independence, and has shown that a preference relation $\succeq$ on L_0 satisfying the axiom of comonotonic independence and other four axioms of EU theory can be represented by the Choquet expected utility (CEU). This extension of the expected utility theory covers situations such as the Allais paradox and the Ellsberg paradox.

In order to weaken the axiom (ii)', Schmeidler introduced the following definition of comotonic acts:

Definition 4.1. *Acts X and Y are said to be comotonic, if for no s and t in Ω, $X(s) \succ X(t)$ and $Y(t) \succ Y(s)$.*

Schmeidler's Comonotonic Independence Axiom now can be stated as:

(ii)"(**Comonotonic Independence**) For all pairwise comonotonic acts X, Y and Z in L_0 and for all α in $(0,1)$: $X \succ Y$ implies $\alpha X + (1-\alpha)Z \succ \alpha Y + (1-\alpha)Z$.

To describe Choquet expected utility, we need still the following axioms.

(iv) (**Monotonicity**) For all X and Y in L_0: If $X(\omega) \succeq Y(\omega)$ on Ω, then $X \succeq Y$;

(v) (**Non-degeneracy**) Not for all Y in L_0: $X \succeq Y$.

The following theorem is due to Schmeidler (1989).

Theorem 4.1. *Suppose that a preference relation $\succeq$ on L_0 satisfies* (*i*), (ii)", (iii), (iv) *and* (v), *then there exists a unique capacity μ on $\mathcal{A}$ and an increasing real-valued function U on $\mathcal{X}$ such that for all X and Y in L_0: $X \succeq Y$ if and only if $\mu(U(X(\cdot))) \geqslant \mu(U(Y(\cdot)))$.*

In view of the equivalent characterization of the uncertainty aversion, given by Schmeidler (1989), we have the following remark.

Remark 4.1. $\succeq$ *exhibits uncertainty aversion (resp. appeal)* $\Longleftrightarrow$ $\mu(U(X)) = \min(\text{resp.}\max)_{\mathbb{P}\in \text{core}(\mu)} \mathbb{E}_{\mathbb{P}}[U(X)]$, *where* $\text{core}(\mu) = \{\mathbb{P} : \mathcal{A} \to [0,1] | \mathbb{P}$ *is additive*, $\mathbb{P}(\Omega) = 1$ and $\mathbb{P}(A) \geqslant \mu(A)$ *for all* $A \in \mathcal{A}\}$.

For some other related works on this topic, we refer the reader to Chateauneuf et al. (2003) and Basset et al. (2004). Recently, Karni (2005) considers a kind of CEU with costly actions.

4.2 Illustrative Examples of CEU

In this subsection, following Narukawa and Murofushi (2004), we will use CEU to solve two famous paradoxes in economics: the Allais paradox and the Ellsberg paradox, which violate the classical expected utility theory.

Paradox 4.1. (**Allais Paradox**) *Let $\Omega = \{1, 2, \cdots, 100\}$ be the state space and $\mathcal{M} = \{0, 100, 200\}$ be the outcome set. Define random payoffs X_1, X_2, X_3 and X_4 by*

$$X_1(x) = 100, \qquad \text{for all } x \in \Omega; \qquad X_2(x) = \begin{cases} 200, & 1 \leqslant x \leqslant 70, \\ 0, & 71 \leqslant x \leqslant 100; \end{cases}$$

$$X_3(x) = \begin{cases} 100, & 1 \leqslant x \leqslant 15, \\ 0, & 16 \leqslant x \leqslant 100; \end{cases} \quad X_4(x) = \begin{cases} 200, & 1 \leqslant x \leqslant 10, \\ 0, & 11 \leqslant x \leqslant 100. \end{cases}$$

Given X_1 and X_2, most people will choose X_1, i.e., $X_2 \prec X_1$. At the same time, most people will choose X_4 when they face X_3 and X_4, i.e., $X_3 \prec X_4$. If the probability measure $\mathbb{P}$ on the elements of Ω are all the same, i.e. $\mathbb{P}(\{i\}) = \frac{1}{100}$ for any $i \in \Omega$, then this preference relation cannot be represented by expected utility. In fact, suppose that there exists a utility function $U : \mathcal{M} \to \mathbb{R}_+$ such that $X \prec Y \iff \mathbb{E}[U(X)] < \mathbb{E}[U(Y)]$. Since $X_2 \prec X_1$, we have $0.7U(200) < U(100)$. On the other hand, it follows from $X_3 \prec X_4$ that $0.15U(100) < 0.1U(200)$. Therefore we have that $1.05U(100) < 0.7U(200) < U(100)$, which is a contradiction.

By decision under uncertainty, we mean, in contrast with decision under risk, situations where no exact objective probability measure on the state space is available to the decision maker. A typical example is the following version of Ellsberg's paradox.

Paradox 4.2. (**Ellsberg's Paradox**) *Assume a urn contains 30 red balls and 60 black and white balls, which numbers are unknown. Now you take a ball from the urn at random. Facing two bets X_R and X_B, where X_R (resp. X_B) means that you will get 100 dollars if you take a red (resp. black) ball, most people will select X_R because there may be few black balls in the urn. On the other hand, facing two bets X_{RW} and X_{BW}, where X_{RW} (resp. X_{BW}) means that you will get 100 dollars if you take a red or white ball (resp. a black or white ball), most people will select X_{BW} because there may be few white balls in the urn. So we have the following preference: $X_B \prec X_R$ and $X_{RW} \prec X_{BW}$. This preference relation cannot be explained by the expected utility theory. In fact, let $\Omega = \{R, B, W\}$ be the state space and $\mathcal{M} = \{0, 100\}$ be the set of outcome. Suppose that there exists a probability measure $\mathbb{P}$ on Ω and* *a utility function U on $\mathcal{M}$ such that $X \prec Y \iff \mathbb{E}[U(X)] < \mathbb{E}[U(Y)]$. It follows from $X_{RW} \prec X_{BW}$ that*

$$\begin{aligned} U(100)\mathbb{P}(\{B\}) &= U(100)(\mathbb{P}(\{B,W\}) - \mathbb{P}(\{W\})) \\ &> U(100)(\mathbb{P}(\{R,W\}) - \mathbb{P}(\{W\})) = U(100)\mathbb{P}(\{R\}), \end{aligned}$$

which contradicts with $X_B \prec X_R$.

Solution to Allais Paradox We will use a distorted probability to solve the Allais paradox. A probability distortion function describes the risk attitude of the decision maker when confront risk. A distortion function γ exhibits *pessimistic* (resp. *optimistic*) if $\gamma \circ \mathbb{P}(X) \leqslant$ (resp. $\geqslant$) $\mathbb{E}[X]$ for all $X \in L_0$, which is equivalent to $\gamma \circ \mathbb{P}(\cdot) \leqslant$ (resp. $\geqslant$) $\mathbb{P}(\cdot)$ (Quiggin, 1993). From the preference relations $X_2 \prec X_1$ and $X_3 \prec X_4$, we may conclude that most people are pessimistic because that most people underestimating the possibility of 200 in X_2 when facing X_1 and X_2. According to this, we can define the distortion as $\gamma(0.1) = 0.08, \gamma(0.15) = 0.1$ and $\gamma(0.7) = 0.45$, then

$$\gamma \circ \mathbb{P}(X_1) = 100 > \gamma \circ \mathbb{P}(X_2) = 90 \text{ and } \gamma \circ \mathbb{P}(X_3) = 10 < \gamma \circ \mathbb{P}(X_4) = 16.$$

Here, we assume that $U(x) = x$.

Solution to Ellsberg's Paradox The preference relations $X_B \prec X_R$ and $X_{RW} \prec X_{BW}$ based on the fact that $\{R\}$ and $\{B,W\}$ have objective probabilities, while the other events are "uncertainty" or have "uncertainty probabilities" (Epstein, 1999). Thus these rankings corresponding to an intuitive notion of uncertainty aversion, i.e. the decision maker underestimating the possibilities of uncertainty event. So we may define the capacity as follows:

$$\begin{aligned} \mu(\{R\}) &= \frac{1}{3}, \quad \mu(\{B\}) = \mu(\{W\}) = \frac{2}{9}, \\ \mu(\{B,W\}) &= \frac{2}{3}, \quad \mu(\{R,W\}) = \mu(\{R,B\}) = \frac{5}{9} \end{aligned}$$

and $\mu(\{R,B,W\}) = 1$ [3]. Then we have the Choquet integral $\mu(\cdot)$ (here we assume that $U(x) = x$) of X^* by the Table 4.1,

Table 4.1

X^*	X_R	X_B	X_{RW}	X_{BW}
$\mu(X^*)$	$\frac{1}{3} \times 100$	$\frac{2}{9} \times 100$	$\frac{5}{9} \times 100$	$\frac{2}{3} \times 100$

which states that $\mu(X_B) < \mu(X_R)$ and $\mu(X_{RW}) < \mu(X_{BW})$. Then there is no contradiction.

[3]The capacities of some events are different from the work of Narukawa and Murofushi (2004), however these are more reasonable in our opinion.

4.3 CEU and Dynamic Consistency

To make CEU more generally applicable, it is desirable to extend it to an intertemporal model. Multi-period decisions also present new problems. Firstly, individuals will receive information as time progresses. It is necessary to model how they update their beliefs as information is received. Secondly, it is not clear whether individuals with non-additive beliefs will be dynamically consistent. In this subsection, we will present a positive answer due to Eichberger et al. (2005).

To begin with we list some notations and assumptions that will be used later.

- $E = \{\omega_1, \cdots, \omega_n\}$: a finite set of states of nature;
- $\mathcal{M} \subset \mathbb{R}^n$: the set of outcomes;
- $L(E)$: the set of all acts, which is a set of functions from E to $\mathcal{M}$;
- $\mathcal{E} = \langle \mathcal{E}_1, \cdots, \mathcal{E}_T \rangle$: a filtration with $\mathcal{E}_T = \langle E_1^T, \cdots E_{K_T}^T \rangle$;
- $L(E_k^T)$: the set of acts available after E_k^T is observed, i.e. $L(E_k^T)$ is a set of functions from E_k^T to $\mathcal{M}$;
- $L(E^t) = \times_{E^T \in \mathcal{E}^T, E^T \subset E^t} L(E^T)$: the set of acts available conditional on event E^t being observed at time t.

The following assumptions will be needed in the proof of the main theorem of this subsection.

- **Assumption 4.1** The utility function $U : \mathcal{M} \to \mathbb{R}$ is continuous;
- **Assumption 4.2** For two acts $X, Y \in L(E)$, if there exists an $\bar{i} \in E$ such that $U(X(\bar{i})) > U(Y(\bar{i}))$ and $U(X(i)) \geqslant U(Y(i))$ for all $i \in E$, then $X \succ Y$;
- **Assumption 4.3** Let μ be a convex capacity on E and let B be an event. Then if μ_B denotes the update of μ conditional on B, we assume that, $\mu(B) + \mu(B^c) = 1$ implies that $\mu_B(A) = \frac{\mu(A \cap B)}{\mu(B)}$ for all $A \subset 2^E$;
- **Assumption 4.4** We assume that the partition $\mathcal{E}_T = \langle E_1^T, \cdots, E_{K_T}^T \rangle$ is non-trivial, i.e. $|E_j^T| \geqslant 2$ for $1 \leqslant j \leqslant K_T$.

The subsequent lemma plays a prominent role in this subsection.

Lemma 4.1. *Let $\{E_i\}_{1 \leqslant i \leqslant K}$ be a partition and let μ be a convex capacity on E such that $\sum_{i=1}^K \mu(E_i) = 1$ then $\mu(B) = \sum_{i=1}^K \mu(B \cap E_i)$ for any $B \subset E$.*

With the aid of the lemma above, we claim that the five updating rules listed in the following satisfy Assumption 4.3, which implies that the result of Eichberger et al. (2005) is very general. In the following we assume that $B \subset A$.

- Bayes' updating rule: $\mu_B(A) = \frac{\mu(A \cap B)}{\mu(B)}$;
- Generalized Bayes' updating rule: $\mu_B(A) = \frac{\mu(A \cap B)}{1 + \mu(A \cap B) - \mu(A \cup B^c)}$;
- Generalized Bayes' updating rule on conjugate capacity: $\mu_B(A) = \frac{\bar{\mu}(A \cap B)}{1 + \bar{\mu}(A \cap B) - \bar{\mu}(A \cup B^c)}$;

• Dempster-Shafer's updating rule: $\mu_B(A) = \frac{\mu(A\cup B^c)-\mu(B^c)}{1-\mu(B^c)}$;

• Dempster-Shafer's updating rule on conjugate capacity: $\mu_B(A) = \frac{\bar{\mu}(A\cup B^c)-\bar{\mu}(B^c)}{1-\bar{\mu}(B^c)}$.

In the following definition, we will present the concepts of conditional CEU and dynamically consistent preference.

Definition 4.2. (i) *We denote CEU preferences conditionally on* $E^t \in \mathcal{E}_t$ *by* $\succeq_{E^t}$. *They are defined by* $X \succeq_{E^t} Y \iff \mu_{E^t}(U(X)) \geqslant \mu_{E^t}(U(Y))$;

(ii) *Preferences are said to be dynamically consistent w.r.t. a filtration* $\mathcal{E}$, *if whenever* $\tau > t$, $X \succeq_{E^\tau} Y$, *for all* $E^\tau \subset E^t$ *implies* $X \succeq_{E^t} Y$.

In the discussion that follows, we are in a position to state the main results of this subsection, which gives a necessary and sufficient condition for CEU preferences to be dynamically consistent.

Theorem 4.2. *Let* $\mathcal{E} = \langle \mathcal{E}_0, \cdots, \mathcal{E}_T \rangle$ *be a given filtration on* E, *which satisfies Assumption 4.4. If a decision maker has CEU preference with belief represented by a convex capacity, which satisfy Assumption 4.1 and Assumption 4.2 and he uses a updating rule satisfying Assumption 4.3 then following conditions are equivalent:*

(i) *he will be dynamically consistent w.r.t.* $\mathcal{E}$;

(ii) $\sum_{E\in\mathcal{E}_T} \mu(E) = 1$.

4.4 Rank-dependent Expected Utility (RDEU)

It is said that RDEU holds for the preference $\succeq$, if there exist a continuous and strictly increasing function U on $\mathbb{R}$ and a distortion γ such that $X \succeq Y \Leftrightarrow V(X) \geqslant V(Y)$ where $V(Z) := \int U(Z)d(\gamma \circ \mathbb{P})$ for a random variable Z. In this case, we will say that RDEU holds for $(\succeq, U)$.

RDEU, as a special case of CEU, is the only natural generalization of EU theory that preserves the standard properties of continuity, transitivity and (first-order) stochastic dominance (Quiggin, 1993). In the discussion that follows, we will focus on studying three aspects of RDEU, axiomatization, the shape of utility function and distortion. For a detailed description on this topic, we refer the reader to Quiggin (1993), Wakker (1994), Schmidt and Zank (2001) and Abdellaoui (2002).

The preference relation $\succeq$ is said to satisfy (first-order) *stochastic dominance* on $\mathcal{X}$ if the following holds:

$$(x_1, p_1; \cdots; x_n, p_n) \succ (y_1, p_1; \cdots; y_n, p_n)$$

whenever $x_i \geqslant y_i$ for all i and $x_i > y_i$ for at least one i with $p_i > 0$.

We say the preference relation has *simple continuity* if the following two sets

$$\{(y_1, \cdots, y_n) : (x_1, p_1; \cdots; x_n, p_n) \succeq (y_1, p_1; \cdots; y_n, p_n)\}$$

and

$$\{(y_1, \cdots, y_n) : (x_1, p_1; \cdots; x_n, p_n) \preceq (y_1, p_1; \cdots; y_n, p_n)\}$$

are closed sets in the Euclidean space $\mathbb{R}^n$ for any lottery $X = (x_1, p_1; \cdots; x_n, p_n)$.

Let $X^* = (x_1, \cdots, x_n)$ represent the lottery $X = (x_1, \frac{1}{n}; \cdots; x_n, \frac{1}{n})$ with equally outcomes. The collection of all this kind of lotteries is denoted by $\mathcal{X}^*$. In what follows, (α, X^*_{-i}) is written for X^* with x_i replaced by α, where α satisfies $x_{i+1} \leqslant \alpha \leqslant x_{i-1}$ for $i \geqslant 2$, $\alpha \geqslant x_2$ for $i = 1$ and $\alpha \leq x_{n-1}$ for $i = n$. Given outcomes α, β, γ and δ, we write $[\alpha; \beta] \succeq^* [\gamma; \delta]$ if

$$(\alpha, X^*_{-i}) \succeq (\beta, Y^*_{-i}) \text{ and } (\gamma, X^*_{-i}) \preceq (\delta, Y^*_{-i}), \text{ for } X^*, Y^* \in \mathcal{X}^*, \quad (4.1)$$

which means that a trade-off α for β is at least as good as a trade-off γ for δ. Furthermore, we write

$$[\alpha; \beta] \succ^* [\gamma; \delta] \text{ if we have } \prec \text{ instead of } \preceq \text{ in (4.1)}$$

and

$$[\alpha; \beta] \sim^* [\gamma; \delta] \text{ if we have } \sim \text{ instead of } \preceq \text{ and } \succeq \text{ in (4.1).}$$

The other relations such as $\prec^*$ and $\preceq^*$ can be defined in the same way.

The preference relation $\succeq$ is said to satisfy *trade-off consistency for equally likely outcomes* (or EL trade-off consistency) if there don't exist outcomes α, β, γ and δ such that $[\alpha; \beta] \succeq^* [\gamma; \delta]$ and $[\alpha; \beta] \prec^* [\gamma; \delta]$.

The following theorem is an axiomatization of RDEU, which is due to Schmidt and Zank (2001).

Theorem 4.3. *For the preference relation $\succeq$ on $\mathcal{X}$, RDEU holds if and only if the following conditions of $\succeq$ are satisfied:*
(i) *weak order;*
(ii) *stochastic dominance;*
(iii) *simple continuity;*
(iv) *EL trade-off consistency.*

Wakker (1994) wanted to separate marginal utility and probabilistic risk aversion by utility function and distortion, respectively. With the help of some suitable definitions, he obtained the following results, which are the analogous results of Pratt-Arrow from another approach.

Theorem 4.4. *Suppose that RDEU holds for $(\succeq_i, U_i), i = 1, 2$. Then the following two statements are equivalent:*

(i) *There exists a continuous, concave (resp. convex), strictly increasing function θ such that $U_2 = \theta \circ U_1$;*

(ii) *$\succeq_2$ exhibits a stronger (resp. weaker) decrease of marginal utility than $\succeq_1$.*

Corollary 4.1. *Under RDEU, U is concave (resp. convex, linear) if and only if $\succeq$ exhibits decreasing (resp. increasing, constant) marginal utility.*

Using the same idea, Abdellaoui (2002) gave a characterization of the shape of distortion in RDEU theory and got the same kind of results, which generalize the corresponding results of Wakker (1994).

5 Choquet Pricing and Portfolio Inertia

In order to capture the presence of friction or the uncertainty aversion of the economic agent, Chateauneuf et al. (1996) proposed to use Choquet integrals as pricing functionals for insurance and finance. The Choquet pricing was further studied by Waegenaere et al. (2003) and Castagnoli et al. (2004). It is well known that the Choquet pricing functional is nonlinear in view of nonlinearity of the Choquet integral. In what follows, we first preset a result of Waegenaere and Wakker (2001), which gives the conditions under which the Choquet pricing functional can be decomposed in a linear part and a non-negative, sub-additive part, and then present a result of Chen and Kulperger (2006), which relates the Choquet pricing to the Maximal (Minimal) pricing. Finally, in view of Dow and Werlang (1992), we present a theoretical foundation of portfolio inertia under Choquet pricing.

5.1 Choquet Pricing versus Linear Pricing

We consider a two-period market model where insurance contracts or financial assets can be traded in the first period, risks occur and contracts pay off in the second period. There are n possible states of the words at the second period, denoted by $i \in E = \{1, \cdots, n\}$. Since we consider a finite state space, a payoff can be represented by a vector $X \in \mathbb{R}^n$, the set of payoff vectors denoted by $L_0(E) \subset \mathbb{R}^n$.

First of all, we call a set function μ on 2^E a *signed set function* if it only satisfies $\mu(\emptyset) = 0$. In this case, the *signed Choquet integral* of X in $L_0(E)$ w.r.t. μ is defined as

$$\mu(X) = \int_E X d\mu = \int_{-\infty}^{0} (\mu([X \geqslant x]) - \mu(E))dx + \int_0^{\infty} \mu([X \geqslant x])dx.$$

The following theorem investigates the conditions under which a pricing functional $\Pi : L_0(E) \to \mathbb{R}$ can be represented as a signed Choquet integral, which follows from Theorem 1 and 2 in Waegenaere and Wakker (2001).

Theorem 5.1. *If Π is continuous, convex, subadditive and comonotonically additive, then there exists a concave set function μ such that $\Pi[X]$ equals the signed Choquet integral of X w.r.t. μ, i.e. $\Pi[X] = \mu(X)$.*

The Choquet pricing functional $\Pi_\mu[X] = \mu(X)$, is fully defined once the class of all set functions can be partitioned in terms of the structure of the pricing functional they generate. The following decomposition was given in the first step.

Theorem 5.2. *A functional $\Pi_\mu[\cdot]$ is a Choquet pricing functional if and only if there exists a vector $\lambda \in \mathbb{R}^n$, and a functional $\Psi : L(E) \to \mathbb{R}_+$ such that*

$$\Pi_\mu[X] = \lambda \cdot X + \Psi[X],$$

where $X \in L_0(E)$ and Ψ is a continuous, subadditive, comonotonic additive functional with $\Psi[X] \geqslant 0$ for all $X \in L_0(E)$ and $\Psi[\lambda I_E] = 0$ for all $\lambda \in \mathbb{R}$.

In general, $\Psi[X]$ is called the *Choquet risk functional*, and λ can be interpreted as a vector of *state prices*. It can be seen from Theorem 5.2, once the state prices and the risk functional are given, the Choquet pricing functional is fully defined. In this sense, in view of Theorem 5.1, fixing a risk functional is equivalent to restrict to Choquet integral w.r.t. a subclass of set functions.

The following example of violation of the Put-call Parity, is an application of Choquet pricing, borrowed from Chateauneuf et al. (1996).

Example 5.1. (**Solution to Violation of the Put-call Parity**) If S is a security, C^S (resp. P^S) denotes the call option (resp. put option) with the exercise price K, and 1 the riskless asset price, then under linear pricing functional $I(\cdot)$ one have the so-called *put-call parity*: $I(P^S) = I(C^S) - I(S) + K$, because at expiration date the same relation holds for payoff: $P^S = C^S - S + K$. However, violations of the put-call parity are observed by many empirical researchers (cf. Sternberg (1994) and references therein).

In the following we show how one can use the Choquet pricing functional to explain violations of the put-call parity. Assume the underlying security S has two states, say ω_1 and ω_2, and can take only two values at the maturity: $K + h$ and $K - h$. Let $\mu(\omega_1)$ and $\mu(\omega_2)$ be the capacities of events ω_1 and ω_2 with $\mu(\omega_1) + \mu(\omega_2) > 1$. Furthermore, assume that the dealer holds asset S which he will sell at price $\mu(S)$, and that he will sell the call option at price $\mu(C^S)$. Now assume an individual needs a put with the exercise price K. He could replicate it by forming portfolio: $C^S - S + K$, which will cost him

$$\mu(C^S) + \mu(-S) + K = h\mu(\omega_1) - K - h + 2h\mu(\omega_2) + K > h\mu(\omega_2).$$

So obviously he will prefer to buy the put directly at the price $\mu(P^S) = h\mu(\omega_2)$. This explains "mispricing" observed when dealers used put-call parity to price puts.

5.2 Choquet Pricing and Maximal(Minimal) Pricing

The g-Expectation, introduced by Peng (1997), is defined via the solution of a backward stochastic differential equation (BSDE) (cf. Pardoux and Peng, 1990). The Choquet integral and the g-expectation are both nonlinear expectations. Maximal(Minimal) pricing (cf. EL Karoui et al, 1997), as a special case of g-expectation, and Choquet pricing have been used in economics, finance and insurance as an alternative to traditional mathematical expectation. In the discussion that follows, we follow Chen and Kulperger (2006) to study the relation between them. For other related results, see Chen et al. 2005(a) and 2005(b).

Let $T \in (0, \infty)$ be a horizon time. Given a filtrated probability space $(\Omega, \mathcal{F}, \{\mathcal{F}_t\}_{0 \leqslant t \leqslant T}, \mathbb{P})$ with $\mathcal{F} = \mathcal{F}_T$ and $\mathcal{F}_t = \sigma(W_s : s \leqslant t)$, where $(W_t)_{0 \leqslant t \leqslant T}$ is a one-dimensional Brownian motion. Define the set

$$\mathcal{P} = \left\{ \mathbb{Q}^v : \frac{d\mathbb{Q}^v}{d\mathbb{P}} = \exp\left\{ -\frac{1}{2} \int_0^T |v_s|^2 ds + \int_0^T v_s dW_s \right\}, \sup_{0 \leqslant t \leq T} |v_s| \leqslant k \right\}.$$

We have the following relations

$$\underline{V}(\xi) \leqslant \underline{\mathcal{E}}(\xi) \leq \overline{\mathcal{E}}(\xi) \leqslant \overline{V}(\xi),$$

where

$$\underline{V}(A) = \inf_{\mathbb{Q} \in \mathcal{P}} \mathbb{Q}(A), \ \overline{V}(A) = \sup_{\mathbb{Q} \in \mathcal{P}} \mathbb{Q}(A),$$

$$\underline{\mathcal{E}}(\xi) = \inf_{\mathbb{Q} \in \mathcal{P}} \mathbb{E}_{\mathbb{Q}}[\xi], \ \overline{\mathcal{E}}(\xi) = \sup_{\mathbb{Q} \in \mathcal{P}} \mathbb{E}_{\mathbb{Q}}[\xi].$$

Note that $\overline{V}$ and $\underline{V}$ are capacities. We call $\overline{V}(\xi)$ (resp. $\underline{V}(\xi)$) Choquet upper-(resp. lower-)pricing of ξ, and $\overline{\mathcal{E}}(\xi)$ (resp. $\underline{\mathcal{E}}(\xi)$) Maximal(resp. Minimal) pricing of ξ.

Suppose that b and $\sigma : [0, T] \times \mathbb{R} \to \mathbb{R}$ are continuous in (t, x) and Lipschitz continuous in x. Let $\{X_s\}$ be the solution of following SDE

$$X_t = x + \int_0^t b(s, X_s) ds + \int_0^t \sigma(s, X_s) dW_s, 0 \leqslant t \leq T.$$

The relations between Choquet pricing and Maximal(Minimal) pricing are stated in the subsequent theorem.

Theorem 5.3. *Assume Φ is a monotonic function such that $\Phi(X_T) \in L^2(\Omega, \mathcal{F}, \mathbb{P})$. Consider the BSDE*

$$y_t = \Phi(X_T) + \int_t^T \mu_s |z_s| ds - \int_t^T z_s dW_s,$$

where μ is an adapted process. Then

$$\underline{V}[\Phi(X_T)] = \underline{\mathcal{E}}[\Phi(X_T)] \text{ and } \bar{\mathcal{E}}[\Phi(X_T)] = \bar{V}[\Phi(X_T)]$$

Assume $\sigma(t,x) > 0$ for all $t \geqslant 0$ and $x \in \mathbb{R}$, then there exist probability measures $\mathbb{Q}_1$ and $\mathbb{Q}_2$ such that

(a) For any Φ that is increasing, we have

$$\bar{V}[\Phi(X_T)] = \mathbb{E}_{\mathbb{Q}_1}[\Phi(X_T)] \text{ and } \underline{V}[\Phi(X_T)] = \mathbb{E}_{\mathbb{Q}_2}[\Phi(X_T)]$$

(b) For any Φ that is decreasing, we have

$$\bar{V}[\Phi(X_T)] = \mathbb{E}_{\mathbb{Q}_2}[\Phi(X_T)] \text{ and } \underline{V}[\Phi(X_T)] = \mathbb{E}_{\mathbb{Q}_1}[\Phi(X_T)]$$

where $\mathbb{Q}_1$ and $\mathbb{Q}_2$ are defined by

$$\frac{d\mathbb{Q}_1}{d\mathbb{P}} = \exp\{-\frac{1}{2}k^2T + kW_T\} \text{ and } \frac{d\mathbb{Q}_2}{d\mathbb{P}} = \exp\{-\frac{1}{2}k^2T - kW_T\}.$$

As a special case of the above theorem, the following corollary is an interesting result.

Corollary 5.1. *Assume that $\{X_t\}$ in the SDE is the price process of a stock. Then Choquet price and Maximal(Minimal) price of European call option $(X_T - K)^+$ are equal, that is*

$$\underline{V}[(X_T - K)^+] = \underline{\mathcal{E}}[(X_T - K)^+] \text{ and } \bar{\mathcal{E}}[(X_T - K)^+] = \bar{V}[(X_T - K)^+],$$

where K is the strike price.

5.3 Portfolio Inertia

Facing a riskless bond with interest rate zero and a risky asset, in the framework of EU theory, the investor will buy (resp. sell) the asset if and only if the mathematical expectation of its future price is higher (resp. lower) than its present price. Consequently, if and only if these two values are equal the investor will hold no position in the asset. Dow and Werlang (1992) generalized this result to the CEU case: there is a range of prices at which the investor has no position in the asset, at price below these, the investor holds a positive amount of the asset and at higher prices he holds a short position, which we call portfolio inertia. In the discussion that follows we will describe this result in detail. Throughout their paper, they assume that all the capacities in question are convex. For some recent developments on this topic we may see Assno (2003).

Definition 5.1. *We say that a capacity* μ_1 *is* at least as uncertainty averse as *another* μ_2 *if for any set* $A \subset \Omega$,

$$u(\mu_1, A) \geqslant u(\mu_2, A),$$

where $u(\mu_i, A) = \bar{\mu}_i(A) - \mu_i(A)$[4] *for* $i = 1, 2$.

For an asset X, $\mu(X)$ (resp. $-\mu(-X)$) denote the selling (resp. buying) price of the asset X by an investor [5]. The subsequent theorem shows that the gap between buying and selling prices of the asset X increases as the degree of uncertainty aversion increases.

Theorem 5.4. *The following statements are equivalent:*
(i) μ_1 *is at least as uncertainty averse as* μ_2;
(ii) *For all measurable functions* X *for which the integrals are finite,*

$$-\mu_1(-X) - \mu_1(X) \geqslant -\mu_2(-X) - \mu_2(X).$$

The following result is the main result of Dow and Werlang (1992).

Theorem 5.5. *Assume the price of the asset is* p, *a risk averse or risk neutral investor will buy (resp. sell) the asset if* $p < \mu(X)$ *(resp.* $p > -\mu(-X)$*) and only if* $p \leqslant \mu(X)$*) (resp.* $p \geqslant -\mu(-X)$*).*

Therefore, the no-trade interval is $(\mu(X), -\mu(-X))$, which depends on the size of the core(μ). Obviously, if the core(μ) is a singleton, then the result will reduce to the EU case. For some illustrative examples and appealing explanations, the reader may consult Dow and Werlang (1992).

6 Application to Risk Measures

6.1 Two Axiomatic Approaches to Characterize Insurance Prices

Yaari (1987) proposes an axiom dual to the independence axiom in EU theory and establishes a dual theory for risk and uncertainty. In Yaari's dual theory, the certainty equivalent of a uniformly bounded economic prospect can be represented as a Choquet integral. Unlike the economic theories of Yaari (1987) and Schmeidler (1989) which concern individual risk preferences, Wang et al.(1997) introduces four axioms to describe

[4]Obviously, we have $u(\mu_i, A) = 1 - \mu_i(A) - \mu_i(A^c)$, which is justly the definition of Dow and Werlang (1992), where it is claimed that $u(\mu_i, A)$ measures the amount of probability "lost" by the presence of uncertainty aversion.

[5]Note that this notation is consist with the fact that selling asset X is the same as buying the short position $-X$

the behavior of insurance prices, and shows that under these axioms the price for an insurance risk has a Choquet integral representation with respect to a distorted probability. Let $\mathcal{X}$ be a set of non-negative random variables defined on a probability space $(\Omega, \mathcal{F}, \mathbb{P})$. Market premiums $H(\cdot)$ can be described as a functional from the set of insurance risks $\mathcal{X}$ to $[0, \infty]$.

The following theorem, which is a kind of characterization of insurance prices under risk, is due to Wang et al.(1997).

Theorem 6.1. *Assume that $\mathcal{X}$ contains all the Bernoulli(u) random variables, $0 < u < 1$. Then H satisfies the comonotonic additivity, monotonicity, the following continuity:*

$$\lim_{a \to 0} H(X - X \wedge a) = H(X), \quad \lim_{b \to \infty} H(X \wedge b) = H(X),$$

and the conditional state independence (i.e. $H(X)$ depends only on the distribution of X), if and only if there is a unique distortion function γ such that

$$H[X] = H[1] \int X d(\gamma \circ \mathbb{P}) = H[1] \int_0^\infty \gamma(\mathbb{P}(X > x)) dx.$$

The following theorem, which is a kind of characterization of insurance prices under uncertainty, is due to Cardin and Ferretti (2001).

Theorem 6.2. *A premium calculation principle H satisfies compatibility ($H(X) \geqslant \mathbb{E}_{\mathbb{P}}[X]$), monotonicity, positive homogeneity, no unjustified loading ($H(I_\Omega) = 1$), weak comonotonic-additivity ($H(X) = H(X \wedge a) + H(X - X \wedge a), a \in \mathbb{R}_+$) and continuity ($\lim_{n\to\infty} H(X_n) = H(X)$ when X_n converges uniformly to X), if and only if there exists a capacity μ on $\mathcal{F}$ such that $\mu(A) \geqslant \mathbb{P}(A)$ for all $A \in \mathcal{F}$ and $H(X) = \mu(X)$ for any $X \in \mathcal{X}$.*

6.2 Characterizations of Risk Measures in Terms of Choquet Integrals

The notion of coherent risk measure was introduced in Artzner *et al.* (1997, 1999) for a probability space with finite states, and further extended to general probability spaces in Delbaen (2001, 2002). In Föllmer and Schied (2002) and Frittelli and Rosazza Gianin (2002) the more general concepts of convex risk measures were introduced. In the context of robust statistics, a result similar to the one in Artzner *et al.* (1997, 1999) was obtained in Huber (1981). Recently, Song and Yan (2006(a))introduced risk measures which are comonotonic subadditive or comonotonic convex, and gave their representations in terms of Choquet integrals. Independently, Heyde et al. (2006) proposed a data based risk measure in which the comonotonic subadditivity is taken as an axiom. Song and Yan (2006(b)) further proposed an axiomatic approach to some new risk measures and gives their representations in terms of Choquet integrals. These risk measures are not only comonotonically subadditive or convex, but also respect stochastic dominance, or stop-loss order, or convex order, and consequently, are law-invariant. These properties have sound economic meaning in insurance.

Let $(\Omega, \mathcal{F})$ be a measurable space, and let $L^\infty(\Omega, \mathcal{F})$ be the space of bounded $\mathcal{F}$-measurable functions, endowed with the supremum norm $\|\cdot\|$. We denote by $\mathcal{M}_{1,m}$ the set of all capacities on $\mathcal{F}$.

We only state the main results of Song and Yan (2006 (a)) concerning the functionals defined on $L^\infty(\Omega, \mathcal{F})$.

Theorem 6.3. *If $\pi : L^\infty(\Omega, \mathcal{F}) \to \mathbb{R}$ satisfies positive homogeneity, monotonicity, translation invariance and comonotonic subadditivity, then*

$$\pi(X) = \max_{\mu \in \mathcal{M}} \mu(X),$$

where

$$\mathcal{M} = \{\mu \in \mathcal{M}_{1,m} : \ \mu(Y) \leqslant \pi(Y), \forall Y \in L^\infty(\Omega, \mathcal{F})\}.$$

Theorem 6.4. *If $\pi : L^\infty(\Omega, \mathcal{F}) \to \mathbb{R}$ satisfies monotonicity, translation invariance and comonotonic convexity, then π is of the form*

$$\pi(X) = \max_{\mu \in \mathcal{M}_{1,m}} (\mu(X) - \alpha(\mu)), \quad X \in L^\infty(\Omega, \mathcal{F}),$$

where

$$\alpha(\mu) = \sup_{\pi(X) \leqslant 0} \mu(X), \quad \mu \in \mathcal{M}_{1,m}.$$

6.3 Risk Measurement with Equivalent Utility Principles

Recently, risk measure is a hot topic in the literature of mathematical finance. There are many kinds of risk measures, such as VaR, Tail-VaR, coherent risk measure, convex risk measure and so on. However, risk measures have been extensively studied in the actuarial literature for more than 30 years, under the guise of premium calculation principle, see Bühlmann (1970) and Gooverts et al. (1984). Bühlmann (1970) introduced the equivalent utility principle in the context of setting insurance premium. Denuit et al. (2006) shows that many well known risk measures can be obtained using some equivalent utility principles under risk and uncertainty. In the discussion that follows, we will explain these equivalent utility principles in detail.

Consider a decision maker with a utility function $U(\cdot)$. Denote $\mathcal{E}(\cdot)$ as the uniform sign of the expected or the non-expected operator. Assume $U(0) = 0$ and $\mathcal{E}(0) = 0$. Facing a random loss X, the decision maker sets $\rho(X)$ as the solution of the equation

$$\mathcal{E}[U(\rho(X) - X)] = 0.$$

Such a rule of setting $\rho(X)$ as the risk measure of X is called an "*equivalent utility principle*".

Before presenting the main results of Denuit et al. (2006), we recall some definitions of familiar risk measures. A risk measure $\rho(X)$ satisfies the properties of monotonicity, positive homogeneity, translation invariance and subadditivity is called a *coherent risk measure.* A risk measure is a *convex risk measure* if it satisfies the property of monotonicity, translation invariance and convexity. For any p in $(0, 1)$, *Value-at-Risk* (VaR) and *Tail Value-at-Risk*(TVaR) for a random variable X with given distribution function F_X is defined by

$$VaR_p(X) = \inf\{x \in \mathbb{R} | F_X(x) \geqslant p\}$$

and

$$TVaR_p(X) = \frac{1}{1-p}\int_p^1 VaR_q(X)dq,$$

respectively.

In Denuit et al. (2006), using different $\mathcal{E}(\cdot)$ and utility functions U and solving the corresponding equations, they obtained the corresponding risk measures which are listed in the sequel.

EU theory: $\rho(X)$ is the solution of $\mathbb{E}[U(\rho(X) - X)] = 0$. As an example, put $U(x) = \frac{1-\exp(-cx)}{c}$, where $x \geqslant 0$ and $c > 0$, then simple calculations yield

$$\rho(X) = \frac{1}{c}\ln \mathbb{E}[\exp(cX)].$$

This is the so-called exponential principle in insurance, which possesses an appealing interpretation in terms of ruin theory.

DU theory: $\rho(X)$ is the solution of $\mu(\rho(X) - X) = 0$, where $\mu = \gamma \circ \mathbb{P}$, which implies that

$$\rho(X) = \bar{\mu}(X),$$

where $\bar{\mu} = \bar{\gamma} \circ \mathbb{P}$, $\bar{\gamma}(x) = 1 - \gamma(1 - x)$. For example, for $p \in (0, 1)$, VaR_p corresponds to

$$\bar{\gamma}(x) = 1, \text{ if } x > 1 - p, \text{ and } \bar{\gamma}(x) = 0, \text{ otherwise, for } 0 \leqslant x \leqslant 1,$$

and $TVaR_p$ corresponds to

$$\bar{\gamma}(x) = \text{Min}(\frac{x}{1-p}, 1).$$

CEU theory: $\rho(X)$ is the solution of $\mu[U(\rho(X) - X)] = 0$, where μ is a capacity. For example, if we let $U(x) = x$, we obtain

$$\rho(X) = \bar{\mu}(X).$$

Furthermore, if we assume μ is convex, then

$$\rho(X) = \sup_{\mathbb{P} \in \text{core}(\mu)} \mathbb{E}_{\mathbb{P}}[X],$$

which is a coherent risk measure.

MEU theory: $\rho(X)$ is the solution of $\inf_{\mathbb{P} \in \mathcal{P}} \mathbb{E}_{\mathbb{P}}[U(\rho(X) - X)] = 0$, where $\mathcal{P}$ is a closed and convex set of additive probability measures. Here MEU stands for *Maximin Expected Utility.* As an example, if we let $U(x) = x$, we obtain

$$\rho(X) = \sup_{\mathbb{P} \in \mathcal{P}} \mathbb{E}_{\mathbb{P}}[X],$$

which is also a coherent risk measure.

RDEU theory: $\rho(X)$ is the solution of $\mu[U(\rho(X) - X)] = 0$, where $\mu = \gamma \circ \mathbb{P}$. Especially, put $U(x) = \frac{1-\exp(-cx)}{c}$, where $x \geqslant 0$ and $c > 0$, we obtain

$$\rho(X) = \frac{1}{c} \ln \left(\int_0^\infty \bar{\gamma}(\mathbb{P}(\exp(cX) > t)dt \right).$$

For some appealing explanations and convincing examples on these results, the reader may consult Denuit et al. (2006).

References

[1] Abdellaoui, M. (2002). A genuine rank-dependent generalization of the von Neumann-Morgenstern exepcted utility theorem. *Econometrica* 70, 717–736.

[2] Anscombe, F. J., and R. J. Aumann (1963). A definition of subjective probability. *Annals of Mathematical Statistics* 34, 199–205.

[3] Artzner, P., F. Delbaen, J. M. Eber, and D. Heath (1997). Thinking coherently. *RISK* 10, 68–71.

[4] Artzner, P., F. Delbaen, J. M. Eber, and D. Heath (1999). Coherent measures of risk. *Math. Finance* 9(3), 203–228.

[5] Asano, T. (2003). Portfolio inertia under ambiguity. Preprint.

[6] Basset, G., R. Koenker and G. Kordas (2004). Pessimistic portfolio allocation and Choquet expected utility. *Journal of Financial Econometrics* 2(4), 477–492.

[7] Bühlmann, H. (1970). *Mathematical Methods in Risk Theory.* Springer Verlag, New York.

[8] Cardin, M. and P. Ferretti (2001). On the use of capacities in representing premium calculations principle. *Decision in Economics and Finance* 24, 71–77.

[9] Castagnoli, E., F. Maccheroni and M. Marinacci (2004). Choquet insurance pricing: a caveat. *Mathematical Finance* 14, 481–485.

[10] Chateauneuf, A., R. Kast and A. Lapied (1996). Choquet pricing for financial markets with frictions. *Mathematical Finance* 6, 323–330.

[11] Chaueauneuf, A. and M. Cohen (2000). Choquet expected utility model: A new approach to individual behavior under uncertainty and to social welfare, in: *Fuzzy Measures and Integrals*, M. Grabisch et al. (Editors), Physica-Verlag, 289–313.

[12] Chateauneuf, A., R. Kast and A. Lapied (2001). Conditioning capacities and Choquet integrals: the role of comonotony. *Theory and Decision* 51, 367–386.

[13] Chateauneuf, A., J. Eichberger and S. Grand (2003). A simple axiomatization and constructive representation proof for Choquet expected utility. *Economic Theory* 22, 907–915.

[14] Chen, Z., R. Kulperger and G. Wei (2005(a)). A comonotonic theorem of BSDEs and its applications. *Stochastic Processes and their Applications* 115 , 41–54.

[15] Chen, Z., T. Chen and M. Davison (2005(b)). Choquet expectation and Peng's g-expectation. *Annals of Probability* 33 , 1179–1199.

[16] Chen, Z. and R. Kulperger (2006). Minimax pricing and Choquet pricing. *Insurance: Mathematics and Economics* 38(3), 518–528.

[17] Choquet, G. (1953). Theory of Capacity. *Ann. Inst. Fourier. (Grenoble)* 5, 131–295.

[18] Cohen, M., I. Gilboa, J. Jaffray and D. Schmeidler (1997). An experimental study of updating ambiguous beliefs. W.P. CEME, University of Paris 1.

[19] Delbaen, F. (2001). *Coherent risk measures.* Lecture Notes, Pisa.

[20] Delbaen, F. (2002). Coherent measures of risk on general probability spaces. In: *Advancees in Finance and Stochastics.* Essays in Honour of Dieter Sondermann, Springer-Verlag, 1–37.

[21] Dempster, A. (1967). Upper and lower probability induced by a multi-valued mapping. *Annals of Mathematical Statistics* 38, 325–339.

[22] Denneberg, D. (1994). *Non-additive Measure and Integral.* Kluwer Academic Publishers.

[23] Denneberg, D. and P. P. Wakker (2001). Nonmonotonic Choquet integrals. *Journal of Mathematical Economics* 36, 45–60.

[24] Denneberg, D. (2002). Conditional expectation for monotone measures, the discrete case. *Journal of Mathematical Economics* 37(2), 105–121.

[25] Denuit, M., J. Dhaene, M. Goovaerts, R. Kaas and R. Laeven (2006). Risk measurement with equivalent utility principles. *Statistics and Decisions* 24, 1–25.

[26] Dow, J. and S. R. C., Werlang (1992). Uncertainty aversion, risk aversion and the optimal choice of portfolio. *Econometrica*, 197–204.

[27] Eichberger, J. and D. Kelsey (1994). *Non-additive Beliefs and Game Theory.*University of Melbourne.

[28] Eichberger, J., S. Grant and D. Kelsey (2005). CEU preference and dynamic consistency. *Mathematical Social Science* 49, 143–151.

[29] El Karoui, N., S. Peng and M. C. Quenez (1997). Backward stochastic differential equations in finance. *Mathematical Finance* 7 , 1–71.

[30] Epstein, L.G. (1999). A definition of uncertainty aversion. *Review of Economic Studies* 66, 579–608.

[31] Fagin, R., J. Halpern, Y. Moses and M. Vardi (1995). *Reasoning About Knowledge.* MIT Press. Cambridge.

[32] Föllmer, H. and A. Schied (2002). Convex measures of risk and trading constraints. *Finance and Stochastics* 6(4), 429–447.

[33] Frittelli, M. and E. Rosazza Gianin (2002). Putting order in risk measures. *Journal Banking Finance* 26(7), 1473–1486.

[34] Ghirardato, P. (1994). Agency theory with non-additive uncertatinty. *Preprint.*

[35] Ghirardato, P. (1997). On independence for non-additive measures, with a Fubini theorem. *Journal of Economic Theory* 73(2), 261–291.

[36] Gilboa, I. and D. Schmeidler (1989). Maxmin expected utility with non-unique prior. *Journal of Mathematical Economics* 18(2), 141–153.

[37] Gilboa, I. and D. Schmeidler (1993). Updating ambiguous beliefs. *Journal of Econonomic Theory* 59, 33–49.

[38] Goovaerts, M. J., F. De Vijlder and J. Haezendonck (1984). *Insurance Premiums.* North-Holland, Amsterdam.

[39] Grabisch, M. (1996). The application of fuzzy integrals in multicriteria decision making. *European Journal of Operational Research* 89, 445–456.

[40] Grabisch, M. (1997). k-order additive discrete fuzzy measures and their representation. *Fuzzy Sets and Systems* 92(2), 167–189.

[41] Graf, S. (1980). A Radon-Nikodym theorem of capacity. *Journal für die Reine und Angewandte Mathematik* 75, 192–214.

[42] Greco, G. (1982). Sulla rappresentazione di funzionali mediante integrali. *Rend. Sem. Math. Univ. Padova* 66, 21–42.

[43] Havrda, J. and F. Charvat (1967). Quantification method in classification processes: concept of structural α-entropy. *Kybernetika* 3, 30–35.

[44] Heyde, C.C., S.G. Kou, and X.H. Peng (2006). What is a good risk measure: bridging the gaps between data, coherent risk measure, and insurance risk measure. *Preprint.*

[45] Hu, Y. (2005). On Jensen's inequality for g-expectation and for nonlinear expectation. *Archiv der Mathematik* 85(6), 572–580.

[46] Huber, P. (1973). The use of Choquet capacity in statistics. *Bull. Institut. Internat. Stat.* 45, 181–191.

[47] Huber, P. and U. Strassen (1973). Minimax tests and the Neyman-Pearson Lemma for capacities. *Annals of Statistics* 1, 251–263.

[48] Huber, P. J. (1981) *Robust Statistics.* Wiley, New York.

[49] Jaynes, E. (1957). Information theory and statistical mechanics. *Phys. Rev.* 106, 620–630.

[50] Karni, E. (2005). Subjective expected utility theory with costly actions. *Games and Economic Behavior* 50(1), 28–41.

[51] Kojadinovic, I. (2005). Minimum variance capacity identification. *Preprint.*

[52] Kojadinovic, I., J. Marichal and M. Roubens (2005). An axiomatic approach to the definition of the entropy of a discrete Choquet capacities. *Information Sciences* 172, 131–153.

[53] Lapied, A. and R. Kast (2005). Updating Choquet valuation and discounting information arrivals. *Preprint.*

[54] Lehrer, E. (2005) Updating non-additive probabilities – a geometric approach. *Games and Economic Behavior* 50(1), 42–57.

[55] Maccheroni, F. and M. Marinacci (2005). A strong law of large numbers of capacities. *Annals of Probability* 33(3), 1171–1178.

[56] Marichal, J. (2000). An axiomatic approach of the discrete Choquet integral as a tool to aggregate interacting criteria. *IEEE Transactions on Fuzzy Systems* 8(6), 800–807.

[57] Marichal, J. (2002). Entropy of discrete Choquet capacities. *European Journal of Operational Research* 137, 612–624.

[58] Narukawa, Y. and T. Murofushi (2004). Decision modeling using the Choquet integral. *Proc. Modeling Decisions for Artificial Intelligence*, Lecture Notes in Computer Science 3131 (2004), 183–193.

[59] Pardoux, E. and S. Peng (1990). Adapted solution of a backward stochastic differential equation. *System Control Letters* 14 , 55–61.

[60] Peng, S. (1997). Backward SDE and related g-expectation. *Backward Stochastic DEs. Pitman* 364, 141–159.

[61] Quiggin, J. (1993). *Generalized Expected Utility Theory: The Rank-Dependent Model.* Kluwer Academic Publishers, Boston / Dordrecht / London.

[62] Schmeidler, D. (1986). Integral representation without additivity. *Proceeding of the American Mathematical Society* 92, 255–261.

[63] Schmeidler, D. (1989). Subjective probability and expected utility without additivity. *Econometrica* 57, 571–587.

[64] Schmidt, U. and H. Zank (2001). A new axiomatization of rank-dependent expected utility with tradeoff consistency for equally likely outcomes. *Journal of Mathematical Economics* 35, 483–491.

[65] Shafer, G. (1976). *A Mathematical Theory of Evidence.* Princeton Univ. Press. Princeton.

[66] Shapley, L. (1953). A Value for N-person Games. In "Contribution to the Theory of Games" (Kuhn H. and A. Tucker, Eds.) Princenton Univ. Press. Princeton.

[67] Song, Y. and J. A. Yan (2006 (a)) The representations of two types of functionals on $L^{\infty}(\Omega, \mathcal{F})$ and $L^{\infty}(\Omega, \mathcal{F}, \mathbb{P})$, *Science in China*, Series A-Mathematics, Vol. 49, No. 10, 1376–1382.

[68] Song, Y. and J. A. Yan (2006 (b)). Risk measures with comonotonic subadditivity or convexity and respecting stochastic orders. *Preprint.*

[69] Sternberg, J. (1984). A re-examination of put-call parity on index futures. *Journal of Futures Markets* 14, 79–101.

[70] Waegenaere, D. and P. P. Wakker (2001). Nonmonotonic Choquet integrals. *J. Math. Econom.* 36(1), 45–60.

[71] Waegenaere, D., R. Kast and A. Lapied (2003). Choquet pricing and equilibrium. *Insurance: Mathematics and Economics* 32(3), 359–370.

[72] Wakker, P. P. (1994). Separating marginal utility and probabilistic risk aversion. *Theory and Decision* 36, 1–44.

[73] Wang, S., V. Young and H. Panjer (1997). Axiomatic characterization of insurance prices. *Insurance: Mathematics and Economics* 21, 173–183.

[74] Wasserman, L. and J. Kadane (1990). Bayes' theorem for Choquet capacity. *Annals of Statistics* 18(3), 1328–1339.

[75] Wasserman, L. (1990). Belief function and statistical inference. *The Canadian Journal of Statistics* 18, 183–196.

[76] Yaari, M. E. (1987). The dual theory of choice under risk. *Econometrica* 55(1), 95–115.

[77] Yan, J. A. (2005). A short presentation of the Choquet integral, *Preprint.*

[78] Zhou, L. (1998). Integral representation of continuous comonotonically additive functionals. *Transactions of American Mathematical Society* 350(5), 1811–1822.

Some Recent Developments in Actuarial Science*

Hailiang Yang
Department of Statistics and Actuarial Science
The University of Hong Kong
Pokfulam Road, Hong Kong, China
E-mail: hlyang@hkusua.hku.hk

Abstract

In this paper, we will provide a brief review on some recent developments in actuarial science. In particular, in view of the changing insurance business nowadays, this paper will illustrate how modern finance theory can be applied to some actuarial science problems, such as pricing of participating products and equity-index annuities. We will also summarize some recent work on risk theory.

1 Introduction

Nowadays, insurance business is changing, the changes are mainly due to the development of modern finance and mathematical finance, in particular the rapid growth of derivatives market. There are many products which exhibit option-embedded features sold by the insurance companies, and such kinds of insurance products, including variable annuities, participating life insurance contracts, equity-index annuities (EIAs) or equity-linked annuities (ELAs), guarantee annuity options (GAOs) and segregated funds, etc., become very popular recently. Earlier work on exploring the interplay between the option pricing theory and life insurance products can be dated back to Boyle and Schwartz (1977) and Brennan and Schwartz (1976, 1979). These works investigate the use of modern option pricing theory and its techniques for the valuation of equity-linked products and life insurance policies with an guarantee on asset value.

*The work described in this paper was supported by a grant from the Research Grants Council of the Hong Kong Special Administrative Region, China (Project No. HKU 7426/06H).

Ruin theory has been of a major interest in mathematical insurance, and have been investigated by many authors. The early work on this problem can track back to Lundberg (1903, 1926) and Cramér (1930). There is a huge amount of literature on ruin theory. Some representative monographs in this area include Bühlmann (1970), Gerber (1979), Grandell (1991), Rolski et al. (1999) and Asmussen (2000).

In the classical risk theory, the compound Poisson process has widely been used to model the surplus process for an insurance company. In particular, it has been used to describe the aggregate claim amounts up to a certain time. The classical model has been generalized in many aspects. Now, the perturbed risk model described by a compound Poisson process with a diffusion term, renewal process models, point process models and Levy process models are also used by many researchers to describe the surplus process of an insurance company. The objective of ruin theory is to obtain closed form or estimations for the ruin probability and related quantities. Closed form solutions are possible only for some very special cases. Most of the works in this area are focus on either upper and lower bounds or asymptotic results for the ruin probability and related quantities.

Recently, many researchers in risk theory are interested in Gerber-Shiu function, models with investment incomes, and models with dividends. Optimal dividend policy has become another popular subject in actuarial science these days. We will summarize some of these developments in this paper. For the purpose of illustration and easy to understand, we will only present the simplest models and refer to the original research papers for various different models and detailed general results.

2 Insurance products with embedded options

In this section, we will give a brief introduction to some of the insurance products with embedded options.

2.1 Variable annuities

In actuarial science, in particular life insurance, life annuities play an important role. A life annuity is a series of payments made continuously or at equal intervals while a given life survives. For a detailed discussion on annuities (or fixed annuities), the readers can see the standard text book by Bowers et al. (1997). Due to the development of derivatives market, the traditional annuities have become less popular, and variable annuities and other equity linked annuities have become more and more popular. For example, in the U.S. annuities market, the portion of annual sale of the traditional annuities was about one half in 1994, and dropped to about one sixth in 2005.

A variable annuity is a contract between an individual and an insurance company, under which the insurer agrees to make periodic payments to the individual, beginning either immediately or at some future date, and the payment will continue until the death of the individual. The variable annuity offers many investment options to the policyholders, and the policyholder can choose the investment options. The investment options for a variable annuity are typically mutual funds that invest in stocks, bonds, money market instruments, or some combination of the three. The payments to the policyholder will linked with the performance of his/her investment options. In addition, variable annuities have a death benefit. The individual purchases a variable annuity contract by making either a single purchase payment or a series of purchase payments. One of the reasons which make the variable annuities so popular is that the payoff of the variable annuity is linked with the performance of the equity market, and another reason is that variable annuities are tax-deferred. The holders of variable annuities pay no taxes on the income and investment gains from the annuity until the holders of the variable annuities withdraw their money. The policyholder can transfer his/her money from one investment option to another within a variable annuity without paying tax at the time of the transfer.

Since the investment options of variable annuities are mutual funds, so, the same as mutual fund, variable annuities also have investment risks. Variable annuities are long-term investment products, they are mainly designed to meet retirement and other long-range goals. If the policyholder withdraw his/her money early, the insurance company will charge certain fee and, of course, the policyholder needs to pay tax. Therefore, variable annuities are not suitable for short-term investment.

2.2 Equity indexed annuities

The first equity indexed annuity product was introduced in 1995 by Keyport Life Insurance Co. An equity indexed annuity can be considered as a combination of a fixed and a variable annuity. EIAs appeal to investors because the payment of the equity indexed annuities has a minimum guarantee, but it also linked with the performance of the equity index. This makes it possible for the policy holders to benefit from potential gains in a stock market index, while still receiving a guaranteed minimum return. Consistent with high risk high return principle, the equity indexed annuity offers higher return, but potentially more risk than a fixed annuity, and less risk, but potentially lower return than a variable annuity.

There are several indexing methods for EIAs. Typically, an equity indexed annuity is linked to a market index such as the Standard & Poor's 500 stock index. If the index goes up, the equity indexed annuity is credited with a return based on a percentage of the gain in that index. The product designs of EIAs are many and can be very complex. In this paper, we only use a very simple example to illustrate the ideas. Tiong (2000) presents a comprehensive discussion on the equity-indexed annuities and discussed the pricing of three of the most common product designs. For more detailed discussions on the EIAs and related products see the monograph by Hardy (2003). Lin and Tan (2003) consider the EIA model with a stochastic interest rate and mortality risk. Lee (2002) provided an introduction to EIAs and introduced some new type of EIAs. Boyle and Hardy (2003) study the guaranteed annuity options. In the first part of this paper, we overview the recent development of insurance products with option-embedded features and the valuation of these products.

2.3 Participating life insurance contracts

Participating life insurance products can be considered investment plans with associated life insurance benefits, a specified benchmark return, a guarantee of an annual minimum rate of return and a specified rule of the distribution of annual excess investment return above the guaranteed return. When policyholders buy the contracts, they pay annual premiums to an insurance company. The insurance company will then manage and invest the funds in a specified reference portfolio. One key feature of these investment plans is the sharing of profits from an investment portfolio between the policyholders and the insurer. The specified rule of surplus distribution commonly used by insurance company is known as reversionary bonus, which is employed to credit interest at or above a specified guaranteed rate to the policyholders every year. The policyholders can receive an additional bonus at the maturity of the contract, namely, the terminal bonus, if the terminal surplus of the fund is positive at the maturity. If the insurer defaults at the maturity of the policy, the policyholders can only receive the outstanding assets. For more comprehensive discussion on various features of participating policies, refer to Grosen and Jørgensen (2000), Ballotta et al. (2006) and Siu et al. (2006). In recent years, participating life insurance products become more and more popular in major insurance and finance markets around the world.

2.4 Segregated funds

Segregated Funds are a kind of investment, but also a kind of insurance contract. When an investor invest in a segregated fund, he/she is actually buying an insurance contract. The insurance company invests the money from the investor in an underlying mutual fund, but the investor does not own any units of mutual fund. The return of the investor is closely linked with the performance of the underlying mutual fund. The segregated fund combines the growth potential of mutual funds with some kind of guarantee from the insurance company. The investor will be guaranteed to get back certain percentage of his/her investment even the price of the underlying mutual fund drops to close to 0.

The maturity date of the segregated funds is fixed. If the investor sells the segregated funds prior to the maturity date, he/she will receive the market value, which may be less than his/her original investment. Following the investor's maturity date or upon his/her death, the investor or his/her beneficiaries will receive either the guaranteed amount or the market value of his/her investment, whichever is higher.

3 Valuation of insurance products with embedded options

In this section we will illustrate how the option pricing methodology can be used to value the insurance products with embedded options. For illustration purpose, we will use equity-index annuity as an example.

We consider a financial model consisting of a risk-free money market account and a reference equity index. We suppose that the market value of the reference index is governed by a geometric Brownian motion model. We assume that the payoff of the EIAs not only depends on the reference index, but also depends on the mortality of the policy holder. For the mortality risk, we use some actuarial approach to price it. Therefore, in some sense, the price of EIAs in this paper is different from the price in finance and is a kind of risk measure. In the sequel, we introduce the set up of our model.

First, we fix a complete probability space $(\Omega, \mathcal{F}, P)$, where P is the real-world probability measure. Let $\mathcal{T}$ denote the time index set $[0, T]$ of an economy in which all economic activities take place. Let r be the instantaneous market interest rate of a bank account or a money market account.

Let $\{W_t\}_{t\in\mathcal{T}}$ denote a standard Brownian Motion on $(\Omega, \mathcal{F}, P)$ with respect to the P-augmentation of its natural filtration $\mathcal{F}^W := \{\mathcal{F}^W_t\}_{t\in\mathcal{T}}$. Let $X_t := \ln(S_t/S_0)$ denote the logarithmic return from the reference index S over the time interval $[0, t]$, where $t \in \mathcal{T}$. Then, we suppose that the dynamic of $\{X_t\}_{t\in\mathcal{T}}$ is given by a Brownian motion with drift:

$$X_t = X_0 + \left(\mu - \frac{1}{2}\sigma^2\right)t + \sigma W_t \, . \tag{3.1}$$

Then, the market values S of the reference index is given by

$$S_t = S_0 \exp(X_t)\ . \tag{3.2}$$

The equity index model here is the standard Black-Scholes model. We can use Girsanov theorem to change the probability measure from P to the risk-neutral probability measure Q. In actuarial science there is another popular method to deal with the changing measure, the Esscher transform, which is proposed by Gerber and Shiu (1994). The advantage of Esscher transform is that even the market is incomplete, the Esscher transform still gives us a unique risk-neutral measure, and this risk neutral measure can be justified by utility principle. In the sequel, we follow Gerber and Shiu (1994) and define the Esscher transform to determine an equivalent martingale measure. Let $\mathcal{M}_W(\theta)_t := E_P[e^{-\theta W_t}] = e^{\frac{1}{2}\theta^2 t}$. Let $\{\Lambda_t^\theta\}_{t\in\mathcal{T}}$ denote a $\mathcal{F}^W$-adapted stochastic process defined as below:

$$\Lambda_t^\theta := \frac{e^{-\theta W_t}}{\mathcal{M}_W(\theta)_t} = \exp\left(-\theta W_t - \frac{1}{2}\theta^2 t\right)\ ,\ t \in \mathcal{T}. \tag{3.3}$$

Then, it is not difficult to check that $\{\Lambda_t^\theta\}_{t\in\mathcal{T}}$ is a P-martingale.

Let

$$\theta^* = \frac{\mu - r}{\sigma}, \tag{3.4}$$

and define a new probability measure Q_{θ^*} by

$$\frac{dQ_{\theta^*}}{dP}|_{\mathcal{F}_t} = \Lambda_t^{\theta^*} = \frac{e^{-\theta^* W_t}}{\mathcal{M}_W(\theta^*)_t}. \tag{3.5}$$

It is easy to check that under Q_{θ^*}, $\{e^{-rt}S_t\}$ is a martingale and,

$$dS_t = rS_t dt + \sigma S_t d\bar{W}_t, \tag{3.6}$$

where $\bar{W} = W_t + \theta^* t$ is a standard Brownian motion under the martingale measure Q_{θ^*}.

Now, we present a simple point-to-point design. There are various EIA designs in the market. In the general case, we have to relay on numerical methods to calculate the price of the EIA.

For the simple point-to-point design, the policy holder receives the realized return on the index over a certain period of time at a prescribed participation rate, but with a minimum guarantee. Let α denote the participation rate. Suppose that at time T, $T > 0$, the policy pays $e^{\alpha X(T)}, \alpha > 0$, or $S_0 e^{gT}$ (that is, at a fixed minimum guarantee rate g), whichever is higher. Hence, we can express the value of this policy as:

$$E_{Q_{\theta^*}}[e^{-rT}\max(e^{\alpha X_T}, S_0 e^{gT})]. \tag{3.7}$$

Therefore, if we assume that the policy holder will survive to time T, it can be shown that the value of the policy is given by:

$$\begin{aligned} V_{pp}(S_0) &= E_{Q_{\theta^*}}\left\{\exp[-rT]\max(e^{\alpha X_T}, S_0 e^{gT})\right\} \\ &= \exp\left(-r(1-\alpha)T + \alpha X_0 + \frac{1}{2}\alpha(\alpha-1)\sigma^2 T\right)\Phi(d_1) \\ &- \exp(-rT+gT)S_0\Phi(d_2)\ , \end{aligned} \tag{3.8}$$

where $\Phi(x)$ is the distribution function of standard normal random variable,

$$d_1 = \frac{(r-\frac{1}{2}\sigma^2+\alpha\sigma^2)T + X_0 - gT/\alpha}{\sigma\sqrt{T}}\ ,$$

and

$$d_2 = \frac{gT/\alpha - X_0 - (r-\frac{1}{2}\sigma^2)T}{\sigma\sqrt{T}}\ .$$

If we assume there is mortality risk for the policyholders. We assume that the financial market is independent of the policyholder's mortality risk. The same as before, we use T to denote the deferred time period and assume that T is an integer, and g to be the minimum guaranteed rate of return. Let $T(x)$ be the future lifetime of a life aged x. We assume that $T(x)$ is independent of X_t. The payoff of the point-to-point design in this case is:

$$\begin{aligned} &\max(e^{\alpha X_T}, S_0 e^{gT}), && \text{if } T(x) > T \\ &\max(e^{\alpha X_s}, S_0 e^{gs}), && \text{if } s-1 < T(x) \leqslant s \end{aligned} \tag{3.9}$$

for $s = 1, 2, \cdots, T$. Due to the mortality risk, and the policy is not tradable now, so the value of the EIA is a kind of risk measure rather than the arbitrage free price in Black-Scholes sense.

Since we assume that $T(x)$ is independent of X_t, when we change the probability measure from P to Q_{θ^*}, we can keep the distribution of $T(x)$ under Q_{θ^*} the same as it was under P. Therefore the value of the

EIA in this case can be written as

$$
\begin{aligned}
&V_{pp}(S_0)\\
&= E_{Q_{\theta^*}}\left[e^{-rT}\max(e^{\alpha X_T}, S_0e^{gT}); T(x) > T\right]\\
&\quad+\sum_{s=1}^{T} E_{Q_{\theta^*}}\left[e^{-rs}\max(e^{\alpha X_s}, S_0e^{gs}); s-1 < T(x) \leqslant s\right]\\
&= E_{Q_{\theta^*}}\left[e^{-rT}\max(e^{\alpha X_T}, S_0e^{gT})\right] Q_{\theta^*}[T(x) > T]\\
&\quad+\sum_{s=1}^{T} E_{Q_{\theta^*}}\left[e^{-rs}\max(e^{\alpha X_s}, S_0e^{gs})\right] Q_{\theta^*}[s-1 < T(x) \leqslant s]\\
&= E_{Q_{\theta^*}}\left[e^{-rT}\max(e^{\alpha X_T}, S_0e^{gT})\right] P[T(x) > T]\\
&\quad+\sum_{s=1}^{T} E_{Q_{\theta^*}}\left[e^{-rs}\max(e^{\alpha X_s}, S_0e^{gs})\right] P[s-1 < T(x) \leqslant s]\\
&= E_{Q_{\theta^*}}\left[e^{-rT}\max(e^{\alpha X_T}, S_0e^{gT})\right] \,{}_Tp_x\\
&\quad+\sum_{s=1}^{T} E_{Q_{\theta^*}}\left[e^{-rs}\max(e^{\alpha X_s}, S_0e^{gs})\right] \,{}_{s-1}p_x \cdot q_{x+s-1},
\end{aligned}
\tag{3.10}
$$

where $E_{Q_{\theta^*}}\left[e^{-rs}\max(e^{\alpha X_s}, S_0e^{gs})\right]$ is given by equation (3.8) with T replaced by s.

In practice, most EIAs are much more complex than the simple point-to-point design. One of the most popular EIA designs is the so called cliquet design. Also in real world, the EIA products often involve exotic options. For detailed discussions on the pricing of EIAs, we refer to Tiong (1999), Lin and Tan (2003) and references therein. Here we would emphasis that, when we consider the mortality risk in the model, the model is incomplete and cannot be completed by adding new securities. The price or value of the EIA here is just a kind of risk measure. It is different from the no-arbitrage price in the Black-Scholes model.

In practice, the risk management and hedging of the EIAs and other insurance products with embedded options are very important and challenging problems. Unlike the derivatives in financial market, even under a very simple model, as long as we include the mortality risk, the market is incomplete and perfect hedging is impossible. The theme of the Society of Actuaries second annual equity-based guarantees conference, which was held in Chicago in October 2006, is risk management and hedging for the equity-based guarantees.

4 Ruin theory

Ruin theory has been one of the main research topics in actuarial science for long time. The most well known model is the compound Poisson model: Let $\{U_t; t \geqslant 0\}$ denote the surplus of the insurance company at time t, $U_0 = u$ be the initial surplus. Then

$$U_t = u + ct - \sum_{i=1}^{N_t} Y_i \tag{4.1}$$

where $c > 0$ is a constant, it represents the premium rate, $\{N_t; t \geqslant 0\}$ is the number of claims up to time t, while the sequence $\{Y_1, Y_2, \cdots\}$ denote the size of the claim, and are independent and identically distributed (i.i.d.) random variables with the same distribution $P(x)$ which has mean ν, and $\{Y_1, Y_2, \cdots\}$ is independent of $\{N_t; t \geqslant 0\}$, N_t is a homogeneous Poisson process with intensity λ.

Let $\tau = \inf\{t \geqslant 0 : U_t < 0\}$ be the ruin time. Define

$$\psi(u) = P\{\tau < \infty | U_0 = u\} \tag{4.2}$$

be the probability of ruin with initial surplus u.

The main classical results about ruin probability for the classical risk model are due to Lundberg (1926) and Cramér (1930). We list some of the main results here:

$$\psi(0) = \frac{\lambda\nu}{c}, \tag{4.3}$$

and $\psi(u)$ satisfies the following integro equation:

$$\psi(u) = \frac{\lambda}{c}\int_u^{\infty}(1 - P(x))\,dx + \frac{\lambda}{c}\int_0^u \psi(u-x)(1-P(x))\,dx. \tag{4.4}$$

If we assume that the moment generator function of $P(x)$ exits, we have

$$\lim_{u\to\infty} e^{Ru}\psi(u) = \frac{\rho\nu}{M_Y'(R) - \nu(1+\rho)}, \tag{4.5}$$

where $\rho = \frac{c}{\lambda\nu} - 1$ is the safety loading, $M_Y(t) = E[e^{tY}] = \int_0^\infty e^{tx} dP(x)$ is the moment generating function of the claim size random variable Y, (4.5) is called Cramér-Lundberg approximation, R satisfies

$$\frac{\lambda}{c}\int_0^{\infty} e^{Rx}(1 - P(x))\,dx = 1, \tag{4.6}$$

and R is called adjustment coefficient (or Lundberg exponent). One of the most well known result in ruin theory is the following Lundberg inequality:

$$\psi(u) \leqslant e^{-Ru}. \tag{4.7}$$

When $P(x)$ is an exponential distribution, $\psi(u)$ has a closed form expression:

$$\psi(u) = \frac{1}{1+\rho}\exp(-\frac{\rho u}{\nu(1+\rho)}), \tag{4.8}$$

where ρ is the safety loading.

4.1 Gerber-Shiu function

In a seminal paper, Gerber and Shiu (1998) propose to consider the following function:

$$\begin{aligned}\phi(u) &= E[w(U(\tau-), |U(\tau)e^{-\delta\tau}I(\tau<\infty)|U(0)=u] \\ &= \int_0^\infty\int_0^\infty\int_0^\infty w(x,y)e^{-\delta t}f(x,y,t|u)dtdxdy \\ &= \int_0^\infty\int_0^\infty w(x,y)f(x,y|u)dxdy, \end{aligned} \tag{4.9}$$

where $w(x,y)$ is a nonnegative function of $x>0$ and $y>0$, δ can be interpreted as a force of interest or as a dummy variable of the Laplace transform, I denotes the indicator function, $f(x,y,t|u)$ denotes the joint probability density function of $U(\tau-)$, $|U(\tau)|$ and τ, and

$$f(x,y|u) = \int_0^\infty e^{-\delta t}f(x,y,t|u)dt.$$

Note that $U(\tau-)$ is the surplus immediately before ruin and $|U(\tau)|$ is the deficit at ruin. The function $\phi(u)$ is called Gerber-Shiu function in the actuarial literature. By specifying the function $w(x,y)$, This function includes many ruin functions as special cases. For example, let $w(x,y) = 1$ and $\delta=0$, then $\phi(u)=\psi(u)$ is the ruin probability; if $w(x,y) = I(y\leqslant z)$ and $\delta=0$, then $\phi(u)=\phi(u,z)$ is the distribution of deficit at ruin; etc. Under the classical compound Poisson model, Gerber and Shiu (1998) derive the integro-differential equations satisfied by $\phi(u)$:

$$\begin{aligned} &c\phi'(u) - (\delta+\lambda)\phi(u) + \lambda\int_0^u \phi(u-x)p(x)dx \\ &+\lambda\int_0^\infty w(u,x)p(u+x)dx, \end{aligned} \tag{4.10}$$

where $p(x)=P'(x)$ is the claim size density function. In their paper, Gerber and Shiu also obtain the asymptotic formulas for $\phi(u)$ by using the renewal theorem, provide closed form results for zero initial surplus, very large initial surplus, and arbitrary initial surplus if the claim amount distribution is exponential or mixture of exponentials.

The function $\phi(u)$ has been studied by many authors under various insurance risk models. In Willmot and Dickson (2002), the Gerber-Shiu function is discussed under a renewal insurance risk model. Li and Garrido (2005) further analyze the Gerber-Shiu function under a class of renewal insurance risk process model. In Pavlova and Willmot (2004), the Gerber-Shiu function is considered under a discrete stationary renewal insurance risk model. In Lin and Pavlova (2005), Gerber-Shiu function is discussed under a compound Poisson model with dividend barriers. In Garrido and Morales (2006), the Gerber-Shiu function is considered under a Levey process model. The Gerber-Shiu function has become a so popular topic these days, there was a conference on Gerber-Shiu function which was held in August 2006 at Concordia University, Montreal.

4.2 Model with investment income

In the classical ruin theory, we assume that there is no investment income and the value of the money is the same at any time. It is well know that if the net profit condition holds, the surplus process of the classical model will converge to $+\infty$ with probability 1, as time goes to ∞. This indicates that if the insurance company cannot get anything from investment and can borrow money with 0 interest rate, the company needs only to worry about the net profit condition.

When we introduce the investment incomes into the insurance risk model, the story changes. When the interest rate is positive, we know that it is possible that the surplus will never come back to positive if the surplus falls below a critical value, this probability is called the absolute ruin probability. Gerber (1971) is the first paper considered the absolute ruin probability. Under the compound Poisson model, the Lundberg type upper bound and some expressions for absolute ruin probability are obtained, and a closed form solution for the absolute ruin probability is obtained in the case of exponential claims. Dassios and Embrechts (1989) and Sundt and Teugels (1995) are two papers on ruin theory under an insurance risk model with interest rate. Dassios and Embrechts (1995) study the absolute ruin probability using the martingale approach. Sundt and Teugels (1995) consider the classical compound Poisson model with constant interest force, ruin probability is discussed in the paper. Yang and Zhang (2001a, 2001b), by using the method in Sundt and Teugels (1995), analyze the joint distribution of surplus before and after ruin and obtain similar results to those in Sundt and Teugels (1995). Yang (1999) obtains both exponential and non-exponential upper bounds for the ruin probability under a discrete model with constant interest rate. Paulsen (1998) and Norberg (1999) use a diffusion models with stochastic interest incomes to model the surplus process and ruin probability is discussed. Paulsen and Gjessing (1997a) provide some results for a model with stochastic investment incomes. Ng et al. (2004) extend the model in Delbaen and Haezendonck (1987) from a deterministic discount factor to a random discount factor. The model in Ng et al. (2004) can be considered as the compound Poisson model with a stochastic interest rate.

4.3 Model with dividend payments

As mentioned in the previous section, in the classical model, when the net profit condition is true, the surplus of the insurance company will go to infinity with probability 1. This indicates that the classical ruin model is not realistic, so we have considered the model with investment incomes in the previous section. On the other hand, this also indicates that the model is too conservative. De Finetti (1957) proposed an insurance risk model which pays dividend when the surplus reaches to a certain barrier. De Finetti's model is a discrete discrete-time model, he considered the problem that the company gains $+1$ with probability $\pi > 1/2$ and -1 with probability $1 - \pi$. Gerber (1973, 1979) considers the classical insurance risk model with a linear barrier dividend policy. The upper bounds for ultimate ruin probability is obtained.

Recently, the optimal dividend problem has become a hot topic in actuarial science, the insurance risk models with dividends are becoming popular too. Albrecher and Kainhofer (2002) extend the model in Gerber (1979) to the case of nonlinear dividend barrier. Lin et al. (2003) analyze the classical insurance risk model with a constant barrier. Gerber-Shiu function is studied and a dividends-penalty identity is derived. Li and Garrido (2004) consider the renewal risk model with a constant barrier. Albrecher et al. (2005) investigate the Gerber-Shiu function under a model that allows dividend payments according to a time-dependent linear barrier strategy. Lin and Pavlova (2005) consider a classical compound Poisson model with a threshold dividend strategy, that is the model assumes that there is no dividend payment if the surplus is below a constant barrier, and pays a constant rate (less than the premium rate) when the surplus is over the constant barrier. The Gerber-Shiu function is studies. Zhu and Yang (2006) study the Markov modulated insurance risk model with a threshold dividend strategy. Ruin probability and related quantities are studied.

5 Optimal dividend strategy

The optimal dividend problem proposed by de Finetti (1957) is to find the dividend payment strategy that maximizes the expected discounted value of dividends which are paid to the shareholders until the company is bankrupt. In Bühlmann (1970), the problem is analyzed in the continuous time model of collective risk theory. In the absence of dividends, the surplus of the company at time t follows a compound Poisson process. Optimal barrier strategy is discussed. Gerber (1972, 1973, 1981) provide a detailed study and show that the optimal strategy for this model is the barrier strategy. After about 20 years, this problem becomes a hot topic in actuarial science again.

To illustrate the idea and the problem, we use the diffusion model. Let Z_t denote the surplus of the company, if no dividends are paid. Then the surplus process follows the following dynamics:

$$dZ_t = (\mu + \rho Z_t)dt + \sigma dW_t, \quad t \geqslant 0, \tag{5.1}$$

where $\{W_t, t \geqslant 0\}$ is a standard Brownian motion, $\mu > 0$ is the drift, $\rho > 0$ is the credit interest rate, and $\sigma > 0$ is the volatility (note that, since this model is an approximation to the insurance risk model, the σ here is much large than the stock volatility).

As in Gerber and Shiu (2004), we assume that dividends are paid to the shareholders according to a barrier strategy, say with parameter $b > 0$. Whenever the (modified) surplus is about to go above the level b, the excess (or "overflow") will be paid as dividends. Let D_t denote the aggregate dividends by time t, and

$$\tau = \inf\{t : Z_t - D_t = 0\} \tag{5.2}$$

be the time of ruin in the model with dividend payments. Due to the barrier dividend policy, the ruin is certain in this case. Now

$$D = \int_0^{\tau} e^{-\delta t} dD_t, \tag{5.3}$$

is the discounted value of the dividends until ruin. We are interested in $V(x; b)$, $0 \leqslant x \leqslant b$, the expectation of D, considered a function of the initial surplus x.

By standard argument, see Gerber and Shiu (2004), it can be shown that $V(x; b)$ satisfies the following second-order differential equation,

$$\frac{\sigma^2}{2} V''(x; b) + (\mu + \rho x)V'(x; b) - \delta V(x; b) = 0, \quad 0 < x < b, \tag{5.4}$$

in combination with the boundary conditions

$$\begin{cases} V(0; b) = 0, \\ V'(b; b) = 1. \end{cases} \tag{5.5}$$

It is possible to write $V(x;b)$ in a more transparent form. Let the function $g(x)$ be a solution of the differential equation

$$\frac{\sigma^2}{2}g''(x) + (\mu + \rho x)g'(x) - \delta g(x) = 0, \quad x > 0, \tag{5.6}$$

combined with the boundary condition

$$g(0) = 0. \tag{5.7}$$

The function $g(x)$ is unique apart from a constant factor and we have the factorization formula,

$$V(x;b) = \frac{g(x)}{g'(b)}, \quad 0 \leqslant x \leqslant b. \tag{5.8}$$

Thus the determination of the function $g(x)$ is a central problem.

The solution of (5.6) can be obtained using the so called confluent hypergeometric functions. See Cai et al. (2005) for details.

It is not difficult to see that

$$V(x;b) = x - b + V(b;b), \ x > b, \tag{5.9}$$

which has the following interpretation: if the initial surplus x exceeds the dividend barrier b, the difference is immediately paid out as dividends.

When $\rho = 0$, $V(x;b)$ can be expressed as, see Gerber and Shiu (2004),

$$V(x;b) = \frac{\sigma^2}{2\mu}(e^{2\mu b/\sigma^2} - e^{2\mu(b-x)/\sigma^2}), \quad 0 \leqslant x \leqslant b. \tag{5.10}$$

The optimal dividend barrier is then given by:

$$b^* = \frac{\sigma^2}{\sqrt{\mu^2 + 2\delta\sigma^2}} \ln\left(\frac{\mu + \sqrt{\mu^2 + 2\delta\sigma^2}}{\mu - \sqrt{\mu^2 + 2\delta\sigma^2}}\right). \tag{5.11}$$

In Gerber and Shiu (2004), many interesting properties and related problems are discussed.

Gerber and Shiu (2006) consider the optimal dividend strategies under the compound Poisson model. Both barrier strategy and threshold strategy are studied. In Paulsen and Gjessing (1997b) a more general model is analyzed. The surplus process before investment is the sum of a shifted compound Poisson process and an independent Brownian motion, and the stochastic rate of return is an independent process of the same type. In their paper, barrier strategies are examined. It is shown that the expected discounted dividends are the solution of an integro-differential equation in conjunction with two boundary conditions. Azcue and Muler (2005) consider the optimal reinsurance and dividend policy that maximizes the cumulative expected discounted dividend payouts. The paper uses the optimal control and the viscosity solution to tackle the problem. In Avanzi et al. (2006), the optimal dividend is studied under a dual model. Using Laplace transforms, the optimal barrier strategy is obtained. Many papers have been published on this topic recently. Most of the papers consider the barrier strategy, however as mentioned in Gerber et al. (2006), if the surplus process is not a Markov process, it is very likely that the optimal dividend strategy is not the barrier strategy.

6 Conclusion remarks

In this paper, we have briefly summarized some recent developments in actuarial science. We hope that this will help colleagues in this area, in particular the young researchers. We are not claiming that the topics listed in this paper are a complete list of actuarial science research topics. There is a huge amount literature on the topics in this paper. The papers cited here are far away from a complete reference list. Here we quote from Hickman and Young (1994): "Our knowledge is limited, but our curiosity is not."

References

[1] H. Albrecher and R. Kainhofer, Risk theory with a nonlinear dividend barrier. Computing 68 (2002), 289–311.

[2] H. Albrecher, J. Hartinger and R. E. Tichy, On the distribution of dividend payments and the discounted penalty function in a risk model with linear dividend barrier. Scandinavian Actuarial Journal 2005 (2005), no. 2, 103–126.

[3] S. Asmussen, Ruin Probabilities. World Scientific, Singapore 2000.

[4] B. Avanzi, H. U. Gerber and E. S. W. Shiu, Optimal dividends in the dual model. Insurance: Mathematics and Economics (2006). To appear.

[5] P. Azcue and N. Muler, Optimal reinsurance and dividend distribution policies in the Cramer-Lundberg model. Mathematical Finance 15 (2005), 261–308.

[6] L. Ballotta, S. Haberman and N. Wang, Guarantees in with-profit and unitised with-profit life insurance contracts: Fair valuation problem in presence of the default option. Journal of Risk and Insurance 73(1), (2006), 97–121.

[7] N. L. Bowers, H. U. Gerber, J. C. Hickman, D. A. Jones and C. J. Nesbitt, Actuarial Mathematics, Society of Actuaries, 2nd edition, 1997.

[8] P. P. Boyle and E. S. Schwartz, Equilibrium Prices of guarantees under equity-linked contracts. Journal of Risk and Insurance 44 (1977), 639–660.

[9] P. P. Boyle and M. R. Hardy, Guaranteed Annuity Options. ASTIN Bulletin 33(2) (2003), 125–152.

[10] M. J. Brennan and E. S. Schwartz, The pricing of equity-linked life insurance policies with an asset value guarantee. Journal of Financial Economics 3 (1976), 195–213.

[11] M. J. Brennan and E. S. Schwartz, Alternative investment strategies for the issuers of equity-linked life insurance with an asset value guarantee. Journal of Business 52 (1979), 63–93.

[12] H. Bühlmann, Mathematical Methods in Risk Theory. Springer Verlag, Heidelberg 1970.

[13] J. Cai, H. U. Gerber and H. Yang, Optimal Dividends in an Ornstein-Uhlenbeck Type Model with Credit and Debit Interest. North American Actuarial Journal 10 (2006), no. 2, 94–108; Discussions, 109–119.

[14] H. Cramér, On the Mathematical Theory of Risk. Skandia Jubilee Volume, Stockholm 1930.

[15] A. Dassios and P. Embrechts, Martingales and insurance risk. Communications in Statistics - Stochastic Models 5 (1989), 181–217.

[16] Bruno De Finetti, Su un' impostazione alternativa dell teoria collettiva del rischio. Transactions of the XVth International Congress of Actuaries 2 (1957), 433–443.

[17] F. Delbaen and J. Haezendonck, Classical risk theory in an economic environment. Insurance: Mathematics and Economics 6 (1987), 85–116.

[18] J. Garrido and M. Morales, On expected discounted penalty function for Levy risk processes. North American Actuarial Journal 10 (2006), no. 4, 196–216; Discussions, 216–218.

[19] H. U. Gerber, Der Einfluss von Zins auf die Ruin wahrscheinlichkeit. Sonderabdruck aus den Mitteilungen der Vereinigung schweizerischer Versicherungsmathematiker 71 (1971), 63–70.

[20] H. U. Gerber, Games of economic survival with discrete and continuous-income process. Operations Research 20 (1972), 37–45.

[21] H. U. Gerber, Martingales in risk theory. Mitteilungen der Vereinigung schweizerischer Versicherungsmathematiker (1973), 205–216.

[22] H. U. Gerber, An Introduction to Mathematical Risk Theory, S.S. Huebner Foundation Monograph Series No. 8, R. Irwin, Homewood, IL 1979.

[23] H. U. Gerber, On the probability of ruin in the presence of a linear dividend barrier. Scandinavian Actuarial Journal 1981 (1981), 105–115.

[24] H. U. Gerber and E. S. W. Shiu, Option pricing by Esscher transforms (with discussions). Transactions of the Society of Actuaries 46 (1994), 99–191.

[25] H. U. Gerber and E. S. W. Shiu, On the time value of ruin. North American Actuarial Journal 2(1) (1998), 48–72.

[26] H. U. Gerber and E. S. W. Shiu, Optimal dividends: analysis with Brownian motion. North American Actuarial Journal 8 (2004), 1–20.

[27] H. U. Gerber and E. S. W. Shiu, On optimal dividend strategies in the compound Poisson model. North American Actuarial Journal 10(2) (2006), 76–93.

[28] H. U. Gerber, X. S. Lin and H. Yang, A note on the dividends-penalty identity and the optimal dividend barrier. ASTIN Bulletin 36 (2006), no. 2, 489–503.

[29] J. Grandell, Aspects of Risk Theory, Springer-Verlag 1991.

[30] A. Grosen and P. L. Jørgensen, Fair valuation of life insurance liabilities: The impact of interest rate guarantees, surrender options, and bonus policies. Insurance: Mathematics and Economics 26 (2000), 37–57.

[31] M. R. Hardy, Investment Guarantees: Modeling and Risk Management for Equity-Linked Life Insurance, Wiley & Sons Inc, Hoboken, New Jersey 2003.

[32] J. C. Hickman and V. Young, Discussion of Gerber and Shiu "Option pricing by Esscher transforms," Transactions of the Society of Actuaries 46 (1994), 154–157.

[33] H. Lee, Pricing exotic options with applications to equity-indexed annuities, Ph.D. thesis, University of Iowa 2002.

[34] S. Li and J. Garrido, On a class of renewal risk models with a constant dividend barrier. Insurance: Mathematics and Economics 35 (2004) 691–701.

[35] S. Li and J. Garrido, On a general class of renewal risk processes: analysis of the Gerber-Shiu function. Advances in Applied Probability 37 (2005) 836–856.

[36] X. S. Lin and K. S. Tan, Valuation of equity-indexed annuities under stochastic interest rates. North American Actuarial Journa 7(4) (2003), 72–91.

[37] X. S. Lin and K. P. Pavlova, The compound Poisson risk model with a threshold dividend strategy. Insurance: Mathematics and Economics (2005).

[38] X. S. Lin, G. E. Willmot and S. Drekic, The classical risk model with a constant dividend barrier: analysis of the Gerber-Shiu discounted penalty function. Insurance: Mathematics and Economics 33 (2003), 551–566.

[39] F. Lundberg, Approximerad Framstallning av Sannolikhetsfunktionen. Aterforsakring af kollektivrisker. Akademisk afhandling. Uppsala 1903.

[40] F. Lundberg, Forsakringsteknisk Riskutjamning. F. Englunds boktryckeri A.B., Stockholm 1926.

[41] K. W. Ng, H. Yang and L. Zhang, Ruin probability in a compound poisson model with random discount rate. Probability in the Engineering and Informational Sciences 18 (2004), no. 1, 55–70.

[42] R. Norberg, Ruin Problems with Asset and Liabilities of Diffusion Type. Stochastic Processes and Their Applications 81 (1999), 255–269.

[43] J. Paulsen, Ruin theory with compounding assets-a survey. Insurance: Mathematics and Economics 22 (1998), 3–16.

[44] J. Paulsen and H. K. Gjessing, Ruin theory with stochastic return on investments. Advance in Applied Probability 29 (1997a), 965–985.

[45] J. Paulsen and H. K. Gjessing, Optimal choice of dividend barriers for a risk process with stochastic return on investments. Insurance: Mathematics and Economics 20 (1997b), 215–223.

[46] K. P. Pavlova and G. E. Willmot, The discrete stationary renewal risk model and the Gerber-Shiu discounted penalty function. Insurance: Mathematics and Economics 35 (2004), 267–277.

[47] T. Rolski, H. Schmidli, V. Schmidt and J. Teugels, Stochastic Processes for Insurance and Finance. Wiley 1999.

[48] T. K. Siu, J. W. Lau and H. Yang, Pricing participating products under a generalized jump-diffusion with a Markov-switching compensator, Submitted (2006).

[49] B. Sundt and J. L. Teugels, Ruin estimates under interest force. Insurance: Mathematics and Economics 16 (1995), 7–22.

[50] S. Tiong, Valuing equity-indexed annuities. North American Actuarial Journal 4 (2000), no. 4, 149–163, discussions, 163–179.

[51] G. E. Willmot and D. Dickson, The Gerber-Shiu discounted penalty function in the statinary renewal risk model (2002).

[52] H. Yang, Non-exponential bounds for ruin probability with interest effect included. Scandinavian Actuarial Journal 1 (1999), 66–79.

[53] H. Yang and L. Zhang, On the distribution of surplus immediately after ruin under interest force", Insurance: Mathematics and Economics 29 (2001a), no. 2, 247–255.

[54] H. Yang and L. Zhang, The joint distribution of surplus immediately before ruin and the deficit at ruin under interest force. North American Actuarial Journal 5 (2001b), no. 3, 92–103.

[55] J. Zhu and H. Yang, Ruin theory for a Markovian regime switching model under a threshold dividend strategy, submitted (2006).